D1714903

INTERNATIONAL UNION OF THEORETICAL AND APPLIED MECHANICS

MECHANICS OF GENERALIZED CONTINUA

PROCEEDINGS OF THE IUTAM-SYMPOSIUM ON THE GENERALIZED COSSERAT CONTINUUM AND THE CONTINUUM THEORY OF DISLOCATIONS WITH APPLICATIONS

FREUDENSTADT AND STUTTGART (GERMANY) 1967

EDITOR

E. KRÖNER

WITH 64 FIGURES

SPRINGER-VERLAG NEW YORK INC. 1968

Professor Dr. rer. nat. EKKEHART KRÖNER
Institut für Theoretische Physik der
Technischen Universität Clausthal

All rights reserved
No part of this book may be translated or reproduced in
any form without written permission from Springer-Verlag
© by Springer-Verlag, Berlin/Heidelberg 1968
Printed in Germany
Library of Congress Catalog Card Number 68-22401

The use of general descriptive names, trade names, trade marks, etc. in this publication even if the former are not especially identified, is not to be taken as a sign that such names, as understood by the Trade Marks and Merchandise Marks Act, may accordingly be used freely by anyone
Title No. 1492

PHYSICS
QA
808.2
I12
1967

1127221

In Memoriam
of the three great French scientists

EUGÈNE and FRANÇOIS COSSERAT

who in 1909 developed the mechanical theory
of a continuum of oriented particles,
now known as the Cosserat continuum, and

ÉLIE CARTAN

who in 1922 introduced the notion of the torsion
of a space, a notion which 30 years later was
recognized to be isomorphic to the notion
of the crystal dislocation.

Preface

The symposium was held in Freudenstadt from 28th to 31st of August 1967 and in Stuttgart from 1st to 2nd of September 1967.

The proposal to hold this symposium originated with the German Society of Applied Mathematics and Mechanics (GAMM) late in 1964 and was examined by a committee of IUTAM especially appointed for this purpose. The basis of this examination was a report in which the present situation in the field and the possible aims of the symposium were surveyed. Briefly, the aims of the symposium were stated to be

1. the unification of the various approaches developed in recent years with the aim of penetrating into the microscopic world of matter by means of continuum theories;

2. the bridging of the gap between microscopic (or atomic) research on mechanics on one hand, and the phenomenological (or continuum mechanical) approach on the other hand;

3. the physical interpretation and the relation to actual material behaviour of the quantities and laws introduced into the new theories, together with applications;

4. the further development of the theories, where necessary, and the clarification of open questions;

5. a stocktaking of present achievements and the prognosis for future developments.

The committee agreed unanimously that the topic of the symposium represented an important phase of current developments in continuum mechanics, from the purely theoretical point of view as well as in connection with possible applications to actual materials. The committee further agreed that the symposium would be desirable in order to clarify the present issues by personal discussions among those most active in the field. The committee hoped in particular that such a symposium would contribute to the physical interpretation of available mathematical theories.

The committee agreed to describe the scope of the symposium as follows:

1. Theory of the Cosserat continuum and its various generalizations.

2. Kinematics and dynamics of dislocations and their description by means of continuum models.

3. Applications in fluid mechanics and solid mechanics, in particular with reference to the dislocation theory of creep and plasticity.

The report of the special IUTAM committee was accepted by the Permanent Bureau of IUTAM. Two working committees were then formed, one of them being an International Scientific Committee consisting of A. E. Green, W. T. Koiter, K. Kondo, E. Kröner (chairman), R. D. Mindlin, L. I. Sedov, the other being a Local Organising Committee consisting of K.-H. Anthony (secretary), W. Günther, A. Seeger (chairman), U. Wegner and K. Zoller.

Since the scientific background of the participants was somewhat heterogeneous, including areas such as mathematics, mechanics and solid state physics, it seemed desirable to have a fairly large number of general lectures (60 minutes) giving a broad survey of the various aspects of the entire field. In addition to these survey lectures there were included several special lectures (45 minutes) in which certain topics of particular interest were presented. Finally, a large number of research lectures (30 minutes) and brief communications (10 minutes) reflected recent progress in the field of the symposium.

Alongside the scheduled lectures two extra discussion sessions (about two hours each) were arranged under the chairmanship of A. E. Green and P. M. Naghdi and of A. Seeger and R. Bullough.

The reader himself may judge whether or not the symposium was a success. Of course, nobody expected that all open questions would be solved during the week in Freudenstadt and Stuttgart. Perhaps even more important than the various results reported in the field covered by the symposium was that the various groups which all strive for progress in mechanics, but who pursue this aim in quite different ways, were united in public and private discussions for a whole week.

An essential aim of the symposium has been to bring physics and mathematics closer together. This remains the main task also after the symposium, notwithstanding the progress which has been achieved in this respect. In particular it was shown by several workers that the mathematical theories of various lattice models, upon use of suitable continuisation procedures exactly result in theories which had been developed earlier on a pure continuum basis, thereby allowing physical interpretations of some of the mathematical continuum theories.

As for the meeting being a success, this was due to the common efforts of the 70 participants from 18 nations all over the world. Before and during the symposium most helpful assistance was provided by the members of the International Scientific Committee and of the Local Organising Committee. Special thanks are owed to U. Wegner and to K.-H. Anthony, the secretary of the meeting. Wegner not only was one of the initiators of the first proposal to hold the symposium; he also took upon himself with greatest energy the difficult task of procuring the needed financial support. As every participant knows, Anthony has done so much for the functioning of the symposium that his merits cannot be enumerated here.

The meeting could not have taken place without generous financial help by the German Bundesminister für Wissenschaftliche Forschung, and by the Kultusminister of the Land Baden-Württemberg. To both of them we wish to express our deep gratitude. Thanks are also due to the members of the Permanent Bureau of IUTAM and to the Presidium of GAMM for their assistance in various phases of this enterprise. Certainly no one attending the symposium will forget the reception party — arranged on behalf of GAMM — which from the beginning made people feeling like one big family.

The editor of this volume wishes to acknowledge having received the most pleasant cooperation from Dr. Mayer-Kaupp of the Springer-Verlag in all phases of the preparation and the printing of this book. Thanks are also due to the printers, who, in spite of the sometimes extremely complicated mathematics, succeeded in giving the text such an appealing appearance. Last but not least the editor is obliged to Dr. B. K. Datta for helpful assistance with the proof-reading.

A final word may be directed towards the reader. Science is becoming increasingly more complex. Less and less can the scientist afford to work out his own ideas, looking neither to right nor left. Steadily increasing is the importance of exchanges of ideas. The exchange of ideas between mathematicians mechanicians, and physicists was the great concern of this symposium. A tender plant has started to grow in Freudenstadt and Stuttgart. Let us all resolve to support its further growth and continued health.

Clausthal-Zellerfeld, June 1968

E. Kröner

Contributors to this Volume

GL = General Lecture, SL = Special Lecture, RL = Research Lecture, BC = Brief Communication

ACHENBACH, J. D.	Department of Civil Engineering, Northwestern University, Evanston, Ill. 60201, U.S.A.	RL
ADOMEIT, G.	Lehrstuhl für Mechanik, Technische Hochschule Aachen, 5100 Aachen, Germany	BC
ALBLAS, J. B.	Department of Mathematics, Technische Hogeschool Eindhoven, Eindhoven, Holland	BC
AMARI, S.	Department of Mathematical Engineering and Instrumentation Physics, University of Tokyo, Tokyo, Japan	BC
ANTHONY, K.-H.	Institut für Physik, Max-Planck-Institut für Metallforschung, 7000 Stuttgart, Germany	BC
BERDITCHEVSKI, V. L.	USSR National Committee of Theoretical and Applied Mechanics, USSR Academy of Sciences, Moscow, U.S.S.R.	GL
BILBY, B. A.	Department of the Theory of Materials, University of Sheffield, Elmfield, Northumberland, Sheffield 10, England	GL
BRINKMAN, J. A.	Science Center, North American Rockwell Corporation Thousand Oaks, Calif., U.S.A.	BC
BULLOUGH, R.	Theoretical Physics Division, Atomic Energy Research Establishment, Harwell, England	RL
COWIN, S. C.	Department of Mechanical Engineering, Tulane University, New Orleans, La.	BC
CROCKER, A. G.	Department of Physics, University of Surrey, London S.W. 11, England	BC
DEWIT, R.	Metallurgy Division, National Bureau of Standards, U.S. Department of Commerce, Washington, D. C. 20234, U.S.A.	RL
DIKMEN, M.	Faculty of Engineering, Middle East Technical University, Ankara, Turkey	BC
ERINGEN, A. C.	Department of Aerospace and Mechanical Sciences, Princeton University, Princeton, N.J. 08540, U.S.A.	SL
ESSMANN, U.	Institut für Physik, Max-Planck-Institut für Metallforschung, 7000 Stuttgart, Germany	BC
FOX, N.	Department of Applied Mathematics and Computing Science, University of Sheffield, Sheffield 10, England	BC

GREEN, A. E.	Department of Mathematics, University of Oxford, Oxford, England	RL
GRIOLI, G.	Seminario Matematico, University of Padova, Padova, Italy	RL
GÜNTHER, H.	Institut für Reine Mathematik, Deutsche Akademie der Wissenschaften zu Berlin, X 1199 Berlin, Germany (D.D.R.)	BC
HEHL, F.	Institut für Theoretische Physik, Technische Universität Clausthal, 3392 Clausthal-Zellerfeld, Germany	BC
HERRMANN, G.	Department of Civil Engineering, Northwestern University, Evanston, Ill. 60201, U.S.A.	RL
INDENBOM, W. L.	Institute of Crystallography, USSR Academy of Sciences, Moscow, U.S.S.R.	GL
KESSEL, S.	Lehrstuhl für Theoretische Mechanik, Universität (Technische Hochschule) Karlsruhe, 7500 Karlsruhe, Germany	RL
KONDO, K.	Department of Mathematical Engineering and Instrumentation Physics, University of Tokyo, Tokyo, Japan	GL
KRÖNER, E.	Institut für Theoretische Physik, Technische Universität Clausthal, 3392 Clausthal-Zellerfeld, Germany	RL
KROUPA, F.	Institute of Physics, Czechoslovak Academy of Sciences, Prague 2, C.S.S.R.	BC
KRUMHANSL, J. A.	Laboratory of Atomic and Solid State Physics, Cornell University, Ithaca, N. Y. 14851, U.S.A.	SL
KUNIN, I. A.	Institute of Thermophysics, USSR Academy of Sciences, Novosibirsk, U.S.S.R.	RL
LAWS, N.	Department of Mathematics, University of Newcastle, Newcastle upon Tyne, England	RL
MINAGAWA, S.	Institute for Strength and Fracture of Materials, Tohoku University, Sendai, Japan	BC
MINDLIN, R. D.	Department of Civil Engineering, Columbia University, New York, N.Y., U.S.A.	RL
MIŞICU, M.	Center of Mechanics of Solids, Academy of Rumanian Socialist Republic, Bucharest, Rumania	RL
MURA, T.	Department of Civil Engineering, Northwestern University, Evanston, Ill. 60201, U.S.A.	RL
NAGHDI, P. M.	Division of Applied Mechanics, University of California, Berkeley, Calif. 94720, U.S.A.	SL
NEUBER, H.	Lehrstuhl für Technische Mechanik und Mechanisch-Technisches Laboratorium, Technische Universität (Hochschule) München, 8000 München 2, Germany	RL
NOLL, W.	Department of Mathematics, Carnegie-Mellon University, Pittsburgh, Pa. 15213, U.S.A.	SL
ORLOV, A. N.	Institute of Technical Physics, USSR Academy of Sciences, Leningrad, U.S.S.R.	GL
PERRIN, R. C.	Theoretical Physics Division, Atomic Energy Research Establishment, Harwell, England	RL
PLAVŠIĆ, M.	Faculty of Sciences, Department of Mechanics, University of Belgrade, Belgrade, Yougoslavia	BC
REISSNER, E.	Department of Mathematics, Massachusetts Institute of Technology, Cambridge, Mass. 02139, U.S.A.	BC

Rivlin, R. S.	Center for the Application of Mathematics, Lehigh University, Bethlehem, Pa. 18015, U.S.A. GL
Satake, M.	Faculty of Engineering, Tohoku University, Sendai, Japan BC
Schaefer, H.	Institut für Mechanik, Technische Universität Braunschweig, 3300 Braunschweig, Germany RL
Sedov, L. I.	USSR National Committee of Theoretical and Applied Mechanics, USSR Academy of Sciences, Moscow, U.S.S.R. GL
Seeger, A.	Institut für Physik, Max-Planck-Institut für Metallforschung, 7000 Stuttgart, Germany BC
Sternberg, E.	Division of Engineering and Applied Sciences, California Institute of Technology, Pasadena, Calif. 91109, U.S.A. SL
Stojanović, R.	Faculty of Sciences, Department of Mechanics, University of Belgrade, Belgrade, Yougoslavia BC
Teodorescu, P. P.	Institute of Mathematics, Academy of Rumanian Socialist Republic, Bucharest 9, Rumania BC
Teodosiu, C.	Institute of Mathematics, Academy of Rumanian Socialist Republic, Bucharest 9, Rumania BC
Toupin, R. A.	International Business Machines Corporation, Research Laboratory Zürich, 8803 Rüschlikon, Switzerland GL
Träuble, H.	Max-Planck-Institut für Physikalische Chemie, 3400 Göttingen, Germany BC
Tucker, M. O.	Department of Physics, University of Surrey, London, England BC
Wan, F. Y. M.	Department of Mathematics, Massachusetts Institute of Technology, Cambridge, Mass. 02139, U.S.A. BC
Wang, C.-C.	Department of Mechanics, The Johns Hopkins University, Baltimore, Ma. 21218, U.S.A. BC
Wesolowski, Z.	Institute of the Basic Technical Problems, Polish Academy of Sciences, Warsaw, Poland BC
Zorski, H.	Institute for Fundamental Engineering Research, Polish Academy of Sciences, Warsaw, Poland RL

Contents

Chapter 1: The Cosserat Continuum and its Generalizations

Chapter 2: Continuous Distributions of Dislocations

Chapter 3: Lattice Structure and Continuum Mechanics

Chapter 4: Applications to other Branches of Physics

Generalized Mechanics of Continuous Media

By

R. S. Rivlin

Center for the Application of Mathematics
Lehigh University, Bethlehem, Pa.

1. Introduction

In the last few years there has been a flurry of interest in continuum-mechanical theories in which the deformation is described not only by the usual vector displacement field, but by other vector or tensor fields as well. This interest was stimulated mainly by the papers of Ericksen [1—3], Toupin [4], and Mindlin and Tiersten [5]. Although it is generally stated that these theories apply to materials with structure, it is not usually made very clear what features of the structure are incorporated into the theory and how the results of the theory are to be interpreted in terms of the structure.

It is one of the objects of the present paper to underline some of the problems which arise in this connection.

Our starting point is an explicit physical model of the body. This consists of a number of "particles", each of which consists of ν (say) mass-points. The kinematic state of the system is described by the vector positions of the centers of mass of the particles with respect to a fixed origin and of the mass-points relative to the centers of mass. The forces applied to the body are considered to act on the various mass-points of which the particles consist. Expressions for the rate at which work is done by the external forces and for the kinetic energy are written down.

Passage to a continuum model is achieved by assuming that, at a given time, the deformation of a particle and the force system applied to it differs little from one particle to its neighbor, except possibly in

Acknowledgment. This work was supported under Office of Naval Research, U.S. Navy, Contract N 00014-67-A-0370-0001 with Lehigh University.

a thin layer at the surface of the body. We first assume that, in this layer, the deformation and applied force system vary but little as we pass from one particle to its neighbor in a direction tangential to the surface, but may vary considerably as we pass from one particle to its neighbor in a direction normal to the surface. With these assumptions, we replace the discrete values of the vectors describing the deformation for the particles and the disrcete values of the forces applied to them by continuous vector fields defined throughout the volume of the continuum model and by similar fields defined on its surface. The latter fields are not necessarily the limits of the former. Summations over the particles in the expression for the rate at which the applied forces do work are replaced by the sum of a volume integral (the rate at which the body forces do work) and a surface integral (the rate at which the surface forces do work). Similarly, the summations in the expression for kinetic energy are replaced by the sum of a volume integral (the volume kinetic energy) and a surface integral (the surface kinetic energy).

In order to highlight other aspects of the theory, we limit our subsequent discussion to deformations which are such that the surface deformation fields are the limits of the volume deformation fields. We also neglect the surface kinetic energy, the surface internal energy and the generation of heat in the surface layer.

With these assumptions, we obtain the thermo-mechanical field equations of the theory by the systematic application of the first and second laws of thermodynamics. In this analysis considerations of invariance play a central role.

The theory is developed on the basis of a particular way of describing the deformation of the particle system, which serves as our starting point, and of the force system acting on it. Many other methods of describing these are possible and, using substantially the same procedure as that adopted in the present paper, we could in each case obtain the appropriate field equations. While these might formally appear quite different from those of the present paper, they would nevertheless have essentially the same physical content. On the other hand, the equations derived in the present paper may well apply to systems other than the particle system which we have taken as our starting point.

The methods employed in this paper are essentially those employed earlier by GREEN and RIVLIN [6, 7] in constructing other theories in which the deformation is described by other fields in addition to the usual displacement field. The description adopted here of the deformation of the particle model and of the force system acting on it is a particular case of the description by generalized coordinates and conjugate generalized forces discussed elsewhere by GREEN and RIVLIN [8].

2. The Physical Model

We consider a mechanical system which consists of N sub-systems, which we call *particles*. Each sub-system consists of ν mass-points. (As a particular case it may consist of a single mass-point.) We can describe the kinematic state of a sub-system, say the Pth., by the vector positions of its mass-points expressed as functions of time t. Let $\boldsymbol{x}_\alpha^{(P)}$ $(\alpha = 1, \ldots, \nu)$ be the vector positions of the ν mass-points of the Pth. particle, at time t, with respect to its center of mass. Let $\boldsymbol{x}^{(P)}$ be the vector position of the center of mass of the Pth. particle with respect to a fixed origin.

We consider now the rate at which the system of forces applied to the Pth. particle do work. We denote this by $\mathfrak{P}_p$. If $m_\alpha^{(P)}$ is the mass of the αth. mass point of the Pth. particle and $\boldsymbol{f}_\alpha^{(P)}$ is the force acting on this mass-point, per unit mass, $\mathfrak{P}_p$ is given by

$$\mathfrak{P}_p = \sum_{\alpha=1}^{\nu} m_\alpha^{(P)} \boldsymbol{f}_\alpha^{(P)} \cdot (\dot{\boldsymbol{x}}^{(P)} + \dot{\boldsymbol{x}}_\alpha^{(P)}), \tag{2.1}$$

where the dot denotes derivation with respect to time. Eq. (2.1) can be re-written as

$$\mathfrak{P}_p = m^{(P)} \boldsymbol{f}^{(P)} \cdot \dot{\boldsymbol{x}}^{(P)} + \sum_{\alpha=1}^{\nu} m_\alpha^{(P)} \boldsymbol{f}_\alpha^{(P)} \cdot \dot{\boldsymbol{x}}_\alpha^{(P)}, \tag{2.2}$$

where

$$m^{(P)} = \sum_{1=\alpha}^{\nu} m_\alpha^{(P)}, \qquad m^{(P)} \boldsymbol{f}^{(P)} = \sum_{\alpha=1}^{\nu} m_\alpha^{(P)} \boldsymbol{f}_\alpha^{(P)}. \tag{2.3}$$

We note that $m^{(P)}$ is the total mass of the Pth. particle and $\boldsymbol{f}^{(P)}$ is the resultant force acting on the Pth. particle, per unit mass.

Bearing in mind that the vectors $\boldsymbol{x}_\alpha^{(P)}$ are measured with respect to the center of mass of the particle, we have

$$\sum_{\alpha=1}^{\nu} m_\alpha^{(P)} \dot{\boldsymbol{x}}_\alpha^{(P)} = 0. \tag{2.4}$$

The rate at which work is done by the forces applied to the whole system, which we denote $\mathfrak{P}$, is given by

$$\mathfrak{P} = \sum_{P=1}^{N} \mathfrak{P}_p = \sum_{P=1}^{N} m^{(P)} \boldsymbol{f}^{(P)} \cdot \dot{\boldsymbol{x}}^{(P)} + \sum_{P=1}^{N} \sum_{\alpha=1}^{\nu} m_\alpha^{(P)} \boldsymbol{f}_\alpha^{(P)} \cdot \dot{\boldsymbol{x}}_\alpha^{(P)}. \tag{2.5}$$

From the point-of-view of classical analytical mechanics, the components in a fixed coordinate system of the vectors $\boldsymbol{x}^{(P)}$ and $\boldsymbol{x}_\alpha^{(P)}$ $(\alpha = 1, \ldots, \nu;\ P = 1, \ldots, N)$ represent a particular choice of the generalized coordinates describing the kinematic state of the system.

The corresponding components of $m^{(P)} \boldsymbol{f}^{(P)}$ and $m_\alpha^{(P)} \boldsymbol{f}_\alpha^{(P)}$ are then the generalized forces conjugate to these generalized coordinates.

The kinetic energy of the system, which we denote $\mathfrak{T}$ is given by

$$\mathfrak{T} = \sum_{P=1}^{N} \sum_{\alpha=1}^{\nu} \tfrac{1}{2} m_\alpha^{(P)} (\dot{\boldsymbol{x}}^{(P)} + \dot{\boldsymbol{x}}_\alpha^{(P)}) \cdot (\dot{\boldsymbol{x}}^{(P)} + \dot{\boldsymbol{x}}_\alpha^{(P)}), \tag{2.6}$$

which, with $(2.3)_1$ and (2.4), yields

$$\mathfrak{T} = \sum_{P=1}^{N} \tfrac{1}{2} m^{(P)} \dot{\boldsymbol{x}}^{(P)} \cdot \dot{\boldsymbol{x}}^{(P)} + \sum_{P=1}^{N} \sum_{\alpha=1}^{\nu} \tfrac{1}{2} m_\alpha^{(P)} \dot{\boldsymbol{x}}_\alpha^{(P)} \cdot \dot{\boldsymbol{x}}_\alpha^{(P)}. \tag{2.7}$$

We now pass from a system of particles to a continuum, by replacing the sums in the expressions for $\mathfrak{P}$ and $\mathfrak{T}$ by integrals. In order to do this, we consider our particle system to occupy a domain $V + \partial V$ at some reference time t_0 and a domain $v + \partial v$ at time t. We consider the applied force system to vary sufficiently slowly from particle to particle, throughout the domain $v + \partial v$, except possibly in the thin layer ∂v at its boundary, so that sums over the particles in v at time t, may be replaced by integrals over v. Within the layer ∂v, we assume that the forces vary sufficiently slowly from particle to particle, as we move in a direction tangential to the surface, in a sense that will be evident from the analysis given below. We make analogous assumptions regarding the vectors $\boldsymbol{x}_\alpha^{(P)}$. From (2.5), we have

$$\mathfrak{P} = \sum_{v} m^{(P)} \boldsymbol{f}^{(P)} \cdot \dot{\boldsymbol{x}}^{(P)} + \sum_{v} \sum_{\alpha=1}^{\nu} m_\alpha^{(P)} \boldsymbol{f}_\alpha^{(P)} \cdot \dot{\boldsymbol{x}}_\alpha^{(P)} +$$
$$+ \sum_{\partial v} m^{(P)} \boldsymbol{f}^{(P)} \cdot \dot{\boldsymbol{x}}^{(P)} + \sum_{\partial v} \sum_{\alpha=1}^{\nu} m_\alpha^{(P)} \boldsymbol{f}_\alpha^{(P)} \cdot \dot{\boldsymbol{x}}_\alpha^{(P)}, \tag{2.8}$$

where $\sum_{v}$ and $\sum_{\partial v}$ denote summation over the particles in v and ∂v respectively. Similarly, we re-write (2.7) as

$$\mathfrak{T} = \sum_{v} \tfrac{1}{2} m^{(P)} \dot{\boldsymbol{x}}^{(P)} \cdot \dot{\boldsymbol{x}}^{(P)} + \sum_{v} \sum_{\alpha=1}^{\nu} \tfrac{1}{2} m_\alpha^{(P)} \dot{\boldsymbol{x}}_\alpha^{(P)} \cdot \dot{\boldsymbol{x}}_\alpha^{(P)} +$$
$$+ \sum_{\partial v} \tfrac{1}{2} m^{(P)} \dot{\boldsymbol{x}}^{(P)} \cdot \dot{\boldsymbol{x}}^{(P)} + \sum_{\partial v} \sum_{\alpha=1}^{\nu} \tfrac{1}{2} m_\alpha^{(P)} \dot{\boldsymbol{x}}_\alpha^{(P)} \cdot \dot{\boldsymbol{x}}_\alpha^{(P)}. \tag{2.9}$$

We define smooth time-dependent vector fields $\dot{\boldsymbol{x}}$ and $\dot{\boldsymbol{x}}_\alpha$ throughout v such that for any portion v' of v large enough to contain many particles

$$\int_{v'} \varrho\, \dot{\boldsymbol{x}} \cdot \dot{\boldsymbol{x}}\, dv = \sum_{v'} m^{(P)} \dot{\boldsymbol{x}}^{(P)} \cdot \dot{\boldsymbol{x}}^{(P)}$$

and (2.10)

$$\int_{v'} \varrho_\alpha\, \dot{\boldsymbol{x}}_\alpha \cdot \dot{\boldsymbol{x}}_\alpha\, dv = \sum_{v'} m_\alpha^{(P)} \dot{\boldsymbol{x}}_\alpha^{(P)} \cdot \dot{\boldsymbol{x}}_\alpha^{(P)},$$

where ϱ is the mass-density of particles in v, ϱ_α is the mass-density of αth. mass-points in v; i.e.

$$\int_{v'} \varrho \, dv = \sum_{v'} m^{(P)}$$

and

$$\int_{v'} \varrho_\alpha \, dv = \sum_{v'} m_\alpha^{(P)}. \tag{2.11}$$

In (2.10) and (2.11), $\int_{v'}$ denotes integration throughout v' and $\sum_{v'}$ denotes summation over the particles in v'. We can evidently choose the fields $\dot{\boldsymbol{x}}$ and $\dot{\boldsymbol{x}}_\alpha$ in a variety of ways and we choose them in such a way that a rigid motion of the system of particles causes the fields $\boldsymbol{x}$ and $\boldsymbol{x}_\alpha$ to change in precisely the same way as do $\boldsymbol{x}^{(P)}$ and $\boldsymbol{x}_\alpha^{(P)}$. It is evident that if we superpose on the assumed deformation a rigid rotation described by the orthogonal matrix $\boldsymbol{R}$ and a translation described by the vector $\boldsymbol{c}$, the center of mass of the Pth. particle moves to $\boldsymbol{R}\,\boldsymbol{x}^{(P)} + \boldsymbol{c}$ and the vector position of its αth. mass-point relative to this center of mass changes to $\boldsymbol{R}\,\boldsymbol{x}_\alpha^{(P)}$. We accordingly choose the vector fields $\boldsymbol{x}$ and $\boldsymbol{x}_\alpha$ so that they are changed to $\boldsymbol{R}\,\boldsymbol{x} + \boldsymbol{c}$ and $\boldsymbol{R}\,\boldsymbol{x}_\alpha$ respectively if we superpose on the assumed deformation a rotation $\boldsymbol{R}$ and translation $\boldsymbol{c}$.

We note that if the relations (2.10) are to remain valid when a time-dependent rigid motion is superposed on the assumed deformation, $\boldsymbol{x}$ and $\boldsymbol{x}_\alpha$ must be chosen to satisfy the following conditions[1]:

$$\left.\begin{aligned}
\int_{v'} \varrho\, \boldsymbol{x}\, dv &= \sum_{v'} m^{(P)}\, \boldsymbol{x}^{(P)},\\
\int_{v'} \varrho\, x_j\, x_k\, dv &= \sum_{v'} m^{(P)}\, x_j^{(P)}\, x_k^{(P)},\\
\int_{v'} \varrho\, x_j\, \dot{x}_k\, dv &= \sum_{v'} m^{(P)}\, x_j^{(P)}\, \dot{x}_k^{(P)},\\
\int_{v'} \varrho_\alpha\, x_j^{(\alpha)}\, x_k^{(\alpha)}\, dv &= \sum_{v'} m_\alpha^{(P)}\, x_j^{(\alpha P)}\, x_k^{(\alpha P)},\\
\int_{v} \varrho_\alpha\, x_j^{(\alpha)}\, \dot{x}_k^{(\alpha)}\, dv &= \sum_{v'} m_\alpha^{(P)}\, x_j^{(\alpha P)}\, \dot{x}_k^{(\alpha P)}.
\end{aligned}\right\} \tag{2.12}$$

Now, let a be the boundary of v and let a' be a part of a. Let $\partial v'$ be the part of $\partial v'$ which is a right cylinder based on a'. We shall assume that a' is large enough so that $\partial v'$ contains many particles. We define surface vector fields $\boldsymbol{y}$ and $\boldsymbol{y}_\alpha$ on a such that

$$\int_{a'} \sigma\, \dot{\boldsymbol{y}} \cdot \dot{\boldsymbol{y}}\, da = \sum_{\partial v'} m^{(P)}\, \dot{\boldsymbol{x}}^{(P)} \cdot \dot{\boldsymbol{x}}^{(P)}$$

and

$$\int_{a'} \sigma_\alpha\, \dot{\boldsymbol{y}}_\alpha \cdot \dot{\boldsymbol{y}}_\alpha\, da = \sum_{\partial v'} m_\alpha\, \dot{\boldsymbol{x}}_\alpha^{(P)} \cdot \dot{\boldsymbol{x}}_\alpha^{(P)}, \tag{2.13}$$

---

[1] $x_i^{(P)}$, $x_i^{(\alpha P)}$, x_i and $x_i^{(\alpha)}$ denote the ith components of the vectors $\boldsymbol{x}^{(P)}$, $\boldsymbol{x}_\alpha^{(P)}$, $\boldsymbol{x}$ and $\boldsymbol{x}^{(P)}$ respectively.

where

$$\int_{a'} \sigma \, da = \sum_{\partial v'} m^{(P)}.$$

and (2.14)

$$\int_{a'} \sigma_\alpha \, da = \sum_{\partial v'} m_\alpha^{(P)}$$

Again, we note that we can choose the fields $\dot{\boldsymbol{y}}$ and $\dot{\boldsymbol{y}}_\alpha$ in a variety of ways and we choose them in such a way that the superposition of a rigid rotation $\boldsymbol{R}$ and translation $\boldsymbol{c}$ on the assumed motion of the particle system changes them to $\boldsymbol{R}\,\boldsymbol{y} + \boldsymbol{c}$ and $\boldsymbol{R}\,\boldsymbol{y}_\alpha$ respectively. We note that if the relations (2.13) are to remain valid when a time-dependent rigid motion is superposed on the assumed deformation, $\boldsymbol{y}$ and $\boldsymbol{y}_\alpha$ must be chosen to satisfy the following conditions

$$\left.\begin{aligned}
\int_{a'} \sigma \, \boldsymbol{y} \, da &= \sum_{\partial v'} m^{(P)} \boldsymbol{x}^{(P)}, \\
\int_{a'} \sigma \, y_j \, y_k \, da &= \sum_{\partial v'} m^{(P)} x_j^{(P)} x_k^{(P)}, \\
\int_{a'} \varrho \, y_j \, \dot{y}_k \, da &= \sum_{\partial v'} m^{(P)} x_j^{(P)} \dot{x}_k^{(P)}, \\
\int_{a'} \varrho_\alpha \, y_j^{(\alpha)} \, y_k^{(\alpha)} \, da &= \sum_{\partial v'} m_\alpha^{(P)} x_j^{(\alpha P)} x_k^{(\alpha P)}, \\
\int_{a'} \varrho_\alpha \, y_j^{(\alpha)} \, \dot{y}_k^{(\alpha)} \, da &= \sum_{\partial v'} m_\alpha^{(P)} x_j^{(\alpha P)} \dot{x}_k^{(\alpha P)}.
\end{aligned}\right\} \qquad (2.15)$$

This ensures that the relations (2.13) remain valid when a rigid motion is superposed on the assumed deformation.

Introducing (2.10) and (2.11) into (2.9), we obtain

$$\begin{aligned}
\mathfrak{T} = \tfrac{1}{2} \int_v \varrho \, \dot{\boldsymbol{x}} \cdot \dot{\boldsymbol{x}} \, dv + \sum_{\alpha=1}^{\nu} \tfrac{1}{2} \int_v \varrho_\alpha \, \dot{\boldsymbol{x}}_\alpha \cdot \dot{\boldsymbol{x}}_\alpha \, dv \, + \\
+ \tfrac{1}{2} \int_a \sigma \, \dot{\boldsymbol{y}} \cdot \dot{\boldsymbol{y}} \, da + \sum_{\alpha=1}^{\nu} \tfrac{1}{2} \int_a \sigma_\alpha \, \dot{\boldsymbol{y}}_\alpha \cdot \dot{\boldsymbol{y}}_\alpha \, da.
\end{aligned} \qquad (2.16)$$

We shall also subject our choices of $\dot{\boldsymbol{x}}_\alpha$ and $\dot{\boldsymbol{y}}_\alpha$ to the conditions [cf. Eqs. (2.4)]

$$\sum_{\alpha=1}^{\nu} \varrho_\alpha \, \dot{\boldsymbol{x}}_\alpha = 0 \quad \text{and} \quad \sum_{\alpha=1}^{\nu} \sigma_\alpha \, \dot{\boldsymbol{y}}_\alpha = 0. \qquad (2.17)$$

We now define smooth vector fields $\boldsymbol{f}$ and $\boldsymbol{f}_\alpha$ throughout the volume v in such a way that

$$\int_{v'} \varrho \, \boldsymbol{f} \cdot \dot{\boldsymbol{x}} \, dv = \sum_{v'} m^{(P)} \boldsymbol{f}^{(P)} \cdot \dot{\boldsymbol{x}}^{(P)}$$

and (2.18)

$$\int_{v'} \varrho_\alpha \boldsymbol{f}_\alpha \cdot \dot{\boldsymbol{x}}_\alpha \, dv = \sum_{v'} m_\alpha^{(P)} \boldsymbol{f}_\alpha^{(P)} \cdot \dot{\boldsymbol{x}}_\alpha^{(P)}.$$

Similarly, we define smooth surface vector fields $\boldsymbol{t}$ and $\boldsymbol{t}_\alpha$ on a in such a way that

$$\int_{a'} \boldsymbol{t}\cdot\dot{\boldsymbol{y}}\,da = \sum_{\partial v'} m^{(P)}\boldsymbol{f}^{(P)}\cdot\dot{\boldsymbol{x}}^{(P)}$$

and (2.19)

$$\int_{a'} \boldsymbol{t}_\alpha\cdot\dot{\boldsymbol{y}}_\alpha\,da = \sum_{\partial v'} m_\alpha^{(P)}\boldsymbol{f}_\alpha^{(P)}\cdot\dot{\boldsymbol{x}}_\alpha^{(P)}.$$

We note that, since the superposition on the assumed deformation of a rigid constant translational velocity $\dot{\boldsymbol{c}}$ leaves $\boldsymbol{f}^{(P)}$ and $\boldsymbol{f}_\alpha^{(P)}$ unchanged, the requirement that the relations (2.18) and (2.19) remain satisfied implies that $\boldsymbol{f}$, $\boldsymbol{f}_\alpha$, $\boldsymbol{t}$ and $\boldsymbol{t}_\alpha$ be chosen in such a way that

$$\int_{v'}\varrho\boldsymbol{f}\,dv = \sum_{v'} m^{(P)}\boldsymbol{f}^{(P)}, \qquad \int_{v'}\varrho_\alpha\boldsymbol{f}_\alpha\,dv = \sum_{v'} m_\alpha^{(P)}\boldsymbol{f}_\alpha^{(P)},$$
$$\int_{a'}\boldsymbol{t}\,da = \sum_{\partial v'} m^{(P)}\boldsymbol{f}^{(P)}, \qquad \int_{a'}\boldsymbol{t}_\alpha\,da = \sum_{\partial v'} m_\alpha^{(P)}\boldsymbol{f}_\alpha^{(P)}. \tag{2.20}$$

Introducing (2.18) and (2.19) into (2.8), we obtain

$$\mathfrak{P} = \int_v \varrho\boldsymbol{f}\cdot\dot{\boldsymbol{x}}\,dv + \sum_{\alpha=1}^{\nu}\int_v \varrho_\alpha\boldsymbol{f}_\alpha\cdot\dot{\boldsymbol{x}}_\alpha\,dv +$$
$$+\int_a \boldsymbol{t}\cdot\dot{\boldsymbol{y}}\,da + \sum_{\alpha=1}^{\nu}\int_a \boldsymbol{t}_\alpha\cdot\dot{\boldsymbol{y}}_\alpha\,da. \tag{2.21}$$

We shall call the vector field $\boldsymbol{x}$ the *simple deformation field* and the vector fields $\boldsymbol{x}_\alpha$, *director deformation fields*. $\boldsymbol{y}$ and $\boldsymbol{y}_\alpha$ are the simple and director deformation fields at the surface of the body.

The vector fields $\boldsymbol{f}$ and $\boldsymbol{f}_\alpha$ ($\alpha = 1, \ldots, \nu$) will be called the *simple body force field* and *director body force fields* respectively. The fields $\boldsymbol{t}$ and $\boldsymbol{t}_\alpha$ ($\alpha = 1, \ldots, \nu$) will be called the *simple surface force field* and *director surface force fields* respectively, measured per unit of surface area at time t.

Suppose that the particles which occupy the volume v' at time t, occupy the volume V' at the reference time t_0, where v' and V' contain many particles. Then, we define $\varrho^{(0)}$ and $\varrho_\alpha^{(0)}$ by

$$\int_{V'}\varrho^{(0)}\,dV = \int_{v'}\varrho\,dv,$$

and (2.22)

$$\int_{V'}\varrho_\alpha^{(0)}\,dV = \int_{v'}\varrho_\alpha\,dv.$$

Similarly, suppose A and A' are the surfaces at time t_0 corresponding to the surfaces a and a' at time t. We define $\sigma^{(0)}$, $\sigma_\alpha^{(0)}$, $\boldsymbol{T}$ and $\boldsymbol{T}_\alpha$ by

$$\int_{A'}\sigma^{(0)}\,dA = \int_{a'}\sigma\,da, \qquad \int_{A'}\sigma_\alpha^{(0)}\,dA = \int_{a'}\sigma_\alpha\,da,$$

and (2.23)

$$\int_{A'}\boldsymbol{T}\,dA = \int_{a'}\boldsymbol{t}\,da, \qquad \int_{A'}\boldsymbol{T}_\alpha\,dA = \int_{a'}\boldsymbol{t}_\alpha\,da.$$

Then, we may re-write Eqs. (2.16) and (2.21) as

$$\mathfrak{T} = \tfrac{1}{2}\int_V \varrho^{(0)}\, \dot{\boldsymbol{x}} \cdot \dot{\boldsymbol{x}}\, dV + \tfrac{1}{2}\sum_{\alpha=1}^{\nu}\int_V \varrho_\alpha^{(0)}\, \dot{\boldsymbol{x}}_\alpha \cdot \dot{\boldsymbol{x}}_\alpha\, dV + \\ + \tfrac{1}{2}\int_A \sigma^{(0)}\, \dot{\boldsymbol{y}} \cdot \dot{\boldsymbol{y}}\, dA + \tfrac{1}{2}\sum_{\alpha=1}^{\nu}\int_A \sigma_\alpha^{(0)}\, \dot{\boldsymbol{y}}_\alpha \cdot \dot{\boldsymbol{y}}_\alpha\, dA,$$

and (2.24)

$$\mathfrak{P} = \int_V \varrho^{(0)}\, \boldsymbol{f} \cdot \dot{\boldsymbol{x}}\, dV + \sum_{\alpha=1}^{\nu}\int_V \varrho_\alpha^{(0)}\, \boldsymbol{f}_\alpha \cdot \dot{\boldsymbol{x}}_\alpha\, dV + \\ + \int_A \boldsymbol{T} \cdot \dot{\boldsymbol{y}}\, dA + \sum_{\alpha=1}^{\nu}\int_A \boldsymbol{T}_\alpha \cdot \dot{\boldsymbol{y}}_\alpha\, dA.$$

From (2.24), we may recover (2.16) and (2.21) by taking the reference time t_0 to coincide with the time t.

We note that the director body forces $\boldsymbol{f}_\alpha$ are measured per unit mass of αth. mass-points. The director body forces, per unit mass of material, are, of course, given by $(\varrho_\alpha^{(0)}/\varrho^{(0)})\, \boldsymbol{f}_\alpha = (\varrho_\alpha/\varrho)\, \boldsymbol{f}_\alpha$.

In an analogous fashion we may express the total internal energy $\mathfrak{U}$ of the body as the sum of a volume and a surface integral. Thus,

$$\mathfrak{U} = \int_v \varrho\, u\, dv + \int_a w\, da = \int_V \varrho^{(0)}\, u\, dV + \int_A W\, dA. \qquad (2.25)$$

u is then the internal energy per unit mass and w and W are internal energies, per unit area of surface measured at times t and t_0 respectively.

Again, the rate at which heat is added to the body, $\mathfrak{Q}$, is given by an expression of the form

$$\mathfrak{Q} = \int_v \varrho\, r\, dv + \int_a h\, da - \int_q da \\ = \int_V \varrho^{(0)}\, r\, dV + \int_A H\, dA - \int_A Q\, da. \qquad (2.26)$$

r is the rate of generation of heat per unit mass; h and H are the rates of generation of heat per unit area of surface, measured at times t_0 and t respectively[1]; q and Q are the rates at which heat leaves the body across its surface, again measured at per unit area at times t and t_0 respectively.

In this paper, we shall assume that $\dot{\boldsymbol{y}}$ and $\dot{\boldsymbol{y}}_\alpha$, at a point on the surface, are the limits of $\dot{\boldsymbol{x}}$ and $\dot{\boldsymbol{x}}_\alpha$ as we approach the point from the interior of the body. Also, we shall assume that the surface integrals in the expression for the kinetic energy of the body are negligible and that the rate of generation of heat (h or H) in the surface is negligible. We shall also neglect the surface internal energy (w or W). Eqs. (2.24), $(2.25)_2$

[1] In terms of the particle model $\int_a h\, da$ is the rate at which heat is generated in the volume ∂v and $\int_a w\, da$ is the internal energy of the particles in the volume ∂v.

and $(2.26)_2$ then become

$$
\begin{aligned}
\mathfrak{T} &= \tfrac{1}{2}\int_V \varrho^{(0)}\, \dot{\boldsymbol{x}} \cdot \dot{\boldsymbol{x}}\, d\,V + \tfrac{1}{2}\sum_{\alpha=1}^{\nu}\int_V \varrho_\alpha^{(0)}\, \dot{\boldsymbol{x}}_\alpha \cdot \dot{\boldsymbol{x}}_\alpha\, d\,V\,, \\
\mathfrak{P} &= \int_V \varrho_\alpha^{(0)}\, \boldsymbol{f} \cdot \dot{\boldsymbol{x}}\, d\,V + \sum_{\alpha=1}^{\nu}\int_V \varrho_\alpha^{(0)}\, \boldsymbol{f}_\alpha \cdot \dot{\boldsymbol{x}}_\alpha\, d\,V \\
&\quad + \int_A \boldsymbol{T} \cdot \dot{\boldsymbol{x}}\, dA + \sum_{\alpha=1}^{\nu}\int_A \boldsymbol{T}_\alpha \cdot \dot{\boldsymbol{x}}_\alpha\, dA\,, \\
\mathfrak{U} &= \int_V \varrho^{(0)}\, u\, d\,V\,, \\
\mathfrak{Q} &= \int_V \varrho^{(0)}\, r\, d\,V - \int_A Q\, dA\,.
\end{aligned}
\tag{2.27}
$$

These expressions for $\mathfrak{T}$, $\mathfrak{P}$, $\mathfrak{U}$ and $\mathfrak{Q}$ form the basis for the subsequent development in this paper.

3. The First Law of Thermodynamics

The first law of thermodynamics states that the sum of the rates of change of the internal energy and kinetic energy of the body is equal to the sum of the rates at which work is done by the applied force system and at which heat is added to the body; i.e.

$$\dot{\mathfrak{U}} + \dot{\mathfrak{T}} = \mathfrak{P} + \mathfrak{Q}. \tag{3.1}$$

Employing the expressions (2.27) in (3.1), we obtain

$$
\begin{aligned}
&\int_V \varrho^{(0)}\, \dot{u}\, d\,V + \int_V \varrho^{(0)}\, \dot{\boldsymbol{x}} \cdot \ddot{\boldsymbol{x}}\, d\,V + \sum_{\alpha=1}^{\nu}\int_V \varrho_\alpha^{(0)}\, \dot{\boldsymbol{x}}_\alpha \cdot \ddot{\boldsymbol{x}}_\alpha\, d\,V \\
&\quad = \int_V \varrho^{(0)}\, \boldsymbol{f} \cdot \dot{\boldsymbol{x}}\, d\,V + \sum_{\alpha=1}^{\nu}\int_V \varrho_\alpha^{(0)}\, \boldsymbol{f}_\alpha \cdot \dot{\boldsymbol{x}}_\alpha\, d\,V \\
&\quad + \int_A \boldsymbol{T} \cdot \dot{\boldsymbol{x}}\, dA + \sum_{\alpha=1}^{\nu}\int_A \boldsymbol{T}_\alpha \cdot \dot{\boldsymbol{x}}_\alpha\, dA + \\
&\quad + \int_V \varrho^{(0)}\, r\, d\,V - \int_A Q\, dA\,.
\end{aligned}
\tag{3.2}
$$

This equation applies to the whole body or, with suitable interpretation of V and A, to any portion of it. We shall apply it to an elementary material tetrahedron in the body which, at time t_0, has three of its faces parallel to the coordinate planes of a rectangular cartesian coordinate system x. Let this tetrahedron be $OABC$ (see Fig. 1a). Let N_A be the components, in the system x, of the unit normal to the face ABC of the tetrahedron. Let the tetrahedron become $oabc$ at time t (see Fig. 1b). In general, $oabc$ will not have its faces parallel to the coordinate planes. We assume the tetrahedron to be sufficiently small, so that the heat flux Q, and the simple and director deformation and surface force vectors are substantially constant over it.

We denote[1] by π_{Bi} and $\pi_{Bi}^{(\alpha)}$ the values of T_i and $T_i^{(\alpha)}$ acting, at time t, on the face of the tetrahedron which was normal to the axis x_B at time t_0. We preserve the notation T_i and $T_i^{(\alpha)}$ for the simple and director surface forces acting, at time t, on the face abc of the tetrahedron. We denote by Q_B the value of Q, at time t, for the face of the tetrahedron which was normal to the axis x_B at time t_0. We preserve the notation Q for the heat flux, at time t, across the face abc of the tetrahedron.

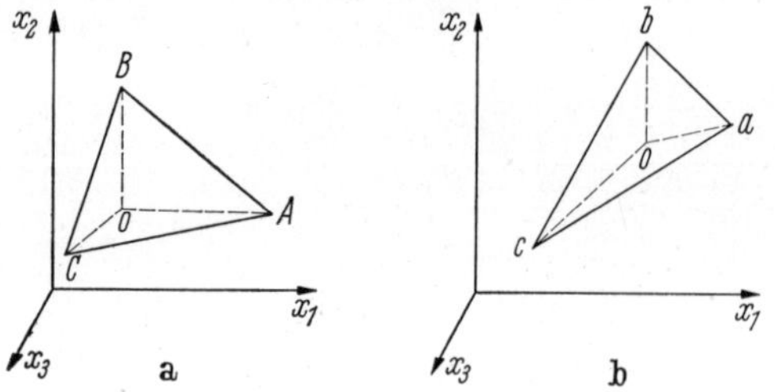

Fig. 1.

Let ΔA be the area of the face ABC of the elementary tetrahedron $OABC$ and ΔA_B that of the face normal to the x_B-axis. Then,

$$\Delta A_B = N_B \Delta A. \tag{3.3}$$

Applying (3.2) to the elementary tetrahedron, we obtain

$$\left(T_i \dot{x}_i + \sum_{\alpha=1}^{\nu} T_i^{(\alpha)} \dot{x}_i^{(\alpha)} - Q\right) \Delta A$$
$$= \left(\pi_{Bi} \dot{x}_i + \sum_{\alpha=1}^{\nu} \pi_{Bi}^{(\alpha)} \dot{x}_i^{(\alpha)} - Q_B\right) \Delta A_B. \tag{3.4}$$

With (3.3), we obtain

$$(T_i - \pi_{Bi} N_B) \dot{x}_i + \sum_{\alpha=1}^{\nu} (T_i^{(\alpha)} - \pi_{Bi}^{(\alpha)} N_B) \dot{x}_i^{(i)} - (Q - Q_B N_B) = 0. \tag{3.5}$$

We now consider a deformation which differs from the assumed deformation, at time t, by a constant velocity $\dot{\boldsymbol{c}}$, say. We note that $\dot{\boldsymbol{x}}$ is changed to $\dot{\boldsymbol{x}} + \dot{\boldsymbol{c}}$, while $T_i - \pi_{Bi} N_B$, and the second and third terms in (3.5) are unaltered. It follows that

$$T_i = \pi_{Bi} N_B$$

and (3.6)

$$\sum_{\alpha=1}^{\nu} (T_i^{(\alpha)} - \pi_{Bi}^{(\alpha)} N_B) \dot{x}_i^{(\alpha)} = Q - Q_B N_B.$$

---

[1] We use the notation T_i and $T_i^{(\alpha)}$ to denote the components in the system x of the vectors $\boldsymbol{T}$ and $\boldsymbol{T}_\alpha$. Analogous notation will be used for the components of other vectors.

From (3.6), we obtain, using the Divergence Theorem,

$$\int_A T_i \dot{x}_i \, dA = \int_A \pi_{Bi} \dot{x}_i N_B \, dA = \int_V (\pi_{Bi,B} \dot{x}_i + \pi_{Bi} \dot{x}_{i,B}) \, dV$$

and (3.7)

$$\int_A \left(\sum_{\alpha=1}^{\nu} T_i^{(\alpha)} \dot{x}_i^{(\alpha)} - Q \right) dA = \int_A \left(\sum_{\alpha=1}^{\nu} \pi_{Bi}^{(\alpha)} \dot{x}_i^{(\alpha)} - Q_B \right) N_B \, dA = \int_V \left[\sum_{\alpha=1}^{\nu} (\pi_{Bi,B}^{(\alpha)} \dot{x}_i^{(\alpha)} + \pi_{Bi}^{(\alpha)} \dot{x}_{i,B}^{(\alpha)}) - Q_{B,B} \right] dV.$$

Introducing the relations (3.7) into (3.2) and applying the resulting equation to an infinitesimal volume element, we obtain

$$(\pi_{Bi,B} + \varrho^{(0)} f_i - \varrho^{(0)} \ddot{x}_i) \dot{x}_i + \sum_{\alpha=1}^{\nu} (\pi_{Bi,B}^{(\alpha)} + \varrho_\alpha^{(0)} f_i^{(\alpha)} - \varrho_\alpha^{(0)} \ddot{x}_i^{(\alpha)}) \dot{x}_i^{(\alpha)} + \pi_{Bi} \dot{x}_{i,B} + \sum_{\alpha=1}^{\nu} \pi_{Bi}^{(\alpha)} \dot{x}_{i,B}^{(\alpha)} - Q_{B,B} + \varrho^{(0)} r - \varrho^{(0)} \dot{u} = 0. \tag{3.8}$$

We again consider a deformation which differs from the assumed deformation, at time t, by a constant translational velocity $\dot{\boldsymbol{c}}$, say. It follows from (3.8) that

$$\pi_{Bi,B} + \varrho^{(0)} f_i - \varrho^{(0)} \ddot{x}_i = 0 \tag{3.9}$$

and

$$\sum_{\alpha=1}^{\nu} G_i^{(\alpha)} \dot{x}_i^{(\alpha)} + \pi_{Bi} \dot{x}_{i,B} + \sum_{\alpha=1}^{\nu} \pi_{Bi}^{(\alpha)} \dot{x}_{i,B}^{(\alpha)} - Q_{B,B} + \varrho^{(0)} r - \varrho^{(0)} \dot{u} = 0, \tag{3.10}$$

where

$$G_i^{(\alpha)} = \pi_{Bi,i}^{(\alpha)} + \varrho_\alpha^{(0)} f_i^{(\alpha)} - \varrho_\alpha^{(0)} \ddot{x}_i^{(\alpha)}. \tag{3.11}$$

We now consider a new deformation which differs from the deformation already considered by a time-dependent rigid rotation, but is such that the configuration of the body is the same at time τ apart from a constant angular velocity. Let $\bar{x}_i(\tau)$ and $\bar{x}_i^{(\alpha)}(\tau)$ be the vectors at time τ, which, for this deformation correspond to the vectors $x_i(\tau)$ and $x_i^{(\alpha)}(\tau)$ for the original deformation. Then, at all times τ,

$$\bar{x}_i(\tau) = R_{ij}(\tau) \, x_j(\tau), \quad \bar{x}_i^{(\alpha)}(\tau) = R_{ij}(\tau) \, x_j^{(\alpha)}(\tau), \tag{3.12}$$

where R_{ij} satisfies the orthogonality condition

$$R_{ij}(\tau) R_{ik}(\tau) = R_{ji}(\tau) R_{ki}(\tau) = \delta_{jk}, \qquad |R_{ij}(\tau)| = 1. \tag{3.13}$$

We also have

$$R_{ij} = R_{ij}(t) = \delta_{ij}. \tag{3.14}$$

From (3.13) and (3.14), we obtain

$$\dot{R}_{ij} + \dot{R}_{ji} = 0. \tag{3.15}$$

From (3.12) and (3.13), we obtain, with the notation $x_i = x_i(t)$, $x_i^{(\alpha)} = x_i^{(\alpha)}(t)$, etc.,

$$\begin{aligned} \dot{\bar{x}}_i^{(\alpha)} &= \dot{R}_{ij}\, x_j^{(\alpha)} + \dot{x}_i^{(\alpha)}, \\ \dot{\bar{x}}_{i,B} &= \dot{R}_{ij}\, x_{j,B} + \dot{x}_{i,B}, \\ \dot{\bar{x}}_{i,B}^{(\alpha)} &= \dot{R}_{ij}\, x_{j,B}^{(\alpha)} + \dot{x}_{i,B}^{(\alpha)}. \end{aligned} \tag{3.16}$$

For the new deformation, we see from their definitions, that π_{Bi}, $\pi_{Bi}^{(\alpha)}$ and $Q_{B,B} + \varrho^{(0)}\, r - \varrho^{(0)}\, \dot{u}$ remain unchanged. Also, $G_i^{(\alpha)}$ remains unchanged since in order that, apart from a rigid rotation, the deformations shall be the same, the applied forces $f_i^{(\alpha)}$ must be changed to an extent which exactly balances the change of the inertial forces $\varrho^{(0)}\, \ddot{x}_i^{(\alpha)}$. Thus, for the new deformation, we have, instead of (3.10),

$$\sum_{\alpha=1}^{\nu} G_i^{(\alpha)}\, \dot{\bar{x}}_i^{(\alpha)} + \pi_{Bi}\, \dot{\bar{x}}_{i,B} + \sum_{\alpha=1}^{\nu} \pi_{Bi}^{(\alpha)}\, \dot{\bar{x}}_{i,B}^{(\alpha)} - Q_{B,B} + \varrho^{(0)}\, r - \varrho^{(0)}\, \dot{u} = 0. \tag{3.17}$$

Subtracting (3.10) from (3.17) and using (3.16), we obtain

$$P_{ij}\, \dot{R}_{ij} = 0, \tag{3.18}$$

where

$$\begin{aligned} P_{ij} &= \sum_{\alpha=1}^{\nu} (G_i^{(\alpha)}\, x_j^{(\alpha)} + \pi_{Bi}^{(\alpha)}\, x_{j,B}^{(\alpha)}) + \pi_{Bi}\, x_{j,B} \\ &= \sum_{\alpha=1}^{\nu} (\pi_{Bi,B}^{(\alpha)} + \varrho_\alpha^{(0)}\, f_i^{(\alpha)} - \varrho_\alpha^{(0)}\, \ddot{x}_i^{(\alpha)})\, x_j^{(\alpha)} + \\ &\quad + \sum_{\alpha=1}^{\nu} \pi_{Bi}^{(\alpha)}\, x_{j,B}^{(\alpha)} + \pi_{Bi}\, x_{j,B}. \end{aligned} \tag{3.19}$$

Since the relation (3.18) is valid for all $\dot{R}_{ij}$ satisfying (3.15), we obtain

$$P_{ij} = P_{ji}. \tag{3.20}$$

We now consider the results that can be obtained from $(3.6)_2$ by considering the effect of the additional rotation described by Eqs. (3.12) to (3.14). We note that $Q - Q_B\, N_B$ and the coefficients of $\dot{x}_i^{(\alpha)}$ are unaltered by this rotation. Consequently, we obtain from $(3.6)_2$ and $(3.16)_1$,

$$\sum_{\alpha=1}^{\nu} (T_i^{(\alpha)} - \pi_{Bi}^{(\alpha)}\, N_B)\, \dot{R}_{ij}\, x_j^{(\alpha)} = 0. \tag{3.21}$$

Since this result is valid for all $\dot{R}_{ij}$ satisfying (3.15), we have

$$(T_i^{(\alpha)} - \pi_{Bi}^{(\alpha)}\, N_B)\, x_j^{(\alpha)} = \sum_{\alpha=1}^{\nu} (T_j^{(\alpha)} - \pi_{Bj}^{(\alpha)}\, N_B)\, x_i^{(\alpha)}. \tag{3.22}$$

4. The Second Law of Thermodynamics

We shall write the second law of thermodynamics in the form of the Clausius-Duhem inequality

$$\int_V \varrho^{(0)}\,\dot{s}\,dV \geqq \int_V \frac{\varrho^{(0)}\,r}{\theta}\,dV - \int_A \frac{Q}{\theta}\,dA\,, \tag{4.1}$$

where s is the entropy per unit mass and θ is the absolute temperature. This relation is applicable to the whole body or to any portion of it. We shall apply it first to an elementary tetrahedron and then to an arbitrary volume element in a manner paralleling the development from the first law of thermodynamics in § 3.

Applying the inequality (4.1) to the elementary tetrahedron described in § 3, we obtain

$$Q - Q_B\,N_B \geqq 0\,. \tag{4.2}$$

We now consider, instead of the tetrahedron $OABC$, a similar tetrahedron, the unit normal to the slant surface of which has components $-N_A$ at time t_0. For this tetrahedron we must replace Q by $-Q$ and N_A by $-N_A$ in (4.2). We thus obtain

$$-Q + Q_B\,N_B \geqq 0\,. \tag{4.3}$$

From (4.2) and (4.3), it follows that

$$Q = Q_B\,N_B\,. \tag{4.4}$$

Introducing (4.4) into $(3.6)_2$, we obtain

$$\sum_{\alpha=1}^{\nu} (T_i^{(\alpha)} - \pi_{Bi}^{(\alpha)}\,N_B)\,\dot{x}_i^{(\alpha)} = 0\,. \tag{4.5}$$

We now use (4.4) to substitute for Q in (4.1) and apply the resulting expression to an infinitesimal volume element of the body. We then obtain

$$\varrho^{(0)}\,\theta\,\dot{s} \geqq \varrho^{(0)}\,r - Q_{B,B} + \frac{Q_B}{\theta}\,\theta_{,B}\,. \tag{4.6}$$

Using (3.10) to substitute for $\varrho^{(0)}\,r - Q_{B,B}$ in (4.6), we obtain

$$\varrho^{(0)}\,(\dot{u} - \theta\,\dot{s}) \leqq \sum_{\alpha=1}^{\nu} (\pi_{Bi,B}^{(\alpha)} + \varrho_\alpha^{(0)}\,f_i^{(\alpha)} - \varrho_\alpha^{(0)}\,\ddot{x}_i^{(\alpha)})\,\dot{x}_i^{(\alpha)} + \\ + \pi_{Bi}\,\dot{x}_{i,B} + \sum_{\alpha=1}^{\nu} \pi_{Bi}^{(\alpha)}\,\dot{x}_{i,B}^{(\alpha)} - \frac{Q_B}{\theta}\,\theta_{,B}\,. \tag{4.7}$$

Denoting the Helmholtz free-energy per unit mass by e, we have

$$e = u - \theta\,s\,. \tag{4.8}$$

Then, if the deformation takes place under isothermal conditions, i.e. $\dot{\theta} = 0$, we obtain, from (4.7) and (4.8),

$$\varrho^{(0)} \dot{e} \leqq \sum_{\alpha=1}^{\nu} (\pi_{Bi,B}^{(\alpha)} + \varrho^{(0)} f_i^{(\alpha)} - \varrho_\alpha^{(0)} \ddot{x}_i^{(\alpha)}) \dot{x}_i^{(\alpha)} + \\ + \pi_{Bi} \dot{x}_{i,B} + \sum_{\alpha=1}^{\nu} \pi_{Bi}^{(\alpha)} \dot{x}_{i,B}^{(\alpha)} - \frac{Q_B}{\theta} \theta_{,B} . \tag{4.9}$$

We have used the Clausius-Duhem inequality as though it were valid whether or not the process involved is reversible. Within the framework of classical thermodynamics (i.e. of thermostatics) it is regarded as valid only when stated in the form of a restriction on the total change of entropy in passing from one state of equilibrium to another, even though this change is effected by an irreversible path. The entropy of the body then has to be defined only for equilibrium conditions. In its application to the thermomechanics of an elastic body, which is taken up in the next section, its use is justified, even in discussing changes from one non-equilibrium state to another, by the assumptions which are made regarding the variables on which the thermodynamic potentials and the stresses depend. For example, in the case of isothermal deformations, in which the temperature throughout the body is uniform, these quantities are regarded as depending on the instantaneous configuration of the body and not on its rate of change. In the case when the deformation is isothermal and the temperature throughout the body is not uniform, the application of the second law of thermodynamics, in the form of the Clausius-Duhem inequality and still within the framework of classical thermodynamics, is justified by the assumption that the thermodynamic potentials, stresses, etc. are independent of the temperature gradient.

The application of the Clausius-Duhem inequality to inelastic materials is much more questionable. It should, however, be realized that the results obtained from much application are, in the main, not very strong. Apart from the inequalities (4.7) or (4.9), the main result is the relation (4.4), which enables us to obtain from $(3.6)_2$ the result (4.5). If we did not allow the application of the Clausius-Duhem inequality in this more general context, we would have to remain content with the single relation $(3.6)_2$.

5. Elastic Materials

In this section, we shall consider the further results which can be obtained if we assume that the material is elastic. In the interests of simplicity, we shall limit our discussion to the case of isothermal deformations. In this case, we assume that the Helmholtz free-energy depends on the deformation through the deformation gradients $x_{i,C}$, the directors $x_i^{(\alpha)}$ $(\alpha = 1, \ldots, \nu)$ and their spatial derivatives $x_{i,C}^{(\alpha)}$. We assume also that Q_B and the coefficients of $\dot{x}_i^{(\alpha)}$, $\dot{x}_{i,B}$ and $\dot{x}_{i\,B}^{(\alpha)}$ on the right-hand side of the inequality (4.9) depend on the deformation only through these quantities. We then have

$$\dot{e} = \frac{\partial e}{\partial x_{i,B}} \dot{x}_{i,B} + \sum_{\alpha=1}^{\nu} \left(\frac{\partial e}{\partial x_i^{(\alpha)}} \dot{x}_i^{(\alpha)} + \frac{\partial e}{\partial x_{i,B}^{(\alpha)}} \dot{x}_{i,B}^{(\alpha)} \right). \tag{5.1}$$

Introducing (5.1) into (4.9), we obtain

$$L_{Bi} \dot{x}_{i,B} + \sum_{\alpha=1}^{\nu} (M_i^{(\alpha)} \dot{x}_i^{(\alpha)} + N_{Bi}^{(\alpha)} \dot{x}_{i,B}^{(\alpha)}) + \frac{Q_B}{\theta} \theta_{,B} \leqq 0, \tag{5.2}$$

where

$$\left.\begin{aligned} L_{Bi} &= \varrho^{(0)} \frac{\partial e}{\partial x_{i,B}} - \pi_{Bi}, \\ M_i^{(\alpha)} &= \varrho^{(0)} \frac{\partial e}{\partial x_i^{(\alpha)}} - (\pi_{Bi,B}^{(\alpha)} + \varrho_\alpha^{(0)} f_i^{(\alpha)} - \varrho_\alpha^{(0)} \ddot{x}_i^{(\alpha)}), \\ N_{Bi}^{(\alpha)} &= \varrho^{(0)} \frac{\partial e}{\partial x_{i,B}^{(\alpha)}} - \pi_{Bi}^{(\alpha)}. \end{aligned}\right\} \tag{5.3}$$

As a result of the condition $(2.17)_1$, the variables $\dot{x}_i^{(\alpha)}$ $(\alpha = 1, \ldots, \nu)$ and the variables $\dot{x}_{i,B}^{(\alpha)}$ $(\alpha = 1, \ldots, \nu)$ are related by

$$\sum_{\alpha=1}^{\nu} \varrho_\alpha^{(0)} \dot{x}_i^{(\alpha)} = 0 \quad \text{and} \quad \sum_{\alpha=1}^{\nu} \varrho_\alpha^{(0)} \dot{x}_{i,B}^{(\alpha)} = 0. \tag{5.4}$$

Using Lagrange undetermined multipliers λ_i and χ_{Bi}, we obtain from (5.2) and (5.4)

$$L_{Bi} = M_i^{(\alpha)} + \varrho_\alpha^{(0)} \lambda_i = N_{Bi}^{(\alpha)} + \varrho_\alpha^{(0)} \chi_{Bi} = 0 \tag{5.5}$$

and

$$Q_B \theta_{,B} \leqq 0. \tag{5.6}$$

If, furthermore, the temperature is constant throughout the body, the equality sign in (5.6) applies.

Since, for an elastic material, T_i, π_{Bi}, Q and Q_A in $(3.6)_2$, are independent of $\dot{x}_i^{(\alpha)}$, we have, bearing in mind the relation $(5.4)_1$,

$$T_i^{(\alpha)} - \pi_{Bi}^{(\alpha)} N_B + \varrho_0^{(\alpha)} \mu_i = 0, \tag{5.7}$$

where μ_i are undetermined multipliers, and

$$Q = Q_B N_B. \tag{5.8}$$

We note that the results (5.7) and (5.8) do not depend on the use of the Clausius-Duhem inequality. In the case when the material is not necessarily elastic, we were able to obtain the relation (5.8) from the Clausius-Duhem inequality, assuming its validity in this case.

Finally, we note that no increase in generality would be obtained by assuming that e depends also on higher order spatial gradients of the directors. For example, suppose e depends also on $x^{(\alpha)}_{i,BC}$. Then, we would have to replace (5.1) by

$$\dot{e} = \frac{\partial e}{\partial x_{i,B}} \dot{x}_{i,B} + \sum_{\alpha=1}^{\nu} \left[\frac{\partial e}{\partial x^{(\alpha)}_i} \dot{x}^{(\alpha)}_i + \frac{\partial e}{\partial x^{(\alpha)}_{i,B}} \dot{x}^{(\alpha)}_{i,B} + \frac{1}{2} \left(\frac{\partial e}{\partial x^{(\alpha)}_{i,BC}} + \frac{\partial e}{\partial x^{(\alpha)}_{i,CB}} \right) \dot{x}^{(\alpha)}_{i,BC} \right]. \tag{5.9}$$

Repeating the analysis of this section, we would obtain, in addition to the results (5.5),

$$\frac{1}{2} \varrho^{(0)} \left(\frac{\partial e}{\partial x^{(\alpha)}_{i,BC}} + \frac{\partial e}{\partial x^{(\alpha)}_{i,CB}} \right) + \varrho^{(0)}_{\alpha} \, \Phi_{BCi} = 0, \tag{5.10}$$

where $\Phi_{BCi} = \Phi_{CBi}$ are Lagrange undetermined multipliers. This is equivalent to the restriction

$$\frac{1}{\varrho^{(0)}_{\alpha}} \left(\frac{\partial e}{\partial x^{(\alpha)}_{i,BC}} + \frac{\partial e}{\partial x^{(\alpha)}_{i,CB}} \right) = \frac{1}{\varrho^{(0)}_{\beta}} \left(\frac{\partial e}{\partial x^{(\beta)}_{i,BC}} + \frac{\partial e}{x^{(\beta)}_{i,CB}} \right), \quad (\alpha, \beta = 1, \ldots, \nu) \tag{5.11}$$

on the manner in which e can depend on the second spatial derivatives of the various directors.

6. Alternative Formulations

It has already been remarked in § 2 that the choice of the vectors $\boldsymbol{x}^{(P)}$ and $\boldsymbol{x}^{(P)}_{\alpha}$ ($\alpha = 1, \ldots, \nu$; $P = 1, \ldots, N$) to describe the kinematic states of the system of particles discussed in § 2 represents one choice of the generalized coordinates of the system. A wide variety of other choices is possible and, for each choice, the force system acting on the system of particles may be described by the appropriate conjugate generalized forces. A procedure analogous to that adopted in §§ 3, 4 and 5 may then be used to arrive at the equations of the corresponding continuum-mechanical theory.

For example, we might choose to describe the kinematic state of the Pth. particle at time t by the vector positions $\boldsymbol{z}^{(P)}_{\alpha}$ of its ν mass-points with respect to a fixed origin. Then, the expressions (2.5) and (2.7) for $\mathfrak{P}$ and $\mathfrak{T}$ would be replaced by

$$\mathfrak{P} = \sum_{P=1}^{N} \sum_{\alpha=1}^{\nu} m^{(P)}_{\alpha} \boldsymbol{f}^{(P)}_{\alpha} \cdot \dot{\boldsymbol{z}}^{(P)}_{\alpha}$$

and (6.1)

$$\mathfrak{T} = \sum_{P=1}^{N} \sum_{\alpha=1}^{\nu} \tfrac{1}{2} m^{(P)}_{\alpha} \dot{\boldsymbol{z}}^{(P)}_{\alpha} \cdot \dot{\boldsymbol{z}}^{(P)}_{\alpha}.$$

Again, we might describe the kinematic state of the Pth. particle at time t in the following way. Let $\boldsymbol{X}_{\alpha}$ be the vector position of the

αth. mass point at the reference time t_0 with respect to the center of mass of the particle. Then, we may express its vector position $\boldsymbol{x}_\alpha$ at time t by an expression of the form

$$x_i^{(\alpha P)} = \sum_{\lambda=0}^{\chi} y_{iA_1\ldots A_\lambda} X_{A_1}^{(\alpha P)} \ldots X_{A_\lambda}^{(\alpha P)}, \tag{6.2}$$

where the y's are independent of α. Then, the kinematic state of the Pth. particle is specified by the vector position $\boldsymbol{x}^{(P)}$ of its center of mass with respect to a fixed origin and by the tensors $y_{iA_1\ldots A_\lambda}^{(P)}$ $(\lambda = 0, \ldots, \chi)$. The expressions (2.5) and (2.7) for $\mathfrak{P}$ and $\mathfrak{T}$ are then replaced by

$$\mathfrak{P} = \sum_{P=1}^{N} \sum_{\alpha=1}^{\nu} m_\alpha^{(P)} f_i^{(\alpha P)} \dot{x}_i^{(P)} + \sum_{P=1}^{N} \sum_{\lambda=1}^{\chi} m^{(P)} F_{iA_1\ldots A_\lambda} \dot{y}_{iA_1\ldots A_\lambda}^{(P)}, \tag{6.3}$$

and

$$\mathfrak{T} = \sum_{P=1}^{N} \tfrac{1}{2} m^{(P)} \dot{x}_i^{(P)} \dot{x}_i^{(P)} + $$
$$+ \sum_{P=1}^{N} \sum_{\lambda=0}^{\chi} \sum_{\bar{\lambda}=0}^{\chi} \left(\tfrac{1}{2} \sum_{\alpha=1}^{N} m_\alpha^{(P)} X_{A_1}^{(\alpha P)} \ldots X_{A_\lambda}^{(\alpha P)} X_{B_1}^{(\alpha P)} \ldots X_{B_{\bar{\lambda}}}^{(\alpha P)} \right) \times$$
$$\times \dot{y}_{iA_1\ldots A_\lambda}^{(P)} \dot{y}_{iB_1\ldots B_{\bar{\lambda}}}^{(P)}, \tag{6.4}$$

where

$$m^{(P)} = \sum_{\alpha=1}^{\nu} m_\alpha^{(P)}, \qquad m^{(P)} F_{iA_1\ldots A_\lambda}^{(P)} = \sum_{\alpha=1}^{\nu} m_\alpha^{(P)} f_i^{(\alpha P)} X_{A_1}^{(\alpha P)} \ldots X_{A_\lambda}^{(\alpha P)}. \tag{6.5}$$

References

[1] Ericksen, J. L.: Arch. Rat. Mech. Anal. **4**, 231 (1960).
[2] Ericksen, J. L.: Trans. Soc. Rheol. **4**, 29 (1960).
[3] Ericksen, J. L.: Trans. Soc. Rheol. **5**, 23 (1960).
[4] Toupin, R.: Arch. Rat. Mech. Anal. **11**, 385 (1962).
[5] Mindlin, R., and H. F. Tiersten: Arch. Rat. Mech. Anal. **11**, 415 (1962).
[6] Green, A. E., and R. S. Rivlin: Arch. Rat. Mech. Anal. **16**, 325 (1964).
[7] Green, A. E., and R. S. Rivlin: Arch. Rat. Mech. Anal. **17**, 113 (1964).
[8] Green, A. E., and R. S. Rivlin: Proc. IUTAM Conference, Vienna 1966 (in press).

Mechanics of Micromorphic Continua[1]

By

A. C. Eringen

Department of Aerospace and Mechanical Sciences
Princeton University, Princeton, N. J.

Abstract. The theory of micromorphic continua, developed by Eringen and his co-workers, is recapitulated and extended. Master equations are obtained in the form of integral operators from which all order volume and surface statistical moments are derivable leading to a hierarchy of balance laws. Micropolar and indeterminate couple stress theories are summarized as special cases. The relation to other recent microstructure theories is discussed briefly.

1. Introduction

In several previous papers, Eringen et al., [1964]—[1967], we presented continuum theories with the hope of explaining certain physical phenomena which are inherently due to the granular and molecular nature of the materials. Under the influence of external effects, the constituents of material bodies are set into motion. The observed outcome is always an average of the individual motions of this substructure. A physical theory is designed with the hope that it will exhibit some important details of these individual motions in an average sense. We may enter the world of the molecular chaos either from the molecular and atomic side or from the continuum end of the spectrum. The theory of micromorphic continua is intended for the latter approach.

Presently there exist several approaches to the formulation of microcontinuum mechanics. Some of these theories are very general in nature but incomplete and not closed while others are concerned with special types of media. Fundamental ideas contained in some of these theories can be traced all the way back to Bernoulli and Euler in connection with their work on beam theories. In elementary beam theory, with a section of the bar, there are associated two sets of kine-

[1] The present work was sponsored by the Office of Naval Research.

matical quantities (a deformation vector and a rotation vector) and two sets of surface loads (tractions and couples). In plate theory, the situation is similar. The existence of surface and body couples independent of traction is fundamental to these theories. With the remarkable monograph of E. and F. COSSERAT [1909], this concept was extended to a three-dimensional continuum where each point of the continuum is supplied with a triad (directors). An oriented continuum of this type was noted earlier by VOIGT [1887] in connection with polar molecules in crystallography. In fact, such a picture of the ether was also introduced by MACCULLAGH [1839] in connection with his theory of optics. LORD KELVIN [1879] went further in constructing a mechanical model of this "quasi-rigid" ether.

Some fifty years later the question of an oriented continuum was rediscovered and/or reopened in various special forms by GRIOLI [1960], AERO and KUVSHINSKII [1961], SCHÄFER [1962], and GÜNTHER [1958] who also remarked on its connection to the theory of dislocations. An incomplete couple stress theory included in TRUESDELL and TOUPIN [1960] was corrected and completed by MINDLIN and TIERSTEN [1962], TOUPIN [1962], and ERINGEN [1962, Arts. 32, 40] and became known as the "indeterminate" couple stress theory.

ERINGEN and SUHUBI [1964a, b] and ERINGEN [1964a], [1964b] introduced a new general theory of nonlinear microcontinua in which the balance laws are supplemented with additional ones, and the intrinsic deformations and motions of microconstituents of the body are taken into account. This theory, in a special case, contains the Cosserat continuum and indeterminate couple stress theory. Independently a linear microstructure theory of elasticity was published by MINDLIN [1964] and a multipolar theory by GREEN and RIVLIN [1964]. Both of these theories, as well as the Cosserat continuum recapitulated by PALMOV [1964], have intimate contacts with the micromorphic theory (cf. Art. 9). Following these works, an intense activity began and literature now contains several hundred papers in this and in related fields. A proper assessment of these works with appropriate references is beyond the scope of this article. Indeed, this task is to be shared by those attending this conference.

The present article is concerned with a discussion of our early work, its various extensions and the establishment of contacts with other theories. It is hoped that the result will be a sensible organization of various approaches to a meaningful structure.

In Arts. 2—6, we present a systematic theory of kinematics and balance laws of micromorphic continua. The basic laws are developed in the forms of integral operators as volume and surface averages over the microcontinuum. In Art. 7, constitutive equations are sketched for various micromorphic continua (grade 1), and specifically stated for the nonlinear micromorphic theory of elasticity. Art. 8 deals with special theories of micropolar continua. In Art. 9, we establish contacts with other theories of microstructure. Unfortunately the space available does not permit a thorough discussion on any aspect of the theory.

2. Deformation and Motion

A micromorphic body is a collection of microcontinua called microelements having volume $\Delta V^{(\alpha)}$ with surface $\Delta S^{(\alpha)}$, $\alpha = 1, 2, \ldots, N$. A material point in $\Delta V^{(\alpha)} + \Delta S^{(\alpha)}$ in a reference state (e.g., undeformed configuration) may be located by its position vector $\boldsymbol{X}^{(\alpha)}$ (or rectangular

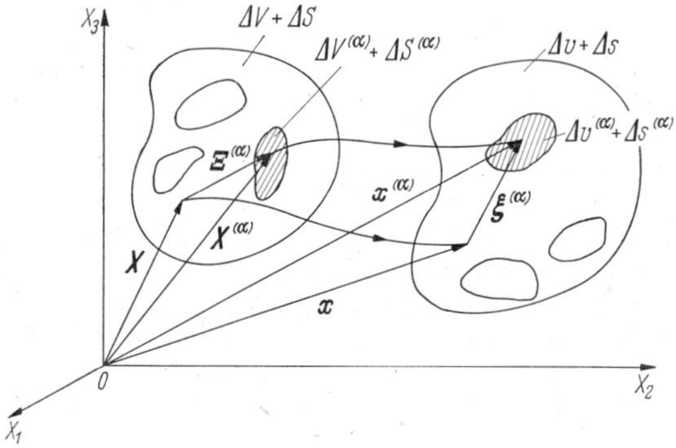

Fig. 1. Motion of microelements.

coordinates $X_K^{(\alpha)}$, $K = 1, 2, 3$). Under the influence of external loads, the microelements move, deform and occupy a new volume $\Delta v^{(\alpha)}$ having surface $\Delta s^{(\alpha)}$. The spatial place of the material point marked by $X_K^{(\alpha)}$ in the undeformed configuration is denoted by $\boldsymbol{x}^{(\alpha)}$ (or by its rectangular coordinates $x_k^{(\alpha)}$, $k = 1, 2, 3$), Fig. 1. The theory of micromorphic continua is concerned with a continuum characterization of the motions of large collections of microelements, $\Delta V^{(\alpha)} + \Delta S^{(\alpha)}$, $\alpha = 1, 2, \ldots, N$ (N large) contained in any macroelement $\Delta V + \Delta S$. To this end, it is assumed that the motion and deformation of $\Delta V + \Delta S$ take place in two steps, namely, the center of mass, $\boldsymbol{X}$, of $\Delta V + \Delta S$ moves to a spatial place, $\boldsymbol{x}$, and any point of $\Delta V^{(\alpha)} + \Delta S^{(\alpha)}$ having a relative position vector, $\boldsymbol{\Xi}^{(\alpha)}$, with respect to $\boldsymbol{X}$ moves to a new relative spatial position, $\boldsymbol{\xi}^{(\alpha)}$, with respect to $\boldsymbol{x}$ in $\Delta v^{(\alpha)} + \Delta s^{(\alpha)}$. Mathematically

$$\boldsymbol{x}^{(\alpha)} = \boldsymbol{x}(\boldsymbol{X}, t) + \boldsymbol{\xi}^{(\alpha)}(\boldsymbol{X}, \boldsymbol{\Xi}^{(\alpha)}, t). \tag{2.1}$$

Conversely we can trace the motion of $\boldsymbol{x}^{(\alpha)}$ back to its original location by

$$\boldsymbol{X}^{(\alpha)} = \boldsymbol{X}(\boldsymbol{x}, t) + \boldsymbol{\Xi}^{(\alpha)}(\boldsymbol{x}, \boldsymbol{\xi}^{(\alpha)}, t). \tag{2.2}$$

We assume that the motion $\boldsymbol{X}(\boldsymbol{x}, t)$ is the unique inverse of the centroidal motion (macromotion) $\boldsymbol{x}(\boldsymbol{X}, t)$, for all times t, in some macro-neighborhood of $\boldsymbol{X}$, except possibly at some singular surfaces, lines, and points. Thus

$$\det\left(\frac{\partial x_k}{\partial X_K}\right) \neq 0. \tag{2.3}$$

Similarly $\boldsymbol{\Xi}^{(\alpha)}(\boldsymbol{x}, \boldsymbol{\xi}^{(\alpha)}, t)$ is the inverse motion to $\boldsymbol{\xi}^{(\alpha)}$ in a neighborhood of $\boldsymbol{\Xi}^{(\alpha)}$ in $\Delta V^{(\alpha)} + \Delta S^{(\alpha)}$ for each $\boldsymbol{X}$. Thus

$$\det\left(\frac{\partial \xi_k^{(\alpha)}}{\partial \Xi_K^{(\alpha)}}\right) \neq 0 \quad \text{for } \boldsymbol{X} \text{ and } t \text{ fixed.} \tag{2.4}$$

Following ERINGEN and SUHUBI [1964a, b] and ERINGEN [1964a, b], we study the special case in which $\boldsymbol{\xi}^{(\alpha)}$ is linear in $\boldsymbol{\Xi}^{(\alpha)}$. For clarity in this exposition, we call such materials micromorphic bodies of degree 1. The generalization to the case when $\boldsymbol{\xi}^{(\alpha)}$ is a polynomial in $\boldsymbol{\Xi}^{(\alpha)}$ was indicated by ERINGEN [1964b]. Other more general cases including a six-dimensional theory are reserved for a forthcoming work.

Thus we assume *affine relative micromotion*

$$\boldsymbol{\xi}^{(\alpha)} = \boldsymbol{\chi}_K(\boldsymbol{X}, t)\, \Xi_K^{(\alpha)} \quad \text{or} \quad \xi_k^{(\alpha)} = \chi_{kK}(\boldsymbol{X}, t)\, \Xi_K^{(\alpha)}, \tag{2.5}$$

$$\boldsymbol{\Xi}^{(\alpha)} = \mathbf{X}_k(\boldsymbol{x}, t)\, \xi_k^{(\alpha)} \quad \text{or} \quad \Xi_K^{(\alpha)} = \mathsf{X}_{Kk}(\boldsymbol{x}, t)\, \xi_k^{(\alpha)} \tag{2.6}$$

where three vectors $\boldsymbol{\chi}_K$ and $\mathbf{X}_k$ are subject to

$$\chi_{kK}\, \mathsf{X}_{Kl} = \delta_{kl}, \qquad \mathsf{X}_{Kk}\, \chi_{kL} = \delta_{KL}. \tag{2.7}$$

Here and throughout, the Latin repeated indices are summed over 1, 2, 3. The summation convention is suspended over an index whenever it is underscored or if it is a Greek index marking a microelement. For these we employ the summation sign when necessary.

From (2.4) it follows that

$$\det \chi_{kK} \neq 0 \quad \text{or conversely } \det \mathsf{X}_{Kk} \neq 0. \tag{2.8}$$

A directed line element in $\Delta v^{(\alpha)} + \Delta s^{(\alpha)}$ is calculated by

$$d\boldsymbol{x}^{(\alpha)} = (\boldsymbol{C}_K + \mathfrak{C}_K)\, dX_K + \boldsymbol{\chi}_K\, d\Xi_K^{(\alpha)} \tag{2.9}$$

where

$$\boldsymbol{C}_K(\boldsymbol{X}, t) \equiv \frac{\partial \boldsymbol{x}}{\partial X_K}, \qquad \mathfrak{C}_K(\boldsymbol{X}, \boldsymbol{\Xi}^{(\alpha)}, t) \equiv \frac{\partial \boldsymbol{\xi}^{(\alpha)}}{\partial X_K} = \boldsymbol{\chi}_{L,K}\, \Xi_L^{(\alpha)},$$
$$\boldsymbol{\chi}_K(\boldsymbol{X}, \boldsymbol{\Xi}^{(\alpha)}, t) \equiv \frac{\partial \boldsymbol{\xi}^{(\alpha)}}{\partial \Xi_K^{(\alpha)}} = \boldsymbol{\chi}_K(\boldsymbol{X}, t). \tag{2.10}$$

The square of the arc length in a microelement is therefore given by

$$(ds^{(\alpha)})^2 = (C_{KL} + 2\Gamma_{KML}\Xi_M^{(\alpha)} + \Gamma_{RMK}\Gamma_{SNL}\overset{-1}{C}_{RS}\Xi_M^{(\alpha)}\Xi_N^{(\alpha)})\, dX_K\, dX_L +$$
$$+ (2\Psi_{KL} + 2\Psi_{SL}\Gamma_{RMK}\overset{-1}{C}_{RS}\Xi_M^{(\alpha)})\, dX_K\, d\Xi_L^{(\alpha)} +$$
$$+ \Psi_{ML}\Psi_{NK}\overset{-1}{C}_{MN}\, d\Xi_K^{(\alpha)}\, d\Xi_L^{(\alpha)}. \tag{2.11}$$

where C_{KL}, Ψ_{KL} and Γ_{KLM} are the deformation tensors of this theory defined by

$$C_{KL}(\boldsymbol{X}, t) \equiv x_{k,K}\, x_{k,L},$$
$$\Psi_{KL}(\boldsymbol{X}, t) \equiv x_{k,K}\, \chi_{kL}, \qquad \Gamma_{KLM}(\boldsymbol{X}, t) \equiv x_{k,K}\, \chi_{kL,M}. \tag{2.12}$$

Of these, C_{KL} is the classical lagrangian deformation tensor and Ψ_{KL} and Γ_{KLM} are the microdeformation tensors. Also $\overset{-1}{C}_{KL}$ is the inverse matrix to C_{KL}.

The material strain measures are defined by

$$E_{KL} = \tfrac{1}{2}(C_{KL} - \delta_{KL}), \qquad \mathfrak{E}_{KL} \equiv \Psi_{KL} - \delta_{KL}, \qquad \Gamma_{KLM}. \tag{2.13}$$

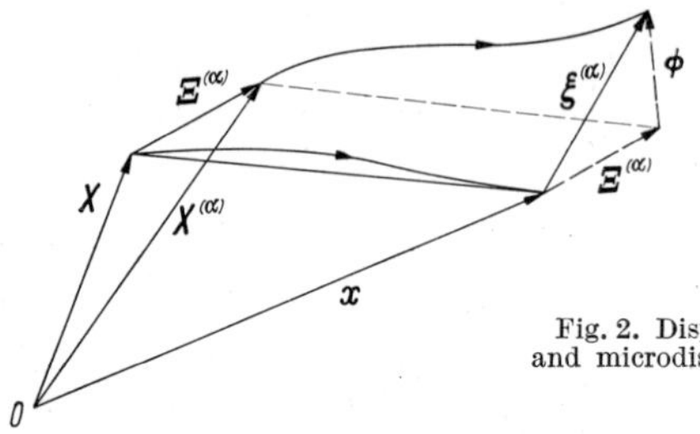

Fig. 2. Displacement and microdisplacement.

The macrodisplacement vector $\boldsymbol{u}$ and microdisplacement vector $\boldsymbol{\phi}$ are introduced through

$$\boldsymbol{u} \equiv \boldsymbol{x} - \boldsymbol{X}, \qquad \boldsymbol{\phi} \equiv \boldsymbol{\xi}^{(\alpha)} - \boldsymbol{\Xi}^{(\alpha)} \tag{2.14}$$

so that the spatial place $\boldsymbol{x}^{(\alpha)}$ can be located now by, Fig. 2,

$$\boldsymbol{x}^{(\alpha)} = \boldsymbol{X} + \boldsymbol{\Xi}^{(\alpha)} + \boldsymbol{u}(\boldsymbol{X}, t) + \boldsymbol{\phi}(\boldsymbol{X}, \boldsymbol{\Xi}^{(\alpha)}, t). \tag{2.15}$$

Upon carrying (2.14) into (2.10) and the result into (2.12) and (2.13), we obtain

$$\left.\begin{aligned} 2E_{KL} &= U_{K,L} + U_{L,K} + U_{M,K}\, U_{M,L}, \\ \mathfrak{E}_{KL} &= \Phi_{KL} + U_{L,K} + U_{M,K}\, \Phi_{ML}, \\ \Gamma_{KLM} &= \Phi_{KL,M} + U_{N,K}\, \Phi_{NL,M}. \end{aligned}\right\} \tag{2.16}$$

The program for micromorphic continuum mechanics is: given external effects and the initial state of the body, determine the motion and micromotion

$$\boldsymbol{x} = \boldsymbol{x}(\boldsymbol{X}, t), \qquad \boldsymbol{\chi}_K = \boldsymbol{\chi}_K(\boldsymbol{X}, t) \tag{2.17}$$

alternatively, the displacements and microdisplacements

$$U_K = U_K(\boldsymbol{X}, t), \qquad \Phi_{KL} = \Phi_{KL}(\boldsymbol{X}, t). \tag{2.18}$$

Since there are 12 independent scalar components in each set, ultimately we need 12 partial differential equations and accompanying boundary and initial conditions. Of course, thermodynamic, electrical and chemical variables when included increase this number further.

Classical continuum mechanics supplies us with three partial differential equations from the momentum balance laws. Thus nine new equations are needed in addition to those provided by the classical momentum balance laws. These do not include mass and energy balance equations.

3. Volume and Surface Averages

For the derivation of balance equations, the following two operational forms of averages over the collection of microvolume elements $\Delta v^{(\alpha)}$ in a macrovolume Δv, and over the open microsurface elements $\Delta s^{(\beta)}$ constituting a part of the surface, Δs, of Δv are essential, Fig. 3

$$\int_{\Delta v} \langle \phi\, \varrho \rangle\, dv \equiv \sum_{\alpha} \int_{\Delta v^{(\alpha)}} \phi^{(\alpha)}\, \varrho^{(\alpha)}\, dv^{(\alpha)}, \qquad \Delta v^{(\alpha)} \subset v, \tag{3.1}$$

$$\int_{\Delta s} \langle \psi_{\underline{k}} \rangle^*\, ds_{\underline{k}} \equiv \sum_{\beta} \int_{\Delta s^{(\beta)}} \psi_{\underline{k}}^{(\beta)}\, da_{\underline{k}}^{(\beta)}, \qquad \Delta s^{(\beta)} \subset s, \tag{3.2}$$

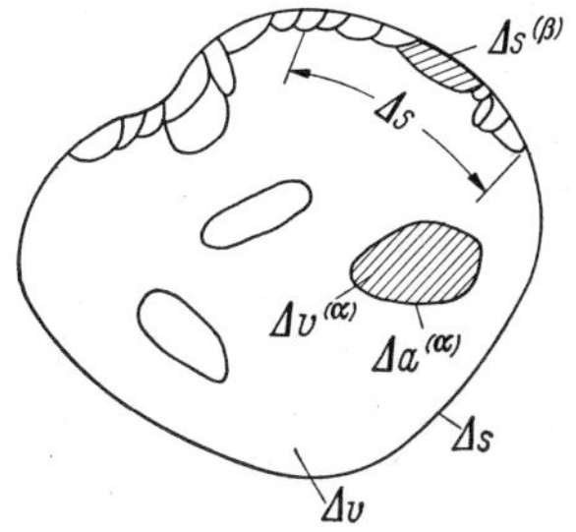

Fig. 3. Volume and surface elements.

where $\langle\,\rangle$ is a function of $\boldsymbol{x}$ only, and to be regarded as an average of its kernel. Quantities $\varrho^{(\alpha)}$ and $dv^{(\alpha)}$ are respectively the mass density and the microvolume element at a point of the microvolume $\Delta v^{(\alpha)}$, and $\Delta s_k^{(\beta)}$ is that portion of the surface of $\Delta v^{(\alpha)}$ that constitutes Δs. The functions $\phi^{(\alpha)}(\boldsymbol{x}, \boldsymbol{\xi}^{(\alpha)}, t)$ and $\psi_k^{(\beta)}(\boldsymbol{x}, \xi^{(\beta)}, t)$ are arbitrary density functions. The sum on α is over all microelements in Δv and the sum on β is over a collection of open surface elements that make up the open surface Δs.

Through the use of the mean value theorem, (3.1) and (3.2) may be expressed as

$$\langle \bar{\phi}\, \bar{\varrho} \rangle\, \Delta \bar{v} = \sum_{\alpha} \bar{\phi}^{(\alpha)}\, \bar{\varrho}^{(\alpha)}\, \Delta \bar{v}^{(\alpha)}, \qquad \Delta \bar{v}^{(\alpha)} \subset \Delta \bar{v}, \tag{3.3}$$

$$\langle \bar{\psi}_{\underline{k}} \rangle^*\, \Delta \bar{s}_{\underline{k}} = \sum_{\beta} \bar{\psi}_{\underline{k}}^{(\beta)}\, \Delta \bar{s}_{\underline{k}}^{(\beta)}, \qquad \Delta \bar{s}_k^{(\beta)} \subset \Delta \bar{s}_k . \tag{3.4}$$

These equations are none other than the expressions of averages over the number of microelements in Δv and on Δs_k. Hence they provide physical meanings for (3.1) and (3.2). For example, if we select $\phi^{(\alpha)} \equiv \phi = 1$, then (3.1) and (3.3) give respectively

$$\int_{\Delta v} \varrho\, dv \equiv \sum_{\alpha} \int_{\Delta v^{(\alpha)}} \varrho^{(\alpha)}\, dv^{(\alpha)}, \qquad \bar{\varrho}\, \Delta \bar{v} = \sum_{\alpha} \bar{\varrho}^{(\alpha)}\, \Delta \bar{v}^{(\alpha)} \tag{3.5}$$

where $\varrho(\boldsymbol{x}, t) \equiv \langle \varrho \rangle$ defines a mass density for the macrovolume element. For large α, a continuous probability density $P(\boldsymbol{x}, \xi, t)$ may be introduced replacing $(3.5)_2$ by

$$\varrho = \int_{\Delta v} P(\boldsymbol{x}, \xi, t)\, dv(\xi). \tag{3.6}$$

If we wish, we can normalize P by writing $p(\boldsymbol{x}, \xi, t) = P(\boldsymbol{x}, \xi, t)/\varrho$.

A similar interpretation is valid for (3.2) and (3.4). Thus, for example, we calculate the surface stress average by

$$\psi_k^{(\beta)} \equiv t_{kl}^{(\beta)}; \qquad \langle \psi_k \rangle \equiv t_{kl}, \tag{3.7}$$

$$\int_{\Delta s} t_{kl}\, da_{\underline{k}} \equiv \sum_{\beta} \int_{\Delta s^{(\beta)}} t_{\underline{k}l}^{(\beta)}\, da_{\underline{k}}^{(\beta)} \tag{3.8}$$

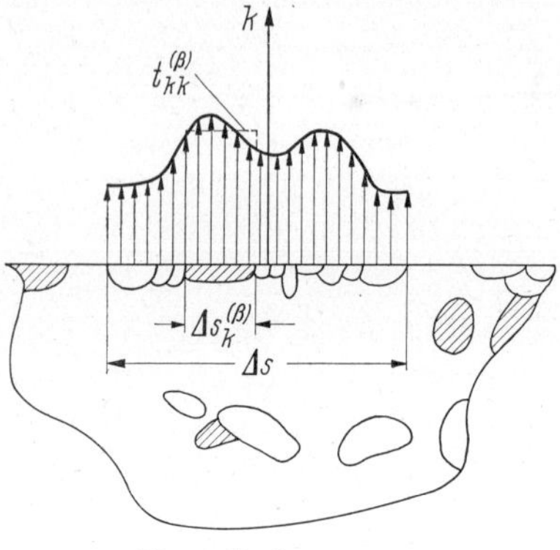

Fig. 4. Surface average.

where $t_{kl}^{(\beta)}$ and t_{kl} are respectively the stress tensor for the microelements β and the average stress tensor.

For continuous and discrete β, the last integral may be replaced respectively by

$$t_{\underline{k}l}\, ds_{\underline{k}} = \int_{\Delta s} t_{\underline{k}l}(\boldsymbol{x}, \xi, t)\, ds_{\underline{k}}(\xi), \tag{3.9}$$

$$\bar{t}_{\underline{k}l}\, \Delta \bar{s}_{\underline{k}} = \sum_{\beta} \bar{t}_{\underline{k}l}^{(\beta)}\, \Delta s_{\underline{k}}^{(\beta)}. \tag{3.10}$$

A graphical illustration of (3.10) is shown on Fig. 4.

4. Balance Laws

We assume that the balance laws of continuum mechanics are valid for each microelement. Thus for the conservation of micromass, balance of micromenta, and conservation of energy, we have respectively

$$\varrho_0^{(\alpha)}\, dv_0^{(\alpha)} = \varrho^{(\alpha)}\, dv^{(\alpha)}, \tag{4.1}$$

$$t_{kl,k}^{(\alpha)} + \varrho^{(\alpha)} (f_l^{(\alpha)} - \dot{v}_l^{(\alpha)}) = 0, \tag{4.2}$$

$$t_{kl}^{(\alpha)} = t_{lk}^{(\alpha)}, \tag{4.3}$$

$$\varrho^{(\alpha)}\, \dot{\varepsilon}^{(\alpha)} = t_{kl}^{(\alpha)}\, v_{l,k}^{(\alpha)} + q_{k,k}^{(\alpha)} + \varrho^{(\alpha)}\, h^{(\alpha)} \tag{4.4}$$

where $\boldsymbol{v}^{(\alpha)}$, $\boldsymbol{q}^{(\alpha)}$, $\varepsilon^{(\alpha)}$ and $h^{(\alpha)}$ are respectively the velocity vector, the heat vector, the internal energy density per unit mass, and the heat source per unit mass of the microelement $\Delta v^{(\alpha)} + \Delta s^{(\alpha)}$. The time rate $(\dot{\ }) \equiv D(\ )/Dt$ of various quantities is defined by

$$(\dot{\ }) = \frac{D(\ )}{Dt} = \left.\frac{\partial(\ )}{\partial t}\right|_{\boldsymbol{X}, \Xi^{(\alpha)}}. \tag{4.5}$$

Thus, for example,

$$\boldsymbol{v}^{(\alpha)} = \frac{\partial \boldsymbol{x}}{\partial t}\bigg|_{\boldsymbol{X}} + \frac{\partial \boldsymbol{\chi}_K}{\partial t}\bigg|_{\boldsymbol{X}} \Xi_K^{(\alpha)}. \tag{4.6}$$

By use of (2.5) to (2.7), we can show that

$$v_l^{(\alpha)} = \dot{x}_l + \dot{\xi}_l^{(\alpha)} = v_l + \nu_{l\,m}\,\xi_m^{(\alpha)} \tag{4.7}$$

where v_l and $\nu_{l\,m}$ are respectively the velocity vector and the gyration tensor defined by

$$v_l(\boldsymbol{x}, t) \equiv \dot{x}_l, \qquad \nu_{l\,m}(\boldsymbol{x}, t) \equiv \dot{\chi}_{lM}\,\mathsf{X}_{M\,m}. \tag{4.8}$$

To obtain the balance laws of microcontinua, we multiply these equations by $\phi^{(\alpha)}(\boldsymbol{x}, \xi^{(\alpha)}, t) = \phi_0^{(\alpha)}(\boldsymbol{X}, \boldsymbol{\Xi}^{(\alpha)})$, integrate over the microvolume, and sum over all elements in $\Delta v + \Delta s$. Thus

$$\sum_\alpha \int_{\Delta v_0^{(\alpha)}} \phi_0^{(\alpha)}\,\varrho_0^{(\alpha)}\,dv_0^{(\alpha)} = \sum_\alpha \int_{\Delta v^{(\alpha)}} \phi^{(\alpha)}\,\varrho^{(\alpha)}\,dv^{(\alpha)}, \tag{4.9}$$

$$\sum_\alpha \int_{\Delta v^{(\alpha)}} \phi^{(\alpha)}\,[t_{kl,k}^{(\alpha)} + \varrho^{(\alpha)}(f_l^{(\alpha)} - \dot{v}_l^{(\alpha)})]\,dv^{(\alpha)} = 0, \tag{4.10}$$

$$\sum_\alpha \int_{\Delta v^{(\alpha)}} \phi^{(\alpha)}[-\varrho^{(\alpha)}\,\dot{\varepsilon}^{(\alpha)} + t_{kl}^{(\alpha)}\,v_{l,k}^{(\alpha)} + q_{k,k}^{(\alpha)} + \varrho^{(\alpha)}\,h^{(\alpha)}]\,dv^{(\alpha)} = 0. \tag{4.11}$$

Using (3.1), we obtain for (4.9)

$$\int_{\Delta v_0} \langle\phi_0\,\varrho_0\rangle\,dv_0 = \int_{\Delta v} \langle\phi\,\varrho\rangle\,dv. \tag{4.12}$$

If this is to be valid for every Δv, we must have

$$\langle\phi_0\,\varrho_0\rangle\,dv_0 = \langle\phi\,\varrho\rangle\,dv. \tag{4.13}$$

This is the master equation for the balance of mass of a micromorphic continuum. It may be stated as:

Theorem 1. *In a micromorphic continuum of degree* 1, *the micromass is conserved if and only if* (4.13) *is satisfied.*

We can use the same procedure for (4.10). For this case, we first write (4.10) in the equivalent form

$$\sum_\alpha \int_{\Delta v^{(\alpha)}} [(\phi^{(\alpha)}\,t_{kl}^{(\alpha)})_{,k} - \phi_{,k}^{(\alpha)}\,t_{kl}^{(\alpha)} + \phi^{(\alpha)}\,\varrho^{(\alpha)}(f_l^{(\alpha)} - \dot{v}_l^{(\alpha)})]\,dv^{(\alpha)} = 0$$

and convert the first term to a surface integral by use of the Green-Gauss theorem. Thus

$$\sum_\alpha \left\{\oint_{\Delta s^{(\alpha)}} \phi^{(\alpha)}\,t_{kl}^{(\alpha)}\,da_k^{(\alpha)} + \int_{\Delta v^{(\alpha)}} [-\phi_{,k}^{(\alpha)}\,t_{kl}^{(\alpha)} + \phi^{(\alpha)}\,\varrho^{(\alpha)}(f_l^{(\alpha)} - \dot{v}_l^{(\alpha)})]\,dv^{(\alpha)}\right\} = 0.$$

Here the surface integrals sum to zero for all internal surfaces whether they are in contact or not. (When microelements are in contact, the surface tractions are balanced from each side; when they form a cavity, the resultant force and its moments on the cavity vanish.) Thus the summation of surface integrals reduces to an integral over the surface of the macroelement. By use of (3.1) and (3.2), we may now write

$$\oint_{\Delta s} \langle \phi\, t_{k l} \rangle^* \, d s_k + \int_{\Delta v} [-\langle \phi_{,k}\, t_{k l} \rangle + \langle \phi\, \varrho (f_l - \dot{v}_l) \rangle]\, d v = 0 . \quad (4.14)$$

Applying the Green-Gauss theorem to the surface integral, we see that the necessary and sufficient condition for (4.14) to be valid for arbitrary $\Delta v + \Delta s$ is

$$\langle \phi\, t_{k l} \rangle^*{}_{,k} - \langle \phi_{,k}\, t_{k l} \rangle + \langle \phi\, \varrho (f_l - \dot{v}_l) \rangle = 0 \text{ in } v \quad (4.15)$$

$$[\langle \phi\, t_{k l} \rangle^*]\, n_k = 0 \quad \text{on } s_d \quad (4.16)$$

where a bold face bracket indicates jumps across a discontinuity surface s_d. Hence

Theorem 2. *In a micromorphic continuum of degree* 1, *the micro-momenta are balanced if and only if* (4.15) *and* (4.16) *are satisfied.*

A similar (but longer) procedure applied to (4.11) leads to the equation for energy balance

$$\begin{aligned} \langle \phi\, \varrho\, \varepsilon \rangle^{\cdot} - \dot{\varrho}\, \varrho^{-1} \langle \phi\, \varrho\, \varepsilon \rangle - \langle \dot{\phi}\, \varrho\, \varepsilon \rangle &= \langle \phi\, t_{k l} \rangle^* \, v_{l,k} + \langle \phi\, t_{k l}\, \xi_m \rangle^* \, \nu_{l m,k} + \\ &+ [\langle \phi t_{k l} \xi_m \rangle^*{}_{,k} - \langle \phi_{,k} t_{k l} \xi_m \rangle + \langle \phi\, \varrho\, (f_l - \dot{v}_l)\, \xi_m \rangle]\, \nu_{l m} + \\ &+ \langle \phi\, q_k \rangle^*{}_{,k} - \langle \phi_{,k}\, q_k \rangle + \langle \phi\, \varrho\, h \rangle \quad \text{in } v, \end{aligned} \quad (4.17)$$

$$[\langle \phi\, t_{k l} \rangle^* \, v_l + \langle \phi\, t_{k l}\, \xi_m \rangle^* \, \nu_{l m} + \langle \phi\, q_k \rangle^*]\, n_k = 0 \quad \text{on } s_d \quad (4.18)$$

and the theorem

Theorem 3. *The necessary and sufficient conditions for the balance of microenergy of a micromorphic continuum of degree* 1 *are* (4.17) *and* (4.18).

Equations (4.13) and (4.15) to (4.17) are the master balance equations from which we can derive various order theories by use of statistical moments.

5. Micromorphic Continuum, Grade 1

We now take three special values for ϕ.

$$\phi = \phi_0 = 1, \quad (5.1)$$

$$\phi = x_k + \xi_k^{(\alpha)}, \quad \phi_0 = x_k + \chi_{k K}\, \Xi_K^{(\alpha)}, \quad (5.2)$$

$$\phi = (x_k + \xi_k^{(\alpha)})\, (x_l + \xi_l^{(\alpha)}), \quad \phi_0 = (x_k + \chi_{k K}\, \Xi_K^{(\alpha)})\, (x_l + \chi_{l L}\, \Xi_L^{(\alpha)}) . \quad (5.3)$$

A micromorphic body of degree 1 is assigned a grade 1 if (5.1) to (5.3) are used in the mass balance (4.13), (5.1) and (5.2) in the momentum balance (4.15), and (5.1) in the energy balance (4.17). According to this, the first three moments of mass balance, the first two moments of momentum balance, and only the average of the energy constitute the complete set of mechanical balance laws.

Using (5.1) to (5.3) in (4.13), we get respectively

$$\varrho\, dv = \varrho_0\, dv_0\,, \tag{5.4}$$

$$\langle \varrho\, \xi_k \rangle\, dv = \langle \varrho_0\, \Xi_K \rangle\, \chi_{kK}\, dv_0 = 0\,, \tag{5.5}$$

$$i_{kl} = I_{KL}\, \chi_{kK}\, \chi_{lL} \tag{5.6}$$

where

$$i_{kl} \equiv \langle \varrho\, \xi_k\, \xi_l \rangle, \qquad I_{KL} \equiv \langle \varrho_0\, \Xi_K\, \Xi_L \rangle \tag{5.7}$$

are called microinertia tensors of the deformed and undeformed bodies. We note that (5.5) vanishes because X_K was taken as the center of mass of a macrovolume, thus, in addition to the classical equation of continuity (5.4), we have two theorems:

Theorem 4. *In a micromorphic body, degree* 1 *and grade* 1, *the motion carries the center of mass of the undeformed body to the center of mass of the deformed body.*

Theorem 5. *In a micromorphic body, degree* 1 *and grade* 1, *the microinertia is conserved* [Eq. (5.6)].

By taking the material derivative of (5.6) and using $(4.8)_2$, we also obtain, Eringen [1964a],

$$\frac{\partial i_{kl}}{\partial t} + i_{kl,m}\, v_m - i_{km}\, \nu_{lm} - i_{ml}\, \nu_{km} = 0\,. \tag{5.8}$$

Of course the time rate of (5.4) gives the classical continuity equation

$$\frac{\partial \varrho}{\partial t} + (\varrho\, v_k)_{,k} = 0\,. \tag{5.9}$$

Eqs. (5.9) and (5.10) are of course equivalent to (5.4) and (5.6) respectively.

Substitution of (5.1) and (5.2) into (4.15) and (4.16) gives

$$t_{kl,k} + \varrho\,(f_l - \dot v_l) = 0\,, \tag{5.10}$$

$$t_{klm,k} + t_{ml} - \bar t_{lm} + \varrho\,(f_{lm} - a_{lm}) = 0\,, \quad \text{in } v\,, \tag{5.11}$$

$$[t_{kl}]\, n_k = 0\,, \qquad [t_{klm}]\, n_k = 0 \quad \text{on } s_d \tag{5.12}$$

where

$$\left.\begin{aligned}
t_{\underline{k}l}\, da_{\underline{k}} &\equiv \langle t_{\underline{k}l} \rangle^*\, da_{\underline{k}} \equiv \int_{da} t_{\underline{k}l}(\boldsymbol{x}, \boldsymbol{\xi}, t)\, da_{\underline{k}}(\boldsymbol{\xi})\,,\\
\bar t_{kl}\, dv &\equiv \langle t_{kl} \rangle\, dv \equiv \int_{dv} t_{kl}(\boldsymbol{x}, \boldsymbol{\xi}, t)\, dv(\boldsymbol{\xi})\,,\\
t_{klm} &\equiv \langle t_{kl}\, \xi_m \rangle^*, \qquad \varrho\, f_{lm} \equiv \langle \varrho\, f_l\, \xi_m \rangle,\\
\varrho\, a_{lm} &\equiv \langle \varrho\, \dot v_l\, \xi_m \rangle = i_{km}(\dot\nu_{lk} + \nu_{lr}\, \nu_{rk})\,.
\end{aligned}\right\} \tag{5.13}$$

Similarly using (5.1) in the master equation of energy (4.17) and (4.18), we get

$$\varrho\,\dot{\varepsilon} = t_{kl}\,v_{l,k} + (\bar{t}_{kl} - t_{kl})\,v_{lk} + t_{klm}\,v_{lm,k} + q_{k,k} + \varrho\,h \quad \text{in } v, \tag{5.14}$$

$$[t_{kl}\,v_l + t_{klm}\,v_{lm} + q_k]\,n_k = 0 \quad \text{on } s_d \tag{5.15}$$

where

$$\varrho\,\varepsilon \equiv \langle\varrho\,\varepsilon\rangle, \qquad q_k \equiv \langle q_k\rangle, \qquad \varrho\,h \equiv \langle\varrho\,h\rangle.$$

Eqs. (5.8), (5.9), (5.10), (5.11) and (5.14) are the local balance laws of a micromorphic continuum, degree 1 and grade 1. Eqs. (5.12) and (5.15) are the jump conditions of this theory. They also give the boundary conditions on stress averages, stress moments, and the heat vector on the surface of the body. We state these results as:

Theorem 6. *In a micromorphic continuum, degree* 1 *and grade* 1, *the momentum is balanced if and only if* (5.10) *and* $(5.12)_1$ *are satisfied.*

Theorem 7. *In a micromorphic continuum, degree* 1 *and grade* 1, *the first moments of momentum are balanced if and only if* (5.11) *and* $(5.12)_2$ *are satisfied.*

Theorem 8. *In a micromorphic continuum, degree* 1 *and grade* 1, *the energy is balanced if and only if* (5.14) *and* (5.15) *are satisfied.*

6. Higher Grades of Micromorphic Continua

A collection of moments up to (not including) order N of the energy equation, and up to order $N+1$ of the momentum equation and order $N+2$ of the continuity equation constitute a micromorphic continuum of grade $N > 0$. For example, for a micromorphic continuum of grade 2, we add to the equations obtained in Art. 5 the third-order moments of the mass, the second-order moments of the momentum balance, and the first-order moments of the energy balance. These are

$$i_{klm} = I_{KLM}\,\chi_{kK}\,\chi_{lL}\,\chi_{mM}, \tag{6.1}$$

$$t_{klmn,k} + t_{mln} + t_{nlm} - \bar{t}_{mln} - \bar{t}_{nlm} + \varrho\,(f_{lmn} - a_{lmn}) = 0, \tag{6.2}$$

$$\begin{aligned}\varrho\,\dot{\varepsilon}_n - \varrho\,\varepsilon_l\,v_{nl} = {} & t_{kln}\,v_{l,k} + t_{klmn}\,v_{lm,k} + \\ & + (\bar{t}_{mln} - t_{mln})\,v_{lm} + q_{kn,k} \\ & + \bar{q}_n + \varrho\,h_n.\end{aligned} \tag{6.3}$$

Various moments are defined as before, e.g.,

$$\varrho\,\varepsilon_n \equiv \langle\varrho\,\varepsilon\,\xi_n\rangle, \qquad \bar{q}_k\,dv \equiv \int_{dv} q_k\,dv.$$

Each of the above sets represents a balance law for the second grade micromorphic continua. Corresponding to each, we also have a set of jump conditions. It is interesting to note the appearance of a micro-energy vector, ε_n, and microheat flux tensor, q_{kn}, in addition to fourth-order stress moments t_{klmn} and third-order spin inertia a_{lmn}. We do not pursue this subject any further here.

7. Constitutive Equations

A general constitutive theory for memory dependent thermo-mechanical micromorphic materials can be constructed by assuming that various stress moments, heat vector, internal energy density, and entropy are functionals of the history of the macrodeformation gradient, microdeformation tensors, temperature, and material position over a range of time $-\infty < \tau \leqq t$. The axiom of objectivity can then be used to show that these functionals depend on the difference histories of strain measures and temperature from their values at $\tau = t$, the microinertia and the material position. For micromorphic continua of degree 1 and grade 1, one of these equations has the form

$$t_{kl} = \underset{\tau=-\infty}{\overset{t}{F_{kl}}}[C_{KL}(t-\tau), \Psi_{KL}(t-\tau), \Gamma_{KLM}(t-\tau), \theta(t-\tau), I_{KL}, X_K]. \tag{7.1}$$

Similar functional equations are written for $\bar{t}_{kl}$, t_{klm}, ε and η. The second law of thermodynamics and the axiom of admissibility, ERINGEN [1966c], and objectivity can then be used to reduce these equations further. Below we reproduce the constitutive equations for a micro-elastic solid obtained by ERINGEN and SUHUBI [1964b].

$$t_{kl} = \frac{\varrho}{\varrho_0}\left[2\frac{\partial\Sigma}{\partial C_{KL}} x_{k,K}\, x_{l,L} + \left(\frac{\partial\Sigma}{\partial\Psi_{KL}}\chi_{lL} + \frac{\partial\Sigma}{\partial\Gamma_{KLM}}\chi_{lL,M}\right) x_{k,K}\right] \tag{7.2}$$

$$\bar{t}_{kl} = 2\frac{\varrho}{\varrho_0}\left[\frac{\partial\Sigma}{\partial C_{KL}} x_{k,K}\, x_{l,L} + \frac{\partial\Sigma}{\partial\Psi_{KL}} x_{(k,K}\chi_{l)L} + \frac{\partial\Sigma}{\partial\Gamma_{KLM}} x_{(k,K}\chi_{l)L,M}\right], \tag{7.3}$$

$$t_{klm} = \frac{\varrho}{\varrho_0}\frac{\partial\Sigma}{\partial\Gamma_{KLM}} x_{k,M}\, x_{l,K}\, \chi_{mL} \tag{7.4}$$

where the potential Σ is the free energy multiplied by ϱ_0.

$$\varrho_0\,\psi \equiv \Sigma = \Sigma(C_{KL}, \Psi_{KL}, \Gamma_{KLM}, \theta, X_K). \tag{7.5}$$

The linear and other approximate forms of these equations for an isotropic medium are to be found in the above cited work. ERINGEN [1964]—[1967] also developed theories of microfluids, various micro-elastic materials, and a special class of materials called micropolar continua. Some of the fundamental results of linear micropolar elasticity and micropolar fluids are condensed in the following section.

8. Linear, Isotropic Micropolar Elasticity[1]

A micropolar continuum, roughly speaking, is a collection of elongated rigid microelements. It emanates as a special continuum from the micromorphic theory by setting

$$\begin{aligned} \Phi_{KL} &= -\Phi_{LK} \equiv -\varepsilon_{KLM}\,\Phi_M, & \nu_{kl} &\equiv \varepsilon_{rkl}\,\nu_r, \\ t_{klm} &= -t_{kml} = -\tfrac{1}{2}\varepsilon_{lmr}\,m_{kr}, & f_{kr} &\equiv -\tfrac{1}{2}\varepsilon_{rkl}\,l_r \end{aligned} \tag{8.1}$$

where ε_{KLM} and ε_{klm} are the alternating tensors.

In this case, for the location of the spatial position of a point $\boldsymbol{X}^{(\alpha)} + \boldsymbol{\Xi}^{(\alpha)}$, in the microelement we have

$$\boldsymbol{x}^{(\alpha)} = \boldsymbol{X} + \boldsymbol{\Xi}^{(\alpha)} + \boldsymbol{u} + \boldsymbol{\Phi} \times \boldsymbol{\Xi}^{(\alpha)} \tag{8.2}$$

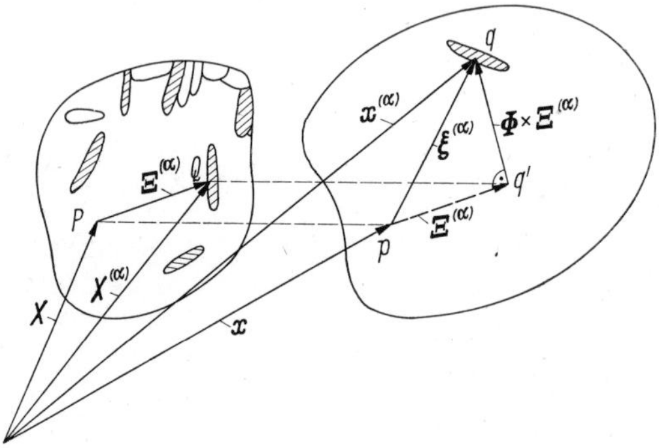

Fig. 5. Micropolar continuum (elongated rigid microelements).

where $\boldsymbol{\Phi}(\boldsymbol{X}, t)$ is the material microrotation vector. Thus the microdeformation is perpendicular to $\boldsymbol{\Xi}^{(\alpha)}$, Fig. 5. In this case, the balance laws can be shown to reduce to

$$\frac{\partial \varrho}{\partial t} + (\varrho\, v_k)_{,k} = 0 \quad \text{(mass)}, \tag{8.3}$$

$$t_{kl,k} + \varrho\,(f_l - \dot{v}_l) = 0 \quad \text{(momentum)}, \tag{8.4}$$

$$m_{rk,r} + \varepsilon_{klr}\,t_{lr} + \varrho\,(l_k - j\,\dot{\nu}_k) = 0 \quad \text{(moment of momentum)}, \tag{8.5}$$

$$\varrho\,\dot{\varepsilon} = t_{kl}(v_{l,k} - \varepsilon_{klr}\,\nu_r) + m_{kl}\,\nu_{l,k} + q_{k,k} + \varrho\,h \quad \text{(energy)}, \tag{8.6}$$

$$\varrho\left(\dot{\eta} - \frac{\dot{\varepsilon}}{\theta}\right) + \frac{1}{\theta}\,t_{kl}(v_{l,k} - \varepsilon_{klr}\,\nu_r) + \frac{1}{\theta}\,m_{kl}\,\nu_{l,k} + \frac{q_k\,\theta_{,k}}{\theta^2} \geqq 0.$$

(Clausius-Duhem inequality) (8.7)

[1] The full theory was first given by ERINGEN and SUHUBI [1964b] and later, ERINGEN [1965], [1966a, b], was recapitulated with additional results under the present title.

The linear constitutive equations for an isotropic micropolar elastic solid are

$$t_{kl} = \lambda\, \varepsilon_{rr}\, \delta_{kl} + (\mu + \varkappa)\, \varepsilon_{kl} + \mu\, \varepsilon_{lk}, \tag{8.8}$$

$$m_{kl} = \alpha\, \phi_{r,r}\, \delta_{kl} + \beta\, \phi_{k,l} + \gamma\, \phi_{l,k}, \tag{8.9}$$

$$\varepsilon_{kl} \equiv u_{l,k} + \varepsilon_{lkm}\, \phi_m, \qquad i_{kl} = -\tfrac{1}{2} j\, \delta_{kl}, \tag{8.10}$$

where $\phi_k(\boldsymbol{x}, t)$ is the spatial microrotation vector. Eringen [1965], [1966a] proved that the internal energy is non negative if and only if the above elastic constants satisfy the inequalities

$$\begin{aligned} &0 \leqq 3\lambda + 2\mu + \varkappa, \qquad && 0 \leqq \mu, \qquad && 0 \leqq \varkappa, \\ &0 \leqq 3\alpha + 2\gamma, && -\gamma \leqq \beta \leqq \gamma, && 0 \leqq \gamma. \end{aligned} \tag{8.11}$$

Substitution of (8.9) to (8.11) into (8.4) and (8.5) gives the field equations

$$(\lambda + 2\mu + \varkappa)\, \nabla\nabla \cdot \boldsymbol{u} - (\mu + \varkappa)\, \nabla \times \nabla \times \boldsymbol{u} + \varkappa\, \nabla \times \boldsymbol{\phi} + \varrho(\boldsymbol{f} - \ddot{\boldsymbol{u}}) = \boldsymbol{0}, \tag{8.12}$$

$$(\alpha + \beta + \gamma)\, \nabla\nabla \cdot \boldsymbol{\phi} - \gamma\, \nabla \times \nabla \times \boldsymbol{\phi} + \varkappa\, \nabla \times \boldsymbol{u} - 2\varkappa\, \boldsymbol{\phi} + \varrho(\boldsymbol{l} - j\, \ddot{\boldsymbol{\phi}}) = \boldsymbol{0}. \tag{8.13}$$

The uniqueness theorem for the solution of (8.12) and (8.13) has been proved. A set of usual boundary and initial conditions is

$$\begin{aligned} \boldsymbol{u}(\boldsymbol{x}', t) &= \boldsymbol{u}', \quad && \boldsymbol{\phi}(\boldsymbol{x}', t) = \boldsymbol{\phi}', \quad && \text{on } s_u, \\ t_{lk}\, n_l &= t'_k, && m_{lk}\, n_l = m'_k, && \text{on } s - s_u, \end{aligned} \tag{8.14}$$

$$\begin{aligned} \boldsymbol{u}(\boldsymbol{x}, 0) &= \boldsymbol{u}_0(\boldsymbol{x}), \quad && \dot{\boldsymbol{u}}(\boldsymbol{x}, 0) = \boldsymbol{v}_0(x), \\ \boldsymbol{\phi}(\boldsymbol{x}, 0) &= \boldsymbol{\phi}_0(\boldsymbol{x}), && \dot{\boldsymbol{\phi}}(\boldsymbol{x}, 0) = \boldsymbol{\nu}_0(x) \quad \text{in } v. \end{aligned} \tag{8.15}$$

The field equations of micropolar fluids are similar to (8.12) and (8.13) with the exception that $\boldsymbol{u}$ now represents the velocity field and $\ddot{\boldsymbol{u}}$ and $\ddot{\boldsymbol{\phi}}$ are respectively replaced by $\dot{\boldsymbol{u}}$ and $\dot{\boldsymbol{\nu}}$. For an extensive account of micropolar elasticity see also Eringen [1967c]. For a discussion of anisotropic fluids and micropolar viscoelasticity, the reader is referred to Eringen [1964b] and [1967a].

9. Relation to Other Theories

In this section, we indicate briefly the relation of micromorphic theory to several other theories.

a) Indeterminate couple stress theory. In several of our previous works [1964b], [1966a, b], we have shown that the indeterminate couple stress theory is obtained from the micropolar theory when the motion is constrained so that the macrorotations and microrotations coincide, i.e.,

$$\phi_k = r_k = \tfrac{1}{2} \varepsilon_{klm}\, u_{m,l}. \tag{9.1}$$

In this case, we write

$$t_{kl} = t_{(kl)} + t_{[kl]} \tag{9.2}$$

and use (8.8) to calculate $t_{(kl)}$ and (8.5) to calculate $t_{[kl]}$

$$\begin{aligned} t_{(kl)} &= \lambda\, u_{r,r}\, \delta_{kl} + \left(\mu + \frac{\varkappa}{2}\right)(u_{k,l} + u_{l,k}), \\ t_{[kl]} &= \frac{\gamma}{2} \nabla^2 u_{[k,l]} - \frac{1}{2} \varrho\,(\varepsilon_{rkl}\, l_r + j\, \ddot{u}_{[k,l]}). \end{aligned} \tag{9.3}$$

When (9.1) and (9.2) are carried into (8.4), we obtain

$$(\lambda + 2\mu + \varkappa)\, \nabla\nabla\cdot\boldsymbol{u} - \left(\mu + \frac{\varkappa}{2} - \frac{\gamma}{4}\nabla^2\right)\nabla\times\nabla\times\boldsymbol{u} + \varrho\left(\boldsymbol{f} - \frac{1}{2}\nabla\times\boldsymbol{l}\right) - \\ - \varrho\left(\boldsymbol{1} + \frac{j}{4}\nabla\times\nabla\times\right)\ddot{\boldsymbol{u}} = 0 \tag{9.4}$$

which becomes identical to the result obtained by Mindlin and Tiersten [1962] if we write μ for $\mu + \varkappa/2$, $\eta = \gamma/4$ and $j \equiv 0$.

b) Asymmetric elasticity. The asymmetric elasticity discussed by Palmov [1964] can be made to coincide with the linear, isotropic micropolar elasticity by some obvious notational changes. Palmov uses the principles of the balance of momenta and virtual displacements, however, with no inertia forces. The inertia forces are introduced later as d'Alembert forces. The theory is linear; a linear rotatory inertia is used and microisotropy is implied. No uniqueness theorem or restrictions on elastic moduli is given.

c) Microstructure elasticity. The microstructure elasticity of Mindlin [1964] can be made to coincide with the linear micromorphic elasticity, of degree 1 and grade 1, if we notice the following correspondence in notation

Micromorphic	*Microstructure*	*Micromorphic*	*Microstructure*
u_i	u_i	$\varrho\, i_{kl}$	$\frac{1}{3}\varrho'\, d_{kl}^2$
ϕ_{ij}	ψ_{ji}	t_{kl}	$\tau_{kl} + \sigma_{kl}$
e_{ij}	ε_{ij}	$s_{kl} \equiv \bar{t}_{kl}$	τ_{kl}
ε_{ij}	$2\varepsilon_{ij} - \gamma_{ji}$	$\lambda_{klm} \equiv t_{klm}$	μ_{kml}
γ_{ijk}	$-\varkappa_{kji}$	$\varrho\, l_{ij}$	Φ_{ji}

Mindlin employs Hamilton's principle to derive the balance of momenta. In this work, the local balance equations of energy, entropy, and microinertia are not given.

d) Multipolar theory. In the space available, it is not possible to discuss fully the relations of the multipolar theory of Green and Rivlin [1964] to the micromorphic theory. We draw attention to a few crucial points: in mechanics, displacement is a well-defined quantity and forces are postulated. GREEN and RIVLIN, however, postulate both multipolar displacement and forces, their inner products being energies. By use of the postulated invariance of energy to rigid translations and rotations, they obtain two balance equations—one for linear momentum and one for energy. Two other sets of fundamental equations are introduced as *definitions*, namely,

$$\bar{\sigma}_{ij_1\cdots j_\beta} \equiv \bar{F}_{ij_1\cdots j_\beta} + \sigma_{kij_1\cdots j_\beta,k},$$

$$\sigma'_{im} = \sigma'_{mi} \equiv \sigma_{im} - \sum_{\beta=1}^{\nu} \bar{\sigma}_{ij_1\ldots j_\beta} x_{mj_1\ldots j_\beta} - \sum_{\beta=1}^{\nu} \sigma_{kij_1\ldots j_\beta} x_{mj_1\ldots j_\beta,k}.$$

Constitutive equations are then written for $\bar{\sigma}_{ij_1\ldots j_\beta}$ and $\sigma_{ij_1\ldots j_\beta}$ which elevate the above definitions to balance laws. This a priori choice needs support on physical grounds if it is to be regarded a physical theory.

The particle model they introduce to provide a contact with physics possesses certain limitations also. For example, the 2^β-pole displacement $x_{iB_1\ldots B_\beta}(\tau)$ introduced by (17.22) is completely symmetrical with respect to all indices except i. Thus the number of independent components is grossly reduced. For example, for the 2^2-pole displacement $x_{iB_1B_2}$, we only have 18 independent components in three dimensions instead of 27. In multipolar theory, no balance laws for multipolar inertia are provided.

GREEN [1965] by redefining and/or reinterpreting some of the abstract notions of multipolar theory has brought the dipolar case in contact with the micromorphic theory (of degree 1 and grade 1).

e) Non-holonomic geometry. Suppose that the condition (2.4) for the existence of the inverse micromotion $\boldsymbol{\Xi} = \boldsymbol{\Xi}(\boldsymbol{x}, \boldsymbol{\xi}, t)$ is violated, i.e.,

$$\det(\partial \xi_k^{(\alpha)}/\partial \Xi_K^{(\alpha)}) = 0.$$

In this case, the microelements slip with respect to each other and split and/or recombine to make new elements. The structure of material is now *torn*. In this case, we can assume

$$d\boldsymbol{\xi}^{(\alpha)} = \boldsymbol{B}_K^{(\alpha)}(\boldsymbol{X}, \boldsymbol{\Xi}^{(\alpha)}, t)\, dX_K.$$

By selecting $\boldsymbol{A}_K^{(\alpha)} \equiv \boldsymbol{B}_K^{(\alpha)} + \partial \boldsymbol{x}^{(\alpha)}/\partial X_K$, (2.1) may be written as

$$d\boldsymbol{x}^{(\alpha)} = \boldsymbol{A}_K^{(\alpha)}\, dX_K.$$

Here $d\boldsymbol{x}^{(\alpha)}$ is *not* a total differential when dX_K is. The deformation tensors are now given by

$$C_{KL}(\boldsymbol{X}, \boldsymbol{\Xi}^{(\alpha)}, t) \equiv \boldsymbol{A}_K^{(\alpha)} \cdot \boldsymbol{A}_L^{(\alpha)}.$$

Thus we obtain a *Finslerian metric* with directors $\Xi_K^{(\alpha)}$. Displacements $\boldsymbol{x}^{(\alpha)} = \boldsymbol{x}^{(\alpha)}(\boldsymbol{X}, \boldsymbol{\Xi}^{(\alpha)}, t)$ cannot be determined since $d\boldsymbol{x}^{(\alpha)}$ is *not* integrable. This is the geometry used by KONDO and his co-workers in connection with plasticity and yielding, c.f., KONDO [1962, Division D] and other earlier references therein.

Finally a six-dimensional geometry can be used encompassing all these theories. A separate publication on this topic is intended in the near future.

References

AERO, E. L., and E. V. KUVSHINSKII [1960]: Fundamental equations of the theory of elastic media with rotationally interacting particles. Fiziko Tverdogo Tela **2**, 1399–1409; Translation, Soviet Physics Solid State **2**, 1272–1281 (1961).

COSSERAT, E., and F. COSSERAT [1909]: Théorie des Corps Déformables. Paris: A. Hermann.

ERINGEN, A. C. [1962]: Nonlinear Theory of Continuous Media. New York: McGraw-Hill.

ERINGEN, A. C., and E. S. SUHUBI [1964a]: Nonlinear Theory of Simple Micro-elastic Solids. I. Int. J. Engng. Sci. **2**, 2, 189–203.

ERINGEN, A. C., and E. S. SUHUBI: [1964b]: Nonlinear Theory of Simple Micro-elastic Solids. II. Int. J. Engng. Sci., **2**, 4, 389–404.

ERINGEN, A. C. [1964a]: Simple Micro-fluids. Int. J. Engng. Sci., **2**, 2, 205–217.

ERINGEN, A. C. [1964b]: Mechanics of Micromorphic Materials. Proc. of the Eleventh International Congress of Applied Mechanics. Berlin/Heidelberg/New York: Springer 1966, pp. 131–138.

ERINGEN, A. C. [1965]: Theory of Micropolar Continua. Proc. 9th Midwestern Mechanics Congress, 3, part 1. Wiley 1967, pp. 23–40.

ERINGEN, A. C. [1966a]: Linear Theory of Micropolar Elasticity. J. Math. & Mech. **15**, 6, 909–924.

ERINGEN, A. C. (1966b]: Theory of Micropolar Fluids. J. Math. & Mech., **16**, 1, 1–18.

ERINGEN, A. C. [1966c]: A Unified Theory of Thermomechanical Materials. Int. J. Engng. Sci., **4**, 2, 179–202.

ERINGEN, A. C. [1967a]: Linear Theory of Micropolar Viscoelasticity. Int. J. Engng. Sci., **5**, 2, 191–204.

ERINGEN, A. C. [1967b]: Theory of Micropolar Plates. J. Appl. Math. & Phys., **18**, 1, 12–30.

ERINGEN, A. C. [1967c]: Theory of Micropolar Elasticity. Contribution to Treatise on Fracture edited by H. LIEBOWITZ. To be published by Academic Press.

GREEN, A. E., and R. S. RIVLIN [1964]: Multipolar Continuum Mechanics. Arch. Rat. Mech. Anal., **17**, 113–147.

GREEN, A. E. [1965]: Micro-Materials and Multipolar Continuum Mechanics. Int. J. Engng. Sci., **3**, 5, 533–537.

GRIOLI, G. [1960]: Elasticity Asimmetrica. Ann. di Mat. Pura ed Appl. Ser. IV, **50**, 389–417.

GÜNTHER, W. [1958]: Zur Statik und Kinematik des Cosseratschen Kontinuums. Abh. d. Braunschweigischen Wiss. Ges. 10, 195–213.

KONDO, K. [1962]: RAAG Memoirs of the Unifying Study of Basic Problems in Engineering and Physical Sciences by Means of Geometry, III.

MacCullagh, J. [1839]: An Essay Towards a Dynamical Theory of Crystalline Reflection and Refraction. Trans. Roy. Irish Acad. Sci., **21**, 17—50.

Mindlin, R. D. [1964]: Micro-structure in Linear Elasticity. Arch. Rat. Mech. Anal., **16**, 51—78.

Mindlin, R. D., and H. F. Tiersten [1962]: Effects of Couple-Stresses in Linear Elasticity. Arch. Rat. Mech. Anal., **11**, 415—448.

Palmov, V. A. [1964]: Fundamental Equations of the Theory of Asymmetric Elasticity. PMM, **28**, 401—408.

Schäefer, H. [1962]: Versuch einer Elastizitätstheorie des zweidimensionalen ebenen Cosserat-Continuums. Miszellaneen der angewandten Mechanik. Akademie-Verlag, pp. 277—292.

Thomson, W. (Lord Kelvin), and P. G. Tait [1867]: Treatise on Natural Philosophy. First ed., Oxford University Press.

Toupin, R. A. [1962]: Elastic Materials with Couple-Stresses. Arch. Rat. Mech. Anal., **11**, 385—414.

Truesdell, C., and R. A. Toupin [1960]: The Classical Field Theories. Handbuch der Physik III/1. Berlin/Göttingen/Heidelberg: Springer 1960.

Voigt, W. [1887]: Theoretische Studien über die Elastizitätsverhältnisse der Krystalle. Abh. Ges. Wiss. Göttingen 34. [1894]: Über Medien ohne innere Kräfte und eine durch sie gelieferte mechanische Deutung der Maxwell-Hertzschen Gleichungen. Gött. Abh. 72.79.

The Cosserat Surface

By

A. E. Green and **P. M. Naghdi**

University of Oxford

University of California
Berkeley, Calif.

Abstract. An account of a general theory of a Cosserat surface is presented and some special cases, including some features of the linear theory of an elastic Cosserat plate, are discussed.

1. Introduction

Historically, the concept of "directed" or "oriented" media was originated by DUHEM [1] and a first systematic development of theories of oriented media in one, two and three-dimensions (the first two being motivated by rods and shells) was carried out by E. and F. COSSERAT [2]. In their work, the Cosserats represented the orientation of each point of their continuum by a set of mutually perpendicular rigid vectors. More recently, the purely kinematical aspects of oriented bodies characterized by ordinary displacement and the independent deformation of *n directors* (i.e. deformable vectors) in *n* dimensional space has been discussed by ERICKSEN and TRUESDELL [3]. General nonlinear theories for three dimensional continua in which directors are admitted as basic kinematical ingredients have been given by ERICKSEN [4] and by GREEN, NAGHDI and RIVLIN [5]. Also TOUPIN [6] has discussed, among other developments, a theory of couple-stress with directors for isothermal elastic materials. References to a number of related recent works on the subject may be found in the papers already cited.

Consider now a surface embedded in a Euclidean 3-space to every point of which one or more deformable directors are assigned. Among recent contributions which deal with some aspects of nonlinear and linear theories of such a surface we cite here references [7—11]. In this paper we present an account of the theory of a deformable surface with a single director, called a *Cosserat surface*, based on references

Acknowledgment. The work of one of us (P.M.N.) was supported by the U.S. Office of Naval Research under Contract Nonr-222(69) with the University of California, Berkeley.

[7, 9—11]. This is a nonlinear theory and is not limited to elastic surfaces.

After deducing the basic field equations, we briefly consider the constitutive equations for an elastic Cosserat surface and also discuss a special case of the general theory which is of interest in certain applications. The remainder of the paper is confined to a linear theory of an initially flat Cosserat surface (or a Cosserat plate) in which the directors are initially coincident with the normals to the surface. In particular, we examine the linear (isothermal) theory of an elastic Cosserat plate whose constitutive equations imitate those of a transversely isotropic material and which are such that the differential equations of the complete theory separate into those for extensional and those for flexural deformations.

2. Preliminaries. Kinematics

A surface to every point of which a single director—not necessarily along the normal to the surface—is assigned will be called a *Cosserat surface*. Although we confine attention here to a surface with a single director, the inclusion of additional assigned directors at each point of the surface is possible and the theory can be developed in a similar manner.

Let a surface $\mathfrak{s}$, embedded in a Euclidean 3-space, be defined by its position vector relative to a fixed origin and let a director (i.e., a deformable vector) be assigned to every point of $\mathfrak{s}$; the director remains invariant in length under superposed rigid body motions. Further, let x^α ($\alpha = 1, 2$) be convected coordinates defining points of the surface. The position vector of a typical point of $\mathfrak{s}$ at time t and the director displacement at the same point will be designated by $\boldsymbol{r}$ and $\boldsymbol{d}$, respectively, where

$$\boldsymbol{r} = \boldsymbol{r}(x^\alpha, t), \qquad \boldsymbol{d} = \boldsymbol{d}(x^\alpha, t). \tag{2.1}$$

If $\boldsymbol{a}_\alpha$ are the base vectors along the x^α-curves on $\mathfrak{s}$, then

$$\begin{gathered} \boldsymbol{a}_\alpha = \boldsymbol{r}_{,\alpha}, \qquad \boldsymbol{a}_\alpha \cdot \boldsymbol{a}_\beta = a_{\alpha\beta}, \qquad a = \det(a_{\alpha\beta}) > 0, \\ \boldsymbol{a}^\alpha \cdot \boldsymbol{a}_\beta = \delta^\alpha_\beta, \qquad a^{\alpha\gamma} a_{\gamma\beta} = \delta^\alpha_\beta, \qquad \boldsymbol{a}^\alpha = a^{\alpha\gamma} \boldsymbol{a}_\gamma, \end{gathered} \tag{2.2}$$

and the unit normal $\boldsymbol{a}_3$ to $\mathfrak{s}$ may be defined by

$$\boldsymbol{a}_\alpha \cdot \boldsymbol{a}_3 = 0, \qquad \boldsymbol{a}_3 \cdot \boldsymbol{a}_3 = 1, \qquad \boldsymbol{a}^3 = \boldsymbol{a}_3, \qquad [\boldsymbol{a}_1\, \boldsymbol{a}_2\, \boldsymbol{a}_3] > 0, \tag{2.3}$$

where a comma denotes partial differentiation with respect to x^α, $a_{\alpha\beta}$ is the first fundamental form of the surface, $a^{\alpha\beta}$ is its conjugate and δ^α_β is the Kronecker symbol in 2-space. For later reference, we also recall the well-known formulae

$$\boldsymbol{a}_{\alpha|\beta} = b_{\alpha\beta}\, \boldsymbol{a}_3, \qquad \boldsymbol{a}_{3,\beta} = -b^\alpha_\beta\, \boldsymbol{a}_\alpha, \qquad b_{\alpha\beta|\gamma} = b_{\alpha\gamma|\beta}, \tag{2.4}$$

where a vertical line stands for covariant differentiation with respect to $a_{\alpha\beta}$ and $b_{\alpha\beta}$ is the second fundamental form of the surface.

In the above equations and throughout the paper, all Greek indices take the values 1, 2, Latin indices take the values 1, 2, 3, and the usual summation convention is employed. We denote the initial values of $\boldsymbol{r}$ and $\boldsymbol{d}$ at time $t = 0$ by $\boldsymbol{R}$ and $\boldsymbol{D}$ and refer to the initial (undeformed) surface by $\mathfrak{S}$. Also, we designate the initial values of $\boldsymbol{a}_i, a_{\alpha\beta}, b_{\alpha\beta}$ by $\boldsymbol{A}_i, A_{\alpha\beta}, B_{\alpha\beta}$, respectively, and note that results similar to (2.2) to (2.4) hold also for the surface $\mathfrak{S}$.

Let $\boldsymbol{v}$ and $\boldsymbol{w}$ denote, respectively, the velocity of a point of $\mathfrak{s}$ and the director velocity at time t. Then,

$$\boldsymbol{v} = \dot{\boldsymbol{r}}, \qquad \boldsymbol{w} = \dot{\boldsymbol{d}}, \tag{2.5}$$

where a superposed dot stands for the material derivative with respect to t, holding x^α fixed. When referred to the base vectors $\boldsymbol{a}_i = \{\boldsymbol{a}_\alpha, \boldsymbol{a}_3\}$, the velocity is

$$\boldsymbol{v} = v^i \boldsymbol{a}_i = v^\alpha \boldsymbol{a}_\alpha + v^3 \boldsymbol{a}_3 = v_i \boldsymbol{a}^i, \tag{2.6}$$

where the lowering and raising of subscripts and superscripts of space tensor functions such as v^i in (2.6) and $\psi_{\mathrm{k}\,i}$ in (2.8) is accomplished with the help of the metric tensor $a_{i\mathrm{k}} = \boldsymbol{a}_i \cdot \boldsymbol{a}_\mathrm{k}$. Since the coordinate curves on $\mathfrak{s}$ are convected, it follows that $\dot{\boldsymbol{a}}_\alpha = \boldsymbol{v}_{,\alpha}$ and

$$\boldsymbol{v}_{,\alpha} = \boldsymbol{v}_{|\alpha} = (v_{\beta|\alpha} - b_{\beta\alpha} v_3)\, \boldsymbol{a}^\beta + (v_{3,\alpha} + b^\beta_\alpha v_\beta)\, \boldsymbol{a}^3. \tag{2.7}$$

We now record without detailed proofs certain kinematical results given by GREEN, NAGHDI and WAINWRIGHT [7]. Thus

$$\dot{\boldsymbol{a}}_i = (\eta_{\mathrm{k}\,i} + \psi_{\mathrm{k}\,i})\, \boldsymbol{a}^\mathrm{k}, \tag{2.8}$$

$$\dot{\boldsymbol{a}}^\alpha = a^{\alpha\lambda} (\psi_{\mathrm{k}\lambda} - \eta_{\mathrm{k}\lambda})\, \boldsymbol{a}^\mathrm{k}, \qquad \dot{\boldsymbol{a}}^3 = \dot{\boldsymbol{a}}_3 = \psi_{\mathrm{k}3}\, \boldsymbol{a}^\mathrm{k}, \tag{2.9}$$

where

$$\begin{aligned} 2\eta_{\alpha\beta} &= v_{\alpha|\beta} + v_{\beta|\alpha} - 2 b_{\alpha\beta} v_3 = 2\eta_{\beta\alpha}, \\ \eta_{3\alpha} &= \eta_{\alpha 3} = 0, \qquad \eta_{33} = 0, \end{aligned} \tag{2.10}$$

$$\begin{aligned} 2\psi_{\alpha\beta} &= v_{\alpha|\beta} - v_{\beta|\alpha} = -2\psi_{\beta\alpha}, \\ \psi_{3\alpha} &= v_{3,\alpha} + b^\beta_\alpha v_\beta = -\psi_{\alpha 3}, \qquad \psi_{33} = 0, \end{aligned} \tag{2.11}$$

and $\eta_{\alpha\beta}$ and $\psi_{\alpha\beta}$ are the surface deformation rate tensor and the surface spin tensor, respectively. For future reference, we define here

$$2 e_{\alpha\beta} = a_{\alpha\beta} - A_{\alpha\beta} \tag{2.12}$$

as a measure of surface strain and also note that $\dot{e}_{\alpha\beta} = \eta_{\alpha\beta}$.

The director displacement $\boldsymbol{d}$ referred to $\boldsymbol{a}^i$ is

$$\boldsymbol{d} = d_i \boldsymbol{a}^i = d_\alpha \boldsymbol{a}^\alpha + d_3 \boldsymbol{a}^3. \tag{2.13}$$

Hence, by $(2.5)_2$, the director velocity may be written in the various forms

$$\begin{aligned} \boldsymbol{w} &= w_{\mathrm{k}}\, \boldsymbol{a}^{\mathrm{k}} \\ &= \boldsymbol{\Gamma} + d^{\alpha}(\boldsymbol{v}_{,\alpha} - 2\boldsymbol{\eta}_{\alpha}) + d^3\, \psi_{\mathrm{k}\,3}\, \boldsymbol{a}^{\mathrm{k}} \\ &= [\dot{d}_{\mathrm{k}} + d^i(\psi_{\mathrm{k}\,i} - \eta_{\mathrm{k}i})]\, \boldsymbol{a}^{\mathrm{k}}, \end{aligned} \tag{2.14}$$

where[1]

$$\boldsymbol{\eta}_{\alpha} = \eta_{k\alpha}\, \boldsymbol{a}^k, \qquad \boldsymbol{\Gamma} = \dot{d}_i\, \boldsymbol{a}^i. \tag{2.15}$$

If we put

$$\boldsymbol{d}_{,\alpha} = \lambda_{i\alpha}\, \boldsymbol{a}^i, \quad \lambda_{i\alpha} = \boldsymbol{a}_i \cdot \boldsymbol{d}_{,\alpha}, \tag{2.16}$$

then

$$\boldsymbol{w}_{,\alpha} = \boldsymbol{\Gamma}_{:\alpha} + \lambda^{i}_{.\alpha}\, \psi_{\mathrm{k}\,i}\, \boldsymbol{a}^{\mathrm{k}} - \lambda^{\beta}_{.\alpha}\, \boldsymbol{\eta}_{\beta}, \tag{2.17}$$

where

$$\begin{aligned} &\lambda_{\beta\alpha} = d_{\beta|\alpha} - b_{\alpha\beta}\, d_3, \quad && \lambda^{\beta}_{.\alpha} = a^{\beta\gamma}\, \lambda_{\gamma\alpha}, \quad && \lambda_{3\alpha} = d_{3,\alpha} + b^{\beta}_{\alpha}\, d_{\beta}, \\ &\lambda^{3}_{.\alpha} = \lambda_{3\alpha}, && \dot{\lambda}_{i\alpha} = \boldsymbol{a}_i \cdot \boldsymbol{\Gamma}_{:\alpha}. && \end{aligned} \tag{2.18}$$

3. Outline of the Theory

Let σ, an area of $\mathfrak{S}$ at time t, be bounded by a closed curve c, let $\boldsymbol{\nu} = \nu_{\alpha}\, \boldsymbol{a}^{\alpha}$ be the outward unit normal to c lying in the surface and let ϱ stand for the mass per unit area of $\mathfrak{S}$. If, for all arbitrary velocities $\boldsymbol{v}$ and all arbitrary director velocities $\boldsymbol{w}$, the scalars $\boldsymbol{N} \cdot \boldsymbol{v}$ and $\boldsymbol{M} \cdot \boldsymbol{w}$ are rates of work per unit length of c, then $\boldsymbol{N}$ and $\boldsymbol{M}$ are called, respectively, a curve force vector (or a force vector) and the director force vector per unit length. Similar definitions can be provided for the assigned surface force $\boldsymbol{F}$ and the assigned surface director force $\boldsymbol{L}$, per unit mass, through their (scalar) rate of work per unit area of $\mathfrak{S}$.

We assume that the assigned forces $\boldsymbol{F}$ and $\boldsymbol{L}$, per unit mass, act throughout σ (an arbitrary area of $\mathfrak{S}$ at time t) and that the force $\boldsymbol{N}$ and the director force $\boldsymbol{M}$, each per unit length, act across c (a boundary curve of σ). We also assume the existence of an internal energy function U per unit mass, an entropy function S per unit mass, a temperature function $T(>0)$, a heat supply function r per unit mass and per unit time and a heat flux h per unit length and per unit time. Then, the equation for the balance of energy and the Clausius-Duhem inequality may be stated in the forms

$$\frac{D}{Dt}\int_{\sigma} \varrho\, E\, d\sigma$$

$$= \int_{\sigma} \varrho[r + \boldsymbol{F}\cdot\boldsymbol{v} + \boldsymbol{L}\cdot\boldsymbol{w}]\, d\sigma + \int_{c} [\boldsymbol{N}\cdot\boldsymbol{v} + \boldsymbol{M}\cdot\boldsymbol{w} - h]\, dc, \tag{3.1}$$

---

[1] The notation for $\boldsymbol{\eta}_{\alpha}$ in $(2.15)_1$ is the negative of the corresponding quantity in Ref. [7].

and

$$\frac{D}{Dt}\int\limits_{\sigma} \varrho\, S\, d\sigma - \int\limits_{\sigma} \varrho \frac{r}{T}\, d\sigma + \int\limits_{c} \frac{h}{T}\, dc \geqq 0, \tag{3.2}$$

where D/Dt stands for the material derivative, E is the sum of the internal and kinetic energies per unit mass and is given by

$$E = U + K, \quad K = \tfrac{1}{2}(\boldsymbol{v}\cdot\boldsymbol{v} + \alpha\, \boldsymbol{w}\cdot\boldsymbol{w}), \tag{3.3}$$

K represents contributions to the kinetic energy due to ordinary and director velocities, and the coefficient α in $(3.3)_2$ is assumed to be a function of surface coordinates x^α, but independent of t.

We assume that ϱ, U, r, h, $\boldsymbol{N}$, $\boldsymbol{M}$, $(\boldsymbol{F} - \dot{\boldsymbol{v}})$ and $\bar{\boldsymbol{L}}$ defined by

$$\bar{\boldsymbol{L}} = \boldsymbol{L} - \alpha\, \dot{\boldsymbol{w}}, \tag{3.4}$$

remain unchanged by arbitrary superposed uniform rigid body translational velocities and consider (3.1) under such motions. We can then show that the local equation of conservation of mass is

$$\frac{D}{Dt}(\varrho\, a^{1/2}) = 0 \tag{3.5}$$

and also deduce the integral form of the equations of motion, namely

$$\int\limits_{\sigma} \varrho(\dot{\boldsymbol{v}} - \boldsymbol{F})\, d\sigma - \int\limits_{c} \boldsymbol{N}\, dc = 0. \tag{3.6}$$

The force vector $\boldsymbol{N}$ and the director force $\boldsymbol{M}$, when referred to the base vectors $\boldsymbol{a}_i$ can be expressed as

$$\boldsymbol{N} = N^i \boldsymbol{a}_i = N^\alpha \boldsymbol{a}_\alpha + N^3 \boldsymbol{a}_3, \quad \boldsymbol{M} = M^i \boldsymbol{a}_i = M^\alpha \boldsymbol{a}_\alpha + M^3 \boldsymbol{a}_3, \tag{3.7}$$

and the assigned force vectors $\boldsymbol{F}$ and $\boldsymbol{L}$ may be expressed similarly. The (physical) curve force vector over a curve with unit normal $\boldsymbol{\nu}$ is $\boldsymbol{N}$. If $\boldsymbol{n}^\alpha$ are the physical force vectors over each coordinate line, then applying (3.6) to a curvilinear triangle on $\mathfrak{s}$ bounded by coordinate curves through the point x^α and by c (with unit normal ν_α), we obtain

$$\boldsymbol{N} = \sum_{\alpha} \nu_\alpha\, \boldsymbol{n}^\alpha (a^{\alpha\alpha})^{1/2} = \boldsymbol{N}^\alpha \nu_\alpha, \quad \boldsymbol{N}^\alpha = \boldsymbol{n}^\alpha (a^{\alpha\alpha})^{1/2}. \tag{3.8}$$

Hence, $\boldsymbol{N}^\alpha$ transforms as a contravariant surface vector and we can put[1]

$$\boldsymbol{N}^\alpha = N^{\alpha i} \boldsymbol{a}_i = N^{\alpha\gamma} \boldsymbol{a}_\gamma + N^{\alpha 3} \boldsymbol{a}_3, \quad N^i = N^{\alpha i} \nu_\alpha, \tag{3.9}$$

where $(3.7)_1$ has been used and where $N^{\alpha\gamma}$ and $N^{\alpha 3}$ are surface tensors under transformation of surface coordinates. With the help of (3.8) and under the usual smoothness assumptions we deduce from (3.6)

[1] The order of indices in $N^{\alpha i}$ differs from that used previously in [7, 9—11], but is in agreement with the more usual notation, e.g., Green and Naghdi [12].

the force equations of motion. The component forms of these equations are

$$N^{\alpha\beta}{}_{|\alpha} - b^{\beta}_{\alpha} N^{\alpha 3} + \varrho F^{\beta} = \varrho c^{\beta}, \qquad N^{\alpha 3}{}_{|\alpha} + b_{\alpha\beta} N^{\alpha\beta} + \varrho F^{3} = \varrho c^{3}, \tag{3.10}$$

where $c^i = \{c^\beta, c^3\}$ are the components of acceleration.

Before returning to the energy Eq. (3.1), it is expedient to consider the entropy production inequality (3.2). Recalling that h is the flux of heat across c per unit length and denoting by h^α the value of the flux of heat across the x^α-curves, then by application of (3.2) to a curvilinear triangle we can show that

$$h = \boldsymbol{q} \cdot \boldsymbol{v}, \qquad \boldsymbol{q} = q^\alpha \boldsymbol{a}_\alpha, \qquad q^\alpha = h^\alpha (a^{\alpha\alpha})^{1/2}, \tag{3.11}$$

provided c is not a curve of discontinuity.

In view of (3.5), (3.6) and $(3.11)_1$, the energy Eq. (3.1) assumes the simpler form

$$\int_\sigma \{\varrho[r - \dot{U} + \bar{\boldsymbol{L}} \cdot \boldsymbol{w}] - q^\alpha_{|\alpha} + \boldsymbol{N}^\alpha \cdot \boldsymbol{v}_{,\alpha}\}\, d\sigma + \int_c \boldsymbol{M} \cdot \boldsymbol{w}\, dc = 0. \tag{3.12}$$

If $\boldsymbol{m}^\alpha$ is the (physical) director force vector over x^α-curves, then the application of (3.12) to a curvilinear triangle on $\mathfrak{S}$ yields[1]

$$\overline{\boldsymbol{M}} \cdot \boldsymbol{w} = 0, \tag{3.13}$$

where we have set

$$\overline{\boldsymbol{M}} = \boldsymbol{M} - \boldsymbol{M}^\alpha \nu_\alpha, \qquad \boldsymbol{M}^\alpha = \boldsymbol{m}^\alpha (a^{\alpha\alpha})^{1/2}. \tag{3.14}$$

If we assume that $\overline{\boldsymbol{M}}$ is an (invariant) vector under the transformation of surface coordinates, then it follows from $(3.14)_1$ that $\boldsymbol{M}^\alpha$ transforms as a contravariant vector under transformation of surface coordinates. Moreover, if we set[2]

$$\boldsymbol{M}^\alpha = M^{\alpha i} \boldsymbol{a}_i = M^{\alpha\gamma} \boldsymbol{a}_\gamma + M^{\alpha 3} \boldsymbol{a}_3, \tag{3.15}$$

$M^i = M^{\alpha i} \nu_\alpha$ by $(3.7)_2$ and $M^{\gamma\alpha}$ and $M^{\alpha 3}$ are surface tensors. Now put $\boldsymbol{m} = \boldsymbol{M}^\alpha{}_{|\alpha} + \varrho \bar{\boldsymbol{L}}$ or in component forms

$$M^{\alpha\beta}{}_{|\alpha} - b^{\beta}_{\alpha} M^{\alpha 3} + \varrho \bar{L}^{\beta} = m^{\beta}, \qquad M^{\alpha 3}{}_{|\alpha} + b_{\alpha\beta} M^{\alpha\beta} + \varrho \bar{L}^{3} = m^{3}, \tag{3.16}$$

where $\bar{L}^i = \bar{\boldsymbol{L}} \cdot \boldsymbol{a}^i$ and $m^i = \boldsymbol{m} \cdot \boldsymbol{a}^i$. We note that in general constitutive equations are required for m^i. With the help of (3.14), the vector form of (3.16) and many of the kinematical results in Sect. 2 and using a familiar argument about uniform superposed rigid body

[1] In certain special cases, e.g., an elastic Cosserat surface for which the constitutive relations are independent of rates, $\overline{\boldsymbol{M}}$ vanishes identically.

[2] The order of indices in $M^{\alpha i}$ differs from that used previously in [7, 9—11], but is in agreement with the more usual notation, e.g., in [12].

angular velocities when the surface occupies the same position at time t, we deduce from (3.13) and (3.12) the equation

$$\overline{\boldsymbol{M}} \cdot (\boldsymbol{\Gamma} - d^{\alpha}\, \boldsymbol{\eta}_{\alpha}) = 0\,, \tag{3.17}$$

the (local) energy equation

$$\varrho\, r - q^{\alpha}_{|\alpha} - \varrho\, \dot{U} + \boldsymbol{N}^{\alpha} \cdot \boldsymbol{\eta}_{\alpha} + \boldsymbol{m} \cdot (\boldsymbol{\Gamma} - d^{\alpha}\, \boldsymbol{\eta}_{\alpha}) + \\ + \boldsymbol{M}^{\alpha} \cdot (\boldsymbol{\Gamma}_{:\alpha} - \lambda^{\beta}_{.\alpha}\, \boldsymbol{\eta}_{\beta}) = 0\,, \tag{3.18}$$

as well as

$$\boldsymbol{d} \times \overline{\boldsymbol{M}} = 0\,, \qquad \boldsymbol{N}^{\alpha} \times \boldsymbol{a}_{\alpha} + (\boldsymbol{M}^{\alpha} \times \boldsymbol{d})_{|\alpha} + \varrho\, \overline{\boldsymbol{L}} \times \boldsymbol{d} = 0\,. \tag{3.19}$$

We complete the theory by recalling the (local) entropy production inequality

$$\varrho\, T\, \dot{S} - \varrho\, r + q^{\alpha}_{|\alpha} - q^{\alpha} \frac{T_{,\alpha}}{T} \geqq 0 \tag{3.20}$$

which is deduced from (3.2) with the help of $(3.11)_1$.

The basic field equations of the theory consist of: The Eq. (3.5) for conservation of mass, the equations of motion (3.10) and (3.16), the energy Eq. (3.18), and the remaining Eq. (3.17) and (3.19) together with the inequality (3.20). These field equations are valid for a Cosserat surface of any material and must be supplemented by appropriate constitutive equations. Inspection of (3.17) and (3.18) suggests that for a complete theory constitutive equations must be found for q^{α}, U, $\boldsymbol{m}$, $\boldsymbol{N}^{\alpha}$, $\boldsymbol{M}^{\alpha}$ and $\overline{\boldsymbol{M}}$ and these can be reduced to a canonical form with the use of invariance conditions for each equation which keeps the left-hand sides of (3.17) and (3.18) unaltered by all superposed rigid body motions.

4. An Elastic Cosserat Surface. A Special Case

In this section we first summarize the principal results for an elastic Cosserat surface and then consider a special case of the general theory which corresponds to an approximate theory of shells (usually derived from the three dimensional theory of elasticity). We also take this opportunity to indicate the nature of the boundary conditions for this special case.

For an elastic Cosserat surface, it can be shown that $\overline{\boldsymbol{M}}$ vanishes identically and (3.15) follows from (3.14) and $(3.7)_2$. It is convenient to express the energy Eq. (3.18) in the alternative form

$$\varrho\, r - q^{\alpha}_{|\alpha} - \varrho\, (T\dot{S} + \dot{T} S) - \varrho\, \dot{A} + \\ + N'^{\alpha\beta}\, \eta_{\alpha\beta} + m^{i}\, \dot{d}_{i} + M^{\alpha i}\, \dot{\lambda}_{i\alpha} = 0\,, \tag{4.1}$$

where A is the Helmholtz free energy function per unit mass defined by $A = U - T\,S$,

$$N'^{\alpha\beta} = N'^{\beta\alpha} = N^{\alpha\beta} - m^{\alpha}\, d^{\beta} - M^{\gamma\alpha}\, \lambda^{\beta}_{.\gamma}\,, \tag{4.2}$$

and where use has been made of (3.9) and (3.15). We assume that A, S, q^α, $N'^{\alpha\beta}$, m^i, $M^{\alpha i}$ are all functions of[1] T, $e_{\alpha\beta}$, $\lambda_{i\alpha}$, d_i and that q^α depends in addition on $T_{,\alpha}$. In general, the constitutive equations depend also on D_i and $\Lambda_{i\alpha} = \boldsymbol{A}_i \cdot \boldsymbol{D}_{,\alpha}$ (the initial values of d_i and $\lambda_{i\alpha}$, respectively) and their dependence on the initial metric tensor $A_{\alpha\beta}$ is understood. With these constitutive assumptions and from the combination of (4.1) and (3.20), we can then deduce[2]

$$S = -\frac{\partial A}{\partial T}, \tag{4.3}$$

$$N'^{\alpha\beta} = \varrho \frac{\partial A}{\partial e_{\alpha\beta}}, \qquad M^{\alpha i} = \varrho \frac{\partial A}{\partial \lambda_{i\alpha}}, \qquad m^i = \varrho \frac{\partial A}{\partial d_i}, \tag{4.4}$$

$$-q^\alpha T_{,\alpha} \geqq 0. \tag{4.5}$$

In view of (4.3) and (4.4), the energy Eq. (4.1) now simplifies and is an equation for the determination of temperature.

We consider now a special case of the general theory under certain special assumptions. The detailed development of this special theory has been discussed elsewhere [11] with particular reference to the nonlinear theory of an elastic Cosserat surface. Here we merely quote the main results from [11] and then extend the discussion to the appropriate boundary conditions. We specify the conditions

$$D_\alpha = 0, \qquad D_3 = 1, \qquad d_\alpha = 0, \tag{4.6}$$

and, in addition, assume that the component L^3 of the assigned director force is zero and that the director inertia in this direction can be neglected. We also assume that the free energy function is independent of $\lambda_{3\alpha} = d_{3,\alpha}$.

With the help of (4.6) and the component forms of the symmetry restrictions $(3.19)_2$, $(3.16)_1$ may be put in the form

$$M^{*\alpha\beta}{}_{|\alpha} + \varrho\, \bar{L}^{*\beta} = N^{\beta 3}, \tag{4.7}$$

where

$$M^{*\alpha\beta} = d_3 M^{\alpha\beta}, \qquad \bar{L}^{*\alpha} = d_3 \bar{L}^\alpha. \tag{4.8}$$

Let

$$M^{*\alpha\beta} = M^{*(\alpha\beta)} + M^{*[\alpha\beta]} \tag{4.9}$$

with the usual notation for symmetry and anti-symmetry in α, β. Under (4.6) and the further conditions stated after (4.6), and assuming that the constitutive equations are such that the value of $M^{*[\alpha\beta]} = 0$ at each point of $\mathfrak{s}$, we can systematically show that the Helmholtz free energy becomes a different function of the form

$$A = A'(T, e_{\alpha\beta}, \lambda_{\alpha\beta}, B_{\alpha\beta}) \tag{4.10}$$

---

[1] Previously in [7] the constitutive relations were expressed in terms of slightly different, but equivalent, variables,

[2] To avoid ambiguity in evaluating $\dfrac{\partial A}{\partial e_{\alpha\beta}}$, the tensor $e_{\alpha\beta}$ in A is understood to stand for $\frac{1}{2}(e_{\alpha\beta} + e_{\beta\alpha})$.

and that (4.3), $(4.4)_1$ and $(4.4)_2$ reduce to $S = -\partial A'/\partial T$ and

$$N'^{\alpha\beta} = \varrho \frac{\partial A'}{\partial e_{\alpha\beta}}, \qquad M^{*(\alpha\beta)} = \varrho \frac{\partial A'}{\partial \lambda_{\beta\alpha}}, \qquad M^{*[\alpha\beta]} = 0, \tag{4.11}$$

where

$$N'^{\alpha\beta} = N'^{\beta\alpha} = N^{\alpha\beta} + M^{*(\gamma\alpha)} b^{\beta}_{\gamma}, \tag{4.12}$$

and where $\lambda_{\alpha\beta}$ is now given by $\lambda_{\alpha\beta} = -b_{\alpha\beta}$. In addition, $M^{\alpha 3} = 0$, m^{α} which now has no constitutive equation can be determined from one of the equations of motion and there is also an equation for the determination of d_3, but this will not be recorded here. Thus, under (4.6) and the additional assumptions stated after (4.6), the theory may be recast in terms of the variables $N^{\alpha i}$, $M^{*(\alpha\beta)}$ and the governing equations are (3.10), (4.7), (4.10), (4.11) and (4.12) apart from (3.5) and the constitutive equations for S and q^{α}.

While the nature of the boundary conditions in the general theory of Sec. 3 is quite clear and follows from the rate of work terms in the balance of energy (3.1), some care is needed in discussing their counterparts for the special case discussed above under the conditions (4.6). Using (3.9) and (3.15), the rate of work by the force vector $\boldsymbol{N}$ and the director force $\boldsymbol{M}$ in (3.1) may be expressed as

$$R = \int_c \nu_{\alpha} [N^{\alpha i} v_i + M^{\alpha i} w_i]\, dc. \tag{4.13}$$

Under the conditions (4.6), it follows from (2.14) that

$$w_{\beta} = -d_3 (v_{3,\beta} + b^{\gamma}_{\beta} v_{\gamma}), \qquad w_3 = \dot{d}_3. \tag{4.14}$$

Substituting (4.14) into (4.13) and remembering that $M^{\alpha 3} = 0$ (for the special case discussed above), we obtain

$$R = \int_c \nu_{\alpha} \{N^{\alpha i} v_i - M^{*\alpha\gamma} b^{\beta}_{\gamma} v_{\beta} - M^{*\alpha\gamma} v_{3,\gamma}\}\, dc. \tag{4.15}$$

If $\frac{\partial}{\partial c}, \frac{\partial}{\partial \nu}$ denote the derivatives along the tangent and normal to c, then

$$v_{3,\gamma} = \nu_{\gamma} \frac{\partial v_3}{\partial \nu} - \varepsilon_{\gamma\alpha} \nu^{\beta} \frac{\partial v_3}{\partial c}, \tag{4.16}$$

where $\varepsilon_{\gamma\beta} = a^{\frac{1}{2}} \bar{e}_{\alpha\beta}$ and $\bar{e}_{\alpha\beta}$ is a permutation symbol with non-vanishing components $\bar{e}_{12} = -\bar{e}_{21} = 1$. Provided the quantities involved are single-valued on a (sufficiently smooth) closed curve c, with the help of (4.16) and an integration by parts, (4.15) may be reduced to

$$R = \int_c \left\{P^{\beta} v_{\beta} + P^3 v_3 - G \frac{\partial v_3}{\partial \nu}\right\} dc, \tag{4.17}$$

where we have put

$$P^\beta = \nu_\alpha [N^{\alpha\beta} - M^{*(\alpha\gamma)} b^\beta_\gamma], \qquad G = M^{*(\alpha\gamma)} \nu_\alpha \nu_\gamma, \tag{4.18}$$

$$P^3 = \nu_\alpha [M^{*(\beta\alpha)}{}_{|\beta} + \varrho \bar{L}^{*\alpha}] - \frac{\partial}{\partial c} [\varepsilon_{\beta\gamma} M^{*(\alpha\beta)} \nu_\alpha \nu^\gamma], \tag{4.19}$$

and in obtaining (4.19) use has been made of $(4.11)_3$ and (4.7). It is clear from (4.17) that the force boundary conditions which hold pointwise on c are given by (4.18) and (4.19).

5. Linear Theory of a Cosserat Plate

We conclude the present paper with some remarks on the linear theory of an elastic, initially flat, Cosserat surface or a Cosserat plate when the initial director $\boldsymbol{D}$ is coincident with the unit normal to $\mathfrak{S}$, i.e.,

$$D_\alpha = 0, \qquad D_3 = 1. \tag{5.1}$$

For purposes of this Section, it is convenient to refer the initially flat surface to a rectangular Cartesian coordinate system and designate the Cartesian components of $\boldsymbol{r}$ and $\boldsymbol{R}$ by z_i and Z_i, respectively. We quote from the infinitesimal theory of a Cosserat surface discussed in [7, Sec. 6], after specializing the results to an initially flat surface and referring all quantities to rectangular Cartesian coordinates.

We recall that an alternative set of the kinematic variables (instead of $e_{\alpha\beta}$, $\lambda_{i\alpha}$, d_i in Sec. 4) which may be used in the constitutive equations are $e_{\alpha\beta}$, $\varkappa_{i\alpha} = \lambda_{i\alpha} - \Lambda_{i\alpha}$, $\delta_i = d_i - D_i$. When the (ordinary) monopolar displacements and the director displacements are infinitesimal and when the initial director components are specified by (5.1), these kinematic variables become

$$\begin{aligned} 2e_{\alpha\beta} &= u_{\alpha,\beta} + u_{\beta,\alpha}, \qquad \varkappa_{i\alpha} = \bar{\delta}_{i,\alpha}, \\ \delta_\alpha &= d_\alpha = \bar{\delta}_\alpha - \beta_\alpha, \qquad \delta_3 = \bar{\delta}_3 = d_3 - 1, \end{aligned} \tag{5.2}$$

where

$$u_i = z_i - Z_i, \qquad \beta_\alpha = -u_{3,\alpha}, \tag{5.3}$$

and in (5.2) and (5.3) and throughout this Section a comma denotes partial differentiation with respect to Z_α. Also, the (ordinary) monopolar velocity and the director velocity associated with (5.2) and (5.3) are

$$v_i = \dot{u}_i, \qquad w_i = \dot{\bar{\delta}}_i. \tag{5.4}$$

The constitutive equations of the linear theory of an elastic Cosserat plate have the form $S = -\dfrac{\partial A}{\partial T}$ and[1]

$$N'_{\alpha\beta} = N_{\alpha\beta} = \varrho_0 \frac{\partial A}{\partial e_{\alpha\beta}}, \qquad M_{\alpha i} = \varrho_0 \frac{\partial A}{\partial \varkappa_{i\alpha}}, \qquad m_i = \varrho_0 \frac{\partial A}{\partial \delta_i}, \tag{5.5}$$

[1] In recalling (5.5), allowance is made for the change in notations for $N_{\alpha i}$, $M_{\alpha i}$ mentioned earlier.

where ϱ_0 is the initial mass density, T and S are now the infinitesimal temperature difference and the infinitesimal entropy difference from T_0 and S_0 (the constant values of temperature and entropy in the initial undeformed state) and A is a quadratic function of the kinematic variables and temperature[1].

We consider now an explicit form of $\varrho_0 A$ for an elastic Cosserat plate which initially is homogeneous and possesses holohedral isotropy (i.e., isotropy with a center of symmetry) and, in addition, require that A must be invariant under the transformations[2]

$$u_\alpha \to u_\alpha, \; u_3 \to -u_3, \; \delta_\alpha \to -\delta_\alpha, \; \delta_3 \to \delta_3. \tag{5.6}$$

If we limit the discussion to the isothermal case, the resulting linear constitutive relations are

$$\left.\begin{aligned} N_{\alpha\beta} &= N_{\beta\alpha} = \alpha_1 \delta_{\alpha\beta} e_{\gamma\gamma} + 2\alpha_2 e_{\alpha\beta} + \alpha_9 \delta_{\alpha\beta} \delta_3, \\ M_{\alpha\beta} &= \alpha_5 \delta_{\alpha\beta} \varkappa_{\gamma\gamma} + \alpha_6 \varkappa_{\beta\alpha} + \alpha_7 \varkappa_{\alpha\beta}, \qquad m_\alpha = \alpha_3 \delta_\alpha, \\ M_{\alpha 3} &= \alpha_8 \varkappa_{3\alpha}, \qquad m_3 = \alpha_4 \delta_3 + \alpha_9 e_{\gamma\gamma}, \end{aligned}\right\} \tag{5.7}$$

where $\alpha_1, \alpha_2, \ldots$ are constants. Some of the elastic coefficients in (5.7) may be identified by studying known exact solutions of the three dimensional equations of classical linear elasticity.

From an examination of the above constitutive relations, together with the kinematical results (5.2) and (5.3) and the related equations of motion [corresponding to infinitesimal deformation and the initial director components specified by (5.1)], it follows that the differential equations of the linear theory separate into those for extensional and those for flexural motions[3]. These two systems of uncoupled equations contain several features which are absent from the corresponding classical (extensional and flexural) theories of plates and (for isothermal deformation) involve 9 elastic constants, 5 for extensional and 4 for flexural deformations. Apart from the limitation of a linearized theory, these equations of the linear theory of a Cosserat plate [9] are exact. On the other hand, it can be shown that these equations may be obtained as a first approximation to an asymptotic expansion of an exact (linearized) three dimensional theory of a continuum which admits a director [10].

In the rest of this Section, we confine attention to the equilibrium of a Cosserat plate with isothermal deformation. Of particular interest

---

[1] In view of (5.1), the initial values $A_{i\alpha}$ and D_i do not occur explicitly in A for a Cosserat plate under discussion.

[2] The symmetries associated with this additional requirement imitate those of a three dimensional plate which is transversely isotropic with respect to the normals of the plate.

[3] See [9, 10] for a detailed discussion.

is the bending (or flexural) theory whose basic equations are collected below:

$$\varkappa_{\alpha\beta} = \varkappa_{(\alpha\beta)} + \varkappa_{[\alpha\beta]},$$

$$\varkappa_{(\alpha\beta)} = \tfrac{1}{2}(\delta_{\beta,\alpha} + \delta_{\alpha,\beta}) - u_{3,\alpha\beta}, \qquad \varkappa_{[\alpha\beta]} = \tfrac{1}{2}(\delta_{\alpha,\beta} - \delta_{\beta,\alpha}), \tag{5.8}$$

$$M_{(\alpha\beta)} = \alpha_5\,\delta_{\alpha\beta}\,\varkappa_{\gamma\gamma} + (\alpha_6 + \alpha_7)\,\varkappa_{(\alpha\beta)}, \qquad M_{[\alpha\beta]} = (\alpha_7 - \alpha_6)\,\varkappa_{[\alpha\beta]},$$

$$m_\alpha = \alpha_3\,\delta_\alpha, \tag{5.9}$$

$$M_{\alpha\beta,\alpha} + \varrho_0 L_\beta = m_\beta, \qquad N_{\alpha 3,\alpha} + \varrho_0 F_3 = 0, \qquad N_{\alpha 3} = m_\alpha. \tag{5.10}$$

Among the interesting features of the above bending theory we mention the presence of (i) the anti-symmetric $M_{[\alpha\beta]}$ which does not occur in any of the previous bending theories of plates; (ii) a contribution which corresponds to the effect of "transverse shear deformation"; and (iii) four constitutive coefficients in contrast to only two in the classical theory for bending of isotropic plates. Two of these constitutive coefficients, however, can be identified by comparison of solutions for pure bending of a plate deduced from (5.8)—(5.10) and a corresponding known exact result from the three dimensional elasticity. Thus, for an isotropic plate, we may set[1]

$$\alpha_5 = \nu D, \qquad (\alpha_6 + \alpha_7) = (1 - \nu)\,D, \tag{5.11}$$

where D is the flexural rigidity of the plate and ν is Poisson's ratio. We close by recalling one further result from [9]. Subject to the usual smoothness assumptions, put

$$\delta_\alpha = \phi_{,\alpha} + \varepsilon_{\alpha\gamma}\,\psi_{,\gamma}, \qquad \chi = \phi - u_3. \tag{5.12}$$

Then, a convenient representation for the solution of the differential equations of the bending theory, in the absence of L_β, is characterized by

$$\nabla^2\phi = -\frac{p}{\alpha_3}, \qquad \nabla^2\chi = \frac{\alpha_3}{D}\,\phi, \qquad \left(\nabla^2 - \frac{1}{\lambda^2}\right)\psi = 0, \tag{5.13}$$

where $p = \varrho_0 F_3$, ∇^2 is the two dimensional Laplacian and

$$\lambda = \left(\frac{k}{\alpha_3}\right)^{1/2}, \qquad k = \frac{1}{2}\,[(1-\nu)\,D + (\alpha_6 - \alpha_7)], \qquad \dim\lambda = [L]. \tag{5.14}$$

References

[1] Duhem, P.: Ann. Ecole Norm. **10**, 187 (1893).
[2] Cosserat, E., and F. Cosserat: Théorie des Corps déformables. Paris: Hermann 1909.
[3] Ericksen, J. L., and C. Truesdell: Arch. Rational Mech. Anal. **1**, 295 (1958).

---

[1] If we regard the equations of a Cosserat plate as an approximation to the three dimensional equations of classical elasticity, then $M_{[\alpha\beta]}$ would be zero and $\alpha_6 = \alpha_7$. However, we retain $M_{[\alpha\beta]}$ since the theory of a Cosserat plate contains more ingredients than its counterpart derived from the classical theory as is evident from the asymptotic expansion of the director theory in [10].

[4] ERICKSEN, J. L.: Trans. Soc. Rheol. **5**, 23 (1961).
[5] GREEN, A. E., P. M. NAGHDI and R. S. RIVLIN: Intern. J. Engng. Sci. **2**, 611 (1965).
[6] TOUPIN, R. A.: Arch. Rational Mech. Anal. **17**, 85 (1964).
[7] GREEN, A. E., P. M. NAGHDI and W. L. WAINWRIGHT: Arch. Rational Mech. Anal. **20**, 287 (1965).
[8] COHEN, H., and C. N. DESILVA: J. Math. Phys. **7**, 960 (1966).
[9] GREEN, A. E., and P. M. NAGHDI: Proc. Cambridge Philos. Soc. **63**, 537, 922 (1967).
[10] GREEN, A. E., and P. M. NAGHDI: Quart. J. Mech. Appl. Math. **20**, 183 (1967).
[11] GREEN, A. E., and P. M. NAGHDI: Quart. J. Mech. Appl. Math. **21**, (1968) to appear.
[12] GREEN, A. E., and P. M. NAGHDI: Quart. J. Mech. Appl. Math. **18**, 257 (1965).

A General Theory of Rods

By

A. E. Green and **N. Laws**

University of Oxford — University of Newcastle upon Tyne

1. Introduction

Three main methods have been used to develop one dimensional theories of rods, mostly in the context of isothermal linear elasticity. The first consists of ad hoc assumptions which are apparently independent of any general theory. In the second approach one can start with the three-dimensional equations of elasticity and use an expansion or perturbation procedure, see for example Hay [1]. However, most of the authors who adopt the three-dimensional approach employ additional assumptions or variational methods. The third approach may be called the direct approach. This is the method employed by Green and Laws [2] whose work we discuss here. A further contribution has been given by Cohen [3) who uses a variational method to produce a theory of elastic rods, but his idea of a rod differs from that of Green and Laws.

The theory of Green and Laws [2] is an exact thermodynamical theory which is not restricted to elastic rods. We note that some progress has been made by Green, Laws and Naghdi [4] in evaluating the relationship between the theory discussed here and the classical three dimensional theory of continuum mechanics.

2. Kinematics

Throughout the paper, Latin indices have the values 1, 2, 3, Greek indices the values 1, 2, and repeated indices are summed.

Let a curve c, embedded in Euclidean 3-space be defined by the equation

$$\boldsymbol{r} = \boldsymbol{r}(\theta, t), \tag{2.1}$$

where $\boldsymbol{r}$ is the position vector, relative to a fixed origin, of a point on c and t denotes the time. We regard θ as a convected coordinate defining points on the curve. The initial position of c is denoted by $\mathscr{C}$. A rod is defined to be a curve at each point of which there are two assigned directors. Thus, let two directors $\boldsymbol{A}_\alpha$ be assigned to every point of $\mathscr{C}$. The duals of $\boldsymbol{A}_\alpha$ at time t are denoted by $\boldsymbol{a}_\alpha$ and the motion of the rod is given by

$$\boldsymbol{r} = \boldsymbol{r}(\theta, t), \quad \boldsymbol{a}_\alpha = \boldsymbol{a}_\alpha(\theta, t). \tag{2.2}$$

We define $\boldsymbol{a}_3$, $\boldsymbol{A}_3$ through

$$\boldsymbol{a}_3 = \boldsymbol{a}_3(\theta, t) = \partial \boldsymbol{r}/\partial \theta, \quad \boldsymbol{A}_3 = \boldsymbol{a}_3(\theta, 0), \tag{2.3}$$

and assume that

$$[\boldsymbol{a}_1\, \boldsymbol{a}_2\, \boldsymbol{a}_3] > 0.$$

We define a set of reciprocal vectors $\boldsymbol{a}^i$ through

$$\boldsymbol{a}^i \cdot \boldsymbol{a}_j = \delta^i_j,$$

where δ^i_j is the Kronecker delta and introduce the notation

$$\begin{aligned} \boldsymbol{a}_i \cdot \boldsymbol{a}_j = a_{ij}, \quad a^{ij}\, a_{kj} = \delta^i_k, \quad a = \det a_{ij}, \\ \varkappa_{ij} = \boldsymbol{a}_j \cdot \partial \boldsymbol{a}_i/\partial \theta, \quad \varkappa_i^{\cdot j} = a^{sj}\, \varkappa_{is}. \end{aligned} \tag{2.4}$$

If $\boldsymbol{b} = \boldsymbol{b}(\theta, t)$ is a vector function of θ, t we write

$$\boldsymbol{b} = b^i\, \boldsymbol{a}_i = b_i\, \boldsymbol{a}^i, \quad b^i = a^{ij}\, b_j. \tag{2.5}$$

At this point we record that the element of length along the curve e is $ds = \sqrt{(a_{33})}\, d\theta$.

We consider a second motion of e, differing from the previous motion only by a superposed rigid body motion and let quantities associated with the second motion be distinguished by an asterisk. Then

$$\boldsymbol{r}^* = \boldsymbol{Q}\, \boldsymbol{r} + \boldsymbol{r}_0, \tag{2.6}$$

where $\boldsymbol{Q}$ denotes a proper orthogonal linear transformation of 3-space into itself, $\boldsymbol{r}_0$ is an arbitrary vector, and both $\boldsymbol{Q}$ and $\boldsymbol{r}_0$ depend only on t. We assume that

$$\boldsymbol{a}^*_\alpha = \boldsymbol{Q}\, \boldsymbol{a}_\alpha. \tag{2.7}$$

3. Theory of a Rod

Let ϱ be the mass per unit length of e, at time t, U be the internal energy per unit mass, r the heat supply function per unit mass per unit time and h the flux of heat along e. For an arbitrary segment of e, with end points $\theta = \theta_1$, $\theta = \theta_2$, we postulate an equation of energy balance in the form

$$\frac{D}{Dt}\int_{\theta_1}^{\theta_2} \varrho \left(U + \frac{1}{2}\boldsymbol{v}\cdot\boldsymbol{v} + \frac{1}{2} y^{\alpha\beta}\, \boldsymbol{w}_\alpha \cdot \boldsymbol{w}_\beta\right) \sqrt{(a_{33})}\, d\theta$$

$$= \int_{\theta_1}^{\theta_2} \varrho (r + \boldsymbol{f}\cdot\boldsymbol{v} + \boldsymbol{l}^\alpha \cdot \boldsymbol{w}_\alpha) \sqrt{(a_{33})}\, d\theta + [\boldsymbol{n}\cdot\boldsymbol{v} + \boldsymbol{p}^\alpha \cdot \boldsymbol{w}_\alpha - h]_{\theta_1}^{\theta_2}, \tag{3.1}$$

where

$$[\psi(\theta, t)]_{\theta_1}^{\theta_2} = \psi(\theta_2, t) - \psi(\theta_1, t),$$

and D/Dt, or a superposed dot, denotes the material derivative. In (3.1), $\boldsymbol{n}$ and $\boldsymbol{p}^\alpha$ are, respectively, the forces and director forces in the rod. Also $\boldsymbol{f}$ and $\boldsymbol{l}^\alpha$ denote the prescribed force and director forces respectively. We assume that $y^{\alpha\beta}$ is independent of t, and, without loss of generality take $y^{12} = y^{21}$. For later convenience we set

$$\boldsymbol{q}^\alpha = \boldsymbol{l}^\alpha - y^{\alpha\beta} \dot{\boldsymbol{w}}_\beta. \tag{3.2}$$

We now assume that ϱ, U, r, h, $(\boldsymbol{f} - \dot{\boldsymbol{v}}) \cdot \boldsymbol{a}^i$, $\boldsymbol{q}^\alpha \cdot \boldsymbol{a}^i$, $\boldsymbol{n} \cdot \boldsymbol{a}^i$ and $\boldsymbol{p}^\alpha \cdot \boldsymbol{a}^i$ are unaltered when the rod receives a superposed rigid body motion. With the help of these invariance requirements we may deduce, from (3.1), the equation of mass conservation and the equations of motion. Although Green and Laws [2] give equations in vector and component form, only the component forms are given here[1].

The local equation of mass conservation is

$$\varrho \sqrt{a_{33}} = k, \tag{3.3}$$

where $k = k(\theta)$. The equations of motion are

$$\partial n^i/\partial \theta + \varkappa_r^{\cdot i} n^r + k f^i = k \dot{\boldsymbol{v}} \cdot \boldsymbol{a}^i, \tag{3.4}$$

$$\pi^{\alpha\beta} - \pi^{\beta\alpha} + p^{\gamma\beta} \varkappa_\gamma^{\cdot\alpha} - p^{\gamma\alpha} \varkappa_\gamma^{\cdot\beta} = 0, \tag{3.5}$$

$$\pi^{\beta 3} + p^{\alpha 3} \varkappa_\alpha^{\cdot\beta} - p^{\alpha\beta} \varkappa_\alpha^{\cdot 3} - n^\beta = 0, \tag{3.6}$$

where the six quantities $\pi^{\alpha i}$ are defined by

$$\pi^{\alpha i} = k q^{\alpha i} + \partial p^{\alpha i}/\partial \theta + \varkappa_r^{\cdot i} p^{\alpha r}. \tag{3.7}$$

We note that the quantities $\pi^{\alpha i}$ are not the same as the corresponding quantities in the work of Green and Laws [2]. By using these results the energy equation may be reduced to

$$-k \dot{U} + k r + (n^3 - p^{\alpha 3} \varkappa_\alpha^{\cdot 3}) \eta_{33} + 2(n^\beta - p^{\alpha 3} \varkappa_\alpha^{\cdot\beta}) \eta_{\beta 3} +$$

$$+ \frac{1}{2} (\pi^{\alpha\beta} + \pi^{\beta\alpha} - p^{\gamma\beta} \varkappa_\gamma^{\cdot\alpha} - p^{\gamma\alpha} \varkappa_\gamma^{\cdot\beta}) \eta_{\alpha\beta} + p^{\alpha i} \dot{\varkappa}_{\alpha i} - \frac{\partial h}{\partial \theta} = 0, \tag{3.8}$$

where

$$\dot{a}_{ij} = 2\eta_{ij}. \tag{3.9}$$

We complete the general theory by postulating the entropy production inequality

$$\frac{D}{Dt} \int_{\theta_1}^{\theta_2} \varrho S \sqrt{(a_{33})}\, d\theta - \int_{\theta_1}^{\theta_2} \frac{\varrho r}{T} \sqrt{(a_{33})}\, d\theta + \left[\frac{h}{T}\right]_{\theta_1}^{\theta_2} \geqq 0, \tag{3.10}$$

[1] Components are obtained thus: $n^i = \boldsymbol{n} \cdot \boldsymbol{a}^i$, $p^{\alpha i} = \boldsymbol{p}^\alpha \cdot \boldsymbol{a}^i$, etc.

where T is the positive temperature and S is the entropy per unit mass. This inequality may be rewritten

$$k\,\dot{S}\,T - k\,r + \frac{\partial h}{\partial \theta} - \frac{h}{T}\,\frac{\partial T}{\partial \theta} \geqq 0\,. \tag{3.11}$$

Finally we define the Helmholtz free energy per unit mass,

$$A = U - T\,S\,. \tag{3.12}$$

4. An Elastic Rod

We define an elastic rod by the constitutive assumptions that A, S, $\pi^{\alpha i}$, $p^{\alpha i}$, n^i depend upon

$$T,\ \gamma_{ij},\ \sigma_{\alpha i},\ A_{ij},\ \mathrm{K}_{\alpha i}, \tag{4.1}$$

where A_{ij}, $\mathrm{K}_{\alpha i}$ are the values of a_{ij}, $\varkappa_{\alpha i}$ in the reference state, and

$$\gamma_{ij} = a_{ij} - A_{ij}, \qquad \sigma_{\alpha i} = \varkappa_{\alpha i} - \mathrm{K}_{\alpha i}. \tag{4.2}$$

Also h depends upon the variables (4.1) and upon $\partial T/\partial\theta$. Then it follows in the usual way that

$$S = -\frac{\partial A}{\partial T}, \qquad p^{\alpha i} = k\,\frac{\partial A}{\partial \sigma_{\alpha i}}, \tag{4.3}$$

$$n^3 - p^{\alpha 3}\,\varkappa_{\alpha}^{\cdot 3} = 2k\,\frac{\partial A}{\partial \gamma_{33}}, \qquad n^{\beta} - p^{\alpha 3}\,\varkappa_{\alpha}^{\cdot\beta} = k\,\frac{\partial A}{\partial \gamma_{\beta 3}}, \tag{4.4}$$

$$\pi^{\alpha\beta} + \pi^{\beta\alpha} - p^{\gamma\beta}\,\varkappa_{\gamma}^{\cdot\alpha} - p^{\gamma\alpha}\,\varkappa_{\gamma}^{\cdot\beta} = 4k\,\frac{\partial A}{\partial \gamma_{\alpha\beta}}, \tag{4.5}$$

$$-h\,\frac{\partial T}{\partial \theta} \geqq 0, \tag{4.6}$$

and the energy Eq. (3.8) may be reduced to

$$k\,r - k\,T\,\dot{S} - \partial h/\partial\theta = 0\,. \tag{4.7}$$

In Eqs. (4.4) and (4.5), A is to be regarded as a function of $\gamma_{\beta 3}, \gamma_{33}, \gamma_{\alpha\beta}$ where $\gamma_{\alpha\beta}$ is interpreted as $\frac{1}{2}(\gamma_{\alpha\beta} + \gamma_{\beta\alpha})$.

To discuss some idea of material and geometrical symmetry we consider the special case of the above, namely

$$A = A\,(T, \gamma_{ij}, \sigma_{\alpha i}). \tag{4.8}$$

We suppose that the free energy, A, is invariant under the transformations

$$\theta \to \pm\theta, \qquad \boldsymbol{a}_1 \to \pm\boldsymbol{a}_1, \qquad \boldsymbol{a}_2 \to \pm\boldsymbol{a}_2, \tag{4.9}$$

where we may take any combination of $+$ and $-$. The first transformation implies that $\boldsymbol{a}_3 \to \pm\boldsymbol{a}_3$, $\boldsymbol{A}_3 \to \pm\boldsymbol{A}_3$, and with the second and third transformations we must associate $\boldsymbol{A}_1 \to \pm\boldsymbol{A}_1$, $\boldsymbol{A}_2 \to \pm\boldsymbol{A}_2$ respectively. Provided A given by (4.8) is a polynomial in its arguments, it follows that A must reduce to a polynomial in T and the forty-five invariants

$$\left.\begin{array}{c}
\gamma_{11},\ \gamma_{22},\ \gamma_{33},\ \gamma_{12}^2,\ \gamma_{23}^2,\ \gamma_{13}^2,\ \gamma_{12}\gamma_{13}\gamma_{23},\\
\sigma_{11}^2,\ \sigma_{11}\sigma_{22},\ \sigma_{22}^2,\ \sigma_{12}^2,\ \sigma_{12}\sigma_{21},\ \sigma_{21}^2,\ \sigma_{13}^2,\ \sigma_{23}^2,\\
\sigma_{11}\sigma_{12}\sigma_{13}\sigma_{23},\ \sigma_{11}\sigma_{21}\sigma_{13}\sigma_{23},\ \sigma_{12}\sigma_{22}\sigma_{13}\sigma_{23},\ \sigma_{21}\sigma_{22}\sigma_{13}\sigma_{23},\\
\gamma_{12}\sigma_{11}\sigma_{12},\ \gamma_{12}\sigma_{11}\sigma_{21},\ \gamma_{12}\sigma_{12}\sigma_{22},\ \gamma_{12}\sigma_{21}\sigma_{22},\ \gamma_{12}\sigma_{13}\sigma_{23},\\
\gamma_{13}\sigma_{11}\sigma_{13},\ \gamma_{13}\sigma_{22}\sigma_{13},\ \gamma_{13}\sigma_{12}\sigma_{23},\ \gamma_{13}\sigma_{21}\sigma_{23},\\
\gamma_{23}\sigma_{11}\sigma_{23},\ \gamma_{23}\sigma_{22}\sigma_{23},\ \gamma_{23}\sigma_{12}\sigma_{13},\ \gamma_{23}\sigma_{21}\sigma_{13},\\
\gamma_{12}\gamma_{13}\sigma_{11}\sigma_{23},\ \gamma_{12}\gamma_{13}\sigma_{22}\sigma_{23},\ \gamma_{12}\gamma_{13}\sigma_{12}\sigma_{13},\ \gamma_{12}\gamma_{13}\sigma_{21}\sigma_{13},\\
\gamma_{12}\gamma_{23}\sigma_{11}\sigma_{13},\ \gamma_{12}\gamma_{23}\sigma_{22}\sigma_{13},\ \gamma_{12}\gamma_{23}\sigma_{12}\sigma_{23},\ \gamma_{12}\gamma_{23}\sigma_{21}\sigma_{23},\\
\gamma_{13}\gamma_{23}\sigma_{11}\sigma_{12},\ \gamma_{13}\gamma_{23}\sigma_{11}\sigma_{21},\ \gamma_{13}\gamma_{23}\sigma_{12}\sigma_{22},\ \gamma_{13}\gamma_{23}\sigma_{21}\sigma_{22},\\
\gamma_{13}\gamma_{23}\sigma_{13}\sigma_{23}.
\end{array}\right\} \quad (4.10)$$

A similar analysis holds for the heat flux h. When A is invariant under the transformations (4.9), we demand that the kinetic energy of the rod be similarly invariant and this implies that

$$y^{12} = y^{21} = 0. \quad (4.11)$$

When the rod has the symmetries described, it is possible to obtain some general solutions of elementary problems of equilibrium. First let the initial curve $\mathscr{C}$ be a straight line and put

$$\boldsymbol{R} = \theta \boldsymbol{A}_3. \quad (4.12)$$

Also choose the directors $\boldsymbol{A}_\alpha$, which are associated with $\mathscr{C}$, so that $\boldsymbol{A}_i$ are a set of orthonormal vectors which are independent of θ:

$$\boldsymbol{A}_i \cdot \boldsymbol{A}_j = A_{ij} = \delta_{ij}, \qquad \mathsf{K}_{ij} = 0. \quad (4.13)$$

Consider the deformation given by

$$\left.\begin{array}{l}
\boldsymbol{r} = \lambda_3 \boldsymbol{R} = \lambda_3 \theta \boldsymbol{A}_3,\\
\boldsymbol{a}_1 = \lambda_1 \boldsymbol{A}_1 \cos\psi\theta + \lambda_1 \boldsymbol{A}_2 \sin\psi\theta, \quad \boldsymbol{a}_2 = -\lambda_2 \boldsymbol{A}_1 \sin\psi\theta + \lambda_2 \boldsymbol{A}_2 \cos\psi\theta,
\end{array}\right\} \quad (4.14)$$

where λ_1, λ_2, λ_3 and ψ are constants. This deformation consists of a (finite) extension and torsion. It is a straightforward matter to show that when the rod is homogeneous and has the symmetries (4.9), then the equations of equilibrium are satisfied with zero body force.

Next consider the deformation, from the reference state (4.12) and (4.13), specified by

$$\left.\begin{array}{l}
\boldsymbol{r} = b \sin\phi\, \boldsymbol{A}_3 - b(1 - \cos\phi)\, \boldsymbol{A}_1,\\
\boldsymbol{a}_1 = \lambda_1 \boldsymbol{A}_1 \cos\phi + \lambda_1 \boldsymbol{A}_3 \sin\phi, \quad \boldsymbol{a}_2 = \lambda_2 \boldsymbol{A}_2,
\end{array}\right\} \quad (4.15)$$

where

$$\phi = \lambda_3 \theta / b, \quad (4.16)$$

and λ_1, λ_2, λ_3 and b are constants. The deformation (4.15) with (4.16) consists of uniform extension (with extension ratios λ_1, λ_2, λ_3) together with pure flexure into an arc of a circle of radius b. Again, it is readily verified that such a deformation is possible when the rod is homogeneous, has the symmetries (4.9) and is under zero body force. The details of both problems have been given by GREEN, KNOPS and LAWS [5].

We remark that these authors have supplied the theory of small deformations superposed on the large deformation of an elastic rod and discussed some aspects of stability.

To conclude this section we consider an incompressible elastic rod. We define a rod to be incompressible if it is susceptible to only those motions for which

$$\sqrt{a} = [a_1\, a_2\, a_3] \tag{4.17}$$

is independent of t. By introducing a Lagrangian multiplier, p, in the usual way, we may show from (3.8), (3.11) and (3.12) that we still recover (4.3), (4.6) and (4.7) but (4.4) and (4.5) are replaced by

$$\left.\begin{aligned} n^3 - p^{\alpha 3}\,\varkappa_{\alpha}^{\cdot 3} &= -p\,a^{33} + 2k\frac{\partial A}{\partial \gamma_{33}}\,,\\ n^\beta - p^{\alpha 3}\,\varkappa_{\alpha}^{\cdot\beta} &= -p\,a^{\beta 3} + k\frac{\partial A}{\partial \gamma_{\beta 3}}\,,\\ \pi^{\alpha\beta} + \pi^{\beta\alpha} - p^{\gamma\beta}\,\varkappa_{\gamma}^{\cdot\alpha} - p^{\gamma\alpha}\,\varkappa_{\gamma}^{\cdot\beta} &= -2p\,a^{\alpha\beta} + 4k\frac{\partial A}{\partial \gamma_{\alpha\beta}}\,. \end{aligned}\right\} \tag{4.18}$$

5. A Linear Theory of Straight Elastic Rods

The linearization of the theory of Sect. 4, for initially, straight rods been given by GREEN, LAWS and NAGHDI [6]. In order to make some simplification these authors considered a rod with symmetries (4.9). It turns out that the occurrence of these symmetries means that the basic equations separate into four distinct groups, two concerned with flexure, one with torsion and one with extension of the rod. Also temperature effects occur only in the latter group.

As a simple application of the linear theory, GREEN, LAWS and NAGHDI [6] discussed the propagation of waves along an infinite rod. It was found that there were two wave speeds for each kind of flexure, two for torsion and three for isothermal extension and all wave speeds were found to depend upon the wave length. In parallel with a classical result in linear three-dimensional elasticity, these authors noted that the positive definiteness of the quadratic free energy were sufficient to ensure that the wave speeds be real.

We remark that many of the usual equations for the motion of initially straight elastic rods arise naturally, or as special cases of this theory. In particular one can recover the equations given by LOVE [7] and Timoshenko beam theory [8].

6. Ideal Fluid Jets

As a further application of the general theory of Sect. 3 GREEN and LAWS [9] have formulated a theory of ideal incompressible fluid jets. For convenience, we derive the necessary constitutive equations from the work of Sect. 5 rather than use the method of GREEN and LAWS [9].

We recall that incompressibility is defined by (4.17). To obtain a theory of ideal incompressible fluid jets we simply take the free energy A of an elastic incompressible rod to depend only upon the temperature T. Then the constitutive equations are

$$S = -\frac{\partial A}{\partial T}, \tag{6.1}$$

$$p^{\alpha i} = 0, \qquad \pi^{\alpha\beta} + \pi^{\beta\alpha} = -2p\, a^{\alpha\beta}, \tag{6.2}$$

$$n^{\beta} = -p\, a^{\beta 3}, \qquad n^{3} = -p\, a^{33}, \tag{6.3}$$

together with (4.6), (4.7) and a constitutive equation for the heat flux h. With the help of (6.2), Eqs. (3.5) to (3.7) are much simplified. Also, we note that the thermal and mechanical effects are distinct and for the rest of this section we consider only the mechanical problem.

A simple problem in this theory is the steady motion of a straight circular jet. By a simple analysis GREEN and LAWS [9] are able to show that such a jet can move with uniform speed in a straight line and twist in either sense about its axis. Another solution which may be obtained from this theory corresponds to the rigid rotation and translation of the jet. It is noteworthy that the solution obtained agrees with the solution of the same problem in the three-dimensional theory of an ideal incompressible fluid.

7. An Elastic-Plastic Rod

Since the general theory of Sect. 3 requires twelve scalar-valued functions to specify the 'strain' in a rod, it is apparent that a theory of elastic-plastic rods of the same level of generality would be rather complicated. With this in mind, LAWS [10] discussed a special case of the general theory in which some severe restrictions were imposed on the possible motions of the directors. The directors were constrained to remain perpendicular to each other and to $\boldsymbol{a}_3$:

$$\boldsymbol{a}_\alpha \cdot \boldsymbol{a}_\beta = \delta_{\alpha\beta}, \qquad \boldsymbol{a}_3 \cdot \boldsymbol{a}_\beta = 0.$$

When these internal constraints are incorporated in the energy Eq. (3.8), it is evident that the resulting energy equation is much less complicated. From this energy equation LAWS [10] formulated a theory of elastic-plastic rods in the spirit of the three-dimensional theory of GREEN and NAGHDI [11]. In addition, the theory of an elastic-perfectly plastic rod emerges as a limiting case of the general theory. This latter theory lends itself readily to linearization, and, compared with a three-dimensional theory, is relatively easy to handle.

References

[1] HAY, G. E.: Trans. Amer. Math. Soc. **51**, 65 (1942).

[2] GREEN, A. E., and N. LAWS: Proc. Roy. Soc. A **293**, 145 (1966).

[3] COHEN, H.: Int. J. Engng. Sci. **4**, 511 (1966).

[4] GREEN, A. E., N. LAWS and P. M. NAGHDI: forthcoming Proc. Camb. Phil. Soc.

[5] GREEN, A. E., R. J. KNOPS and N. LAWS: forthcoming Int. J. Solids Struct.

[6] GREEN, A. E., N. LAWS and P. M. NAGHDI: Arch. Rat. Mech. Anal. **25**, 285 (1967).

[7] LOVE, A. E. H.: A Treatise on the Mathematical Theory of Elasticity, Cambridge 1927.

[8] TIMOSHENKO, S. P.: Phil. Mag. (Ser. 6) **41**, 744 (1921).

[9] GREEN, A. E., and N. LAWS: forthcoming Int. J. Engng. Sci.

[10] LAWS, N.: Quart. J. Mech. Appl. Math. **20**, 167 (1967).

[11] GREEN, A. E., and P. M. NAGHDI: Arch. Rat. Mech. Anal. **18**, 251 (1965).

The Basic Affine Connection in a Cosserat Continuum

By

H. Schaefer

Institut für Mechanik
Technische Universität Braunschweig

1. The Transport Law of a Motor

More than forty years ago R. v. Mises [1] developed the motor calculus, a calculus of vector fields in rigid bodies. It is well known that a force field in a rigid body can be "reduced" at a point P and represented by a single force vector $\boldsymbol{F}$ through P and a moment vector $\boldsymbol{M}(P)$. If we choose another reduction point Q, then we have the *transport law*

$$\begin{aligned} \boldsymbol{F}(Q) &= \boldsymbol{F}(P), \\ \boldsymbol{M}(Q) &= \boldsymbol{M}(P) + \boldsymbol{F} \times (Q - P). \end{aligned} \tag{1.1}$$

We say the two representations at P and at Q are *equivalent* or *equipollent.*

The same is the case with the velocity field of a rigid body. Let $\overset{1}{\boldsymbol{v}}$ be the angular velocity and $\overset{2}{\boldsymbol{v}}$ the displacement velocity. From P to Q we have the same transport law

$$\begin{aligned} \overset{1}{\boldsymbol{v}}(Q) &= \overset{1}{\boldsymbol{v}}(P), \\ \overset{2}{\boldsymbol{v}}(Q) &= \overset{2}{\boldsymbol{v}}(P) + \overset{1}{\boldsymbol{v}} \times (Q - P). \end{aligned} \tag{1.2}$$

$\overset{2}{\boldsymbol{v}}$ has the character of a moment vector.

A motor $\boldsymbol{v}$ is the compound of two vectors $\overset{1}{\boldsymbol{v}}, \overset{2}{\boldsymbol{v}}$

$$\boldsymbol{v} = \begin{pmatrix} \overset{1}{\boldsymbol{v}} \\ \overset{2}{\boldsymbol{v}} \end{pmatrix}. \tag{1.3}$$

The second vector $\overset{2}{\boldsymbol{v}}$ shall be the moment vector of $\boldsymbol{v}$. To each point $\boldsymbol{x}$ of a Cosserat continuum a number of motors is attached. For example we look at the motor of infinitesimal displacement

$$\boldsymbol{v}(\boldsymbol{x}) = \begin{pmatrix} \boldsymbol{\varphi}(\boldsymbol{x}) \\ \boldsymbol{u}(\boldsymbol{x}) \end{pmatrix} = \begin{pmatrix} \text{rotation vector} \\ \text{displacement vector} \end{pmatrix}. \tag{1.4}$$

If

$$\begin{aligned} \boldsymbol{\varphi}(\boldsymbol{x}+d\boldsymbol{x}) &= \boldsymbol{\varphi}(\boldsymbol{x}), \\ \boldsymbol{u}(\boldsymbol{x}+d\boldsymbol{x}) &= \boldsymbol{u}(\boldsymbol{x}) + \boldsymbol{\varphi}\times d\boldsymbol{x}, \end{aligned} \tag{1.5}$$

or

$$\begin{aligned} d\boldsymbol{\varphi} &= 0, \\ d\boldsymbol{u} - \boldsymbol{\varphi}\times d\boldsymbol{x} &= 0, \end{aligned} \tag{1.6}$$

we know from the transport law that in the neighborhood of $\boldsymbol{x}$ the continuum moves like a rigid body.

2. The Affine Connection in the Motor Space

The two vector Eqs. (1.6) define a *parallel transport* in the sense of differential geometry and the *covariant differential*

$$D\boldsymbol{v} = \begin{pmatrix} d\boldsymbol{\varphi} \\ d\boldsymbol{u} - \boldsymbol{\varphi}\times d\boldsymbol{x} \end{pmatrix} \tag{2.1}$$

measures the deviation of the displacement motor field from rigid body displacement. We have defined rather a simple *affine connection* in the motor space V^6 over the position space E^3. Of course the parallel transport of a motor is integrable and therefore the V^6 is a space of *distant parallelism.* We introduce the skew-symmetric matrix

$$\boldsymbol{\omega} = (\omega_{\alpha\beta}) = (e_{\alpha\varrho\beta}\, dx^{\varrho}). \tag{2.2}$$

Then

$$D\boldsymbol{v} = \begin{pmatrix} d\overset{1}{\boldsymbol{v}} \\ d\overset{2}{\boldsymbol{v}} + \boldsymbol{\omega}\overset{1}{\boldsymbol{v}} \end{pmatrix}, \tag{2.3}$$

In our example we have (in cartesian coordinates x^{ϱ})

$$D\boldsymbol{v} = \begin{pmatrix} \partial_{\varrho}\varphi^{\alpha}\, dx^{\varrho} \\ (\partial_{\varrho}u_{\alpha} + e_{\alpha\varrho\beta}\,\varphi^{\beta})\, dx^{\varrho} \end{pmatrix} = \begin{pmatrix} \varkappa_{\varrho}^{\cdot\alpha}\, dx^{\varrho} \\ \varepsilon_{\varrho\alpha}\, dx^{\varrho} \end{pmatrix}. \tag{2.4}$$

$\varkappa_{\varrho}^{\cdot\alpha}$ and $\varepsilon_{\varrho\alpha}$ are the deformation tensors of the Cosserat continuum. Here they are represented as vector-valued differential one-forms.

3. The Metric in the Motor Space

In the motor space V^6 there is still another geometric structure. The virtual work: force motor $\circ$ displ. motor $=$ force vector $\circ$ displ. vector $+$ moment vector $\circ$ rot. vector is independent of the reduction point and defines a crosswise *scalar product* in V^6:

$$\boldsymbol{u}\circ\boldsymbol{v} = \overset{1}{\boldsymbol{u}}\circ\overset{2}{\boldsymbol{v}} + \overset{2}{\boldsymbol{u}}\circ\overset{1}{\boldsymbol{v}} = \overset{1}{u}{}^{\alpha}\,\overset{2}{v}{}^{\alpha} + \overset{2}{u}{}^{\alpha}\,\overset{1}{v}{}^{\alpha} \quad (\alpha = 1, 2, 3). \tag{3.1}$$

If we introduce a basis $\boldsymbol{e}_1, \ldots, \boldsymbol{e}_6$ in the motor space V^6, then

$$\boldsymbol{u}\circ\boldsymbol{v} = (\overset{1}{u}{}^{\alpha}\,\boldsymbol{e}_{\alpha} + \overset{2}{u}{}^{\alpha}\,\boldsymbol{e}_{\alpha+3}) \circ (\overset{1}{v}{}^{\beta}\,\boldsymbol{e}_{\beta} + \overset{2}{v}{}^{\beta}\,\boldsymbol{e}_{\beta+3}) \tag{3.2}$$

and the scalar products of the basis vectors are

$$\boldsymbol{e}_\alpha \circ \boldsymbol{e}_\beta = \boldsymbol{e}_{\alpha+3} \circ \boldsymbol{e}_{\beta+3} = 0; \qquad \boldsymbol{e}_\alpha \circ \boldsymbol{e}_{\beta+3} = \boldsymbol{e}_{\alpha+3} \circ \boldsymbol{e}_\beta = \delta_{\alpha\beta} \tag{3.3}$$

or

$$\boldsymbol{e}_i \circ \boldsymbol{e}_k = a_{ik} \quad (i, k = 1, \ldots, 6) \tag{3.4}$$

with the 6×6 matrix

$$\boldsymbol{a} = (a_{ik}) = (a^{ik}) = \left(\begin{array}{c|c} \boldsymbol{O} & \boldsymbol{U} \\ \hline \boldsymbol{U} & \boldsymbol{O} \end{array}\right). \tag{3.5}$$

($\boldsymbol{U}$ is the 3×3 unit matrix). The a_{ik} and a^{ik} can be used to lower and rise the indices.

Now we define an *orthonormal basis* in the sense of our scalar product. It is constituted by the column motors of the 6×6 matrix

$$\boldsymbol{e} = \left(\begin{array}{c|c} \boldsymbol{U} & \boldsymbol{O} \\ \hline \boldsymbol{A} & \boldsymbol{U} \end{array}\right), \tag{3.6}$$

where the 3×3 matrix A is given by

$$\boldsymbol{A} = \int_{x_P}^{x} \boldsymbol{\omega} = (A_{\alpha\beta}) = \left(e_{\alpha\varrho\beta}\,(x^\varrho - x^\varrho_P)\right) \tag{3.7}$$

and is independent of path. P is some fixed point. Thus (3.6) is a basis not only at P but also over the whole E^3.

4. The Basic Structure Equations of the Connection in V^6

We define the Frenet-Cartan equations of the connection by

$$\boldsymbol{d}P = \boldsymbol{e}\,\sigma = \boldsymbol{e}_\alpha \overset{1}{\sigma}{}^\alpha + \boldsymbol{e}_{\alpha+3} \overset{2}{\sigma}{}^\alpha, \tag{4.1}$$

$$\boldsymbol{de} = \boldsymbol{\Omega}\,\boldsymbol{e}. \tag{4.2}$$

(See FLANDERS [2], p. 144.)

$\overset{1}{\sigma}{}^\alpha$ and $\overset{2}{\sigma}{}^\alpha$ are vector-valued differential one-forms over the position space E^3. Further

$$\boldsymbol{\Omega} = \left(\begin{array}{c|c} \boldsymbol{O} & \boldsymbol{O} \\ \hline \boldsymbol{\omega} & \boldsymbol{O} \end{array}\right). \tag{4.3}$$

Now we use the exterior calculus of differential forms. $\boldsymbol{dd}P$ is the *torsion motor* of the connection:

$$\boldsymbol{dd}P = \boldsymbol{e}\,d\boldsymbol{\sigma} + \boldsymbol{de} \wedge \boldsymbol{\sigma} = \boldsymbol{e}(d\boldsymbol{\sigma} + \boldsymbol{\Omega} \wedge \boldsymbol{\sigma}) = \boldsymbol{e}\,\boldsymbol{\tau} = \boldsymbol{e}_\alpha \overset{1}{\tau}{}^\alpha + \boldsymbol{e}_{\alpha+3} \overset{2}{\tau}{}^\alpha, \tag{4.4}$$

$$\boldsymbol{\Omega} \wedge \boldsymbol{\sigma} = \left(\begin{array}{c|c} \boldsymbol{O} & \boldsymbol{O} \\ \hline \boldsymbol{\omega} & \boldsymbol{O} \end{array}\right) \wedge \begin{pmatrix} \overset{1}{\boldsymbol{\sigma}} \\ \overset{2}{\boldsymbol{\sigma}} \end{pmatrix} = \begin{pmatrix} \boldsymbol{O} \\ \boldsymbol{\omega} \wedge \overset{1}{\boldsymbol{\sigma}} \end{pmatrix}, \tag{4.5}$$

$$\boldsymbol{\tau} = \begin{pmatrix} \overset{1}{\boldsymbol{\tau}} \\ \overset{2}{\boldsymbol{\tau}} \end{pmatrix} = \begin{pmatrix} d\overset{1}{\boldsymbol{\sigma}} \\ d\overset{2}{\boldsymbol{\sigma}} + \boldsymbol{\omega} \wedge \overset{1}{\boldsymbol{\sigma}} \end{pmatrix} = D\,\boldsymbol{\sigma}. \tag{4.6}$$

The torsion motor $\boldsymbol{\tau}$ of the connection turns out to be the covariant differential of motor-valued one-forms $\boldsymbol{\sigma}$. It is a motor-valued two-form. In the exterior calculus of differential forms $ddx^{\varkappa} = 0$. Then by inspection $\boldsymbol{dde} = 0$; the V^6 has zero-curvature. This shouldn't surprise because the V^6 is a space of distant parallelism.

5. Disclination and Dislocation Density

We set

$$\begin{pmatrix} \overset{1}{\sigma}{}^{\alpha} \\ \overset{2}{\sigma}_{\alpha} \end{pmatrix} = \begin{pmatrix} \varkappa_{\varrho}^{\cdot\alpha}\, dx^{\varrho} \\ \varepsilon_{\varrho\alpha}\, dx^{\varrho} \end{pmatrix}. \tag{5.1}$$

If

$$\begin{pmatrix} \overset{1}{\sigma}{}^{\alpha} \\ \overset{2}{\sigma}{}^{\alpha} \end{pmatrix} = D \begin{pmatrix} \varphi^{\alpha} \\ u_{\alpha} \end{pmatrix}, \quad \text{then} \quad D \begin{pmatrix} \overset{1}{\sigma}{}^{\alpha} \\ \overset{2}{\sigma}_{\alpha} \end{pmatrix} = 0$$

and $D \begin{pmatrix} \varkappa_{\varrho}^{\cdot\alpha}\, dx^{\varrho} \\ \varepsilon_{\varrho\alpha}\, dx^{\varrho} \end{pmatrix} = 0$ are the compatibility conditions of the deformations $\varkappa_{\varrho}^{\cdot\alpha}(\boldsymbol{x})$ and $\varepsilon_{\varrho\alpha}(\boldsymbol{x})$.

If $\varkappa_{\varrho}^{\cdot\alpha}$ and $\varepsilon_{\varrho\alpha}$ are incompatible, then

$$D \begin{pmatrix} \varkappa_{\varrho}^{\cdot\alpha}\, dx^{\varrho} \\ \varepsilon_{\varrho\alpha}\, dx^{\varrho} \end{pmatrix} = \begin{pmatrix} \overset{1}{\tau}{}^{\alpha} \\ \overset{2}{\tau}_{\alpha} \end{pmatrix} \neq 0. \tag{5.2}$$

In detail

$$\begin{aligned} D \begin{pmatrix} \varkappa_{\varrho}^{\cdot\alpha}\, dx^{\varrho} \\ \varepsilon_{\varrho\alpha}\, dx^{\varrho} \end{pmatrix} &= \begin{pmatrix} \partial_{\mu}\varkappa_{\varrho}^{\cdot\alpha}\, dx^{\mu} \wedge dx^{\varrho} \\ \partial_{\mu}\varepsilon_{\varrho\alpha}\, dx^{\mu} \wedge dx^{\varrho} + e_{\alpha\mu\beta}\, dx^{\mu} \wedge \varkappa_{\varrho}^{\cdot\beta}\, dx^{\varrho} \end{pmatrix} \\ &= \begin{pmatrix} \partial_{\mu}\varkappa_{\varrho}^{\cdot\alpha}\, dx^{\mu} \wedge dx^{\varrho} \\ \partial_{\mu}\varepsilon_{\varrho\alpha}\, dx^{\mu} \wedge dx^{\varrho} + e_{\alpha\mu\beta}\, \varkappa_{\varrho}^{\cdot\beta}\, dx^{\mu} \wedge dx^{\varrho} \end{pmatrix} \\ &= \begin{pmatrix} e^{\mu\varrho\gamma}\, \partial_{\mu}\varkappa_{\varrho}^{\cdot\alpha}\, dF_{\gamma} \\ e^{\mu\varrho\gamma}\, \partial_{\mu}\varepsilon_{\varrho\alpha}\, dF_{\gamma} + (\delta_{\alpha}^{\gamma}\varkappa_{\beta}^{\cdot\beta} - \varkappa_{\alpha}^{\cdot\gamma})\, dF_{\gamma} \end{pmatrix} \\ &= \begin{pmatrix} \overset{1}{A}{}^{\gamma\alpha}\, dF_{\gamma} \\ \overset{2}{A}{}^{\gamma}_{\cdot\alpha}\, dF_{\gamma} \end{pmatrix} = \begin{pmatrix} \overset{1}{\tau}{}^{\alpha} \\ \overset{2}{\tau}_{\alpha} \end{pmatrix} \end{aligned} \tag{5.3}$$

with

$$dx^{\mu} \wedge dx^{\varrho} = e^{\mu\varrho\gamma}\, dF_{\gamma}.$$

(See GÜNTHER [3].)

$\overset{1}{\boldsymbol{\tau}}$ = 1. Burgers vector = rotation dislocation (disclination) density,

$\overset{2}{\boldsymbol{\tau}}$ = 2. Burgers vector = (translation) dislocation density.

Physical examples of disclinations were given quite recently by ANTHONY, ESSMANN, SEEGER, TRÄUBLE [4].

6. The Divergence Theorem for Disclinations and Dislocations

$$DD\boldsymbol{v} = D\begin{pmatrix} d\overset{1}{\boldsymbol{v}} \\ d\overset{2}{\boldsymbol{v}} + \boldsymbol{\omega} \wedge \overset{1}{\boldsymbol{v}} \end{pmatrix} = \begin{pmatrix} dd\overset{1}{\boldsymbol{v}} \\ dd\overset{2}{\boldsymbol{v}} + d(\boldsymbol{\omega} \wedge \overset{1}{\boldsymbol{v}}) + \boldsymbol{\omega} \wedge d\overset{1}{\boldsymbol{v}} \end{pmatrix}. \qquad (6.1)$$

But in the exterior calculus $dd\overset{1}{\boldsymbol{v}} = dd\overset{2}{\boldsymbol{v}} = 0$ and $d(\boldsymbol{\omega} \wedge \overset{1}{\boldsymbol{v}}) = d\,\boldsymbol{\omega} \wedge \overset{1}{\boldsymbol{v}} - \boldsymbol{\omega} \wedge d\overset{1}{\boldsymbol{v}} = -\boldsymbol{\omega} \wedge d\overset{1}{\boldsymbol{v}}$, because $\boldsymbol{\omega}$ is a matrix of one-forms with constant coefficients.

Thus

$$DD\boldsymbol{v} = 0. \qquad (6.2)$$

With (5.1) and (5.3) we have

$$DD\begin{pmatrix} \overset{1}{\sigma}{}^{\alpha} \\ \overset{2}{\sigma}{}_{\alpha} \end{pmatrix} = D\begin{pmatrix} \overset{1}{\tau}{}^{\alpha} \\ \overset{2}{\tau}{}_{\alpha} \end{pmatrix} = \begin{pmatrix} \partial_{\varrho} \overset{1}{A}{}^{\gamma\alpha}\, dx^{\varrho} \wedge dF_{\gamma} \\ \partial_{\varrho} \overset{2}{A}{}^{\gamma}_{\cdot\alpha}\, dx^{\varrho} \wedge dF_{\gamma} + e_{\alpha\varrho\beta}\, dx^{\varrho} \wedge \overset{1}{A}{}^{\gamma\beta}\, dF_{\gamma} \end{pmatrix} \qquad (6.3)$$

or

$$(dx^{\varrho} \wedge dF_{\gamma} = \delta^{\varrho}_{\gamma}\, dV), \quad \partial_{\gamma} \overset{1}{A}{}^{\gamma\alpha} = 0, \quad \partial_{\gamma} \overset{2}{A}{}^{\gamma}_{\cdot\alpha} + e_{\alpha\gamma\beta}\, \overset{1}{A}{}^{\gamma\beta} = 0. \qquad (6.4)$$

(GÜNTHER [3].)

7. Some Concluding Remarks

If the motor $\boldsymbol{\sigma}$ is represented by 6 differential p-forms, we write

$$\left.\begin{aligned} D\boldsymbol{\sigma} &= \operatorname{Grad}\boldsymbol{\sigma} \quad \text{for } p = 0, \\ D\boldsymbol{\sigma} &= \operatorname{Rot}\boldsymbol{\sigma} \quad \text{for } p = 1, \\ D\boldsymbol{\sigma} &= \operatorname{Div}\boldsymbol{\sigma} \quad \text{for } p = 2, \end{aligned}\right\} \qquad (7.1)$$

and by (6.2) we have

$$\begin{aligned} DD\boldsymbol{\sigma} &= \operatorname{Rot}\operatorname{Grad}\boldsymbol{\sigma} = 0 \quad \text{for } p = 0, \\ DD\boldsymbol{\sigma} &= \operatorname{Div}\operatorname{Rot}\boldsymbol{\sigma} = 0 \quad \text{for } p = 1. \end{aligned} \qquad (7.2)$$

Thus the torsion motor $\boldsymbol{\tau} = \operatorname{Rot}\boldsymbol{\sigma}$ and $\operatorname{Div}\boldsymbol{\tau} = 0$.

The operation $D\boldsymbol{v}$ works not only in kinematics but also in statics. E.g. if $\overset{1}{\boldsymbol{v}}, \overset{2}{\boldsymbol{v}}$ are the vector-valued force stress and moment stress 2-forms respectively, then $D\boldsymbol{v} = \operatorname{Div}\boldsymbol{v} = 0$ is the equilibrium condition, and $\boldsymbol{v} = D\boldsymbol{s} = \operatorname{Rot}\boldsymbol{s}$ represents the stress 2-forms by stress functions 1-forms $\boldsymbol{s}$. Here I may refer to my paper [5].

References

[1] MISES, R. v.: Motorrechnung, ein neues Hilfsmittel der Mechanik. Anwendungen der Motorrechnung, ZAMM 4, 155—181, 193—213 (1924). Also: Selected Papers of Richard v. Mises, Vol. I, Amer. Math. Soc. 1964.

[2] Flanders, H.: Differential Forms with Applications to the Physical Sciences, New York 1963.

[3] Günther, W.: Zur Statik und Kinematik des Cosseratschen Kontinuums. Abh. d. Braunschweigischen Wiss. Ges. **10** (1958).

[4] Anthony, K., U. Essmann, A. Seeger and H. Träuble: Disclinations and the Cosserat continuum with incompatible rotations. This book, p. 355.

[5] Schaefer, H.: Analysis der Motorfelder im Cosserat-Kontinuum. ZAMM **47**, 319–328 (1967).

On the Thermodynamic Potential of Cosserat Continua

By

G. Grioli

Seminario Matematico
University of Padova

The theory of the linear Cosserat continua has made good progress in the last years. In the case of finite deformations the general statements of the theory have been established but, whatever I know, little progress have been made.

It makes a great difference between the case of finite deformations and the linear one, especially on account of the local rotation associated to every point of the continuum.

Going into full details, one is surprised at seing that only particular kinds of Cosserat continuum are mathematically consistent.

On account of the short time I have to myself, I shall exhibit the results without going into details.

I shall consider a Cosserat continuum in the classic sense.

Let C be a reference configuration of the continuum which I suppose without stress and C' the actual one. I assume that the transformation between C and C' is determined by the vector u joining the typical point P of C with its correspondent P' of C' and by a rotation R, independent of u, of a trihedron T associated to P. As it is known, one may interpret R as a rotation around P' of the material particle P, independent of displacement u.

The vector u and the rotation R are to be considered with reference to a fixed rectangular reference frame, with respect to which I denote by x_i and x'_i the coordinates of P and P'.

Now we consider an infinitesimal transformation from C to $C' + dC'$.

Let du be the displacement vector and $\delta\omega$ the whole local rotation associated to P'. Let $\delta'\omega$ be the local rotation due to the displacement du. Clearly is

$$(\delta\omega)_i = (\delta'\omega)_i + (\delta R)_i \tag{1}$$

with

$$(\delta'\omega)_i = \tfrac{1}{2} e_{itm} (\delta u_m)_{/t}. \tag{2}$$

In (2) e_{itm} is the Ricci alternating tensor and the bar denotes the derivative with respect to the actual coordinate x'_t.

It is to remark that in the case of finite deformations δR does not coincide with the differential of a function of the material coordinates $\varkappa_r$.

In order to get actual results, it is advisable to characterize the rotation R in a explicit manner. It is possible to use the Euler angles respect to T of the trihedron T' which is the actual position of T, but I think that not convenient on account of some singularities which are present.

I think more profitable to characterize the R rotation by the use of the vector $q = 2\, v \tan\frac{\varphi}{2}$, where v is the unit vector of the axis direction of the rotation and φ its amplitude.

However, there exists always a vector q_i such that results

$$(\delta R)_r = \tau_{ri}\, \delta q_i \tag{3}$$

where $|\tau_{ri}|$ is an invertible matrix depending on q_l and δq_l is the variation of q_l corresponding to the transformation from C' to $C' + dC'$.

Two asymmetric matrices, $|X'_{rs}|$ and $|\psi'_{rs}|$ characterize the inner forces: $|X'_{rs}|$ the Cauchy stress, $|\psi'_{rs}|$ the Cosserat couple stress.

It is easy to show that the work per unit volume of C' of the inner forces, corresponding to the transformation from C' to $C' + dC'$, has the following expression:

$$\delta l^{(i)} = X_{rs}(\delta u_r)_{/s} + \psi_{rs}(\delta\omega_r)_{/s} + e_{rpq}\, X_{pq}\, \delta\omega_r. \tag{4}$$

I shall denote by $f_{,r}$ the derivative of a function f with respect to the material coordinate x_r. It is convenient to introduce the matrices $|X_{rs}|$, $|\psi_{rs}|$ defined by the equalities

$$X'_{rs} = \frac{1}{D}\, x'_{r,l}\, x'_{s,m}\, X_{lm}, \qquad \psi'_{rs} = \frac{1}{D}\, x'_{s,l}\, \psi_{rl}, \tag{5}$$

where D denotes the determinant

$$D \equiv |\det x'_{r,s}| > 0. \tag{6}$$

and X_{rs} is the Piola-Kirchhoff stress tensor.

Further, I denote by A_{rs} the cofactor of $x'_{r,s}$ in the matrix $|x_{r,s}|$ and put

$$\mu_{rs} = \frac{e_{rtm}}{2D}\, A_{tp}\, u_{m,ps}. \tag{7}$$

After these preliminaries, one may show that a lagrangian expression, $\delta^* l^{(i)}$, of the work per unit volume of C done by the inner forces, corresponding to expression (4), is the following:

$$\delta^* l^{(i)} = [X_{(pl)}\, x'_{m,l} + B_{rlmp}\, \psi_{rl}]\, \delta u_{m,p} + \psi_{rl}[\delta\mu_{rl} + \tau_{ri}\, \delta q_{i,l}] + \\ + \left[\psi_{rl}\frac{\partial\tau_{ri}}{\partial q_t}\, q_{t,l} + e_{rpq}\, x'_{p,l}\, x'_{q,m}\, \tau_{ri}\, X_{[lm]}\right] \delta q_i, \tag{8}$$

where $X_{(p l)}$ and $X_{[p l]}$ are respectively the symmetric and antisymmetric part of $X_{p l}$ and

$$B_{r l m p} = \frac{u_{\gamma, \sigma l}}{2 D^2} [e_{r t m} A_{t \sigma} A_{\gamma p} - e_{r t \gamma} A_{t p} A_{m \sigma}]. \tag{9}$$

It is easy to show that vector u_r satisfies the equality

$$A_{r l} \mu_{r l} = 0. \tag{10}$$

Then $\delta u_{r, s}$ and $\delta \mu_{r s}$ are not independent, but connected by the relation

$$\delta A_{r l} \mu_{r l} + A_{r l} \delta \mu_{r l} = 0. \tag{11}$$

Now I suppose valid the following hypotheses: 1) there is no inner constraint; 2) the transformation from C' to $C' + d C'$ is an isotermic reversible transformation. Then, denoting by W the isothermic potential density, well known thermodynamic considerations show that the equality

$$\delta^* l^{(i)} = - \delta W \tag{12}$$

subsists for arbitrary values of δq_i, $\delta q_{i,h}$ and for every set of values of $\delta u_{r, s}$, $\delta \mu_{r s}$ which satisfy equality (11). It follows that W depends on the variables $x'_{r, s}$, $\mu_{r s}$, q_r, $q_{r, s}$.

Equalities (11), (12) imply some rather complicated relations between variables $X_{(r s)}$, $X_{[r s]}$, $\psi_{r s}$, the gradient of W and a certain parameter σ. However, it is possible to show that $X_{(r s)}$ do not depend on σ. Further, if we put formally

$$W = W' - \sigma A_{r l} \mu_{r l}, \tag{13}$$

it is possible to show that is

$$X_{(p t)} = \frac{A_{m t}}{D} \left[B_{r l m p} \frac{\partial W'}{\partial \mu_{r l}} - \frac{\partial W'}{\partial x'_{m, p}} \right], \tag{14}$$

$$\psi_{r l} \frac{\partial \tau_{r i}}{\partial q_t} q_{t, l} + e_{r p q} x'_{p, l} x'_{q, m} \tau_{r i} X_{[l m]} = - \frac{\partial W'}{\partial q_i}, \tag{15}$$

$$\psi_{r l} = - \frac{\partial W'}{\partial \mu_{r l}} \tag{16}$$

$$\psi_{r l} \tau_{r i} = - \frac{\partial W'}{\partial q_{i, l}}. \tag{17}$$

It is possible to show that the term of (13) depending on σ has no influence on the right member of (14), (15), (17) and one my substitute W to W' in these equalities.

The relations (14)—(17) imply important restrictions to the analytical structure of W'. First of all it is necessary for the second member of (14) to be symmetric with respect to indexes p, t.

This fact agrees with the principle of material indifference [1] and implies the equations

$$e_{r p t} \left[x'_{p, l} \frac{\partial W'}{\partial x'_{t, l}} + \mu_{p l} \frac{\partial W'}{\partial \mu_{t l}} \right] = 0. \tag{18}$$

The most general expression of W' which satisfies to (18) is an arbitrary function of the right Cauchy-Green matrix $|b_{rs} = x'_{i,r} x'_{i,s}|$, and of the matrices $|m_{rs} = x'_{i,r} \mu_{is}|$, $|q_{r,s}|$, $|q_r|$:

$$W' = F(b_{rs}, m_{rs}, q_{r,s}, q_r). \tag{19}$$

The expression (19) of W', as it is possible to show, satisfies to the principle of material indifference also respect to the couple stress $\psi'_{r,s}$.

If one assumes for F the expression

$$F = L(b_{rs}, m_{rs}, q_{i,r} q_{i,s}, q^2), \tag{20}$$

where L is an arbitrary function of its arguments, then W satisfies also the Euclid invariance principle [2].

However, independently of this principle and of expression (20), Eqs. (16), (17) present some difficulties. Precisely, they imply the compatibility equations

$$\tau_{ri} \frac{\partial W'}{\partial \mu_{rl}} - \frac{\partial W'}{\partial q_{i,l}} = 0. \tag{21}$$

The most general solution of these equations is an arbitrary function of the variables

$$z_{pq} = \mu_{pq} + \tau_{pi} q_{i,q}, \tag{22}$$

and of $x'_{r,s}$, q_r:

$$W' = f(x'_{r,s}, z_{pq}, q_i). \tag{23}$$

Now it is to observe that (19) implies that W' depends on the variables $x'_{r,s}$ only by the matrices b_{rs} and m_{rs}, while (22), (23) show that this is impossible because W' depend on μ_{rs} only through the variables z_{rs}. In other words, Eqs. (18), (21) are inconsistent in general.

On the contrary, in the linear case the term depending on μ_{rs} disappears in Eqs. (18), while the matrix $|\tau_{ri}|$ becomes the identity. Then, Eqs. (18), (21) are consistent and W' is an arbitrary function of the variables b_{rs}, $z_{pq} = \mu_{p,q} + q_{p,q}$ and q_r.

Nevertheless, one is in doubt of the meaning of the linear theory if assume it as a first approximation of the exact theory.

At least it seems that a general Cosserat continuum cannot experiences a reversible transformation and cannot be a hyperelastic material.

It is interesting to observe that if one takes into account only the local rotation due to displacement u, supposing $q_i \equiv 0$, $q_{i,l} \equiv 0$ (constrained rotations), the terms depending on δq_i, $\delta q_{i,l}$ disappear in expression (8), of $\delta^* l^{(i)}$ and Eqs. (15), (17), (21) are not to be considered. In this case the remaining equations are consistent and W' is an arbitrary function of variables b_{rs}, m_{rs}:

$$W' = \varphi(b_{rs}, m_{rs}). \tag{24}$$

This is a particular case, already studied by me [3], TOUPIN [4] and MINDLIN [5] since 1960.

One has another consistent particular case if supposes W' independent of z_{pq}. Now Eqs. (21) are satisfied, W is independent of μ_{rs}, $q_{r,s}$, the couple stress are not present and results $\sigma = 0$.

It seems that these particular cases characterize the only kinds of Cosserat continuum which are consistent in point of fact.

It is possible to realize the above incompatibility keeping in mind that the stress is determined by 18 variables, X_{rs}, ψ_{rs}, while the kinematical response of the continuum by 27 ones: b_{rs}, μ_{rs}, $q_{r,s}$, q_i. Then it is clear that the equality (12) must imply some compatibility equations which may be inconsistent.

On the contrary, in the case that is $q_i \equiv 0$, $\delta q_i \equiv 0$, there are only 15 kinematical variables: b_{rs}, μ_{rs}. In this case the relation (12) does not determine the antisymmetric part of the stress, $X_{[rs]}$, which is determined by the dynamical equations.

Often, in order to obtain some generalizations, the continuum's theory is stated postulating a certain kind of thermodynamic potential and deducing the dynamical equations by means of well known variational principles. So doing no incompatibility is present because the stress variables are defined in a mathematical manner as the components of the potential's gradient.

Nevertheless, may happen that dynamical equations so obtained are not the only equations compatible with the postulated expression of the potential because there are some stress variables whose inner work is always equal to zero. Then these variables may be present in the dynamical equations but disappear in the expression of the inner work.

The former happens, for example, in the particular case of Cosserat continuum corresponding to the hypothesis $R = 0$ (constrained rotation) before recalled by me.

In this case the antisymmetric part of stress $X_{[rs]}$ is present in the dynamical equations but disappears in the expression of the work of the inner forces. Consequently, in this particular case it is impossible to obtain the known dynamical equations by using the energetic variational principles, as it happens instead starting from the integral equations of the Mechanics.

References

[1] Truesdell, C., and W. Noll: The non-Linear Field Theory of Mechanics, Handbuch der Physik III/3. Berlin/Heidelberg/New York: Springer 1965, p. 43.

[2] Truesdell, C., and W. Noll: The non-Linear Field Theory of Mechanics, Handbuch der Physik III/3. Berlin/Heidelberg/New York: Springer 1965, p. 395.

[3] Grioli, G.: Elasticità asimmetrica, Annali di Matematica pura e Applicata (LV), V. 4, p. 389—418 (1960).

GRIOLI, G.: Onde die discontinuità ed elasticità asimmetrica, Acc. Naz. dei Lincei, S. VIII, V. XXIX, fasc. 5, Nov. 1960.

GRIOLI, G.: Mathematical Theory of Elastic Equilibrium (Recent Result), Erg. Angew. Mathem. **7**, 141–160 (1962).

GRIOLI, G.: Sulla Meccanica dei continui a trasformazioni reversibili con caratteristiche di tensione asimmetriche, Seminari dell'Istituto Nazionale di Alta Matematica, 1962–1963.

[4] TOUPIN, R.: Elastic Materials with couple-stress, Arch. Rational Mech. Anal., **11**, 5, December 1962.

[5] MINDLIN, R. D., and H. F. TIERSTEN: Effects of Couple-stress in Linear Elasticity, Arch. Rational Mech. Anal., V. 11, 5, December 1962.

Applications of Theories of Generalized Cosserat Continua to the Dynamics of Composite Materials[1]

By

G. Herrmann and **J. D. Achenbach**

Northwestern University, Evanston, Ill.

Abstract. It is shown that the Cosserat continuum and the theory of elasticity with micro-structure can be interpreted as analytical models describing the dynamic behavior of a composite material. The nonclassical material constants are simply functions of the geometry and the classical constants of the two materials constituting the composite. The study of wave propagation in a laminated composite reveals that a more complex micro-structure needs to be introduced in a continuum in order to describe adequately the dispersive character of (essentially) longitudinal waves.

1. Introduction

The 19th century idea that models of physical bodies should consist not merely of an assemblage of points, but should also include effects of directions associated with the points (oriented bodies), as suggested by Voigt and Duhem, sprang from the desire to describe various phenomena on the microscale which ordinary continuum mechanics is not able to accommodate. E. and F. Cosserat constructed a theory of elasticity corresponding to this idea for a special case, namely, when orientation is specified at each point by a rigid triad, entailing the introduction of the couple per unit area, acting across a surface within a material volume or on its boundary, in addition to the usual force per unit area. A modern derivation of a Cosserat-type theory and a discussion of typical effects of couple stresses within the framework of a linearized form of the couple-stress theory for perfectly elastic, centrosymmetric-isotropic materials were given by Mindlin and Tiersten [1]. It was mentioned by these authors that in their theory the new material constant l, which has the dimension of length and which embodies all the difference between analogous equations or solutions with and without couple stresses, is presumably small in com-

[1] This work was supported by the Office of Naval Research under Contract ONR Nonr. 1228(34) with Northwestern University.

parison with bodily dimensions and wave lengths normally encountered, as there appears to be no conclusive experimental evidence of its existence. Various other aspects of Cosserat-type continua and related theories were discussed by TOUPIN [2, 3], KOITER [4], SCHAEFER [5] and MINDLIN and ESHEL [6].

To incorporate in a continuum theory of mechanics further micro-scale phenomena occurring in a crystal lattice, MINDLIN [7] established a theory of linear elasticity with micro-structure (TEMS) by assuming, in effect, that each leg of the Cosserat triad can stretch and rotate independently of the other two. This model is equivalent to the inclusion, at each point of the macro-medium, of a unit cell of a micro-medium which deforms homogeneously. For a centrosymmetric-isotropic material there are sixteen additional independent material constants which describe the properties of this continuum. If the cell is made rigid, but is allowed to rotate independently of the macro-rotation, one reverts to the Cosserat theory (COST). With the further constraint of the cell having the same rotation as the macro-rotation (Cosserat's "trièdre caché"), one is led to the special theory of elasticity with couple stresses (TECS). Alternatively, the theory can be made more complex, for example, by placing into each cell several mass points, and by specifying interaction forces between mass points in the same cell and in neighboring cells, as was discussed by KUNIN [8].

It appears now, as some recent work by the authors indicates, that the concepts and theories of a Cosserat continuum and its generalizations have broad applicability in describing phenomena which occur on a macro- rather than a microscale. Indeed, if one wishes to describe the dynamic behavior of periodically macro-heterogeneous solids, such as, for example, fiber-reinforced or laminated composites, one can be led to similar mathematical relations. One approach explored by the authors [9, 10] consists in using representative elastic moduli for the binder (soft layers) and combining the elastic and geometric properties of the fibers or the sheets (stiff layers) into "effective stiffnesses". Depending upon certain supplementary kinematical assumptions describing the deformation of reinforcing elements, a continuum theory can be evolved which bears strong resemblance to Mindlin's theory of elasticity with micro-structure or, in its simpler version, to the Cosserat continuum. What is noteworthy, however, is that the nonclassical material constants are now simply functions of the geometric layout and of the classical constants of the two constituent homogeneous materials.

To render the indicated connection specific and precise, the most important concepts and relations of TEMS are set down in Sect. 2. To make the point, it suffices to consider plane deformation and unidirectional structuring, and, for the sake of brevity, boundary conditions are not discussed. Sect. 3 presents the fundamental relations of one version of the effective stiffness theory for laminated composites recently proposed by the authors [9, 10] and identifies the material coefficients of TEMS in terms of the classical material constants and the geometric layout of the composite. This interpretation is followed by a reduction corresponding to COST and TECS.

In Sect. 4, the viability of the proposed theories is discussed by a study of the dispersion characteristics of free plane harmonic waves in the direction of lamination. Comparisons with "exact" dispersion curves obtained by solving the appropriate classical elasticity problem [11] reveal that the lowest (predominantly) transverse mode is rather well described in its strong dispersion, by contrast to the lowest (predominantly) longitudinal mode. The reason for strong dispersion in this mode, as revealed by examining the exact solution, is due primarily to the dispersive properties of the soft layers. Authors' more complicated versions of the effective stiffness theory for laminated media [12] are briefly summarized.

2. Uni-Directional Micro-Structure in Linear Plane Elasticity

Let us assume that in Mindlin's TEMS [7] the deformation is two-dimensional. With a cartesian frame of reference x_1, x_2, the components of displacement u_i are

$$u_1 = u_1(x_1, x_2, t); \qquad u_2 = u_2(x_1, x_2, t) \tag{1}$$

where t is the time. If micro-structure is introduced in one direction only, say x_2, there will be only two nonvanishing components of micro-deformation ψ_{ij}, namely

$$\psi_{21} = \psi_{21}(x_1, x_2, t); \qquad \psi_{22} = \psi_{22}(x_1, x_2, t). \tag{2}$$

The components of macro-strain are

$$\varepsilon_{11} = \partial u_1/\partial x_1; \qquad \varepsilon_{22} = \partial u_2/\partial x_2; \qquad \varepsilon_{12} = (\partial u_2/\partial x_1 + \partial u_1/\partial x_2)/2, \tag{3}$$

and the components of macro-rotation are

$$\omega_{12} = -\omega_{21} = (\partial u_2/\partial x_1 - \partial u_1/\partial x_2)/2. \tag{4}$$

The relative deformation has only two relevant components, namely

$$\gamma_{21} = \partial u_1/\partial x_2 - \psi_{21}; \qquad \gamma_{22} = \partial u_2/\partial x_2 - \psi_{22}. \tag{5}$$

The four nonvanishing components of the micro-deformation gradient are

$$\begin{aligned} \varkappa_{121} &= \partial \psi_{21}/\partial x_1; & \varkappa_{122} &= \partial \psi_{22}/\partial x_1; \\ \varkappa_{221} &= \partial \psi_{21}/\partial x_2; & \varkappa_{222} &= \partial \psi_{22}/\partial x_2. \end{aligned} \tag{6}$$

It is assumed, however, that the gradients in the direction of structuring do not contribute to the potential energy and thus will be ignored in the sequel, i.e., $\varkappa_{221} \equiv \varkappa_{222} \equiv 0$. The reason for this assumption will be discussed in Sect. 3.

The potential energy W is assumed to be a function of the seven variables ε_{11}, ε_{12}, ε_{22}, γ_{21}, γ_{22}, $\varkappa_{121}$ and $\varkappa_{122}$.

The three nonvanishing components of Cauchy stress τ_{ij} are defined as

$$\tau_{ij} = \partial W/\partial \varepsilon_{ij} = \tau_{ji} \quad (i, j = 1, 2). \tag{7}$$

The two nonvanishing components of relative stress σ_{ij} are defined as

$$\sigma_{21} = \partial W/\partial \gamma_{21}; \qquad \sigma_{22} = \partial W/\partial \gamma_{22} \tag{8}$$

and the two nonvanishing components of double stress μ_{ijk} are defined as

$$\mu_{121} \equiv \partial W/\partial \varkappa_{121}; \qquad \mu_{122} \equiv \partial W/\partial \varkappa_{122}. \tag{9}$$

The kinetic energy T is taken in the form

$$T = \tfrac{1}{2}\varrho(\dot{u}_1^2 + \dot{u}_2^2) + \tfrac{1}{6}\varrho' d^2(\dot{\psi}_{21}^2 + \dot{\psi}_{22}^2) \tag{10}$$

where ϱ is the sum of the masses of macro-material and micro-material per unit macro-volume, ϱ' is the mass of the micro-material per unit macro-volume and $2d$ is the characteristic length of the (presently one-dimensional) micro-medium. The dot indicates differentiation with respect to time.

Hamilton's principle for independent variations δu_i and $\delta \psi_{ij}$ leads to the following four stress-equations of motion, in the absence of body forces and body double forces,

$$\left.\begin{aligned}
&\partial \tau_{11}/\partial x_1 + \partial(\tau_{21} + \sigma_{21})/\partial x_2 = \varrho\, \ddot{u}_1,\\
&\partial \tau_{12}/\partial x_1 + \partial(\tau_{22} + \sigma_{22})/\partial x_2 = \varrho\, \ddot{u}_2,\\
&\partial \mu_{121}/\partial x_1 + \sigma_{21} = \tfrac{1}{3}\varrho' d^2 \ddot{\psi}_{21},\\
&\partial \mu_{122}/\partial x_1 + \sigma_{22} = \tfrac{1}{3}\varrho' d^2 \ddot{\psi}_{22}.
\end{aligned}\right\} \tag{11}$$

For an isotropic material and for the restricted deformations presently considered, Mindlin's general potential energy density reduces to

$$\begin{aligned}
W = {} & \tfrac{1}{2}\lambda(\varepsilon_{11}^2 + 2\varepsilon_{11}\varepsilon_{22} + \varepsilon_{22}^2) + \mu(\varepsilon_{11}^2 + \varepsilon_{22}^2 + 2\varepsilon_{12}^2) + \\
& + \tfrac{1}{2}b_1\gamma_{22}^2 + \tfrac{1}{2}b_2(\gamma_{21}^2 + \gamma_{22}^2) + \tfrac{1}{2}b_3\gamma_{22}^2 + \\
& + g_1(\gamma_{22}\varepsilon_{11} + \gamma_{22}\varepsilon_{22}) + g_2(\gamma_{21}\varepsilon_{12} + \gamma_{21}\varepsilon_{21} + 2\gamma_{22}\varepsilon_{22}) + \\
& + \tfrac{1}{2}(a_4 + a_{10} + a_{13})\varkappa_{122}^2 + \tfrac{1}{2}(a_8 + a_{10} + a_{15})\varkappa_{121}^2.
\end{aligned} \tag{12}$$

The substitution of constitutive equations, after replacement of ε_{ij}, γ_{ij} and $\varkappa_{ijk}$ by u_i and ψ_{ij}, into the stress-equations of motion results readily in displacement-equations of motion, which, for the sake of brevity, will not be displayed. The displacement-equations of motion can, alternatively, be derived directly by appropriate reduction of Mindlin's displacement-equations of motion [7], Eqs. (6.1), (6.2).

The equations of motion appropriate to TEMS can be reduced to those of COST and then to TECS in different ways, as discussed in detail by MINDLIN [7]. In rhe follnwing we outline one possible procedure of simplification.

To perform the first step of reduction we let $\sigma_{22} \to 0$, which permits to express γ_{22} in therms of ε_{11} and ε_{22}. The first three equations of motion (11) will thus involve only the unknown functions u_1, u_2 and ψ^{21} and are those of COST. The fourth equation of motion of (11) is ignored.

To carry out the next step of reduction the deformation associated with σ_{21} is made to vanish, i.e.,

$$\psi_{21} = \frac{g_2}{b_2} \frac{\partial u_2}{\partial x_1} + \left(1 + \frac{g_2}{b_2}\right) \frac{\partial u_1}{\partial x_2}. \tag{13}$$

The relative deformation γ_{21} contributing to τ_{12}, on the other hand, is expressed in terms of ε_{21} and σ_{21}, resulting in

$$\tau_{12} = \tau_{21} = \frac{g_2}{b_2} \sigma_{21} + \left(2\mu - 2\frac{g_2^2}{b_2}\right) \varepsilon_{21}. \tag{14}$$

Only two stress-equations of motion remain, namely

$$\begin{gathered}
\frac{\partial \tau_{11}}{\partial x_1} + \frac{\partial \tau_{21}}{\partial x_2} - \frac{\partial \mu_{121}}{\partial x_1 \partial x_2} \\
= \left[\varrho - \frac{1}{3} \varrho' \left(1 + \frac{g_2}{b_2}\right) d^2 \frac{\partial^2}{\partial x_2^2}\right] \ddot{u}_1 - \frac{1}{3} \varrho' d^2 \frac{g_2}{b_2} \frac{\partial^2 \ddot{u}_2}{\partial x_1 \partial x_2} \\
\text{and} \qquad \frac{\partial \tau_{12}}{\partial x_1} + \frac{\partial \tau_{22}}{\partial x_2} = \varrho \, \ddot{u}_2
\end{gathered} \tag{15}$$

where the third equation of motion of (11) has been employed. With

$$\varkappa_{121} = \frac{\partial \psi_{21}}{\partial x_1} = \frac{g_2}{b_2} \frac{\partial^2 u_2}{\partial x_1^2} + \left(1 + \frac{g_2}{b_2}\right) \frac{\partial^2 u_1}{\partial x_1 \partial x_2}, \tag{16}$$

the two displacement-equations of motion, with $d^2 = 0$, are those of TECS and can be written down immediately.

3. Interpretation of TEMS as a Continuum Theory for a Laminated Material

In an earlier paper [9] the authors derived an expression for the potential energy density of a uniformly laminated composite according to what was termed the effective single stiffness theory. In this approximate theory it is assumed that the components of displacement (in plane strain) of the kth reinforcing sheet, whose midplane position is defined by x_2^k, may be expressed in the form

$$\begin{aligned}
u_1^{fk} &= u_1^k(x_1, x_2^k, t) + x_2' \psi_{21}^k(x_1, x_2^k, t), \\
u_2^{fk} &= u_2^k(x_1, x_2^k, t) + x_2' \psi_{22}^k(x_1, x_2^k, t)
\end{aligned} \tag{17}$$

where x_2' is the coordinate in a local coordinate system, and u_i^k are the displacements in the midplanes. The displacement distributions (17) may be used to compute the potential energy V_f^k per unit surface of the kth reinforcing sheet. If there are n reinforcing sheets per unit length in x_2-direction the potential energy stored in the reinforcing sheets is obtained as a summation of V_f^k over the n discrete points x_2^k. The basic premise of the effective stiffness theory is that the sum may be approximated by a weighted integral (smoothing operation)

$$\sum_{k=1}^{n} V_f^k \cong \frac{\eta}{h} \int V_f \, dx_2 \tag{18}$$

where η is the density of the reinforcing sheets

$$\eta = h/(h+H). \tag{19}$$

In Eq. (19) h and H are the thicknesses of the reinforcing sheets and the matrix layers, respectively. By means of the smoothing operation the field variables which were previously defined at discrete points x_2^k, now have become continuously varying functions of x_2 and the superscript k is, henceforth, omitted. The resulting expression for the potential energy density V_f is

$$V_f = \frac{1}{2} D_f \left(\frac{\partial \psi_{21}}{\partial x_1}\right)^2 + \frac{1}{2} G_f \left(\psi_{21} + \frac{\partial u_2}{\partial x_1}\right)^2 + \frac{1}{2} \lambda_f h \left(\frac{\partial u_1}{\partial x_1} + \psi_{22}\right)^2 + \\ + \mu_f h \left[\left(\frac{\partial u_1}{\partial x_1}\right)^2 + \psi_{22}^2\right] + \frac{1}{24} \mu_f h^3 \left(\frac{\partial \psi_{22}}{\partial x_1}\right)^2 \tag{20}$$

where D_f is the bending stiffness

$$D_f = \mu_f h^3/6(1-\nu_f), \tag{21}$$

and G_f is the shear stiffness, which is, as an approximation

$$G_f = \mu_f h. \tag{22}$$

In Eqs. (20)—(22) μ_f, λ_f and ν_f are Lamé's elastic constants and Poisson's ratio of the reinforcing material, respectively. The first two terms in Eq. (20) represent the strain energy of bending and transverse shear of a single reinforcing sheet, respectively, and the remaining three terms represent the strain energy of extension.

By applying a similar smoothing operation to the matrix layers the contribution to the total potential energy density is obtained as $(1-\eta)\, V_m$, where for V_m we write (see [9] for greater detail),

$$V_m = \tfrac{1}{2} \bar{\lambda} (\varepsilon_{11}^2 + 2\varepsilon_{11}\varepsilon_{22} + \varepsilon_{22}^2) + \bar{\mu} (\varepsilon_{11}^2 + \varepsilon_{22}^2 + 2\varepsilon_{12}^2). \tag{23}$$

In Eq. (23) $\bar{\lambda}$ and $\bar{\mu}$ could be the elastic constants of the matrix, but more appropriate values can be assigned based on solutions for wave motion at long wave lengths.

The total potential energy density of the laminated medium may thus be written as

$$V = (\eta/h)\, V_f + (1-\eta)\, V_m \tag{24}$$

where V_f and V_m are defined by Eqs. (20) and (23), respectively.

In terms of the kinematic variables γ_{21}, γ_{22}, $\varkappa_{121}$ and $\varkappa_{122}$ defined by Eqs. (5) and (6) and the usual components of strain ε_{ij} (3), the strain energy V_f may also be written as

$$\begin{aligned} V_f = & \left(\frac{1}{2}\lambda_f + \mu_f\right) h (\varepsilon_{11}^2 + \varepsilon_{22}^2) + \lambda_f h \, \varepsilon_{11} \varepsilon_{22} + \\ & + 2\mu_f h \, \varepsilon_{12}^2 - 2\mu_f h \, \varepsilon_{12} \gamma_{21} - \lambda_f h \, \varepsilon_{11} \gamma_{22} - \\ & - (\lambda_f + 2\mu_f) h \, \varepsilon_{22} \gamma_{22} + \frac{1}{2} \mu_f h \gamma_{21}^2 + \left(\frac{1}{2}\lambda_f + \mu_f\right) h \gamma_{22}^2 + \\ & + \frac{1}{2} D_f \varkappa_{121}^2 + \frac{1}{2} \mu_f \frac{h^3}{12} \varkappa_{122}^2 . \end{aligned} \tag{25}$$

The micro-deformation gradients $\varkappa_{221}$ and $\varkappa_{222}$ do not appear in the above expression because of the underlying assumption that the differences in rotation ψ_{21} and stretch ψ_{22} of two neighboring reinforcing sheets for like x_1 do not contribute to the potential energy of the laminated continuum.

Comparing now the expression $\eta V_f/h + (1-\eta) V_m$ with that for W as given by Eq. (12), we recognize that they can be made identical provided the coefficients in one expression are related to the coefficients in the other by the following

$$\left.\begin{aligned} \lambda = \eta \lambda_f + (1-\eta)\bar{\lambda}; \quad & \mu = \eta \mu_f + (1-\eta)\bar{\mu}; \\ b_1 + b_3 = \eta(\lambda_f + \mu_f); \quad b_2 = \eta \mu_f; \quad & g_1 = -\eta \lambda_f; \quad g_2 = -\eta \mu_f; \\ a_8 + a_{10} + a_{15} = \eta D_f/h; \quad & a_4 + a_{10} + a_{13} = \eta \mu_f h^2/12 . \end{aligned}\right\} \tag{26}$$

Thus the elastic coefficients of the theory of elasticity with microstructure, if interpreted as those of a laminated composite, are seen to be determined in terms of the classical Lamé coefficients of the two constituent materials and the geometric lay-out as described by η and h.

A similar juxtaposition can also be carried out for the kinetic energy. For a laminated composite it was taken by the authors in [9] as

$$T = \frac{\eta}{h} T_f + (1-\eta) T_m \tag{27}$$

where

$$T_f = \frac{1}{2} \varrho_f h (\dot{u}_1^2 + \dot{u}_2^2) + \frac{1}{2} \varrho_f \frac{h^3}{12} (\dot{\psi}_{21}^2 + \dot{\psi}_{22}^2), \tag{28}$$

$$T_m = \tfrac{1}{2} \varrho_m (\dot{u}_1^2 + \dot{u}_2^2) \tag{29}$$

and ϱ_f and ϱ_m are the mass densities of the laminate and the matrix material, respectively. Comparing the above with the expression (10) we find

$$\varrho = \eta \varrho_f + (1-\eta) \varrho_m , \tag{30}$$

$$\varrho' d^2 = \eta \varrho_f h^2/4 . \tag{31}$$

Since, from the definition of ϱ', $\varrho' = \eta\,\varrho_f$, it follows that $h = 2d$ is the length characterizing micro-structure.

The components of Cauchy stress for a composite under consideration then are

$$\left.\begin{aligned}
\tau_{11} &= [\eta(\lambda_f + 2\mu_f) + (1-\eta)(\bar\lambda + 2\bar\mu)]\frac{\partial u_1}{\partial x_1} + (1-\eta)\,\bar\lambda\,\frac{\partial u_2}{\partial x_2} + \\
&\qquad + \eta\,\lambda_f\,\psi_{22},\\
\tau_{12} &= \tau_{21} = \eta\,\mu_f\left(\frac{\partial u_2}{\partial x_1} + \psi_{21}\right) + (1-\eta)\,\bar\mu\left(\frac{\partial u_2}{\partial x_1} + \frac{\partial u_1}{\partial x_2}\right),\\
\tau_{22} &= \eta\left[(\lambda_f + 2\mu_f)\,\psi_{22} + \lambda_f\frac{\partial u_1}{\partial x_1}\right] + (1-\eta)(\bar\lambda + 2\bar\mu)\frac{\partial u_2}{\partial x_2} + \\
&\qquad + (1-\eta)\,\bar\lambda\,\frac{\partial u_1}{\partial x_1}
\end{aligned}\right\} \tag{32}$$

and the components of relative stress are

$$\left.\begin{aligned}
\sigma_{21} &= -\eta\,\mu_f\left(\frac{\partial u_2}{\partial x_1} + \psi_{21}\right),\\
\sigma_{22} &= -\eta(\lambda_f + 2\mu_f)\,\psi_{22} - \eta\,\lambda_f\frac{\partial u_1}{\partial x_1}.
\end{aligned}\right\} \tag{33}$$

The substitution of these constitutive relations into the stress-equations of motion (11) results in displacement-equations of motion identical to Eqs. (38)—(41) of Ref. [9] and are not reproduced here.

For a Cosserat medium (COST) $\sigma_{22} \to 0$, hence

$$\psi_{22} = -[\nu_f/(1-\nu_f)]\,\partial u_1/\partial x_1,$$

and the three remaining displacement-equations of motion are the same as Eqs. (38)—(40) in [9], except that the two terms $(\lambda_f + 2\mu_f) \times \times\, \partial^2 u_1/\partial x_1^2 + \lambda_f\,\partial\psi_{22}/\partial x_1$ in (38) combine into a single term $[E_f/(1-\nu_f^2)]\,\partial^2 u_1/\partial x_1^2$ with coefficient appropriate to compression of a plate in plane strain.

For the restricted Cosserat medium (TECS) τ_{12} should be expressed in terms of σ_{21}, see Eq. (14). Next, σ_{21} is to be made a "reactive stress" through the third equation of motion in (11) and the associated deformation has to vanish, i.e., $\partial u_2/\partial x_1 = -\psi_{21}$. The two remaining equations of motion (15) become in terms of displacements

$$\left.\begin{aligned}
&(1-\eta)\left[(\bar\lambda + 2\bar\mu)\frac{\partial^2 u_1}{\partial x_1^2} + \bar\mu\,\frac{\partial^2 u_1}{\partial x_2^2} + (\bar\lambda + \bar\mu)\frac{\partial^2 u_2}{\partial x_1\,\partial x_2}\right] + \\
&\qquad + \eta\,\frac{E_f}{1-\nu_f^2}\,\frac{\partial^2 u_1}{\partial x_1^2} = \varrho\,\ddot u_1,\\
&(1-\eta)\left[(\bar\lambda + 2\bar\mu)\frac{\partial^2 u_2}{\partial x_2^2} + \bar\mu\,\frac{\partial^2 u_2}{\partial x_1^2} + (\bar\lambda + \bar\mu)\frac{\partial^2 u_1}{\partial x_1\,\partial x_2}\right] - \\
&\qquad - \frac{\eta}{h}\,D_f\,\frac{\partial^4 u_2}{\partial x_1^4} = \varrho\,\ddot u_2 - \frac{\eta}{12}\,\varrho_f\,h^2\,\frac{\partial^2 \ddot u_2}{\partial x_1^2}.
\end{aligned}\right\} \tag{34}$$

These equations are those of TECS, as discussed in Sect. 2, except for the inertia term with h^2. The material constant l, of the dimension of length, discussed by MINDLIN and TIERSTEN [1], is here

$$l^2 = \frac{\eta}{h} \frac{D_f}{(1-\eta)\,\bar{\mu}} = \frac{\eta}{1-\eta} \frac{h^2}{6(1-\nu_f)} \frac{\mu_f}{\bar{\mu}} \tag{35}$$

and depends again, just as the material coefficients of TEMS, only on the classical material properties of the two constituent materials of the composite and on its geometric properties.

4. Wave Motion and More Complex Theories

The equations of motion (38) through (41) of Ref. [9], which are now referred to as those of the effective single stiffness theory, have been used to study the propagation of plane waves in the direction of the layering, focusing attention on the lowest (predominantly) transverse and the lowest (predominantly) longitudinal mode. For a laminated composite of periodic structure it is possible to calculate "exact" dispersion curves of those modes by solving an appropriate eigenvalue problem of the classical theory of elasticity, as was done in [11]. The comparison revealed that the proposed approximate theory of the TEMS-type describes rather well the strong dispersion of the medium in the transverse mode for wave lengths which are large as compared to the lengths which characterize the lamination of the composite, as indicated in Fig. 1, which shows a plot of dimensionless frequency vs. dimensionless wave number.

By contrast, the dispersion predicted by the single stiffness theory advanced in [9] is not too satisfactory for the predominantly longitudinal mode. A detailed examination of the displacement distribution as determined by the "exact" analysis reveals that in this mode the dispersive behavior of the soft matrix is responsible for the discrepancy, which is not accounted for by the version of Ref. [9] of the effective stiffness theory. The authors were thus induced to construct a continuum theory with more complex micro-structure [12], in which the effective stiffnesses of both the laminates and the matrix layers are introduced (effective double stiffness theory) and supplemented by appropriate continuity conditions of perfect bond at the layer interfaces. A first- and a second-order theory of this type were developed, where, in the latter, quadratic terms in the displacement expansions were retained. The effective double stiffness theory corresponds to a theory of elasticity with micro-structure in which the unit cell is allowed to undergo two different micro-deformations, whose weighted sum is proportional to the macro-displacement gradient. Alternatively, the micro-kinematics of this theory can be interpreted as that of two deformable Cosserat triads being defined at each point which are, however, suitably related to the macro-displacement gradient.

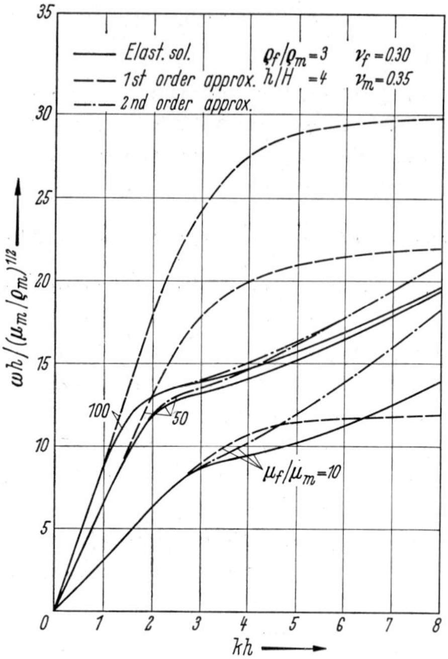

Fig. 1. Lowest transverse mode according to effective single stiffness theory.

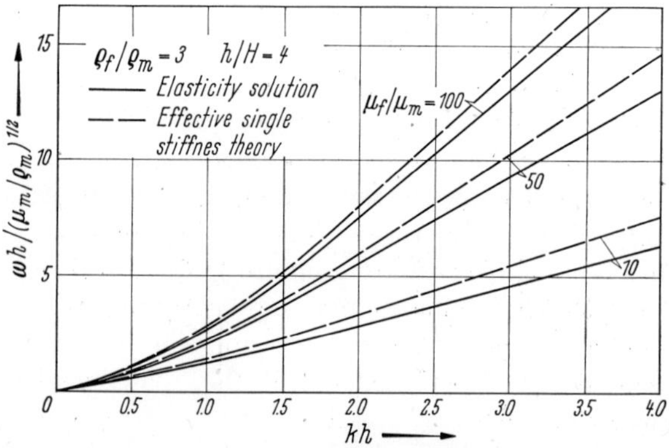

Fig. 2. Lowest longitudinal mode according to effective double stiffness theory.

Fig. 2 gives a plot of dimensionless frequency vs. dimensionless wave number, and it is concluded that the second-order approximation contributes to an improvement of the dispersion characteristics, particularly for large values of the ratio of the shear moduli of the two materials.

Finally it may be remarked that just as TEMS can be reduced to COST and then to TECS, similarly the effective double stiffness theory may thus be reduced by introducing corresponding constraints in the micro-structure.

References

[1] Mindlin, R. D., and H. F. Tiersten: Effects of couple-stresses in linear elasticity. Arch. Rational Mech. Anal. **11**, 415 (1962).

[2] Toupin, R. A.: Elastic materials with couple-stresses. Arch. Rational Mech. Anal. **11**, 385 (1962).

[3] Toupin, R. A.: Theories of elasticity with couple-stress. Arch. Rational Mech. Anal. **17**, 85 (1964).

[4] Koiter, W. T.: Couple-stresses in the theory of elasticity. Proc. Koninkl. Nederl. Akad. Wet., Ser. B. **67**, 17 (1964).

[5) Schaefer, H.: Das Cosserat-Kontinuum. ZAMM (in press).

[6] Mindlin, R. D., and N. N. Eshel: On first strain-gradient theories in linear elasticity. Int. J. of Solids and Structures **4**, 109 (1968).

[7] Mindlin, R. D.: Micro-structure in linear elasticity. Arch. Rational Mech. Anal. **16**, 51 (1964).

[8] Kunin, I. A.: Theory of elasticity with spatial dispersion. One-dimensional complex structure. PMM **30**, 866 (1966).

[9] Herrmann, G., and J. D. Achenbach: On dynamic theories of fiber-reinforced composites. Northwestern Univ. Str. Mech. Lab. Techn. Rep. 67-2 (1967); also Proc. AIAA/ASME Eighth Structures, Structural Dynamics, and Materials Conf., New York: AIAA 1967, pp. 112–118.

[10] Herrmann, G., and J. D. Achenbach: Wave propagation in laminated and fiber-reinforced composites. To be pub. in Proc. of the Int. Conf. on the mechanics of composite materials, Pergamon Press 1968.

[11] Sun, C. T., J. D. Achenbach and G. Herrmann: Time-harmonic waves in a stratified medium propagating in the direction of the layering. J. Appl. Mech., forthcoming.

[12] Sun, C. T., J. D. Achenbach and G. Herrmann: Effective stiffness theory for laminated media. Northwestern Univ. Str. Mech. Lab. Techn. Rep. 67-4 (1967).

Determination of Elastic Constants of a Structured Material[1]

By

G. Adomeit

Lehrstuhl für Mechanik
Technische Hochschule, Aachen

It has been demonstrated in Refs. [1], [2], and [3] that an elastic material with structure may be considered a kind of generalized Cosserat continuum. These investigations arrive at the same results for the linear, bipolar case, the formal procedures employed and also their physical significance are, however, not fully clear. Furthermore no particular structures have been investigated so far leaving unused the possibility of obtaining numerical values of material constants.

Therefore a specific structured material shown in Fig. 1 was chosen as a model. This structure is sufficiently simple to permit computation of its elastic behavior by simple methods of the conventional theory of elasticity. It turns out that this material, when subjected to respective deformations, may be described by the same relations as a homogeneous, linear, bipolar Cosserat continuum. The corresponding material constants may be obtained by comparison.

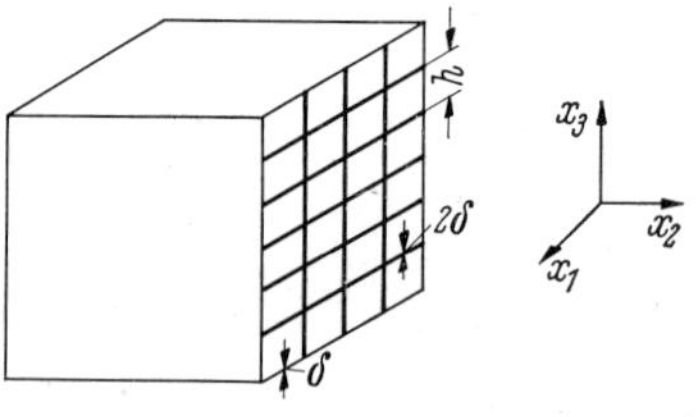

Material composed of cubes

Fig. 1. Model of a structured material.

The linear constitutive equations, as given in [1, 3], have the form

$$\left.\begin{aligned}
t_{ij} &= E^{\varepsilon}_{ijkl}\,\varepsilon_{kl} + E^{\varkappa}_{ijk[lm]}\,\varkappa_{k[lm]} + E^{\varkappa}_{ijk(lm)}\,\varkappa_{k(lm)} + E^{\gamma}_{ijkl}\,\gamma_{kl},\\
s_{ij} &= K^{\varepsilon}_{ijkl}\,\varepsilon_{kl} + K^{\varkappa}_{ijk[lm]}\,\varkappa_{k[lm]} + K^{\varkappa}_{ijk(lm)}\,\varkappa_{k(lm)} + K^{\gamma}_{ijkl}\,\gamma_{kl},\\
\mu_{i[jk]} &= L^{\varepsilon}_{i[jk]lm}\,\varepsilon_{lm} + L^{\varkappa}_{i[jk]l[mn]}\varkappa_{l[mn]} + L^{\varkappa}_{i[ik]l(mn)}\varkappa_{l(mn)} + \\
&\qquad + L^{\gamma}_{i[jk]lm}\,\gamma_{lm},\\
\mu_{i(jk)} &= L^{\varepsilon}_{i(jk)lm}\,\varepsilon_{lm} + L^{\varkappa}_{i(jk)l[mn]}\,\varkappa_{l[mn]} + L^{\varkappa}_{i(jk)l(mn)}\varkappa_{l(mn)} + \\
&\qquad + L^{\gamma}_{i(jk)lm}\,\gamma_{lm}
\end{aligned}\right\} \qquad (1)$$

---

[1] This is a summary of part of the "Habilitation"-Thesis of the author, Technische Hochschule Aachen 1967 [4].

where t_{ij}, s_{ij} denote stress tensors as introduced in [1], μ_{ijk} are the stress moments defined in [3], or in [1] as λ_{ijk}. ε_{ij}, γ_{ij} and $\varkappa_{ijk}$ are the tensors of strain, relative strain and microstrain gradients introduced in [3] and, in different notation, in [1]. The capital letters denote material constants. These equations simplify if symmetry conditions are taken into account. Thus deformations symmetric to the coordinate system used in Fig. 1 must result in symmetric stress fields, because the cubic element is symmetric itself. Consequently, since the stress moment μ_{ijk} is a measure of the antisymmetric part of the microstress, μ_{ijk} must be zero, i.e. $L^{\varepsilon}_{ijklm} = 0$. Similarly, imposing antisymmetric deformations, it follows, that $E^{\varkappa}_{ijklm} = 0$, $K^{\varkappa}_{ijklm} = 0$. For the structure investigated here, relative deformations γ_{ij} do not occur under static loading, hence the corresponding terms may be omitted. The gradients of microstrain then become equal to the gradients of macrostrain, $\varkappa_{ijk} = \partial^2 v_k / \partial x_i \, \partial x_j$ and Eq. (1) becomes

$$\begin{aligned} t_{ij} &= s_{ij} = E_{ijkl}\,\varepsilon_{kl}, \qquad m_{ij} = L_{ijkl}\,\varkappa_{kl} + L_{ijk(lm)}\,\varkappa_{k(lm)}, \\ \mu_{i(jk)} &= L^{\varkappa}_{i(jk)lm}\,\varkappa_{lm} + L^{\varkappa}_{i(jk)l(mn)}\,\varkappa_{l(mn)} \end{aligned} \tag{2}$$

with

$$m_{ij} = e_{jkl}\,\mu_{ikl}, \qquad \varkappa_{ij} = \tfrac{1}{2} e_{jkl}\,\varkappa_{ikl} = \partial\omega_j/\partial x_i, \qquad \omega_i = \tfrac{1}{2} e_{ijk}\,\partial v_k/\partial x_j$$

Here e_{ijk} is the antisymmetric permutation tensor. From the transformation invariances of cubes, the existence of an elastic potential, the symmetries of t_{ij}, ε_{ij}, $\mu_{i(jk)}$ and $\sum\limits_i \varkappa_{ii} = 0$ it follows that E_{ijkl} and L_{ijkl} must be determined by 3 constants each, $L_{ijk(lm)}$ and $L^{\varkappa}_{i(jk)lm}$ together by 4 and $L^{\varkappa}_{i(jk)l(mn)}$ by 3 constants. The relation between t_{ij}, m_{ij} and ε_{ij}, $\varkappa_{ij}$ then becomes

$$\begin{aligned} t_{ij} &= \lambda\,\delta_{ij}\,\varepsilon_{kk} + 2\mu\,\varepsilon_{ij} + (\tau - \lambda - 2\mu)\,\gamma_{ijkl}\,\varepsilon_{lk}, \\ m_{ij} &= 4\eta\left[\frac{\partial\omega_j}{\partial x_i} + \frac{\eta'}{\eta}\frac{\partial\omega_i}{\partial x_j} + \left(d - 1 - \frac{\eta'}{\eta}\right)\gamma_{ijkl}\frac{\partial\omega_l}{\partial x_k}\right]. \end{aligned} \tag{3}$$

The constants of Eqs. (3) were determined by imposing various deformations upon the material of Fig. 1, e.g. uniform compression, tension with restricted lateral contractions, shear, bending, and torsion and comparing then with a material described by Eq. (3). Assuming the wall thickness $\delta \ll h$ one obtains

$$\left.\begin{aligned} &\lambda = \frac{K E_0}{2}\frac{\nu_0}{1-\nu_0^2}, \qquad \mu = \frac{K G_0}{2}, \qquad \tau = \frac{K E_0}{1-\nu_0^2}, \\ &\eta = \frac{h^2}{48}\frac{K E_0}{1-\nu_0^2}, \qquad \frac{\eta'}{\eta} = -\nu_0, \qquad d = 4\frac{\mu}{\tau} = (1-\nu_0), \\ &\frac{\eta}{\mu} = l^2 = \frac{h^2}{12}\frac{1}{1-\nu_0}. \end{aligned}\right\} \tag{4}$$

Here E_0 is YOUNG's modulus, G_0 is the shear modulus, and ν_0 is Poisson's ratio of the wall material of the cubic element. The other quantities of Eq. (4) are defined in Fig. 1.

References

[1] ERINGEN, A. C., and E. S. SUHUBI: Int. J. Engng. Sci. **2**, 189—203, 389—404 (1964).
[2] GREEN, A. E., and R. S. RIVLIN: Arch. Rational Mech. Anal. **17**, 113—147 (1964).
[3] MINDLIN, R. D.: Arch. Rational Mech. Anal. **16**, 51—78 (1964).
[4] ADOMEIT, G.: Habilitationsschrift, Technische Hochschule Aachen 1967.

A Note on Günther's Analysis of Couple Stress

By

E. Reissner and **F. Y. M. Wan**

Department of Mathematics
Massachusetts Institute of Technology, Cambridge, Mass.

The following considerations are concerned with GÜNTHER's form of couple stress theory [1]. While the observations which follow were made without knowledge of the earlier work[1], they are offered here as a supplement to it.

We assume an orthogonal coordinate system x_i with position vector $\boldsymbol{x}$ and with coordinate tangent unit vectors $\boldsymbol{t}_i = \boldsymbol{x}_{,i}/\alpha_i$ where $\boldsymbol{x}_{,i} \cdot \boldsymbol{x}_{,j} = \alpha_i \alpha_j \delta_{ij}$. We designate force stress vectors by σ_i, moment stress vectors by $\boldsymbol{\tau}_i$ and body force and moment intensity vectors by $\boldsymbol{p}$ and $\boldsymbol{q}$. We take as basic relations the two equations of force and moment equilibrium

$$\sum (S_i \boldsymbol{\sigma}_i)_{,i} + V \boldsymbol{p} = 0, \qquad \sum (S_i \boldsymbol{\tau}_i)_{,i} + \boldsymbol{x}_{,i} \times (S_i \boldsymbol{\sigma}_i) + V \boldsymbol{q} = 0 \tag{1}$$

where $V = \alpha_1 \alpha_2 \alpha_3$, $S_1 = \alpha_2 \alpha_3$, etc.

We next introduce force and moment strain vectors $\boldsymbol{\varepsilon}_i$ and $\boldsymbol{\varkappa}_i$, and translational and rotational displacement vectors $\boldsymbol{u}$ and $\boldsymbol{\phi}$. We obtain expressions for $\boldsymbol{\varepsilon}_i$ and $\boldsymbol{\varkappa}_i$ in terms of $\boldsymbol{u}$ and $\boldsymbol{\phi}$ through use of the principle of virtual work, written in the form

$$\int (\sum \boldsymbol{\sigma}_i \cdot \delta \boldsymbol{\varepsilon}_i + \boldsymbol{\tau}_i \cdot \delta \boldsymbol{\varkappa}_i)\, dV = \int (\boldsymbol{p} \cdot \delta \boldsymbol{u} + \boldsymbol{q} \cdot \delta \boldsymbol{\phi})\, dV + \oint (\boldsymbol{\sigma}_n \cdot \delta \boldsymbol{u} + \boldsymbol{\tau}_n \cdot \delta \boldsymbol{\phi})\, dS. \tag{2}$$

In this p and q are taken from (1) and $\boldsymbol{\sigma}_n$ and $\boldsymbol{\tau}_n$ are surface traction vectors. Integration by parts to eliminate derivatives of $\boldsymbol{\sigma}_i$ and $\boldsymbol{\tau}_i$ in (2) and observation that $\boldsymbol{\sigma}_i$ and $\boldsymbol{\tau}_i$ are arbitrary functions in the remaining volume integral leads to the vectorial strain displacement relations

$$\alpha_i \boldsymbol{\varepsilon}_i = \boldsymbol{u}_{,i} + \boldsymbol{x}_{,i} \times \boldsymbol{\phi}, \qquad \alpha_i \boldsymbol{\varkappa}_i = \boldsymbol{\phi}_{,i} \tag{3}$$

Acknowledgment. Preparation of this note has been supported by the Office of Naval Research.

[1] We are indebted to E. KRÖNER for bringing GÜNTHER's paper to our attention.

which are equivalent to relations stated by GÜNTHER [1]. The present approach and Günther's approach differ from each other inasmuch as Günther departs from (3) and obtains (1) by use of the principle of virtual work. It seems to us easier to depart from (1), which is in accordance with elementary principles of dynamics, and avoid the less elementary geometrical considerations which are involved in stipulating (3).

From (3) follows by inspection, as noted previously in [1], a system of six vectorial compatibility equations

$$(\alpha_1 \boldsymbol{\varkappa}_1)_{,2} - (\alpha_2 \boldsymbol{\varkappa}_2)_{,1} = (\alpha_2 \boldsymbol{\varkappa}_2)_{,3} - (\alpha_3 \boldsymbol{\varkappa}_3)_{,2} = (\alpha_3 \boldsymbol{\varkappa}_3)_{,1} - (\alpha_1 \boldsymbol{\varkappa}_1)_{,3} = 0, \quad (4)$$

$$(\alpha_1 \boldsymbol{\varepsilon}_1)_{,2} - (\alpha_2 \boldsymbol{\varepsilon}_2)_{,1} = \boldsymbol{x}_{,1} \times (\alpha_2 \boldsymbol{\varkappa}_2) - \boldsymbol{x}_{,2} \times (\alpha_1 \boldsymbol{\varkappa}_1), \quad (5a)$$

$$(\alpha_2 \boldsymbol{\varepsilon}_2)_{,3} - (\alpha_3 \boldsymbol{\varepsilon}_3)_{,2} = \boldsymbol{x}_{,2} \times (\alpha_3 \boldsymbol{\varkappa}_3) - \boldsymbol{x}_{,3} \times (\alpha_2 \boldsymbol{\varkappa}_2), \quad (5b)$$

$$(\alpha_3 \boldsymbol{\varepsilon}_3)_{,1} - (\alpha_1 \boldsymbol{\varepsilon}_1)_{,3} = \boldsymbol{x}_{,3} \times (\alpha_1 \boldsymbol{\varkappa}_1) - \boldsymbol{x}_{,1} \times (\alpha_3 \boldsymbol{\varkappa}_3). \quad (5c)$$

We complement Günther's formulation by phenomenological stress strain relations as follows. Writing

$$(\boldsymbol{\sigma}_i, \boldsymbol{\tau}_i, \boldsymbol{\varepsilon}_i, \boldsymbol{\varkappa}_i) = \sum (\sigma_{ij}, \tau_{ij}, \varepsilon_{ij}, \varkappa_{ij})\, \boldsymbol{t}_j \quad (6)$$

we stipulate the existence of functions $A(\varepsilon, \varkappa)$ and $B(\sigma, \tau)$ such that

$$\sigma_{ij} = \frac{\partial A}{\partial \varepsilon_{ij}}, \qquad \tau_{ij} = \frac{\partial A}{\partial \varkappa_{ij}}, \qquad \varepsilon_{ij} = \frac{\partial B}{\partial \sigma_{ij}}, \qquad \varkappa_{ij} = \frac{\partial B}{\partial \tau_{ij}}. \quad (7)$$

An appropriate definition of vectorial derivatives allows us to write (7) in the form

$$\boldsymbol{\sigma}_i = \frac{\partial A}{\partial \boldsymbol{\varepsilon}_i}, \qquad \boldsymbol{\tau}_i = \frac{\partial A}{\partial \boldsymbol{\varkappa}_i}, \qquad \boldsymbol{\varepsilon}_i = \frac{\partial B}{\partial \boldsymbol{\sigma}_i}, \qquad \boldsymbol{\varkappa}_i = \frac{\partial B}{\partial \boldsymbol{\tau}_i}. \quad (7')$$

Observing (1), (3) and (7) we readily have as variational principles, generalizing corresponding principles of elasticity without couple stresses, a variational principle for stresses and displacements, $\delta I_{SD} = 0$, and a variational principle for stresses, displacements and strains, $\delta I_{SDS} = 0$, where

$$I_{SD} = \int \left[\sum \boldsymbol{\sigma}_i \cdot \frac{\boldsymbol{u}_{,i} + \boldsymbol{x}_{,i} \times \boldsymbol{\phi}}{\alpha_i} + \boldsymbol{\tau}_i \cdot \frac{\boldsymbol{\phi}_{,i}}{\alpha_i} - \boldsymbol{p} \cdot \boldsymbol{u} - \boldsymbol{q} \cdot \boldsymbol{\phi} - B \right] dV -$$

$$- \int_{S_s} [\bar{\boldsymbol{\sigma}}_n \cdot \boldsymbol{u} + \bar{\boldsymbol{\tau}}_n \cdot \boldsymbol{\phi}]\, dS - \int_{S_d} [(\boldsymbol{u} - \bar{\boldsymbol{u}}) \cdot \boldsymbol{\sigma}_n + (\boldsymbol{\phi} - \bar{\boldsymbol{\phi}}) \cdot \boldsymbol{\tau}_n]\, dS \quad (8)$$

and

$$I_{SDS} = \int \left[\sum \boldsymbol{\sigma}_i \cdot \left(\frac{\boldsymbol{u}_{,i} + \boldsymbol{x}_{,i} \times \boldsymbol{\phi}}{\alpha_i} - \boldsymbol{\varepsilon}_i \right) + \boldsymbol{\tau}_i \cdot \left(\frac{\boldsymbol{\phi}_{,i}}{\alpha_i} - \boldsymbol{\varkappa}_i \right) - \right.$$

$$\left. - \boldsymbol{p} \cdot \boldsymbol{u} - \boldsymbol{q} \cdot \boldsymbol{\phi} + A \right] dV -$$

$$- \int_{S_s} [\bar{\boldsymbol{\sigma}}_n \cdot \boldsymbol{u} + \bar{\boldsymbol{\tau}}_n \cdot \boldsymbol{\phi}]\, dS - \int_{S_d} [(\boldsymbol{u} - \bar{\boldsymbol{u}}) \cdot \boldsymbol{\sigma}_n + (\boldsymbol{\phi} - \bar{\boldsymbol{\phi}}) \cdot \boldsymbol{\tau}_n]\, dS. \quad (9)$$

In δI_{SD} stresses and displacements are varied independently and the Euler differential equations consist of the equations of equilibrium and the stress strain relations. In δI_{SDS} stresses, displacements and strains are varied independently and the Euler differential equations are equilibrium equations, strain displacement relations and stress strain relations[1].

A recent result [4] on a static-geometric analogue of the two-dimensional elastic-shell theory version of the variational principle for stresses and displacements [3] suggests the formulation of a variational principle for strains and stress functions in couple stress elasticity, such that the compatibility Eqs. (4) and (5), together with the stress strain relations, are Euler differential equations.

A small amount of mathematical experimentation indicates that a suitable stress function representation of the solutions of the homogeneous Eqs. (1) is

$$S_1\,\boldsymbol{\sigma}_1 = \boldsymbol{F}_{3,2} - \boldsymbol{F}_{2,3}, \quad S_2\,\boldsymbol{\sigma}_2 = \boldsymbol{F}_{1,3} - \boldsymbol{F}_{3,1}, \quad S_3\,\boldsymbol{\sigma}_3 = \boldsymbol{F}_{2,1} - \boldsymbol{F}_{1,2}, \tag{10}$$

$$S_1\,\boldsymbol{\tau}_1 = \boldsymbol{H}_{3,2} - \boldsymbol{H}_{2,3} + \boldsymbol{x}_{,2} \times \boldsymbol{F}_3 - \boldsymbol{x}_{,3} \times \boldsymbol{F}_2, \tag{11a}$$

$$S_2\,\boldsymbol{\tau}_2 = \boldsymbol{H}_{1,3} - \boldsymbol{H}_{3,1} + \boldsymbol{x}_{,3} \times \boldsymbol{F}_1 - \boldsymbol{x}_{,1} \times \boldsymbol{F}_3, \tag{11b}$$

$$S_3\,\boldsymbol{\tau}_3 = \boldsymbol{H}_{2,1} - \boldsymbol{H}_{1,2} + \boldsymbol{x}_{,1} \times \boldsymbol{F}_2 - \boldsymbol{x}_{,2} \times \boldsymbol{F}_1. \tag{11c}$$

Taking Eqs. (10) and (11) as equations of definition, we find that the variational equation

$$\delta \int [\sum \boldsymbol{\sigma}_i \cdot \boldsymbol{\varepsilon}_i + \boldsymbol{\tau}_i \cdot \boldsymbol{\varkappa}_i - A]\, d\,V = 0 \tag{12}$$

where the $\boldsymbol{F}_i$, $\boldsymbol{H}_i$, $\boldsymbol{\varepsilon}_i$ and $\boldsymbol{\varkappa}_i$ are varied independently does in fact have the compatibility Eqs. (4) and (5) and the stress strain relations (7) as Euler equations. In addition, it is found that the associated Euler boundary conditions are displacement boundary conditions expressed in terms of tangential strain vectors. These conditions are homogeneous conditions which may be made non-homogeneous conditions upon adding in (12) an appropriate surface integral.

We state two further variational principles which we think will eventually be found useful. The first of these is a *mixed* principle, in the sense that it has parts of the equilibrium equations and compatibility equations as Euler equations, while the complementary parts are equations of definition.

---

[1] A restricted form of the variational principle $\delta I_{SDS} = 0$ has previously been stated by NAGHDI [2]. The restriction consists in assuming symmetry conditions $\varepsilon_{ij} = \varepsilon_{ji}$ as equations of definition. As a consequence of this, an additional Euler differential equation $\boldsymbol{\phi} = \frac{1}{2}\operatorname{curl} \boldsymbol{u}$ is obtained, the force stress strain relations become $\frac{1}{2}(\sigma_{ij} + \sigma_{ji}) = \partial A/\partial \varepsilon_{ij} = \partial A/\partial \varepsilon_{ji}$ and the number of associated Euler boundary conditions is five instead of six.

Restricting attention to the case of absent body forces and moments, we take as equations of definition the conditions of force equilibrium in (1), via the stress function Eqs. (10), and the equations for bending strains in (3), these being equivalent to the compatibility Eqs. (4). Additionally, we now take the relations between stresses and strains in the mixed form

$$\boldsymbol{\varepsilon}_i = \frac{\partial B_f}{\partial \boldsymbol{\sigma}_i}, \qquad \boldsymbol{\tau}_i = \frac{\partial A_c}{\partial \boldsymbol{\varkappa}_i} \tag{13}$$

as equations of definition.

We then have that the equation $\delta I_M = 0$ where

$$I_M = \int \left[\frac{\boldsymbol{F}_1}{\alpha_1} \cdot \left(\frac{\boldsymbol{x}_{,2}}{\alpha_2} \times \frac{\boldsymbol{\phi}_{,3}}{\alpha_3} - \frac{\boldsymbol{x}_{,3}}{\alpha_3} \times \frac{\boldsymbol{\phi}_{,2}}{\alpha_2} \right) + \cdots + B_f - A_c \right] dV \tag{14}$$

and where $\boldsymbol{\phi}$ and the $\boldsymbol{F}_i$ are varied independently has as Euler differential equations the conditions of moment equilibrium in (1) together with the force strain compatibility Eqs. (5).

The second additional variational principle is for boundary values. It is a generalization of a principle previously stated for elasticity without couple stresses [5]. Define a functional

$$I_B = \int_{S_s} [\tfrac{1}{2}(\boldsymbol{u} \cdot \boldsymbol{\sigma}_n + \boldsymbol{\phi} \cdot \boldsymbol{\tau}_n) + \Psi_s(\boldsymbol{u}, \boldsymbol{\phi})] \, dS - \\ - \int_{S_d} [\tfrac{1}{2}(\boldsymbol{u} \cdot \boldsymbol{\sigma}_n + \boldsymbol{\phi} \cdot \boldsymbol{\tau}_n) + \Psi_d(\boldsymbol{\sigma}_n, \boldsymbol{\tau}_n)] \, dS \tag{15}$$

where $\boldsymbol{u}$, $\boldsymbol{\phi}$, $\boldsymbol{\sigma}_n$ and $\boldsymbol{\tau}_n$ are the surface values of states of displacement and stress which in the interior satisfy equilibrium and stress displacement relations, subject to the limitation of absent body forces and moments and subject to the limitation that the stress energy function B in (7) is homogeneous of the second degree in the sense that $\sum \boldsymbol{\sigma}_i \cdot \partial B / \partial \boldsymbol{\sigma}_i + \boldsymbol{\tau}_i \cdot \partial B / \partial \boldsymbol{\tau}_i = 2B$. It can then be shown, in extension of what is done in [5] for elasticity without couple stresses, that the Euler equations of $\delta I_B = 0$ are the boundary conditions.

$$\boldsymbol{\sigma}_n + \frac{\partial \Psi_s}{\partial \boldsymbol{u}} = 0, \qquad \boldsymbol{\tau}_n + \frac{\partial \Psi_s}{\partial \boldsymbol{\phi}} = 0 \quad \text{on } S_s, \tag{16}$$

$$\boldsymbol{u} + \frac{\partial \Psi_d}{\partial \boldsymbol{\sigma}_n} = 0, \qquad \boldsymbol{\phi} + \frac{\partial \Psi_d}{\partial \boldsymbol{\tau}_n} = 0 \quad \text{on } S_d. \tag{17}$$

References

[1] GÜNTHER, W.: Abh. d. Braunschweigischen Wiss. Ges. **10**, 195—213 (1958)
[2] NAGHDI, P. M.: J. Appl. Mech. **31**, 647—652 (1964).
[3] REISSNER, E.: Proc. American Soc. Civil Eng. 88 (EM), 23—57 (1962).
[4] WAN, F. Y. M.: J. Math. and Phys. **47** (1968).
[5] REISSNER, E.: Problems of Continuum Mechanics (Muskhelishvili Anniversary Volume), Philadelphia 1961, pp. 371—381.

Note on the Statics and Stability of Polar Hyperelastic Materials

By

M. Dikmen

Middle East Technical University, Ankara

The present Note generalizes two theorems already known to be valid for hyperelastic materials [1], to the case of *polar hyperelastic materials*, as they are described by the theory of Toupin [2].

The notation shall be the same as in [1], unless specified otherwise here.

We consider the boundary-value problem of place and traction in statics of a polar hyperelastic body $\mathscr{B}$, and assume that the deformation is prescribed upon a portion $\mathscr{S}_1$ of the boundary, while the surface tractions (per unit area in the reference configuration) are prescribed upon the remainder $\mathscr{S}_2$, i.e.

$$\boldsymbol{x} = \boldsymbol{x}(\boldsymbol{X}) \quad \text{and} \quad \boldsymbol{d}_{\mathfrak{a}} = \boldsymbol{d}_{\mathfrak{a}}(\boldsymbol{D}_{\mathfrak{b}}) = \boldsymbol{d}_{\mathfrak{a}}(\boldsymbol{X}) \quad \text{given if} \quad \boldsymbol{X} \in \mathscr{S}_1, \tag{1}$$

$$\underset{R}{\boldsymbol{T}}\,\underset{R}{\boldsymbol{n}} = \underset{R}{\boldsymbol{t}} \quad \text{and} \quad \underset{R}{\boldsymbol{T}}{}^{a}\,\underset{R}{\boldsymbol{n}} = \underset{R}{\boldsymbol{t}}{}^{a} \quad \text{given if} \quad \boldsymbol{X} \in \mathscr{S}_2 \tag{2}$$

where $\boldsymbol{x}$ denotes the spatial coordinates, $\boldsymbol{X}$ the material coordinates, $\boldsymbol{d}_{\mathfrak{a}}$ the directors at $\boldsymbol{x}$ ($\mathfrak{a} = 1, \ldots, \mathfrak{p}$), $\boldsymbol{D}_{\mathfrak{a}}$ the directors at $\boldsymbol{X}$, $\underset{R}{\boldsymbol{T}}$ and $\underset{R}{\boldsymbol{T}}{}^{\mathfrak{a}}$ the first Piola-Kirchhoff macrostress and microstress tensors, $\underset{R}{\boldsymbol{t}}$ and $\underset{R}{\boldsymbol{t}}{}^{\mathfrak{a}}$ the corresponding traction vectors in the reference configuration, and $\underset{R}{\boldsymbol{n}}$ the unit normal in the reference configuration.

We assume then the existence of a *density function*

$$L = L(\boldsymbol{x}, \boldsymbol{d}_{\mathfrak{a}}, \boldsymbol{F}, \boldsymbol{F}_{\mathfrak{a}}, \boldsymbol{X}, \boldsymbol{D}_{\mathfrak{a}}) \tag{3}$$

where $\boldsymbol{F}$ and $\boldsymbol{F}_{\mathfrak{a}}$ are the gradients of $\boldsymbol{x}$ and $\boldsymbol{d}_{\mathfrak{a}}$, having the components $x^k_{,\alpha}$ and $d^k_{\mathfrak{a},\alpha}$, respectively, and such that

$$-L_{\boldsymbol{F}} = \underset{R}{\boldsymbol{T}}, \qquad -L_{\boldsymbol{F}_{\mathfrak{a}}} = \underset{R}{\boldsymbol{T}}{}^{\mathfrak{a}}. \tag{4}$$

We furthermore assume that the body macroforce density $\boldsymbol{b}$ and the body microforce densities $\boldsymbol{b}^{\mathfrak{a}}$ have each a potential, so that

$$\boldsymbol{b} \overset{\text{def}}{=} -\operatorname{grad} w, \qquad \boldsymbol{b}^{\mathfrak{a}} \overset{\text{def}}{=} -\operatorname{grad} w^{\mathfrak{a}} \tag{5}$$

where

$$w = w(\boldsymbol{x}), \qquad w^{\mathfrak{a}} = w^{\mathfrak{a}}(\boldsymbol{d}_{\mathfrak{a}}). \tag{6}$$

We then consider the integral

$$J = -\int_{\mathscr{B}} L \, d\underset{R}{v} + \int_{\mathscr{B}} \underset{R}{\varrho}\, w \, d\underset{R}{v} - \int_{\mathscr{S}_2} \boldsymbol{u}\, \underset{R}{\boldsymbol{t}} \, d\underset{R}{s} + \sum_{\mathfrak{a}} \int_{\mathscr{B}} \underset{R}{\varrho}\, w^{\mathfrak{a}} \, d\underset{R}{v} - \int_{\mathscr{S}_2} \boldsymbol{l}_{\mathfrak{a}} \cdot \underset{R}{\boldsymbol{t}}^{\mathfrak{a}} \, d\underset{R}{s} \tag{7}$$

where $\underset{R}{\varrho}$ is the mass density in the reference configuration,

$$\boldsymbol{u} \overset{\text{def}}{=} \boldsymbol{x} - \boldsymbol{X} \quad \text{denotes the displacement,}$$

$$\boldsymbol{l}_{\mathfrak{a}} \overset{\text{def}}{=} \boldsymbol{d}_{\mathfrak{a}} - \boldsymbol{D}_{\mathfrak{a}} \quad \text{denotes the deviation,}$$

and where summation over diagonally repeated gothic indices is understood.

The first variation of the integral J, with respect to the independent variables $\boldsymbol{x}$ and $\boldsymbol{d}_{\mathfrak{a}}$ is formally equivalent to Toupin's action principle for the particular case of a *density of action* L independent of time and body forces having potentials, and yields after suitable transformations and in view of the boundary conditions (1) of place

$$\begin{aligned} \delta J = & \int_{\mathscr{B}} [-L_{\boldsymbol{x}} - \operatorname{Div} \underset{R}{\boldsymbol{T}} - \underset{R}{\varrho}\, \boldsymbol{b}] \cdot \delta \boldsymbol{x} \, d\underset{R}{v} + \\ & + \int_{\mathscr{B}} [-L_{\boldsymbol{d}_{\mathfrak{a}}} - \operatorname{Div} \underset{R}{\boldsymbol{T}}^{\mathfrak{a}} - \underset{R}{\varrho}\, \boldsymbol{b}^{\mathfrak{a}}] \cdot \delta \boldsymbol{d}_{\mathfrak{a}} \, d\underset{R}{v} + \\ & + \int_{\mathscr{S}_2} [\underset{R}{\boldsymbol{T}}\, \underset{R}{\boldsymbol{n}} - \underset{R}{\boldsymbol{t}}] \cdot \delta \boldsymbol{x} \, d\underset{R}{s} + \int_{\mathscr{S}_2} [\underset{R}{\boldsymbol{T}}^{\mathfrak{a}}\, \underset{R}{\boldsymbol{n}} - \underset{R}{\boldsymbol{t}}^{\mathfrak{a}}] \cdot \delta \boldsymbol{d}_{\mathfrak{a}} \, d\underset{R}{s}. \end{aligned} \tag{8}$$

Since the surface integrals in (8) vanish whenever the boundary conditions (2) are fulfilled, the first variation δJ itself vanishes, for all possible δx and $\delta d_{\mathfrak{a}}$, if and only if

$$\operatorname{Div} \underset{R}{\boldsymbol{T}} + \underset{R}{\varrho}\, \boldsymbol{b} = -L_{\boldsymbol{x}}, \qquad \operatorname{Div} \underset{R}{\boldsymbol{T}}^{\mathfrak{a}} + \underset{R}{\varrho}\, \boldsymbol{b}^{\mathfrak{a}} = -L_{\boldsymbol{d}_{\mathfrak{a}}}. \tag{9}$$

The requirements of *Euclid invariance* of L, and equilibrium of all external forces (*principle of solidification*) [3] yield each, in specialization of the Cosserat-Toupin equivalence theorem [1], the necessary and sufficient conditions

$$L_{\boldsymbol{x}} = \boldsymbol{0}, \quad \boldsymbol{K} = \boldsymbol{K}^T, \tag{10}$$

where

$$\boldsymbol{K} \overset{\text{def}}{=} \boldsymbol{p} \otimes L_{\boldsymbol{x}} + \boldsymbol{d}_{\mathfrak{a}} \otimes L_{\boldsymbol{d}_{\mathfrak{a}}} + \boldsymbol{F}\, L_{\boldsymbol{F}}^T + \boldsymbol{F}_{\mathfrak{a}}\, L_{\boldsymbol{F}_{\mathfrak{a}}}^T. \tag{11}$$

tr stands for trace, T for transpose, and $\boldsymbol{p}$ denotes the position vector of place $\boldsymbol{x}$.

A necessary condition for the stability of the equilibrium is obtained by the requirement that, in addition to $\delta J = 0$ and the boundary conditions,

$$\delta^2 J \geqq 0 \tag{12}$$

be satisfied. This yields the criterion

$$\delta^2 J = \int_{\mathscr{B}} \delta(\operatorname{Div} L_{\boldsymbol{F}}) \cdot \delta \boldsymbol{x} \, d\underset{R}{v} + \int_{\mathscr{B}} \delta(\operatorname{Div} L_{\boldsymbol{F}_{\mathfrak{a}}}) \cdot \delta \boldsymbol{d}_{\mathfrak{a}} \, d\underset{R}{v} + \\ + \int_{\mathscr{B}} tr[(\delta L_{\boldsymbol{F}}^T)(\delta \boldsymbol{F})] \, d\underset{R}{v} + \int_{\mathscr{B}} tr[(\delta_{\boldsymbol{F}_{\mathfrak{a}}}^T L)(\delta \boldsymbol{F}_{\mathfrak{a}})] \, d\underset{R}{v} \geqq 0. \tag{13}$$

References

[1] Truesdell, C., and W. Noll: Handbuch der Physik III/3. The Non-Linear Field Theories of Mechanics. Berlin/Heidelberg/New York: Springer 1965, p. 326 and p. 329.

[2] Toupin, R. A.: Arch. Rational Mech. Anal. **17**, 85—112 (1964).

[3] Cosserat, E., and F. Cosserat: Théorie des Corps Déformables, Paris: Hermann 1906, pp. 122—154.

The Characteristic Length of a Polar Fluid

By

S. C. Cowin

Department of Mechanical Engineering
Tulane University, New Orleans, La.

The constitutive equations for a linear polar fluid were introduced in 1952 by GRAD [1] and were motivated by statistical mechanics considerations. Recently polar fluids have been considered from a statistical mechanics viewpoint by DAHLER [2, 3] and from a continuum approach by COWIN [4], JAUNZEMIS and COWIN [5], and ERINGEN [6]. The constitutive equations for a polar fluid relate the stress tensor $\boldsymbol{T}$ and the couple stress tensor $\hat{\boldsymbol{\Lambda}}$ to the rate of deformation tensor $\boldsymbol{D}$, to the angular velocity of the rigid Cosserat triad $\hat{\boldsymbol{G}}$, to the gradient of $\hat{\boldsymbol{G}}$ denoted by $\hat{\boldsymbol{\Psi}}$, and to the difference $\hat{\boldsymbol{H}}$ between $\hat{\boldsymbol{G}}$ and the usual or regional angular velocity $\hat{\boldsymbol{W}}$ ($\hat{\boldsymbol{W}}$ is one half the vorticity $\nabla \times \boldsymbol{v}$),

$$\hat{\boldsymbol{H}} = \hat{\boldsymbol{G}} - \hat{\boldsymbol{W}} = \hat{\boldsymbol{G}} - \tfrac{1}{2}\nabla \times \boldsymbol{v}. \tag{1}$$

Absolute tensors of ranks one, two and three are denoted here in boldface by small Latin letters, capital Latin letters and capital Greek letters, respectively; the circumflex ^ over a tensor symbol denotes an associated axial tensor one rank lower, for example, in cartesian component notation

$$\hat{G}_i = -\tfrac{1}{2} e_{ijk}\, G_{jk}, \quad \hat{\Lambda}_{im} = -\tfrac{1}{2} e_{ijk}\, \Lambda_{jkm}, \tag{2}$$

where e_{ijk} is the alternating symbol. The constitutive equations for the linear polar fluid are

$$\begin{aligned} \boldsymbol{T} + p\,\boldsymbol{1} &= \boldsymbol{1}\,\lambda\, tr\,\boldsymbol{D} + 2\mu\,\boldsymbol{D} - 2\tau\,\boldsymbol{H}, \\ \hat{\boldsymbol{\Lambda}} &= \boldsymbol{1}\,\alpha\, tr\,\hat{\boldsymbol{\Psi}} + (\beta+\gamma)\,\hat{\boldsymbol{\Psi}} + (\beta-\gamma)\,\hat{\boldsymbol{\Psi}}^T, \end{aligned} \tag{3}$$

where the T denotes transpose, $\boldsymbol{1}$ is the unit tensor, p is the pressure, λ and μ are the usual viscosities, τ is a relative rotational viscosity

Acknowledgment. This work was supported in part by Tulane University and, in part, by the U.S. Army Research Office in Durham under Grant DA-ARO-D-31-124-G599 with Tulane University.

and α, β, γ are viscosities associated with the gradient of the total rotation. Thermodynamic arguments may be presented to show that $\mu, \beta, \gamma, \tau, 3\lambda + 2\mu$ and $3\alpha + 2\beta$ must all be greater than or equal to zero. When the constitutive Eqs. (3) are substituted into the expressions for the conservation of linear and angular momentum, the following equations of motion result:

$$(\lambda + \mu)\nabla\nabla \cdot \boldsymbol{v} + \mu \nabla^2 \boldsymbol{v} + 2\tau \nabla \times \hat{\boldsymbol{H}} + \varrho \boldsymbol{b} - \nabla p = \varrho \dot{\boldsymbol{v}}, \tag{4}$$

$$(\alpha + \beta - \gamma)\nabla\nabla \cdot \hat{\boldsymbol{G}} + (\beta + \gamma)\nabla^2 \hat{\boldsymbol{G}} - 4\tau \hat{\boldsymbol{H}} + \varrho \hat{\boldsymbol{C}} = \varrho \dot{\hat{\boldsymbol{M}}}, \tag{5}$$

where $\boldsymbol{b}$ is the body force density, $\hat{\boldsymbol{C}}$ is the body couple density and $\hat{\boldsymbol{M}}$ is the intrinsic angular momentum density. Eq. (4) differs from the Navier Stokes equation only by the term $2\tau \nabla \times \hat{\boldsymbol{H}}$.

The following paragraphs concern a special class of flows of *incompressible* polar fluids defined by the conditions: (i) that $\nabla \cdot \boldsymbol{v}$, $\nabla \cdot \hat{\boldsymbol{G}}$, and $\nabla \times (\dot{\hat{\boldsymbol{M}}} - \hat{\boldsymbol{C}})$ all vanish, (ii) that $\hat{\boldsymbol{G}}$ vanish on all boundaries and that $\boldsymbol{v}$ satisfy the usual "no slip" boundary condition, and (iii) that the flow region be connected. Subject to conditions (i) the equations of motion (4) and (5) take the form

$$\begin{aligned} \mu \nabla^2 \boldsymbol{v} + 2\tau\nabla \times \hat{\boldsymbol{H}} + \varrho \boldsymbol{b} - \nabla p &= \varrho \dot{\boldsymbol{v}}, \\ (\beta + \gamma) \nabla^2 \hat{\boldsymbol{G}} - 2\tau \hat{\boldsymbol{H}} &= \varrho (\dot{\hat{\boldsymbol{M}}} - \hat{\boldsymbol{C}}). \end{aligned} \tag{6}$$

In the study of this special class of flows two significant and new dimensionless parameters occur. To obtain the first of these note from the constitutive Eqs. (3) that the viscosities λ, μ and τ are of the same dimension and that the gradient viscosities α, β and γ are of dimension length squared times the dimension of the viscosity μ, hence material coefficients with the dimension of length squared may be formed in a variety of ways. For the special class of flows considered here the *material characteristic length* l, $l = ((\beta + \gamma)/\mu)^{1/2}$, is most significant. The dimensionless parameter L, called the *length ratio*, is defined as the ratio of a characteristic geometrical length L_0 to the material characteristic length l,

$$L = L_0/l = L_0(\mu/(\beta + \gamma))^{1/2}. \tag{7}$$

The second significant dimensionless parameter is the *coupling number* N, $N = (\tau/(\mu + \tau))^{1/2}$, which characterizes the coupling of the differential Eqs. (4) and (5) governing the velocity and angular velocity fields.

Two important flows that fall into the special class defined above are steady flow between parallel flat plates in relative motion (Couette flow) and steady pipe flow (Poiseuille flow). The velocity distributions for these two flows are presented graphically in Figs. 1 and 2, respectively, and the solutions to these boundary value problems are discussed in detail by PENNINGTON [7]. In Figs. 1 and 2 the value of N^2 has been set at 0.9 and the velocity distributions for L equal to 0, 2, 5 and ∞ are illustrated. The velocity distributions for an L of ∞ cor-

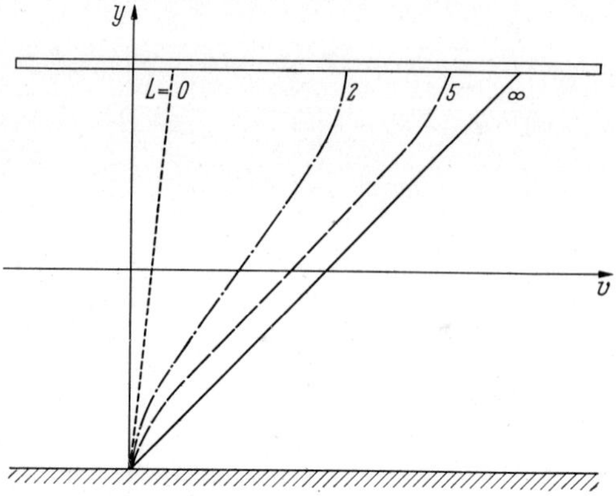

Fig. 1. Velocity distribution for flow between parallel flat plates in relative motion (Couette flow). $N^2 = 0.9$, $L = 0$ (- - - -), $L = 2$ (—-—-—), $L = 5$ (— — —), $L = \infty$ (———).

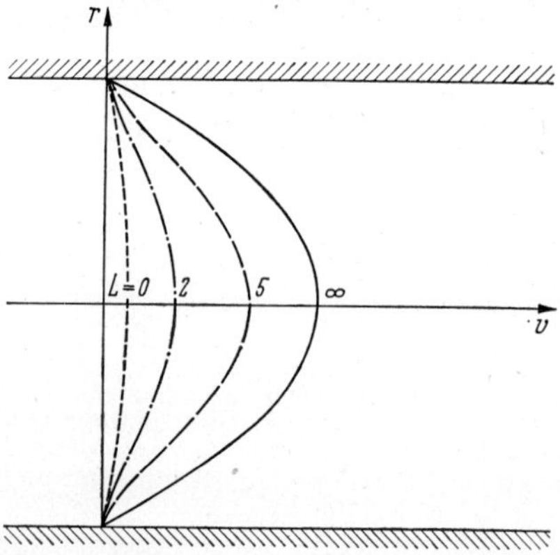

Fig. 2. Velocity distribution for pipe flow. $N^2 = 0.9$, $L = 0$ (- - - -), $L = 2$ (—-—-—), $L = 5$ (— — —), $L = \infty$ (———).

respond to the solution given by the Navier Stokes equations for both flows, that is to say, the velocity distribution is linear for Couette flow and parabolic for pipe flow. An interesting result illustrated by these figures is that the velocity distributions for an L of 0 are also the linear and parabolic ones, respectively, predicted by the Navier Stokes equations, but the viscosity μ in the Navier Stokes equations must be replaced by $\mu + \tau$.

This interesting result is true for the entire special class of flows considered. It is easy to show that, as L tends to ∞, the system of Eqs. (6) reduce for this special class of flows to the single equation

$$\mu \nabla^2 \boldsymbol{v} + \varrho \boldsymbol{b} - \nabla p = \varrho \dot{\boldsymbol{v}} \tag{8}$$

which is the appropriate form of the Navier Stokes equation for this special class of flows. To show that the system of Eqs. (6) for this class of flows reduce to Eq. (8) with μ replaced by $\mu + \tau$ as L tends to zero, it must be assumed that N is not zero or one. For fixed L_0, L tending to zero means that $(\beta + \gamma)^{-1}$ tends to zero, hence $(6)_2$ reduces to the condition that $\hat{\boldsymbol{G}}$ be harmonic. Since $\hat{\boldsymbol{G}}$ is harmonic and zero everywhere on the boundary of a connected region, it follows that $\hat{\boldsymbol{G}}$ is zero everywhere. Eq. (1) then shows that $\hat{\boldsymbol{H}}$ is equal to $-\frac{1}{2}\nabla \times \boldsymbol{v}$. Substitution of this expression for $\hat{\boldsymbol{H}}$ into $(6)_1$ and subsequently employing the incompressibility condition and a vector identity, it is easy to show that $(6)_1$ takes the form of Eq. (8) with μ replaced by $\mu + \tau$. This is the appropriate form of the Navier Stokes equation for this class of flows with a viscosity of $\mu + \tau$ rather than μ.

Another important characteristic of the length ratio L is its function as a measure of the "narrowness" of a geometry for a given polar fluid. The geometry of the flow situation (i.e. the distance between the plates or the pipe diameter) can always be made sufficiently large so that a given polar fluid will behave in the manner of a usual viscous fluid. As the geometry is narrowed from this extreme polar effects will become increasingly significant. When the geometry is narrowed to a point where the geometric characteristic length is very much less than the material characteristic length, the flow will again be that predicted by viscous flow theory, but with an increased viscosity of $\mu + \tau$ rather than μ. There is some experimental evidence that qualitatively agrees with this prediction. A summary of some related experimental literature has been given by Henniker [8] who notes that "... the evidence for abnormally high viscosity in the neighborhood of a solid surface is extensive." Henniker mentions an experiment in which a tenfold increase in viscosity was found within 5000 Å of a solid boundary and one performed by several different investigators in which a ten fold increase in the viscosity of water between glass plates 2500 Å apart was observed.

References

[1] Grad, H.: Communs. Pure and Appl. Math. **5**, 455 (1952).
[2] Dahler, J. S.: Article in Research Frontiers of Fluid Mechanics. New York: Interscience 1965.

[3] Condiff, D. W., and J. S. Dahler: Phys. Fluids **7**, 842 (1964).
[4] Cowin, S. C.: Mechanics of Cosserat Continua, Thesis, The Pennsylvania State University 1962.
[5] Jaunzemis, W., and S. C. Cowin: Proc. Princeton Conf. on Solid Mechanics, 1963.
[6] Eringen, A. C.: Theory of Micropolar Fluids. Techn. Rep. 27, Contract Nonr-1100(23), Purdue University 1965.
[7] Pennington, C. J.: Certain Steady Flows of Polar Fluids, Thesis, Tulane University 1966.
[8] Henniker, J. C.: Rev. Modern Phys. **21**, 322 (1949).

Couple-Stresses and Singular Stress Concentrations in Elastic Solids[1]

By

E. Sternberg

California Institute of Technology, Pasadena, Calif.

Introduction

Continuum-mechanical theories of the type initiated in rudimentary form by the COSSERATS [1], which admit in addition to ordinary stresses the presence of couple-stresses, have—for several and diverse reasons—attracted a renewed and growing interest during recent years. A particularly comprehensive study of the *linearized* Cosserat theory with constrained rotations for elastic solids was published by MINDLIN and TIERSTEN [2][2] in 1962. Since that time various related broader generalizations of conventional elasticity theory have been proposed in investigations that introduce, beyond ordinary and couple-stresses, hyper-stresses of ever increasing complexity and physical elusiveness.

This paper aims not at generality but rather at the exploration of certain implications of the simplest theory of elasticity in which couple-stresses make their appearance. Thus we limit our attention to the *linear* couple-stress theory treated in [2] and, in particular, to mechanically *homogeneous, isotropic* and *centrosymmetric, elastic* solids, as well as to the *equilibrium* case and to conditions of *plane strain*. Within this context we summarize, amplify, and discuss the main results obtained in a sequence of investigations [3—6] concerned with the influence of couple-stresses in problems for which the analogous classical

---

Acknowledgment. The author is indebted to D. B. BOGY and R. MUKI for their assistance in the preparation of this paper.

[1] This paper was prepared under Contract Nonr-220(58) with the Office of Naval Research in Washington, D.C., for presentation at the IUTAM Symposium on the Generalized Cosserat Continuum and the Continuum Theory of Dislocations in Stuttgart and Freudenstadt, August 1967.

[2] A bibliography tracing the history of the theories to which we are alluding, as well as the rapidly expanding related current literature, is beyond the scope and space limitations of the present paper. References up to 1962 may be found in [2].

treatment furnishes locally unbounded stresses or deformations, which are accompanied by unbounded deformation gradients. Our chief interest is directed at the extent to which such pathological predictions of classical elastostatics in *singular* stress-concentration problems are altered, mitigated, or possibly even eliminated once couple-stresses are taken into account. This question is motivated by results in [2] that reveal a reduction, in the departure from the classical theory, of the *regular* stress-concentration around a circular hole in an infinite sheet under a non-isotropic uniform loading at infinity.

1. Plane Strain in the Linearized Couple-Stress Theory

In this section we recall briefly certain pertinent elements of the modified plane-strain theory with which we are concerned[1]. Adhering to the notation introduced in [3], we call $\boldsymbol{u}$ and $\boldsymbol{\omega}$ the displacement and rotation vector fields, denote by $\boldsymbol{e}$ and $\boldsymbol{\varkappa}$ the strain and curvature-twist tensor fields, while we designate by $\boldsymbol{\tau}$ and $\boldsymbol{\sigma}$ the tensor fields of stress and couple-stress. Further, in connection with the corresponding cartesian field components we employ the usual indicial notation, Latin and Greek subscripts having the respective ranges $(1, 2, 3)$ and $(1, 2)$.

Suppose the medium at hand occupies a cylindrical or prismatic domain R of space, let D be the cross-section of R, call ∂D the boundary of D, and choose rectangular cartesian coordinates x_i such that the x_3-axis is parallel to the generators of R. The assumption of plane deformations parallel to the plane $x_3 = 0$ then takes the form

$$u_{\alpha,3} = 0, \quad u_3 = 0 \quad \text{on } R. \tag{1.1}$$

Subjecting the relevant three-dimensional field equations in [2] to the restriction (1.1) and adopting the normalization

$$\sigma_{kk} = 0 \quad \text{on } R \tag{1.2}$$

of the couple-stress field[2], one concludes that all field quantities are independent of x_3 and that throughout D

$$\begin{gathered} \omega_\alpha = e_{3i} = e_{i3} = \varkappa_{3i} = \varkappa_{\alpha\beta} = 0, \\ \tau_{3\alpha} = \tau_{\alpha 3} = 0, \quad \sigma_{33} = 0, \quad \sigma_{\alpha\beta} = 0 \quad (\alpha \neq \beta). \end{gathered} \tag{1.3}$$

With the abridged notation

$$\omega \equiv \omega_3, \quad \varkappa_\alpha \equiv \varkappa_{\alpha 3}, \quad \sigma_\alpha \equiv \sigma_{\alpha 3}, \tag{1.4}$$

---

[1] A direct ad hoc treatment of the plane-strain case was given by MINDLIN [7], whose results were subsequently amplified and reaffirmed in [3, 4] through an appropriate specialization of the three-dimensional theory.

[2] Recall from [2] that the isotropic part of $\boldsymbol{\sigma}$ and the skew-symmetric part of $\boldsymbol{\tau}$ remain indeterminate in the couple-stress theory under consideration. Condition (1.2) serves to remove this indeterminacy.

the resulting system of two-dimensional field equations is represented by the *kinematic relations*

$$e_{\alpha\beta} = u_{(\alpha,\beta)}, \qquad \omega = \tfrac{1}{2}\varepsilon_{\alpha\beta}\, u_{\beta,\alpha}, \qquad \varkappa_\alpha = \omega_{,\alpha} \tag{1.5}$$

together with the *constitutive relations*

$$e_{\alpha\beta} = \frac{1}{2\mu}\left[\tau_{(\alpha\beta)} - \nu\, \delta_{\alpha\beta}\, \tau_{\gamma\gamma}\right], \qquad \varkappa_\alpha = \frac{\sigma_\alpha}{4\mu l^2} \tag{1.6}$$

and the *stress equations of equilibrium*

$$\tau_{\beta\alpha,\beta} = 0, \qquad \varepsilon_{\alpha\beta}\, \tau_{\alpha\beta} + \sigma_{\alpha,\alpha} = 0, \tag{1.7}$$

provided the body-force and body-couple fields are supposed to vanish identically. Here $\delta_{\alpha\beta}$ and $\varepsilon_{\alpha\beta}$ are the KRONECKER delta and the two-dimensional alternator, respectively, whereas the constants μ, ν, and l—in this order—stand for the shear modulus, POISSON's ratio, and the intrinsic length-parameter of the material.

Eqs. (1.5)—(1.7), which must hold on D, are to be accompanied by the appropriate boundary conditions. Let (u^T, u^N) and (t^T, t^N) designate the tangential and normal scalar components of the displacement and ordinary traction vectors on ∂D, and call s the axial component of the couple-traction vector on the boundary. Thus, let

$$\begin{gathered} u^T = \varepsilon_{\alpha\beta}\, u_\beta\, n_\alpha, \qquad u^N = u_\alpha\, n_\alpha, \\ t^T = \varepsilon_{\alpha\beta}\, \tau_{\gamma\beta}\, n_\alpha\, n_\gamma, \qquad t^N = \tau_{\beta\alpha}\, n_\alpha\, n_\beta, \qquad s = \sigma_\alpha\, n_\alpha \end{gathered} \tag{1.8}$$

on ∂D, where $\boldsymbol{n}$ is the unit outer normal of ∂D. Then the *boundary conditions* become

$$\begin{gathered} u^T = \overset{*}{u}{}^T \;\text{ or }\; t^T = \overset{*}{t}{}^T, \qquad u^N = \overset{*}{u}{}^N \;\text{ or }\; t^N = \overset{*}{t}{}^N, \\ \omega = \overset{*}{\omega} \;\text{ or }\; s = \overset{*}{s} \;\text{ on } \partial D, \end{gathered} \tag{1.9}$$

in which letters carrying an asterisk denote functions prescribed on ∂D. Once the fifteen Eqs. (1.5)—(1.7) have been solved for the fifteen unknowns u_α, ω, $e_{\alpha\beta}$, $\varkappa_\alpha$, $\tau_{\alpha\beta}$, and σ_α, subject to (1.9), the only remaining unknowns τ_{33} and $\sigma_{3\alpha}$ follow from

$$\tau_{33} = \nu\, \tau_{\alpha\alpha}, \qquad \sigma_{3\alpha} = 4\eta'\, \varkappa_\alpha \quad \text{on } D, \tag{1.10}$$

if η' stands for the second new elastic constant arising in the couple-stress theory at hand.

The constitutive relations (1.6) are generated, in the sense of

$$\tau_{(\alpha\beta)} = \partial W / \partial e_{\alpha\beta}, \qquad \sigma_\alpha = \partial W / \partial \varkappa_\alpha, \tag{1.11}$$

by the *strain-energy density*

$$W(\boldsymbol{e}, \boldsymbol{\varkappa}) = \mu \left[\frac{\nu}{1-2\nu}\, e_{\alpha\alpha}\, e_{\beta\beta} + e_{\alpha\beta}\, e_{\alpha\beta} + 2 l^2\, \varkappa_\alpha\, \varkappa_\alpha\right], \tag{1.12}$$

which is positive-definite provided

$$\mu > 0, \qquad -1 < \nu < \tfrac{1}{2}, \qquad l > 0. \tag{1.13}$$

If (1.13) hold, the solution to the foregoing boundary-value problem is unique (except possibly for an additive rigid displacement of the entire body) under suitable assumptions regarding the regularity of D and of the admitted solution fields. In the event that D is unbounded, (1.9) need to be supplemented by appropriate regularity conditions on $\boldsymbol{\tau}$ and $\boldsymbol{\sigma}$ at infinity. In the subsequent applications it will be assumed without further mention, unless otherwise specified, that

$$\tau_{\alpha\beta} = o(1), \quad \sigma_\alpha = o(1) \quad \text{as} \quad r \to \infty, \tag{1.14}$$

where r is the distance from the origin.

Elimination of all field quantities other than u_α among (1.5)—(1.7) leads to the *displacement equations of equilibrium*

$$l^2 \, \varepsilon_{\alpha\beta} \, \varepsilon_{\gamma\varrho} \, \nabla^2 u_{\varrho,\beta\gamma} + \nabla^2 u_\alpha + \frac{1}{1-2\nu} u_{\beta,\beta\alpha} = 0, \tag{1.15}$$

if ∇^2 is the Laplacian operator. Elimination of all kinematic fields among (1.5)—(1.7), alternatively, yields a characterization of $\tau_{\alpha\beta}$ and σ_α in terms of the stress equations of equilibrium and compatibility. The system of equations thus obtained, in turn, furnishes the point of departure for an extension of Airy's integration scheme of the classical two-dimensional theory[1].

The couple-stress theory recalled above may be viewed as a generalization of the conventional theory of plane strain, from which it differs in several important respects:

a) The ordinary stress tensor $\boldsymbol{\tau}$ is no longer symmetric in the present circumstances. Further, the modified constitutive relations (1.6), as well as the elastic potential (1.12), involve explicitly—in addition to the strains—the rotation gradients. Consequently, the local state of stress at a material point of the body depends not only on the corresponding local state of deformations, but—to some extent—also on the deformations in a vicinity of the point in question.

b) The presence in (1.6), beyond the pair of conventional elastic constants, of the material length parameter l assures the analytical possibility of size effects, which cannot arise in the classical theory.

c) As l tends to zero, σ_α approaches zero on D according to (1.6), provided $\varkappa_\alpha$ remains bounded in this limit. One thus recovers from (1.5)—(1.8) the classical fields equations of plane strain

$$e_{\alpha\beta} = u_{(\alpha,\beta)}, \quad e_{\alpha\beta} = \frac{1}{2\mu}[\tau_{\alpha\beta} - \nu \, \delta_{\alpha\beta} \, \tau_{\gamma\gamma}], \quad \tau_{\beta\alpha,\beta} = 0, \tag{1.16}$$

together with the conventional mixed boundary conditions

$$u^T = \overset{*}{u}{}^T \quad \text{or} \quad t^T = \overset{*}{t}{}^T, \quad u^N = \overset{*}{u}{}^N \quad \text{or} \quad t^N = \overset{*}{t}{}^N \quad \text{on} \quad \partial D. \tag{1.17}$$

[1] In this connection see [7, 3, 4], as well as Günther [8], Schaeffer [9], and Carlson [10].

Accordingly, the quantitative influence of couple-stresses is bound to depend upon the relative size of the material length-parameter l compared to the characteristic dimensions of the body. As is apparent from (1.15), (1.9), (1.17), the order of the governing partial differential equations is lowered and the number of requisite boundary conditions is diminished in the transition to the classical theory. This fact suggests the emergence of boundary-layer effects in the departure from the conventional theory.

Finally, we cite from [4] a theorem concerning the connection between the modified and the ordinary theory of plane strain, which will be needed later on. Let u_α, $e_{\alpha\beta}$, $\tau_{\alpha\beta}$ be a solution of the classical field equations of plane strain on D and define ω, $\varkappa_\alpha$, σ_α on D through

$$\omega = \tfrac{1}{2}\varepsilon_{\alpha\beta}\, u_{\beta,\alpha}, \qquad \varkappa_\alpha = \omega_{,\alpha}, \qquad \sigma_\alpha = 4\mu\, l^2\, \varkappa_\alpha. \tag{1.18}$$

Then u_α, ω, $e_{\alpha\beta}$, $\varkappa_\alpha$, $\tau_{\alpha\beta}$, and σ_α satisfy the modified field Eqs. (1.5) to (1.7) on D; moreover,

$$s \equiv \sigma_\alpha\, n_\alpha = 4\mu\, l^2 \frac{\partial\omega}{\partial n} \quad \text{on } \partial D. \tag{1.19}$$

The preceding elementary result enables one to generate, on the basis of a known solution to a conventional plane-strain problem, a solution to an associated plane-strain problem in the couple-stress theory; the latter obeys the same ordinary boundary conditions (1.17) met by the original solution but conforms to the usually artificial distribution of couple-tractions (1.19) on the boundary. If, in particular, $\partial\omega/\partial n$ happens to vanish on ∂D, then the associated solution obtained in this manner corresponds to vanishing couple-tractions on ∂D. This physically useful eventuality arises, for example, in the axisymmetric problem of plane strain, whose classical displacement field is irrotational.

2. Results for Singular Stress-Concentration Problems

We turn now to applications of the couple-stress theory reviewed in Sect. 1 to various singular stress-concentration problems of plane strain. In this connection we take up first the problems of the half-plane under concentrated or discontinuously distributed normal and tangential loads, in the absence of boundary couple-tractions (Fig. 1).

Thus let D at present be the open half-plane defined by

$$D = \{\boldsymbol{x} \mid -\infty < x_1 < \infty,\ 0 < x_2 < \infty\}. \tag{2.1}$$

For the sake of brevity we refer to [3], [5] for a detailed formulation of the singular problems to be considered presently. Their solution is defined and deduced by means of the appropriate limit process applied to the solution for the half-plane under regular distributed normal or shearing tractions. This preliminary *regular* problem, in turn, is solved in [3], in integral form, with the aid of the generalized Airy stress functions and the exponential Fourier transform. The limit process just alluded to thus leads to an integral representation for the desired solutions corresponding to the *singular* loadings at hand.

Throughout the remainder of this paper we denote by $\overset{\circ}{\tau}_{\alpha\beta}(\boldsymbol{x})$ and $\overset{\circ}{\omega}(\boldsymbol{x})$ the stress and rotation values at position $\boldsymbol{x}$ of the *classical* solution to the particular problem under consideration, while designating by $\tau_{\alpha\beta}(\boldsymbol{x}, l)$, $\sigma_\alpha(\boldsymbol{x}, l)$, and $\omega(\boldsymbol{x}, l)$ the stress, couple-stress, and rotation values of the corresponding *modified* solution[1]. In what follows we cite from [3][2], with reference to the choice of coordinates appearing in Fig. 1, representative results pertaining to the classical solution and to the asymptotic behavior near the origin of the modified solution for the concentrated-load problems.

Concentrated normal load:

$$\overset{\circ}{\tau}_{11}(\boldsymbol{x}) = -\frac{2P}{\pi}\,\frac{\cos^2\theta\sin\theta}{r}, \ldots, \qquad \overset{\circ}{\omega}(\boldsymbol{x}) = -\frac{(1-\nu)P}{\pi\mu}\,\frac{\cos\theta}{r}. \tag{2.2}$$

As $r \to 0$, for every fixed $l > 0$,

$$\left.\begin{aligned} \tau_{11}(\boldsymbol{x}, l) &= \frac{2(1-2\nu)P}{(3-2\nu)\pi}\,\frac{\cos^2\theta\sin\theta}{r} + O(1), \ldots, \\ \sigma_1(\boldsymbol{x}, l) &= \frac{2(1-\nu)P}{(3-2\nu)\pi}\log r + O(1), \quad \sigma_2(\boldsymbol{x}, l) = O(1), \\ \omega(\boldsymbol{x}, l) &= O(1). \end{aligned}\right\} \tag{2.3}$$

Concentrated tangential load:

$$\overset{\circ}{\tau}_{11}(\boldsymbol{x}) = -\frac{2P}{\pi}\,\frac{\cos^3\theta}{r}, \ldots, \qquad \overset{\circ}{\omega}(\boldsymbol{x}) = \frac{(1-\nu)P}{\pi\mu}\,\frac{\sin\theta}{r}. \tag{2.4}$$

As $r \to 0$, for every fixed $l > 0$,

$$\begin{aligned} \tau_{11}(\boldsymbol{x}, l) &= -\frac{2P}{(3-2\nu)\pi}\,\frac{\cos\theta}{r}\,[1 + (1-2\nu)\sin^2\theta] + O(1), \ldots, \\ \sigma_\alpha(\boldsymbol{x}, l) &= O(1), \quad \omega(\boldsymbol{x}, l) = O(1). \end{aligned} \tag{2.5}$$

The subsequent asymptotic estimates appropriate to discontinuous normal and tangential surface tractions, which are taken from [5], presuppose a load-function p that is antisymmetric and has a finite jump-discontinuity at the origin (Fig. 1).

---

[1] Observe that this notation fails to make explicit the possible dependence of the solution fields on POISSON's ratio.

[2] Although rotation fields were not examined in [3], the results for ω included here are easily deduced from those given in [3].

Discontinuous normal load:

As $r \to 0$,

$$\mathring{\tau}_{\alpha\beta}(\boldsymbol{x}) = O(1), \quad \mathring{\omega}(\boldsymbol{x}) = \frac{2(1-\nu)\,p(0+)}{\pi\,\mu}\log r + O(1). \tag{2.6}$$

As $r \to 0$, for every fixed $l > 0$,

$$\left.\begin{aligned} \tau_{\alpha\beta}(\boldsymbol{x}, l) &= O(1)\ (\tau_{\alpha\beta} \neq \tau_{12}), \\ \tau_{12}(\boldsymbol{x}, l) &= \frac{8(1-\nu)\,p(0+)}{(3-2\nu)\,\pi}\log r + O(1), \\ \sigma_\alpha(\boldsymbol{x}, l) &= o(1), \quad \omega(\boldsymbol{x}, l) = o(1). \end{aligned}\right\} \tag{2.7}$$

Discontinuous tangential load:

As $r \to 0$,

$$\left.\begin{aligned} &\mathring{\tau}_{\alpha\beta}(\boldsymbol{x}) = O(1) \quad (\mathring{\tau}_{\alpha\beta} \neq \mathring{\tau}_{11}), \quad \mathring{\tau}_{11}(\boldsymbol{x}) = \frac{4\,p(0+)}{\pi}\log r + O(1), \\ &\mathring{\omega}(\boldsymbol{x}) = O(1). \end{aligned}\right. \tag{2.8}$$

As $r \to 0$, for every fixed $l > 0$,

$$\left.\begin{aligned} \tau_{\alpha\beta}(\boldsymbol{x}, l) &= O(1) \quad (\tau_{\alpha\beta} \neq \tau_{11}), \\ \tau_{11}(\boldsymbol{x}, l) &= \frac{4\,p(0+)}{(3-2\nu)\,\pi}\log r + O(1), \\ \sigma_1(\boldsymbol{x}, l) &= O(1), \quad \sigma_2(\boldsymbol{x}, l) = o(1), \quad \omega(\boldsymbol{x}, l) = o(1). \end{aligned}\right\} \tag{2.9}$$

Refinements of the estimates (2.6) to (2.9) are to be found in [5].

We proceed next to the problem of an orthogonal wedge, one face of which is subjected to arbitrary (sufficiently regular) shearing tractions, in the absence of other loads (Fig. 2). This problem—hereafter briefly referred to as the "corner problem"—is treated in [6]. In the present instance D is the open quarter-plane characterized by

$$D = \{\boldsymbol{x} \mid 0 < x_1 < \infty,\ 0 < x_2 < \infty\} \tag{2.10}$$

and the governing boundary conditions take the form

$$\left.\begin{aligned} &\tau_{11}(0, x_2) = 0, \quad \tau_{12}(0, x_2) = p(x_2), \quad \sigma_1(0, x_2) = 0 \ \ (0 \leqq x_2 < \infty), \\ &\tau_{21}(x_1, 0) = 0, \quad \tau_{22}(x_1, 0) = 0, \qquad\quad \sigma_2(x_1, 0) = 0 \ \ (0 \leqq x_1 < \infty), \end{aligned}\right. \tag{2.11}$$

in which p is a pre-assigned load-function, with $p(0) \neq 0$. The corresponding classical problem is amenable to a Mellin-transform technique[1], which is not applicable to the modified corner problem. The latter may, however, by means of a dual reflection scheme (involving two auxiliary half-plane problems) be reduced to a regular integral equation from which the required asymptotic behavior of the solution may be inferred. The pertinent results deduced in [6] are given below.

---

[1] An elementary closed solution for the special case in which p is constant was given earlier by Reissner [11].

Corner problem:

As $r \to 0$,

$$\overset{\circ}{\tau}_{11}(\boldsymbol{x}) = \frac{p(0)}{2}[-\pi + \sin(2\theta) + 2\theta] + o(1), \ldots, \qquad (2.12)$$

$$\overset{\circ}{\omega}(\boldsymbol{x}) = -\frac{(1-\nu)}{\mu} p(0) \log r + o(1).$$

As $r \to 0$, for every fixed $l > 0$,

$$\tau_{\alpha\beta}(\boldsymbol{x}, l) = o(1) \quad (\tau_{\alpha\beta} \neq \tau_{12}), \qquad \tau_{12}(\boldsymbol{x}, l) = p(0) + o(1), \qquad (2.13)$$

$$\sigma_\alpha(\boldsymbol{x}, l) = o(1), \qquad \omega(\boldsymbol{x}, l) = o(1).$$

All of the results cited so far refer to *load*-induced singular behavior. Two problems concerning *geometrically* conditioned singular stress concentrations have also been dealt with: the half-plane indented by a smooth flat-ended punch in [3] and the crack in a uniform field of transverse tension in [4]. We omit the punch problem here and turn directly to the physically more interesting crack problem.

Thus D is now the slit entire plane (Fig. 2) given by

$$D = \{\boldsymbol{x} \mid -\infty < x_\alpha < \infty, \ |x_1| > a \quad \text{if} \quad x_2 = 0\}, \qquad (2.14)$$

where $2a$ is the length of the crack. The two faces of the crack are taken to be free of ordinary and couple-tractions, while the assumed loading at infinity obeys

$$\tau_{22} = p_0 + o(1), \quad \tau_{\alpha\beta} = o(1) \ (\tau_{\alpha\beta} \neq \tau_{22}), \quad \sigma_\alpha = o(1) \text{ as } r \to \infty, \qquad (2.15)^1$$

with p_0 a constant. This singular problem is rendered fully determinate by the requirement that its solution must coincide with the appropriate limit of the solution to the corresponding problem for the plane with an elliptic hole. Such a limit treatment of the crack problem, however, offers formidable difficulties. The direct method of attack adopted in [4] may be described as follows. The original problem, by virtue of its symmetry, is equivalent to a singular mixed half-plane problem, which is reduced to a simultaneous system of two pairs of dual integral equations. This system, in turn, is shown to be equivalent to a single integral equation of Fredholm's second kind on the assumption that the stress singularities at the crack-tips are of the same order as in the well-known classical solution. Representative results belonging to the latter, along with typical estimates for the modified solution, all of which refer to the auxiliary polar coordinates appearing in Fig. 2, are given below.

---

[1] The solution for the problem of the uniformly pressurized crack, in the absence of loads at infinity, is obtained by an elementary modification of the results for the present problem (see [4]).

Crack problem:

$$\mathring{\tau}_{11}(\boldsymbol{x}) = \frac{p_0 r}{\sqrt{r_1 r_2}} \left\{\cos[\theta - (\theta_1 + \theta_2)/2] - \frac{a^2 \sin\theta}{r_1 r_2} \sin[3(\theta_1 + \theta_2)/2]\right\} - p_0,$$

$$\ldots, \quad \mathring{\omega}(\boldsymbol{x}) = \frac{p_0(1-\nu) r}{\mu \sqrt{r_1 r_2}} \sin[\theta - (\theta_1 + \theta_2)/2]. \tag{2.16}$$

As $r_\alpha \to 0$, for every fixed $l > 0$,

$$\left.\begin{aligned} \tau_{11}(\boldsymbol{x}, l) &= -\frac{p_0(1-2\nu)}{1-\nu} F(l/a, \nu) \frac{r}{\sqrt{r_1 r_2}} \left\{\cos[\theta - (\theta_1 + \theta_2)/2] - \right. \\ &\quad \left. - \frac{a^2 \sin\theta}{r_1 r_2} \sin[3(\theta_1 + \theta_2)/2]\right\} + O(1), \ldots, \\ \sigma_1(\boldsymbol{x}, l) &= -\frac{p_0 a^2}{2(1-\nu)} \frac{G(l/a, \nu)}{\sqrt{r_1 r_2}} \sin[(\theta_1 + \theta_2)/2] + O(1), \\ &\quad \ldots, \omega(\boldsymbol{x}, l) = O(1). \end{aligned}\right\} \tag{2.17}$$

Here F and G are functions characterized in [4] in terms of the solution of a Fredholm integral equation. From (2.16) one has, as $r_\gamma \to 0$,

$$\mathring{\tau}_{\alpha\beta}(\boldsymbol{x}) = O(r_\gamma^{-1/2}), \quad \mathring{\omega}(\boldsymbol{x}) = O(r_\gamma^{-1/2}), \tag{2.18}$$

while (2.17) imply, as $r_\gamma \to 0$, for every fixed $l > 0$,

$$\tau_{\alpha\beta}(\boldsymbol{x}, l) = O(r_\gamma^{-1/2}), \quad \sigma_\alpha(\boldsymbol{x}, l) = O(r_\gamma^{-1/2}), \quad \omega(\boldsymbol{x}, l) = O(1). \tag{2.19}$$

3. Discussion and Conclusions

The results cited in Sect. 2, as well as additional findings emerging from the analysis contained in the underlying investigations [3—6] of singular plane-strain problems in the couple-stress theory, permit various conclusions which we now attempt to summarize.

With a view toward clarifying the *transition from the modified to the classical theory* in the present context, we let $\bar{D}$ denote the closure of D and call S the set of all singular points[1] on the boundary ∂D. Then, in the examples considered,

$$\begin{gathered} \lim_{l\to 0} \tau_{\alpha\beta}(\boldsymbol{x}, l) = \mathring{\tau}_{\alpha\beta}(\boldsymbol{x}), \\ \lim_{l\to 0} \sigma_\alpha(\boldsymbol{x}, l) = 0, \quad \lim_{l\to 0} \omega(\boldsymbol{x}, l) = \mathring{\omega}(\boldsymbol{x}), \end{gathered} \tag{3.1}$$

at every point $\boldsymbol{x}$ in $\bar{D} - S$, the convergence being uniform (with respect to $\boldsymbol{x}$) on every closed subset of $\bar{D} - S$, but non-uniform on $\bar{D} - S$ itself. Accordingly, as was to be anticipated from related general observations in Sect. 1, each modified solution at all non-singular points of the body passes over continuously into its classical counterpart as

[1] Thus S contains only the origin in all but the last problem taken up in Sect. 2, while S consists of the pair of points with the coordinates $(0, a)$ and $(0, -a)$ in the crack problem. See Figs. 1, 2.

$l \to 0$. On the other hand, the non-uniformity of this transition on $\bar{D} - S$ gives rise to severe boundary-layer effects (in the vicinity of the singular boundary points) which will be discussed further later on.

Turning first to the behavior of the *rotation field*, we observe that in every case for which the conventional solution predicts a locally (on S) unbounded rotation[1], the latter remains bounded throughout $\bar{D}$ in the corresponding modified solution. This fact represents a mitigating influence inherent in the couple-stress theory, which takes account of the rotation gradients in its governing constitutive relations.

The conclusions reached regarding the (ordinary) *stress field* in the vicinity of S are more involved and are in marked contrast to those concerning the rotation field. Whenever the classical stress field becomes unbounded on S, the same is true of the associated modified stress field. This behavior is to be expected (on the basis of elementary equilibrium considerations) in the concentrated-load problems but is not a priori predictable in the remaining problems treated, for which the conventional theory furnishes stress singularities weaker than that induced by a concentrated load.

While the *order* of an unbounded stress singularity is preserved in the departure from the classical theory, its *detailed structure* is altered. Thus the dominating unbounded terms in the stress estimates contained in (2.3), (2.5), (2.9) depend on Poisson's ratio and those in (2.17) involve in addition the length-parameter l; yet $\mathring{\tau}_{\alpha\beta}$ is independent of the elastic constants in each instance. There is, therefore, a discontinuous change of the "singular part" of the solution in the departure from the classical theory, which is reflected in the boundary-layer effect mentioned earlier. To make the preceding remark fully meaningful and to fix ideas, consider the problem of the discontinuous tangential load. Here one draws from (3.1), (2.8), (2.9) that

$$\lim_{r\to 0} \lim_{l\to 0} [\tau_{11}(\boldsymbol{x}, l)/\mathring{\tau}_{11}(\boldsymbol{x})] = 1,$$
$$\lim_{l\to 0} \lim_{r\to 0} [\tau_{11}(\boldsymbol{x}, l)/\mathring{\tau}_{11}(\boldsymbol{x})] = \frac{1}{3-2\nu}. \tag{3.2}$$

The limits of $\tau_{\alpha\beta}(\boldsymbol{x}, l)/\mathring{\tau}_{\alpha\beta}(\boldsymbol{x})$ as $l \to 0$ and as $\boldsymbol{x}$ tends to a point in S are found to be irreversible whenever $\tau_{\alpha\beta}$ and $\mathring{\tau}_{\alpha\beta}$ both become unbounded on S.

Fig. 3, taken from [3], furnishes a quantitative illustration of the preceding observations. This figure refers to the half-plane under uniformly distributed shearing tractions confined to a boundary-segment of width $2a$ (see the inset diagram). The curves shown display

[1] See (2.2), (2.4), (2.6), (2.12), and (2.16); $\mathring{\omega}$ is finite on S only in the problem of the discontinuous tangential load.

the variation with x_1/a of the stress-ratio[1] $\tau_{11}(x_1, 0, l)/\mathring{\tau}_{11}(x_1, 0)$ for $\nu = 1/2$ and various values of the characteristic length-ratio l/a. The family of functions depicted here, as $l/a \to 0$ converges non-uniformly on $[0, a)$ to its classical member $(l/a = 0)$, the limit-function being discontinuous on $[0, a]$. Consistent with (3.2), there is a sharp *decrease* in the stress under discussion compared to its classical values at points sufficiently close to the load-discontinuities, for arbitrarily small positive values of l/a.

Fig. 4 refers to the crack problem and is reproduced from [4]. It shows the dependence on l/a of $\tau_{22}(x_1, 0, l)/\mathring{\tau}_{22}(x_1, 0, l)$ in the limit as $x_1 \to a+$. Superimposed on the same diagram are curves illustrative of the analogous stress-ratio in the problem of the circular hole, based on the results obtained in [2]. Whereas the stress concentration at the hole is mitigated by the admission of couple-stresses, the modified theory is found to predict a discontinuous *aggravation* of the concentration of stress at the crack tips as l/a departs from zero[2].

The asymptotic estimates (2.6), (2.7), appropriate to the discontinuous normal load, lead to the surprising and somewhat disturbing conclusion that a stress field which is wholly bounded in the conventional solution of a singular problem may become unbounded according to the modified theory: here τ_{12} has a logarithmic singularity at the origin.

In contrast, the results (2.12), (2.13) for the corner problem reveal that the finite stress discontinuity at the apex predicted by the classical theory disappears entirely in the modified solution. In this connection we observe that the classical singularity at the corner stems from the incompatibility of the boundary conditions (2.11) with the symmetry of the stress tensor, which no longer prevails in the couple-stress theory.

As far as the character of the *couple-stress field* in the vicinity of S is concerned, $\boldsymbol{\sigma}$ is evidently better behaved than the associated ordinary stress field $\boldsymbol{\tau}$, with the exception of the crack problem in which both fields possess singularities of the same order.

It is clear from (2.2) and (2.16) that the classical rotation fields for the problem of the concentrated normal load and the crack problem conform to

$$\frac{\partial \mathring{\omega}}{\partial n} = 0 \quad \text{on } \partial D; \tag{3.3}$$

the same is true also for the punch problem [3] and the corner problem [6], provided in the latter the shear load is constant[3]. Therefore, and in view of the theorem recalled at the end of Sect. 1, all of the afore-

---

[1] Here and in connection with Fig. 4 we write $\tau_{\alpha\beta}(x_1, x_2, l)$, $\mathring{\tau}_{\alpha\beta}(x_1, x_2)$ in place of $\tau_{\alpha\beta}(\boldsymbol{x}, l)$, $\mathring{\tau}_{\alpha\beta}(\boldsymbol{x})$.

[2] An analogous conclusion is reached for the punch problem in [3].

[3] See Footnote 1, p. 101.

mentioned problems admit alternative solutions in the modified theory, which on $\bar{D} - S$ are given by

$$\tau_{\alpha\beta} = \mathring{\tau}_{\alpha\beta}, \quad \sigma_\alpha = 4\mu\, l^2\, \partial\omega/\partial n, \quad \omega = \mathring{\omega}. \tag{3.4}$$

The existence of more than one solution to presumably the same *singular* problem in no way contradicts the uniqueness theorem mentioned in Sect. 1, which does not apply in the present circumstances.

In deciding which of the two alternative "solutions" needs to be rejected several considerations are pertinent. The elementary solutions (3.4), in contrast to those deduced in [3—6], display no interaction between ordinary and couple-stresses, the former being entirely unaffected by the presence of couple-stresses. The couple-stress field of the competing elementary solution is invariably more strongly singular; indeed, the total strain energy associated with the elementary solution to the problem of the pressurized crack[1] and to the punch problem is no longer finite. Finally, the elementary solution to the problem of the concentrated normal load fails to coincide with the corresponding limit of the solution appropriate to distributed normal tractions, from which

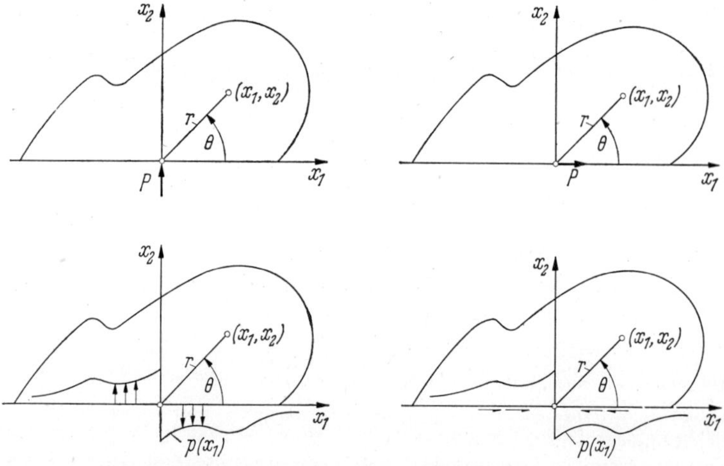

Fig. 1. Half-plane under concentrated or discontinuous normal and tangential loads.

this singular problem derives its very meaning. On all of these grounds the degenerate singular solutions of the form (3.4) must be dismissed as physically irrelevant pseudo-solutions[2].

---

[1] See Footnote 1, p. 102.

[2] The pseudo-solution to the problem of the half-plane under a concentrated normal load was arrived at independently by TIWARI [12], as well as by BERT and APPL [13], but its spurious significance was not recognized in either of these publications.

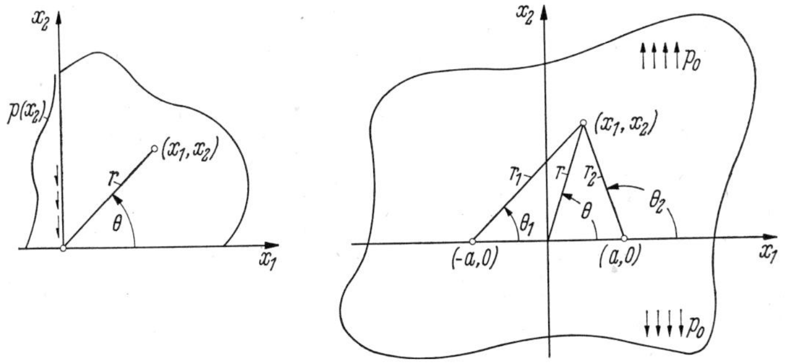

Fig. 2. Corner problem and crack problem.

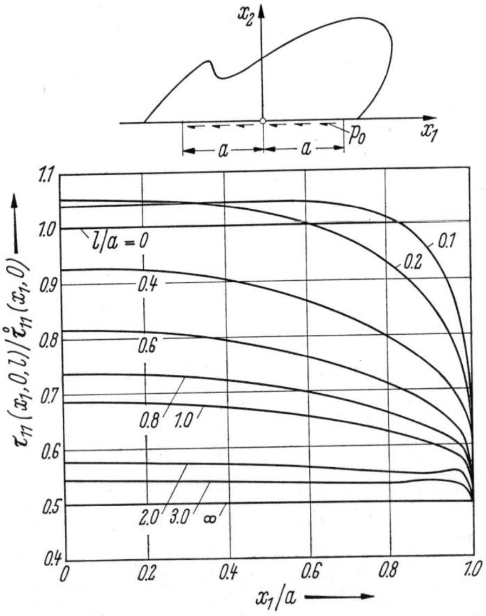

Fig. 3. Discontinuous uniform shear load $\tau_{11}/\mathring{\tau}_{11}$ along boundary for various l/a and $\nu = \frac{1}{2}$.

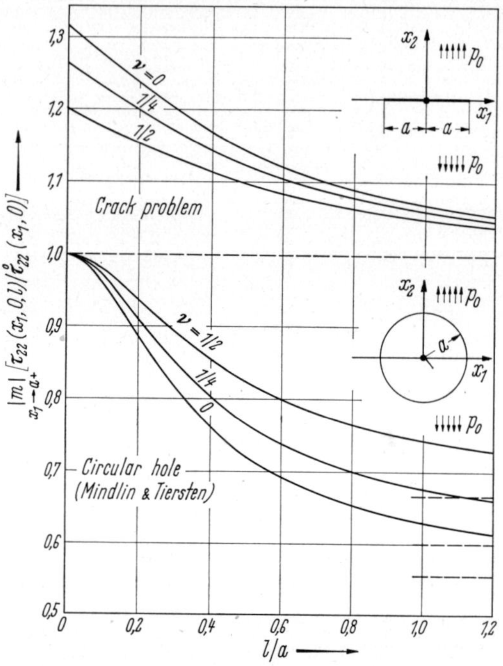

Fig. 4. Crack problem. Dependence on l/a of $\tau_{22}/\overset{\circ}{\tau}_{22}$ at ends of crack.

References

[1] Cosserat, E., and F. Cosserat: Théorie des corps déformables. Paris: Hermann 1909.
[2] Mindlin, R. D., and H. F. Tiersten: Arch. Rat. Mech. Anal. **11**, 415 (1962).
[3] Muki, R., and E. Sternberg: Z. angew. Math. Phys. **16**, 611 (1965).
[4] Sternberg, E., and R. Muki: Int. J. Solids Struct. **3**, 69 (1967).
[5] Bogy, D. B., and E. Sternberg: Int. J. Solids Struct. **3**, 757 (1967).
[6] Bogy, D. B., and E. Sternberg: Int. J. Solids Struct. **4**, 159 (1968).
[7] Mindlin, R. D.: Exper. Mech. 1 (Jan. 1963).
[8] Günther, W.: Abh. d. Braunschweigischen Wiss. Ges. **10**, 195 (1958).
[9] Schaefer, H.: in Miszellaneen der Angewandten Mathematik. Berlin: Akademie-Verlag 1962, pp. 277—292.
[10] Carlson, D. E.: Z. angew. Math. Phys. **17**, 789 (1966).
[11] Reissner, E.: J. Math. Phys. **23**, 192 (1944).
[12] Tiwari, G. R.: J. Science and Engineering Research, Indian Inst. Tech., Kharagpur, **9**, 37 (1965).
[13] Bert, C. W., and F. J. Appl: Proc., Fifth U.S. Nat. Congress Appl. Mech., New York: A.S.M.E., 1966, p. 245.

On the Effect of Stress Concentration in Cosserat Continua

By

H. Neuber

Lehrstuhl für Technische Mechanik und
Mechanisch-Technisches Laboratorium
Technische Universität (Hochschule) München

1. A Restrictive Condition for Continuum Theories

The theories of continuum mechanics use idealised models mostly based on the assumption of a continuous distribution of matter. Only in this way the fundamental definitions can be introduced as there are the continuous and differentiable displacement and rotation vectors, the stress and strain tensors and the necessary material constants characterising the deformation behaviour. All these quantities represent statistical mean values taken over very small ranges of volume and time. Consequently continuum theories cannot give satisfying predictions of the behaviour of the material within very small ranges of volume and time, if high gradients of stress and strain occur; especially these theories cannot describe processes with dominating structure effects since all material qualities producing effects of such kind had been eliminated by introducing the idealised models. In consequence of this situation a restrictive condition must be respected with regard to the applicability of continuum theories in very small ranges of volume and time. There are some analogies to the physical restrictive condition of Heisenberg.

2. The Theory of Stress Mean Value and the Micro-Support-Effect

Within the field of strength of materials the defined restrictive condition becomes very important in connection with problems of high stress concentration as caused by extremely curved surfaces, e.g. at holes, notches or cracks, or by the action of concentrated forces. In such cases the classical theory of elasticity predicts immense values of stress and strain, which—in consequence of the defined restrictive condition—are to be replaced by *mean values*. By such mathematical procedure an approximate representation of a fundamental structure effect becomes possible which may be called *micro-support-effect*: Certain domains of the material have stronger bonds between their particles and therefore are supported as quasi-rigid blocks by their environment. In connection with his *theory of sharply curved notches* the author already introduced in 1936 a *stress mean value* taken over a finite length normal to the surface within the range of high stress concentration [1, 2]. This so-called *fictive length of structure* represents an additional material constant. In connection with the interpretation of a large series of fatigue tests at stress concentration the author modified this *theory of stress mean value* within the last years by introducing the mean value of the so-called *comparison stress* calculated by means of a suitable strength hypothesis, valid for the material. In this way the threeaxiality of the stress tensor can be taken into account more accurately [3, 4]:

$$\bar{\sigma} = \frac{1}{R^*} \int_{y=y_0}^{y_0+R^*} \sigma_c \, dy. \tag{1}$$

Here R^* means the fictive length of structure, σ_c the comparison stress (calculated by means of the classical theory of elasticity using a suitable strength hypothesis) and y the coordinate normal to the surface

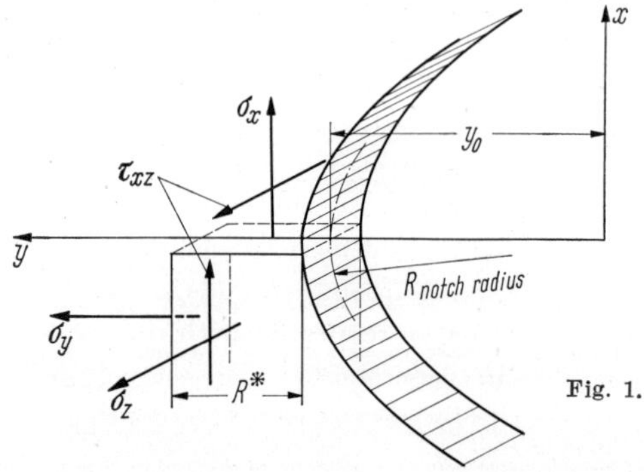

Fig. 1.

(see Fig. 1, the boundary lying at $y = y_0$). As results from this theory the stress $\bar{\sigma}$ at the root of the notch also depends on the fictive length of structure.

In most technical materials the domains in the vicinity of the highly stressed point participate in the force transmission more intensively than according to the linear theory of elasticity, during the endangered point itself is somewhat relieved; this *micro-support-effect* is taken into account by the above formula. For sharply curved notches a fictive radius of curvature R_F can be introduced which differs from the real radius of curvature R by the small length $s\,R^*$:

$$R_F = R + s\,R^*. \tag{2}$$

The fictive length of structure R^* depends on the properties of the material, the methods of machining and other secondary influences. The dimensionless factor s only depends on the type of loading and the applied strength hypothesis (see the table).

Strength Hypothesis	Factor s of Micro-Support-Effect for $R^* \ll R$		
	Tension and Bending		Torsion and Shear
	Flat Plates with Notches, Holes and Fillets	Shafts with Circumferential Notches	
Normal Stress Hypothesis	2	2	1
Shear Hypothesis	2	$\frac{2-\nu}{1-\nu}$	1
Strain Energy Hypothesis	2,5	$\frac{5-2\nu+2\nu^2}{2-2\nu+2\nu^2}$	1

Herein the constant ν is Poisson's ratio. With the new radius R_F the new stress concentration factor $\bar{\alpha} = \alpha_H \sqrt{R/R_F}$ is obtained, which replaces the ideal (Hookian) stress concentration factor α_H.

With regard to fatigue tests the high-cycle fatigue strength at push-pull (reversed) loading of unnotched specimens may be denoted by σ_a and that of notched specimens by σ_{an} (also this value is defined as

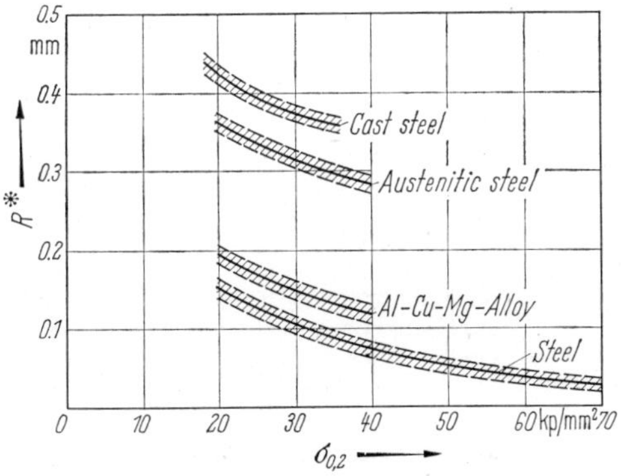

Fig. 2.

mean value over the minimum cross section). Then the ratio σ_a/σ_{an} is the so-called *fatigue strength reduction factor*. By the authors theory this value can be identified with $\bar{\alpha}$. Using the results of numerous fatigue tests the values of the fictive length of structure could be determined, as shown in Fig. 2.

The fictive length of structure mostly is considerably larger than real structural lengths as e.g. the dimension of grains in crystalline materials. This seems to be a consequence of the fact that the mean value procedure and therefore the correction of the error of the linear theory of elasticity too are resfricted to the highly stressed zone and due to the simplicity of the method no further corrections are included with regard to the other domains of the material.

3. The Effect of Stress Concentration in Cosserat Continua

A more accurate continuum model is represented by the Cosserat body with six elastic constants in the linear-isotropic case (one of them with the dimension of length). In the case of sharply curved notches a similar relation for a fictive radius of curvature could be expected as derived by means of the theory of stress mean value. Using his ix-functions-set [5] the author gave solutions of some stress concentraion problems (circular holes and spherical cavities) [6] and of a shear

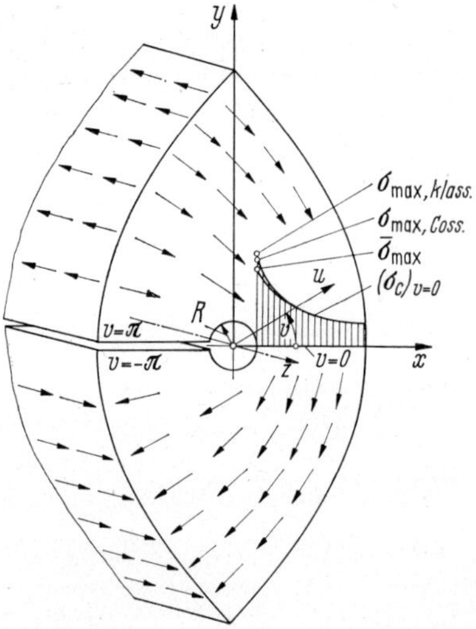

Fig. 3.

strained sharply curved notch with straight flanges and a circular root (Fig. 3) [7]. As the results show the Cosserat length l enters the stress concentration factor within the range $l \ll R$ in the second power. Therefore the results of the theory of stress mean value and those of the Cosserat theory are not identic, though there are qualitative analogies. If the theory of stress mean value is applied to stress concentration problems at holes and notches in the Cosserat continuum then the fictive radius of curvature can be represented in the form

$$R_F = R\left(1 + \sum_{\alpha=1}^{\infty} \sum_{\beta=0}^{\infty} s_{\alpha\beta} R^{*\alpha} l^{\beta} R^{-\alpha-\beta}\right) \tag{3}$$

with $s_{10} = s$ (see the table). This formula may become important in connection with the prediction of some kinds of failure, as e.g. fatigue or brittle fracture.

References

[1] NEUBER, H.: Zur Theorie der technischen Formzahl, Forschung a. d. Geb. d. Ingenieurwes. 7, 271—274 (1936).

[2] NEUBER, H.: Kerbspannungslehre, Grundlagen für genaue Festigkeitsberechnung mit Berücksichtigung von Konstruktionsform und Werkstoff. Berlin/Göttingen/Heidelberg: Springer, 1. Aufl. 1937, 2. Aufl. 1958.

[3] NEUBER, H.: Technische Mechanik, methodische Einführung, zweiter Teil: Festigkeitslehre. Berlin/Heidelberg/New York: Springer i. Vorb.

[4] NEUBER, H.: Über die Berücksichtigung der Spannungskonzentration bei Festigkeitsberechnungen. Konstruktion 1968.

[5] NEUBER, H.: Über die Dauerfestigkeit bei Spannungskonzentration. Sonderheft über die Sitzung d. WGLR-Ausschüsse am 19. u.20. 10.1967, VDI-Verlag 1968.

[6] NEUBER, H.: On the general solution of linear-elastic problems in isotropic and anisotropic Cosserat continua. Proc. Intern. Congr. Appl. Mechanics München 1964, Ed. H. GÖRTLER. Berlin/Heidelberg/New York: Springer 1966, pp. 153—158.

[7] NEUBER, H.: Über Probleme der Spannungskonzentration im Cosseratkörper. Acta Mechanica II (1966) pp. 48—69.

[8] NEUBER, H.: Die schubbeanspruchte Kerbe im Cosseratkörper. ZAMM 47, 313—318 (1967).

Stress Functions and Loading Singularities for the Infinitely Extended, Linear Elastic-Isotropic Cosserat Continuum

By

S. Kessel

Lehrstuhl für Theoretische Mechanik
Universität (Technische Hochschule) Karlsruhe

1. Preliminary Remarks

The equations of kinematic and static in a Cosserat continuum can be written in a clear and concise manner if we introduce the following differential operators [1]

$$\operatorname{Grad}\begin{pmatrix}\boldsymbol{V}\\ \boldsymbol{W}\end{pmatrix} \triangleq \begin{cases}\partial_i V_k\\ \partial_i W_k \overset{*}{-} e_{ikl} V_l\end{cases}, \quad \operatorname{Div}\begin{pmatrix}\mathbf{Q}\\ \mathbf{R}\end{pmatrix} \triangleq \begin{cases}\partial_i Q_{ik}\\ \partial_i R_{ik} \overset{*}{+} e_{klm} Q_{lm}\end{cases}, \tag{1a, b}$$

$$\operatorname{Rot}\begin{pmatrix}\mathbf{Q}\\ \mathbf{R}\end{pmatrix} \triangleq \begin{cases}e_{ikl}\, \partial_k Q_{lm}\\ e_{ikl}(\partial_k R_{lm} \overset{*}{+} e_{mkn} Q_{ln})\end{cases}. \tag{1c}$$

Changing the signs marked by a star we get the operators Grad*, Div*, and Rot*.

Looking at the identities easily to be proved

$$\left.\begin{aligned}
&\operatorname{Div}\operatorname{Grad}^* = \Delta, \quad \operatorname{Div}^*\operatorname{Grad} = \Delta;\\
&\operatorname{Div}\operatorname{Rot} \;\;= 0, \quad \operatorname{Div}^*\operatorname{Rot}^* = 0;\\
&\operatorname{Rot}\operatorname{Grad} = 0;\\
&\operatorname{Rot}\operatorname{Rot}^* \;= \operatorname{Grad}^*\operatorname{Div} - \Delta,
\end{aligned}\right\} \tag{2}$$

the method of notation becomes clear; Δ is the Laplace operator. We use cartesian coordinates throughout this paper.

2. Kinematic and Static Equations

Being $\boldsymbol{u}$ the displacement vector field and $\boldsymbol{\varphi}$ the vector field of small rotations of the "points" of the continuum, the tensors ε and $\varkappa$, describing small deformations of the continuum [2], read

$$\begin{pmatrix}\varkappa\\ \varepsilon\end{pmatrix} = \operatorname{Grad}\begin{pmatrix}\boldsymbol{\varphi}\\ \boldsymbol{u}\end{pmatrix}. \tag{3}$$

The constitutive equations [3] in the linear, elastic-isotropic case

$$\left.\begin{aligned} \varkappa &= \frac{1}{2GL^2}\left[\left(\frac{1}{2}+\frac{1}{c_2}\right)\mu+\left(\frac{1}{2}-\frac{1}{c_2}\right)\mu^T-\frac{c_3}{1+3c_3}\mathbf{E}\,\mu\right] \\ \varepsilon &= \frac{1}{2G}\left[\left(\frac{1}{2}+\frac{1}{c_1}\right)\sigma+\left(\frac{1}{2}-\frac{1}{c_1}\right)\sigma^T-\frac{\nu}{1+\nu}\mathbf{E}\,\sigma\right] \end{aligned}\right\}\begin{pmatrix}\varkappa\\ \varepsilon\end{pmatrix}=\mathbf{M}\begin{pmatrix}\sigma\\ \mu\end{pmatrix} \tag{4}$$

contain 6 material constants and link the force stress tensor σ and the couple stress tensor μ to the deformation tensors ε and $\varkappa$ respectively.

The equations of equilibrium read

$$\operatorname{Div}\begin{pmatrix}\sigma\\ \mu\end{pmatrix}=-\begin{pmatrix}\boldsymbol{X}\\ \boldsymbol{Y}\end{pmatrix} \tag{5}$$

with $\boldsymbol{X}$ the volume force and $\boldsymbol{Y}$ the volume couple.

3. The Stress Function Solution

As SCHAEFER [1] has proved, for each equilibrium stress state we can find a representation

$$\begin{pmatrix}\sigma\\ \mu\end{pmatrix}=\operatorname{Rot}\operatorname{Rot}^*\begin{pmatrix}\mathbf{F}\\ \mathbf{G}\end{pmatrix}+\operatorname{Grad}^*\begin{pmatrix}\boldsymbol{S}\\ \boldsymbol{T}\end{pmatrix} \tag{6}$$

where

$$\Delta\begin{pmatrix}\boldsymbol{S}\\ \boldsymbol{T}\end{pmatrix}=-\begin{pmatrix}\boldsymbol{X}\\ \boldsymbol{Y}\end{pmatrix}. \tag{7}$$

If the deformation state of the continuum is compatible, the stress function tensors $\mathbf{F}$ and $\mathbf{G}$ have to satisfy the condition

$$\mathbf{M}\cdot\left\{\operatorname{Rot}\operatorname{Rot}^*\begin{pmatrix}\mathbf{F}\\ \mathbf{G}\end{pmatrix}+\operatorname{Grad}^*\begin{pmatrix}\boldsymbol{S}\\ \boldsymbol{T}\end{pmatrix}\right\}=\operatorname{Grad}\begin{pmatrix}\boldsymbol{\varphi}\\ \boldsymbol{u}\end{pmatrix}. \tag{8}$$

This equation can be written as

$$\operatorname{Grad}\begin{pmatrix}\boldsymbol{N}^{(1)}\\ \boldsymbol{N}^{(2)}\end{pmatrix}+\begin{pmatrix}\mathfrak{L}^{(1)}(\mathbf{F},\mathbf{G})\\ \mathfrak{L}^{(2)}(\mathbf{F},\mathbf{G})\end{pmatrix}=\operatorname{Grad}\begin{pmatrix}\boldsymbol{\varphi}\\ \boldsymbol{u}\end{pmatrix} \tag{9}$$

with

$$\begin{aligned} \boldsymbol{N}^{(1)} &= \frac{1}{2GL^2}(\nabla\cdot\mathbf{G}+2\boldsymbol{F}+\boldsymbol{T}), \\ \boldsymbol{N}^{(2)} &= \frac{1}{2G}(\nabla\cdot\mathbf{F}+\mathbf{S}) \end{aligned} \tag{10}$$

and two tensor differential operators $\mathfrak{L}^{(1)}$ and $\mathfrak{L}^{(2)}$ applied to the stress function tensors $\mathbf{F}$ and $\mathbf{G}$.

It is obvious to try

$$\boldsymbol{\varphi} = \boldsymbol{N}^{(1)}, \qquad \boldsymbol{u} = \boldsymbol{N}^{(2)}; \tag{11}$$

$$\mathfrak{L}^{(1)}(\mathbf{F}, \mathbf{G}) = 0, \qquad \mathfrak{L}^{(2)}(\mathbf{F}, \mathbf{G}) = 0. \tag{12}$$

The Eqs. (12) are the differential equations for the compatible stress functions. Splitting them up into their symmetric and skew symmetric parts and using appropriate set-ups for the symmetrical parts of the stress function tensors **F** and **G**, it is possible to reduce these complicated equations to harmonic, Poisson- and Helmholtz equations [4]. The calculations become relatively simple if we compute the compatible stress functions for a concentrated force and a concentrated couple acting at the same point of the infinitely extended continuum [5].

4. Kinematical Fields for a Concentrated Force and a Concentrated Couple

With $\boldsymbol{X} = \boldsymbol{K}\,\delta(\boldsymbol{r})$ and $\boldsymbol{Y} = \boldsymbol{M}\,\delta(\boldsymbol{r})$, where $\delta(\boldsymbol{r})$ is Dirac's delta function, we obtain

$$\boldsymbol{S} = \frac{1}{4\pi r}\boldsymbol{K}, \qquad \boldsymbol{T} = \frac{1}{4\pi r}\boldsymbol{M}. \tag{13}$$

The set-ups

$$F_{(ik)} = \delta_{ik}\, f, \qquad G_{(ik)} = \delta_{ik}\, g, \tag{14}$$

introduced into the differential Eqs. (12), lead to the equations

$$\Delta f = -\frac{\nu}{1-\nu}\nabla \cdot \boldsymbol{S}, \tag{15}$$

$$\left(\Delta - \frac{1}{l_2^2}\right) g = -\frac{c_3}{1+c_3}\nabla \cdot \boldsymbol{T}, \qquad l_2^2 = \frac{L^2(1+c_3)}{c_1} \tag{16}$$

and the representations

$$\boldsymbol{F} = -\frac{1+c_3}{2c_3}\nabla g - \frac{1}{2}\boldsymbol{T} + \nabla \times \boldsymbol{A}, \tag{17}$$

$$\boldsymbol{G} = 2\boldsymbol{A} - \frac{L^2(2+c_1)}{2c_1}\Delta \boldsymbol{A} + \frac{L^2(2-c_1)}{2c_1}\boldsymbol{S} - \frac{L^2(2+c_1)}{4c_1}\nabla \times \boldsymbol{T} \tag{18}$$

for the vector fields corresponding to the skew symmetric parts of **F** and **G**. The vector field $\boldsymbol{A}$ is a solution of the differential equation

$$\begin{gathered}\Delta(1 - l_1^2\Delta)\boldsymbol{A} = (1 + l_1^2\Delta)\frac{1}{2}\nabla \times \boldsymbol{T} + \left(1 - \frac{2-c_1}{2+c_1} l_1^2 \Delta\right)\boldsymbol{S}, \\ l_1^2 = \frac{L^2(2+c_1)(2+c_2)}{8c_1}.\end{gathered} \tag{19}$$

Finally we obtain for the kinematical fields

$$\boldsymbol{u} = \frac{1}{16\pi G}\left\{\frac{3-4\nu}{1-\nu}\,\frac{\boldsymbol{K}}{r}\left[\overset{*}{1} + \frac{4c_1(1-\nu)}{(2+c_1)(3-4\nu)}\left(\frac{l_1}{r}\right)^2\left(1-h_2\left(\frac{r}{l_1}\right)\right)\right] + \right.$$

$$+\frac{1}{1-\nu}\,\frac{\boldsymbol{K}\cdot\boldsymbol{r}\,\boldsymbol{r}}{r^3}\left[\overset{*}{1} - \frac{12c_1(1-\nu)}{2+c_1}\left(\frac{l_1}{r}\right)^2\left(1-h_3\left(\frac{r}{l_1}\right)\right)\right] +$$

$$\left. + 2\,\frac{\boldsymbol{M}\times\boldsymbol{r}}{r^3}\left[\overset{*}{1} - h_1\left(\frac{r}{l_1}\right)\right]\right\}, \tag{20a}$$

$$\boldsymbol{\varphi} = \frac{1}{16\pi G}\left\{2\,\frac{\boldsymbol{K}\times\boldsymbol{r}}{r^3}\left[\overset{*}{1} - h_1\left(\frac{r}{l_1}\right)\right] - \right.$$

$$-\frac{\boldsymbol{M}}{r^3}\left[\overset{*}{1} + \frac{2}{c_1}h_1\left(\frac{r}{l_2}\right) - \frac{2+c_1}{c_1}h_2\left(\frac{r}{l_1}\right)\right] +$$

$$\left. + \frac{3\boldsymbol{M}\cdot\boldsymbol{r}\,\boldsymbol{r}}{r^5}\left[\overset{*}{1} + \frac{2}{c_1}h_3\left(\frac{r}{l_2}\right) - \frac{2+c_1}{c_1}h_3\left(\frac{r}{l_1}\right)\right]\right\}, \tag{20b}$$

with the following abbreviations

$$\left.\begin{aligned} h_1\left(\frac{r}{l}\right) &= \left(1+\frac{r}{l}\right)\exp\left(-\frac{r}{l}\right),\\ h_2\left(\frac{r}{l}\right) &= \left(1+\frac{r}{l}+\frac{r^2}{l^2}\right)\exp\left(-\frac{r}{l}\right),\\ h_3 &= \frac{1}{3}\,(2h_1+h_2). \end{aligned}\right\} \tag{21}$$

It is $\boldsymbol{r} = \boldsymbol{R} - \boldsymbol{R}_0$, where $\boldsymbol{R}$ is directed to the field point and $\boldsymbol{R}_0$ to the loading point. The terms marked by a star are the classical solutions.

The fundamental solutions of the field equations of the linear elastic Cosserat continuum have also been given by MINDLIN [6], who used stress functions of the Neuber-Papkowich type.

5. Kinematical Fields for Other Loading Singularities

Starting from the solutions (20a, b) we can by a TAYLOR expansion find the solution for a loading singularity acting at the point $\boldsymbol{R}_0 + \boldsymbol{a}$ instead of the point $\boldsymbol{R}_0$ $(|\boldsymbol{a}|/|\boldsymbol{r}| \ll 1)$. Superposing such solutions we can construct the kinematical fields for a dilatation centre, a torsion centre and a force couple.

a) The dilatation centre.

We superpose the solutions of 6 concentrated forces $-K\boldsymbol{e}_1$, $K\boldsymbol{e}_1$, $-K\boldsymbol{e}_2$, $K\boldsymbol{e}_2$, $-K\boldsymbol{e}_3$, $K\boldsymbol{e}_3$ acting at the points $\boldsymbol{R}_0$, $\boldsymbol{R}_0 + a\,\boldsymbol{e}_1$, $\boldsymbol{R}_0$, $\boldsymbol{R}_0 + a\,\boldsymbol{e}_2$, $\boldsymbol{R}_0$, $\boldsymbol{R}_0 + a\,\boldsymbol{e}_3$, where $\boldsymbol{e}_1$, $\boldsymbol{e}_2$ and $\boldsymbol{e}_3$ are mutually perpendicular unit vectors. In the limit $(a \to 0,\ K \to \infty,\ a\,K = p,\ p$ finite$)$

we obtain

$$\boldsymbol{u} = \frac{1 - 2\nu}{8\pi G(1-\nu)}\, p \frac{\boldsymbol{r}}{r^3}, \tag{22a}$$

$$\boldsymbol{\varphi} = 0. \tag{22b}$$

This result is the same as in the classical case.

b) The torsion centre.

Exchanging the forces K in the dilatation centre by couples M, we get the torsion centre. The kinematical fields are

$$\boldsymbol{u} = 0, \tag{23a}$$

$$\boldsymbol{\varphi} = \frac{1}{8\pi G c_1 l_2^2}\, q \frac{\boldsymbol{r}}{r^3} h_1\left(\frac{r}{l_2}\right) \tag{23b}$$

($a \to 0$, $M \to \infty$, $a\,M = q$, q finite).

c) The force couple.

We superpose now the kinematical fields of 4 concentrated forces $K\,\boldsymbol{e}_1$, $-K\,\boldsymbol{e}_1$, $-K\,\boldsymbol{e}_2$, $K\,\boldsymbol{e}_2$ acting at the points $\boldsymbol{R}_0$, $\boldsymbol{R}_0 + (a/2)\,\boldsymbol{e}_2$, $\boldsymbol{R}_0$, $\boldsymbol{R}_0 + (a/2)\,\boldsymbol{e}_1$ respectively, where $\boldsymbol{e}_1$ and $\boldsymbol{e}_2$ are mutually perpendicular unit vectors.

The result is

$$\boldsymbol{u} = \frac{1}{8\pi G} \frac{\boldsymbol{M}^* \times \boldsymbol{r}}{r^3}\left[1 - \frac{c_1}{2 + c_1} h_1\left(\frac{r}{l_1}\right)\right], \tag{24a}$$

$$\boldsymbol{\varphi} = \frac{1}{16\pi G}\left\{-\frac{\boldsymbol{M}^*}{r^3}\left[1 - h_2\left(\frac{r}{l_1}\right)\right] + 3\,\frac{\boldsymbol{M}^* \cdot \boldsymbol{r}\,\boldsymbol{r}}{r^5}\left[1 - h_3\left(\frac{r}{l_1}\right)\right]\right\}; \tag{24b}$$

($a \to 0$, $K \to \infty$, $\boldsymbol{M}^* = a\,K\,\boldsymbol{e}_1 \times \boldsymbol{e}_2$, $a\,K$ finite).

It is interesting that there is a difference between the kinematical fields of a couple $\boldsymbol{M}$ contained in (20a, b), and the kinematical fields of a force couple $\boldsymbol{M}^*$ (24a, b); ($\boldsymbol{M} \triangleq \boldsymbol{M}^*$)

$$\boldsymbol{u}(\boldsymbol{M}^*) - \boldsymbol{u}(\boldsymbol{M}) = \frac{1}{4\pi G(2 + c_1)} \frac{\boldsymbol{M} \times \boldsymbol{r}}{r^3} h_1\left(\frac{r}{l_1}\right),$$

$$\boldsymbol{\varphi}(\boldsymbol{M}^*) - \boldsymbol{\varphi}(\boldsymbol{M}) \tag{25}$$

$$= \frac{1}{8\pi G c_1}\left\{\frac{\boldsymbol{M}}{r^3}\left[h_1\left(\frac{r}{l_2}\right) - h_2\left(\frac{r}{l_1}\right)\right] - 3\,\frac{\boldsymbol{M} \cdot \boldsymbol{r}\,\boldsymbol{r}}{r^5}\left[h_3\left(\frac{r}{l_2}\right) - h_3\left(\frac{r}{l_1}\right)\right]\right\}.$$

This difference vanishes in the special case $c_1 \to \infty$ which corresponds to the kinematically narrowed continuum where the rotations are given by the curl of the displacement field.

This result makes clear that in a Cosserat continuum we have to distinguish between a force couple $\boldsymbol{M}^*$ and a couple $\boldsymbol{M}$.

The loading singularity of a force couple in a Cosserat continuum has also been discussed by Teodorescu [7] and Weitsman [8].

References

[1] Schaefer, H.: Bull. Acad. Pol. Sci. XV, 1, 63 (1967).
[2] Günther, W.: Abh. der Braunschweigischen Wiss. Ges. X, 195 (1958).
[3] Kessel, S.: Abh. d. Braunschweigischen Wiss. Ges. XVI, 1 (1964).
[4] Kessel, S.: ZAMM **47** (1967).
[5] Kessel, S.: Abh. d. Braunschweigischen Wiss. Ges. XIX, 72 (1967).
[6] Mindlin, R. D.: Int. J. Solids Structures **1**, 265 (1965).
[7] Teodorescu, P. P.: Article contained in this book.
[8] Weitsman, Y.: Quart. Applied Math. XXV, 213 (1967).

On the Action of Concentrated Loads in the Case of a Cosserat Continuum

By

P. P. Teodorescu

Institute of Mathematics
Academy of S. R. Rumania, Bucharest

1. As it is known [4], all concentrated loads can be constructed starting from a single concentrated load, considered as a fundamental load; usually the concentrated force is considered as such a load. The result is valid for all deformable continua with mechanical symmetry for which the principle of the local superposition of effects can be used.

In this way, starting from the *concentrated force* (which one supposes having a given definition, although it does not exist till now—as far as one knows it—a satisfactory definition) one can construct, for instance, *directed moments* and *dipoles of forces*. Superposing two directed moments equal and of the same sense, acting on two orthogonal directions, one obtains a concentrated load which does not have a privileged direction in the plane of the two couples (with axial antisymmetry): the *concentrated moment* or the *centre of rotation*. In the same way, by the superposition of two dipols of forces equal and of the same sense, acting on two orthogonal directions, one obtains another concentrated load which does not have a privileged direction in the plane of the two dipoles (with axial symmetry): the *centre of plane dilatation* (or condensation); superposing another dipol of forces, equal to the first ones and of the same sense, on a direction orthogonal to their corresponding directions, one obtains a concentrated load which does not have any privileged direction (with central symmetry): the *centre of spatial dilatation* (condensation). The interior concentrated loads can be thus represented by the aid of the *body forces*.

2. In the case of a Cosserat continuum appear *body moments*; a problem of relations between these moments and the centres of rotation defined previously is put.

One shall consider the particular case of small deformations of a Cosserat continuum, linearly elastic, isotropic and homogeneous, for which a constitutive law of the form (see, for instance, [3])

$$\sigma_{ij} = \lambda\, \delta_{ij}\, \beta_{hh} + 2\mu\, \beta_{(ij)} + 2\alpha\, \beta_{[ij]}, \tag{1}$$

$$\mu_{ij} = \beta\, \delta_{ij} \varphi_{hh} + 2\gamma\, \varphi_{(ij)} + 2\varepsilon\, \varphi_{[ij]} \tag{1'}$$

is valid; σ_{ij} and μ_{ij} are the components of the stress tensor, respectively of the couple-stress tensor; δ_{ij} is KRONECKER's symbol. The tensors β_{ij} and φ_{ij} (the round and the straight brackets correspond to the symmetric, respectively antisymmetric parts of the tensors) are the characteristics of the deformation and can be put in connection with the displacement and rotation vectors through the relations

$$\varphi_{ij} = \Phi_{j,i}, \tag{2}$$

$$\beta_{ij} = u_{j,i} - \varepsilon_{ijk}\, \Phi_k, \tag{2'}$$

where ε_{ijk} is the total antisymmetric operator, the indices after the comma precising the derivation with respect to the corresponding variable. The coefficients λ and μ are LAMÉ's elastic constants; given the general asymmetry, the constants α, β, γ and ε are also introduced.

The rotations Φ_i are generally free. In particular, for $\alpha \to \infty$ one obtains $\beta_{[ij]} = 0$ (the case of constraint rotations, characterized by the relation

$$\vec{\Phi} = \tfrac{1}{2} \operatorname{curl} \vec{u}). \tag{3}$$

For $\alpha = 0$ one obtains the case of symmetric elasticity.

3. In the case of a *force* $\vec{F} = F_1\, \vec{\imath}_1$, acting along the $0x_1$-axis in the point $(0, 0, 0)$ ($\vec{\imath}_1, \vec{\imath}_2, \vec{\imath}_3$ are the versors of the coordinate-axes), one obtains [5][1]

$$\vec{u} = \frac{1}{16\pi\mu(1-\nu)} \frac{F_1}{R} \left[(3-4\nu)\, \vec{\imath}_1 + \frac{x_1}{R^2} \vec{R}\right] +$$
$$+ \frac{l'^2}{4\pi\mu} \frac{F_1}{R^3} \left\{[1-(1+\chi+\chi^2)\, e^{-\chi}]\, \vec{\imath}_1 - \frac{x_1}{R^2} [3-(3+3\chi+\chi^2)\, e^{-\chi}]\, \vec{R}\right\}, \tag{4}$$

$$\vec{\Phi} = -\frac{1}{8\pi\mu} \frac{F_1}{R^3} [1-(1+\chi)\, e^{-\chi}]\, (x_3\, \vec{\imath}_2 - x_2\, \vec{\imath}_3), \tag{4'}$$

where ν is POISSON's ratio, related to LAMÉ's constants by the relation

$$\nu = \frac{\lambda}{2(\lambda+\mu)}; \tag{5}$$

the adimensional coefficient χ is given by

$$\chi = \frac{R}{l}, \tag{6}$$

R is the radius

$$R = \sqrt{x_1^2 + x_2^2 + x_3^2} \tag{7}$$

[1] Results in condensed form have been given by R. D. MINDLIN [1] for stress functions of Neuber-Papkovič type and by N. ŞANDRU [3] for displacement and rotation vectors.

and l is a length depending on the elastic constants and characterizing the asymmetry; it is precised by the relation

$$l^2 = \frac{(\gamma + \varepsilon)(\mu + \alpha)}{4\mu\alpha}. \tag{8}$$

For $\alpha \to \infty$ one obtains $l \to l'$ ($l \geqq l'$) with

$$l'^2 = \frac{\gamma + \varepsilon}{4\mu}. \tag{8'}$$

4. Taking into account also the force $\vec{F}' = -F_1 \vec{i}_1$, acting in the point $(0, c_2, 0)$, and passing to the limit

$$M_{12} = \lim_{\substack{F_1 \to \infty \\ c_2 \to 0}} F_1 c_2 \tag{9}$$

one obtains the action of the *directed moment* M_{12}; it results

$$\begin{aligned}\vec{u} = &-\frac{1}{16\pi\mu(1-\nu)} \frac{M_{12}}{R^3}\left[(3-4\nu)x_2\vec{i}_1 - x_1\vec{i}_2 + 3\frac{x_1x_2}{R^2}\vec{R}\right] - \\ &-\frac{l'^2}{4\pi\mu}\frac{M_{12}}{R^5}\left\{[3-(3+3\chi+2\chi^2+\chi^3)e^{-\chi}]x_2\vec{i}_1 + \right.\\ &\left. + [3-(3+3\chi+\chi^2)e^{-\chi}]x_1\vec{i}_2 - \frac{x_1x_2}{R^2}[15-(15+15\chi+6\chi^2+\chi^3)e^{-\chi}]\vec{R}\right\},\end{aligned} \tag{10}$$

$$\begin{aligned}\vec{\Phi} = &\frac{1}{8\pi\mu}\frac{M_{12}}{R^3}\left\{[1-(1+\chi)e^{-\chi}]\vec{i}_3 + \right.\\ &\left. + \frac{x_2}{R^2}[3-(3+3\chi+\chi^2)e^{-\chi}](x_3\vec{i}_2 - x_2\vec{i}_3)\right\}.\end{aligned} \tag{10'}$$

Calculating in an analogous manner the displacement state corresponding to the directed moment M_{21} and superposing the effects in the case

$$M_{12} = M_{21} = \frac{M_3}{2} \tag{11}$$

one obtains the displacement state corresponding to a *centre of rotation*

$$\vec{u} = \frac{1}{8\pi\mu}\frac{M_3}{R^3}\left[1 - \frac{\alpha}{\mu+\alpha}(1+\chi)e^{-\chi}\right]r_3\vec{i}_\theta, \tag{12}$$

$$\begin{aligned}\vec{\Phi} = &-\frac{1}{16\pi\mu}\frac{M_3}{R^3}\left\{[1-(1+\chi+\chi^2)e^{-\chi}]\vec{i}_3 - \right.\\ &\left. - \frac{x_3}{R^2}[3-(3+3\chi+\chi^2)e^{-\chi}]\vec{R}\right\},\end{aligned} \tag{12'}$$

where

$$r_3 = \sqrt{x_1^2 + x_2^2}, \tag{13}$$

the versor $\vec{i}_\theta$ (polar coordinates in the plane $0x_1x_2$) being given by

$$-r_3\vec{i}_\theta = x_2\vec{i}_1 - x_1\vec{i}_2. \tag{13'}$$

In the case of a *body moment* $\vec{\mathscr{M}} = \mathscr{M}_3 \vec{i}_3$ one obtains [5][1]

$$\vec{u} = \frac{1}{8\pi\mu} \frac{\mathscr{M}_3}{R^3} [1 - (1+\chi) e^{-\chi}] r_3 \vec{i}_\theta, \tag{14}$$

$$\vec{\Phi} = \frac{1}{16\pi\alpha} \frac{\mathscr{M}_3}{R^3} \left\{[1 - (1+\varkappa) e^{-\varkappa}] \vec{i}_3 - \frac{x_3}{R^2} [3 - (3 + 3\varkappa + \varkappa^2) e^{-\varkappa}] \vec{R}\right\} -$$
$$- \frac{\mu+\alpha}{16\pi\mu\alpha} \frac{\mathscr{M}_3}{R^3} \left\{[1 - (1+\chi+\chi^2) e^{-\chi}] \vec{i}_3 - \frac{x_3}{R^2} [3 - (3 + 3\chi + \chi^2) e^{-\chi} \vec{R}\right\}, \tag{14'}$$

where the adimensional coefficient

$$\varkappa = \frac{R}{h} \tag{15}$$

is used, the length h being given by

$$h^2 = \frac{\beta + 2\gamma}{4\alpha}. \tag{16}$$

For $\alpha \to \infty$ one obtains $h = 0$.

5. Comparing the axial antisymmetric displacement states (12), (12′) and (14), (14′) one observes that the rotation centre M_3 and the body moment $\mathscr{M}_3$ lead to different results for $M_3 = \mathscr{M}_3$. It is thus to be noticed that in the asymmetric linear elasticity with free rotations the body moment $\vec{\mathscr{M}}$ represents a concentrated load which cannot be constructed starting only from the notion of concentrated force as in the case of the centre of rotation M. Thus, from a qualitative point of view one can affirm that in a Cosserat continuum *the body moment* $\vec{\mathscr{M}}$ *is a fundamental load* as well as the body force $\vec{F}$.

Starting from the body moment one can thus construct an infinity of other concentrated loads. Evidently, one can construct also loads of a mixed nature, using simultaneously body forces and body moments.

For $M_3 = \mathscr{M}_3 = 1$ in the formulae (12), (12′), (14), (14′) one observes that the differences $\vec{u}(M_3) - \vec{u}(\mathscr{M}_3)$ and $\vec{\Phi}(M_3) - \vec{\Phi}(\mathscr{M}_3)$ tend to zero only for $\alpha \to \infty$ $(h = 0,\ l = l')$. Thus, only for constrained rotations the centre of rotation and the body moment give the same results

$$\vec{u} = \frac{1}{8\pi\mu} \frac{M_3}{R^3} [1 - (1+\chi') e^{-\chi'}] r_3 \vec{i}_\theta, \tag{17}$$

$$\vec{\Phi} = - \frac{16\pi\mu}{1} \frac{M_3}{R^3} \left\{[1 - (1+\chi'+\chi'^2) e^{-\chi'}] \vec{i}^3 - \right.$$
$$\left. - \frac{x_3}{R^2} [3 - (3 + 3\chi' + \chi'^2) e^{-\chi'}]\vec{R}\right\}, \tag{17'}$$

obtained by R. D. Mindlin and H. F. Tiersten [2], with the notation

$$\chi' = \frac{R}{l'}. \tag{18}$$

The demonstration is valid only for asymmetric elasticity.

[1] See the previous footnote.

6. Considering also the force $\vec{F}'' = -F_1 \vec{i}_1$, acting in the point $(-c_1, 0, 0)$, and passing to limit

$$D_{11} = \lim_{\substack{F_1 \to \infty \\ c_1 \to 0}} F_1 c_1 \tag{19}$$

one obtains the action of the *dipol of forces* D_{11}; it results

$$\vec{u} = \frac{1}{16\pi\mu(1-\nu)} \frac{D_{11}}{R^3} \left[2(1-2\nu) x_1 \vec{i}_1 - \left(1 - 3\frac{x_1^2}{R^2}\right)\vec{R}\right] +$$
$$+ \frac{l'^2}{4\pi\mu} \frac{D_{11}}{R^5} \Big\{[6 - (6 + 6\chi + 3\chi^2 + \chi^3) e^{-\chi}] x_1 \vec{i}_1 +$$
$$+ \Big\{3 - (3 + 3\chi + \chi^2) e^{-\chi} - \frac{x_1^2}{R^2} [15 - (15 + 15\chi + 6\chi^2 + \chi^3) e^{-\chi}]\Big\} \vec{R}\Big\}, \tag{20}$$

$$\vec{\Phi} = -\frac{1}{8\pi\mu} \frac{x_1}{R^5} D_{11} [3 - (3 + 3\chi + \chi^2) e^{-\chi}] (x_3 \vec{i}_2 - x_2 \vec{i}_3). \tag{20'}$$

Calculating in an analogous manner the displacement state corresponding to the dipol D_{22} and superposing the effects in the case

$$D_{11} = D_{22} = \frac{D_3}{2}, \tag{21}$$

one obtains the displacement state with axial symmetry corresponding to the *centre of plane dilatation* D_3

$$\vec{u} = -\frac{1}{32\pi\mu(1-\nu)} \frac{D_3}{R^3} \left[2(1-2\nu) x_3 \vec{i}_3 - \left(3 - 4\nu - 3\frac{x_3^2}{R^2}\right)\vec{R}\right] -$$
$$- \frac{l'^2}{8\pi\mu} \frac{D_3}{R^5} \Big\{[6 - (6 + 6\chi + 3\chi^2 + \chi^3) e^{-\chi}] x_3 \vec{i}_3 +$$
$$+ \Big\{3 - (3 + 3\chi + \chi^2) e^{-\chi} - \frac{x_3^2}{R^2} [15 - (15 + 15\chi + 6\chi^2 + \chi^3) e^{-\chi}]\Big\} \vec{R}\Big\}, \tag{22}$$

$$\vec{\Phi} = \frac{1}{16\pi\mu} \frac{x_3}{R^5} D_3 [3 - (3 + 3\chi + \chi^2) e^{-\chi}] (x_2 \vec{i}_1 - x_1 \vec{i}_2). \tag{22'}$$

Superposing also the action of the dipol D_{33} and admiting that

$$D_{11} = D_{22} = D_{33} = \frac{D}{3}, \tag{23}$$

one obtains the centro-symmetrical displacement state corresponding to a *centre of spatial dilatation* D

$$\vec{u} = \frac{1-2\nu}{24\pi\mu(1-\nu)} \frac{D}{R^3} \vec{R}, \tag{24}$$

$$\vec{\Phi} = 0. \tag{24'}$$

One observes that the results obtained are the same as in the classical elasticity, the effect of mechanical asymmetry being lost.

7. One has shown in [6] that the above obtained results are valid also for the case of a plane state of deformation.

References

[1] Mindlin, R. D.: Int. J. Solids Struct. **1**, 265 (1965).
[2] Mindlin, R. D., and H. F. Tiersten: Arch. Rat. Mech. Anal. **11**, 415 (1962).
[3] Şandru, N.: Int. J. Engng. Sci. **4**, 81 (1966).
[4] Teodorescu, P. P.: Accad. Naz. Lincei, Rend., Cl. Sci. fis., mat. nat. Ser. VIII, **40**, 251 (1966).
[5] Teodorescu, P. P.: Bull. Acad. Pol. Sci., sér. Sci. Techn. **15**, 65 (1967).
[6] Teodorescu, P. P., and N. Şandru: Rev. Roum. Math. Pures et Appl. **12**, 1399 (1967).

Dislocated and Oriented Media

By

R. A. Toupin

IBM Zürich Research Laboratory
Rüschlikon-ZH

Abstract. A continuum model of perfect or dislocated crystals is considered, and it is shown how one can view such a crystalline medium as an oriented medium, and as a material manifold with an irrotational law of distant parallelism. In dislocated crystals with twist, the Burgers vectors of homologous cycles are not independent of the cycles unless the amount of twist is restricted.

1. Introduction

The concepts of a dislocated material medium and of an oriented material medium had different origins, although both stemmed from special aspects of classical elasticity theory. Dislocated elastic media were first considered by VOLTERRA [1], and the first systematic treatment of a broad class of oriented elastic media appears in the memoir of E. and F. COSSERAT [2]. Both concepts, however, are essentially kinematical and independent of the idea of elastic response and the laws of mechanics which we may suppose govern the motion of either kind of material medium. Oriented media provide a unifying concept for various special theories of elastic rods and shells, of elastic media with microstructure, of liquid crystals, and many other special theories. An oriented medium is nothing more nor less kinematically than a continuous medium of dimension one, two, or three, at each point of which and at each instant of time there is defined a set of vectors $\underset{a}{\boldsymbol{d}}$, $a = 1, 2, \ldots, m$. In many considerations, the physical dimension or any other physical significance of each member of this set of vectors is unimportant. Theories of dislocated media have developed principally as a means to understand the mechanism of initial stress and of the macroscopic inelastic deformations and physical properties of imperfect crystals. In 1958, GÜNTHER [3] called attention to the relevance of the Cosserats theory of oriented media to the then newly developing theory of continuously dislocated material media. More recently, N. FOX [4] has established a connection between these seemingly diverse theories, and W. NOLL [5] has shown how the mathematical structure of a continuously dislocated medium flows naturally from the definition of a materially uniform, yet inhomogeneous simple medium. In this note, we show how one can begin with NOLL's general definition of the isotropy group, define perfect and imperfect crystals, and view a dislocated crystal as an oriented medium.

2. The Isotropy Group of a Continuous Medium

We shall consider a set $\overline{\mathscr{B}}$ of objects $X, Y, \ldots$ called *material points*, and we shall call $\overline{\mathscr{B}}$ a *body*. We shall consider only the cases for which $\overline{\mathscr{B}}$ is a *standard 3-manifold* in the sense defined by WHITNEY [6]. Thus $\overline{\mathscr{B}}$ is a compact, connected topological space $\overline{\mathscr{B}}$, and we are given a closed subset $\partial\overline{\mathscr{B}}$ of $\overline{\mathscr{B}}$ called *boundary points*, and a closed subset $\partial_0\overline{\mathscr{B}}$ of $\partial\overline{\mathscr{B}}$ called *edge* and *corner points*. The set $\mathscr{B} = \overline{\mathscr{B}} - \partial\overline{\mathscr{B}}$ of *interior points* is a smooth 3-dimensional manifold, and the set $\partial\overline{\mathscr{B}} - \partial_0\overline{\mathscr{B}}$ is a finite collection of smooth 2-dimensional manifolds. A standard 3-manifold is an abstraction of a polyhedral region in 3-dimensional Euclidean space, or of a smoothly deformed polyhedral region. Let $\mathscr{T}_X$ denote the *tangent space* of $\mathscr{B}$ at the point X. Each $\mathscr{T}_X$ is a 3-dimensional vector space and we call the elements $\boldsymbol{U}, \boldsymbol{V}, \ldots$ of $\mathscr{T}_X$ *material vectors* at X. An assignment of a material vector in each $\mathscr{T}_X$, $X \in \mathscr{B}$ is a *material vector field over* $\mathscr{B}$. The conjugate space $\mathscr{T}_X^*$ of $\mathscr{T}_X$ is the set of real valued linear functions $A : \mathscr{T}_X \to R$ of material vectors. We call the elements of $\mathscr{T}_X^*$, *material covectors at* X. Material covector fields over $\mathscr{B}$, or, more generally, material tensor fields over $\mathscr{B}$ are defined in the obvious way.

Let $\mathscr{E}$ be 3-dimensional Euclidean space with elements $x, y, \ldots$ we shall call *places* or *positions*, and let $\mathscr{V}$ denote the translation space of $\mathscr{E}$. Then $\mathscr{V}$ is a 3-dimensional vector space and $x = y + \boldsymbol{v}$, $(\boldsymbol{v} \in \mathscr{V}$, $x, y \in \mathscr{E})$ is the point y translated by $\boldsymbol{v}$. Every two points $x, y \in \mathscr{E}$ determine a unique translation $\boldsymbol{v} = x - y$. We call the elements of $\mathscr{V}$, *spatial vectors*.

A configuration $\varkappa$ of $\overline{\mathscr{B}}$ is a mapping

$$\varkappa : \overline{\mathscr{B}} \to \mathscr{E} \tag{2.1}$$

which assigns a position to each point of the body. We write $x_\varkappa = \varkappa(X)$ for the position assigned to X in the configuration $\varkappa$.

A configuration $\varkappa$ of $\overline{\mathscr{B}}$ is faithful if a) $\varkappa$ is one-one. b) $\overline{\mathscr{B}}_\varkappa$ is a standard 3-manifold with boundary points $\partial\overline{\mathscr{B}}_\varkappa = \varkappa(\partial\overline{\mathscr{B}})$, and edge and corner points $\partial_0 \overline{\mathscr{B}}_\varkappa = \varkappa(\partial_0 \overline{\mathscr{B}})$. c) $\varkappa$ is continuous in $\overline{\mathscr{B}}$ and smooth in $\mathscr{B}$.

A *motion* of $\overline{\mathscr{B}}$ is a one-parameter family $\varkappa_t$ of configurations of $\overline{\mathscr{B}}$, one configuration for each value of the *time* t. If each configuration $\varkappa_t$ is faithful, the motion is *regular*. A motion of $\overline{\mathscr{B}}$ can fail to be regular in many ways; e.g., the restriction of $\varkappa$ to $\partial\overline{\mathscr{B}}$ may fail to be one-one.

A *local configuration* of $\overline{\mathscr{B}}$ at a point $X \in \mathscr{B}$ is a one-one linear mapping

$$K_X : \mathscr{T}_X \to \mathscr{V} \tag{2.2}$$

which assigns to each material vector at X a unique spatial vector. A field of local configurations over $\mathscr{B}$ is a *reference*. Every faithful configuration $\varkappa$ of $\overline{\mathscr{B}}$ determines a reference defined by

$$K_X = \nabla_X \varkappa, \tag{2.3}$$

where $\nabla_X \varkappa$ denotes the gradient of $\varkappa$ evaluated at the interior point X of $\overline{\mathscr{B}}$. Not every reference is given by such a gradient.

If $X \neq Y$, the tangent spaces $\mathscr{T}_X$ and $\mathscr{T}_Y$ of $\mathscr{B}$ are distinct 3-dimensional vector spaces, and the addition or subtraction of material vectors at different material points is a meaningless operation. Suppose, however, for each pair of points $(X, Y) \in \mathscr{B} \times \mathscr{B}$ we are given a linear transformation

$$\gamma_{XY} : \mathscr{T}_Y \to \mathscr{T}_X, \tag{2.4}$$

and that this set of transformations satisfies the three conditions

$$\left.\begin{aligned} &1)\ \gamma_{XX} = I_X,\\ &2)\ \gamma_{XY} = \gamma_{YX}^{-1},\\ &3)\ \gamma_{XY} \circ \gamma_{YZ} = \gamma_{XZ}, \end{aligned}\right\} \tag{2.5}$$

where I_X denotes the identity map of $\mathscr{T}_X$. Every such set of transformations determines an equivalence relation between material vectors at the same or different points of $\mathscr{B}$ defined by

$$\boldsymbol{V}_X \overset{\gamma}{\sim} \boldsymbol{V}_Y \quad \text{iff} \quad \boldsymbol{V}_X = \gamma_{XY} \cdot \boldsymbol{V}_Y. \tag{2.6}$$

We call such a set $\{\gamma_{XY}\}$ a *material parallelism*. Every reference $\{K_X\}$ of $\overline{\mathscr{B}}$ determines a corresponding material parallelism defined by setting

$$\gamma_{XY} = K_X^{-1} \circ K_Y. \tag{2.7}$$

Thus, in particular, every faithful configuration $\varkappa$ of $\overline{\mathscr{B}}$ determines a material parallelism defined by setting $K_X = \nabla_X \varkappa$ in (2.7).

We say that a material covector $\boldsymbol{A}_X$ at $X \in \mathscr{B}$ is *γ-equivalent* to a material covector $\boldsymbol{A}_Y$ at $Y \in \mathscr{B}$ and write $\boldsymbol{A}_X \overset{\gamma}{\sim} \boldsymbol{A}_Y$ if and only if $\boldsymbol{A}_X \cdot \boldsymbol{V}_X = \boldsymbol{A}_Y \cdot \boldsymbol{V}_Y$ whenever $\boldsymbol{V}_X \overset{\gamma}{\sim} \boldsymbol{V}_Y$.

It is not difficult to show that if $\{\gamma_{XY}\}$ is a material parallelism, then there exists a linearly independent set $\{\underset{a}{\boldsymbol{E}}_X;\ a = 1, 2, 3\}$ of γ-equivalent material vector fields such that

$$\gamma_{XY} = \underset{a}{\boldsymbol{E}}_X \otimes \overset{a}{\boldsymbol{E}}_Y, \tag{2.8}$$

where the material covectors $\overset{a}{\boldsymbol{E}}_X$ are *reciprocal* to the material vectors $\underset{a}{\boldsymbol{E}}_X$;

$$\overset{a}{\boldsymbol{E}}_X \cdot \underset{b}{\boldsymbol{E}}_X = \delta^a_b. \tag{2.9}$$

The fields $\{\underset{a}{\boldsymbol{E}}_X\}$ in the representation (2.8) of $\{\gamma_{XY}\}$ are uniquely determined by $\{\gamma_{XY}\}$ up to a non-singular linear transformation $\underset{a}{\boldsymbol{E}}_X \to L^b_a \underset{b}{\boldsymbol{E}}_X$ which is independent of X.

Let $\mathscr{C}$ be a smooth closed curve in $\mathscr{B}$ with differential element $\boldsymbol{dX} \in \mathscr{T}_X$ at $X \in \mathscr{C}$. Choose the point $Y \in \mathscr{B}$ arbitrarily and consider the line integral

$$\boldsymbol{V}_Y(\mathscr{C}) = \oint_{\mathscr{C}} \gamma_{YX} \cdot \boldsymbol{dX}. \tag{2.10}$$

If $\boldsymbol{V}_Y(\mathscr{C}) = 0$ for some $Y \in \mathscr{B}$, then $\boldsymbol{V}_Y(\mathscr{C}) = 0$ for every $Y \in \mathscr{B}$. If $\boldsymbol{V}_Y(\mathscr{C}) = 0$ for every closed curve $\mathscr{C} \subset \mathscr{B}$, we say that the material parallelism $\{\gamma_{XY}\}$ is *torsionless*. Substituting the representation (2.8) of γ_{XY} into (2.10), one concludes that a necessary and sufficient condition for $\{\gamma_{XY}\}$ to be torsionless is that

$$\oint_{\mathscr{C}} \overset{a}{\boldsymbol{E}}_X \cdot \boldsymbol{dX} = 0, \qquad a = 1, 2, 3. \tag{2.11}$$

In other words, the material parallelism $\{\gamma_{XY}\}$ is torsionless if and only if each *exterior differential 1-form* $\overset{a}{E} = \overset{a}{\boldsymbol{E}}_X \cdot \boldsymbol{dX}$ is *exact*. If $\boldsymbol{V}_X(\mathscr{C}) = 0$ whenever $\mathscr{C}$ is the boundary $\partial\mathscr{A}$ of a standard 2-manifold $\mathscr{A} \subset \mathscr{B}$, the material parallelism $\{\gamma_{XY}\}$ is said to be *irrotational*. Then (2.11) must hold for every 1-cycle which is the boundary of an $\mathscr{A} \subset \mathscr{B}$, and the 1-form $\overset{a}{E}$ must be *irrotational*. But not every 1-cycle $\mathscr{C} \subset \mathscr{B}$ may be a boundary $\partial\mathscr{A}$ of a surface $\mathscr{A} \subset \mathscr{B}$, in which case, there exist irrotational material parallelisms which are not torsionless. It is precisely this circumstance which can occur in dislocated crystalline media, as we shall see.

Let $\mathscr{C}_X$ be the set of all local configurations $K_X : \mathscr{T}_X \to \mathscr{V}$ of $\mathscr{B}$ at the point X. Then, following NOLL, we may view the *constitutive relations* of a simple body as a set of mappings

$$\{\mathfrak{G}_X : \mathscr{C}_X \to \mathbb{R};\quad X \in \mathscr{B}\}, \tag{2.12}$$

where the range $\mathbb{R}$ of each $\mathfrak{G}_X$ is the space of *response descriptors*. The nature of the elements of this space is unimportant for the present discussion. Let $\varkappa$ be a faithful configuration of $\overline{\mathscr{B}}$. It need not be a configuration $\varkappa_t$ experienced by $\overline{\mathscr{B}}$ at any time during its motion. It need not be an "equilibrium" configuration of $\overline{\mathscr{B}}$. Define the functions $\mathfrak{G}_X^{\varkappa}$ by

$$\mathfrak{G}_X^{\varkappa} = \mathfrak{G}_X \circ \nabla_X \varkappa. \tag{2.13}$$

Let $\mathscr{L}(\mathscr{V}, \mathscr{V})$ denote the set of all one-one linear transformations $\mathscr{V} \to \mathscr{V}$. Then, by definition, each $\mathfrak{G}_X^{\varkappa}$ is a mapping

$$\mathfrak{G}_X^{\varkappa} : \mathscr{L}(\mathscr{V}, \mathscr{V}) \to \mathbb{R}. \tag{2.14}$$

The functions $\mathfrak{G}_X^{\varkappa}$ and $\varkappa$ determine the $\mathfrak{G}_X$ uniquely. The constitutive relations $\mathfrak{G}_X^{\varkappa}$ depend on $\varkappa$, but the constitutive relations $\mathfrak{G}_X$ are independent of any configuration of $\overline{\mathscr{B}}$.

Every smooth transformation $\chi : \mathscr{E} \to \mathscr{E}$ determines another faithful configuration $\varkappa'$ of $\overline{\mathscr{B}}$ given by $\varkappa' = \chi \circ \varkappa$. Every pair of faithful configurations $(\varkappa, \varkappa')$ of $\overline{\mathscr{B}}$ determines a smooth transformation $\chi : \overline{\mathscr{B}}_{\varkappa} \to \overline{\mathscr{B}}_{\varkappa'}$ given by $\chi = \varkappa' \circ \varkappa^{-1}$. If $\varkappa' = \chi \circ \varkappa$, then

$$\nabla_X \varkappa' = \nabla_{x_{\varkappa}} \chi \circ \nabla_X \varkappa. \tag{2.15}$$

From this relation, we see that the argument $K_X^{\varkappa} = K_X \circ (\nabla_X \varkappa)^{-1}$ of $\mathfrak{G}_X^{\varkappa}$, for each fixed value of K_X depends on $\varkappa$ as follows

$$K_X^{\varkappa'} = K_X^{\varkappa} \circ (\nabla_{x_{\varkappa}} \chi)^{-1}. \tag{2.16}$$

From the definition of the functions $\mathfrak{G}_X^{\varkappa}$ we have then

$$\mathfrak{G}_X^{\varkappa}(K_X^{\varkappa}) = \mathfrak{G}_X^{\varkappa'}(K_X^{\varkappa'}) = \mathfrak{G}_X(K_X). \tag{2.17}$$

Let L be any element of $\mathscr{L}(\mathscr{V}, \mathscr{V})$. If there exists an L such that

$$\mathfrak{G}_X^{\varkappa}(S) = \mathfrak{G}_Y^{\varkappa}(S \circ L) \tag{2.18}$$

for all S in the domain $\mathscr{L}(\mathscr{V}, \mathscr{V})$ of $\mathfrak{G}_X^{\varkappa}$ and $\mathfrak{G}_Y^{\varkappa}$, we say that the two material points X and Y are equivalent. While this definition of the equivalence of X and Y might appear to depend on $\varkappa$, the relations (2.16) and (2.17) show that if (2.18) holds for $\varkappa$ and L, then it also holds with $\varkappa$ replaced by $\varkappa'$ and L replaced by $L' = L \circ \nabla(x_{\varkappa} \chi)^{-1}$, $\chi = \varkappa' \circ \varkappa^{-1}$. Thus the definition of equivalent material points is independent of the configuration.

The *symmetry group* $\mathscr{G}_X$ of the material point $X \in \mathscr{B}$ is the set of all one-one linear transformations

$$P_X : \mathscr{T}_X \to \mathscr{T}_X \tag{2.19}$$

of the tangent space of $\mathscr{B}$ at X such that

$$\mathfrak{G}_X(K_X \circ P_X^{-1}) = \mathfrak{G}_X(K_X) \tag{2.20}$$

for all K_X in the domain of $\mathfrak{G}_X$.

We readily find that (2.20) holds if and only if

$$\mathfrak{G}_X^\varkappa(S \circ \overset{-1}{P_X^\varkappa}) = \mathfrak{G}_X^\varkappa(S). \tag{2.21}$$

For all $S \in \mathscr{L}(\mathscr{V}, \mathscr{V})$, where $P_X^\varkappa$ is the linear transformation $\mathscr{V} \to \mathscr{V}$ defined by

$$P_X^\varkappa = \nabla_{x_\varkappa} \varkappa \circ P_X \circ (\nabla_{x_\varkappa} \varkappa)^{-1}. \tag{2.22}$$

The set of all $P_X^\varkappa$ with $P_X \in \mathscr{G}_X$ comprises a *representation* $\mathscr{G}_X^\varkappa$ of the symmetry group $\mathscr{G}_X$ by linear transformations of $\mathscr{V}$. The representations $\mathscr{G}_X^\varkappa$ and $\mathscr{G}_X^{\varkappa'}$ are conjugate groups of transformations of $\mathscr{V}$. If X and Y are equivalent material points, the symmetry groups and their representations $\mathscr{G}_X^\varkappa$ and $\mathscr{G}_Y^\varkappa$ are conjugate.

$$\mathscr{G}_X^\varkappa = L \circ \mathscr{G}_X^\varkappa \circ L^{-1}. \tag{2.23}$$

If every pair of material points of a body is equivalent, the body is *uniform*. If there exists a faithful configuration $\varkappa$ of a uniform body $\overline{\mathscr{B}}$ such that the representations $\mathscr{G}_X^\varkappa$ and $\mathscr{G}_Y^\varkappa$ are not only conjugate for every pair of points, but *identical* ($\mathscr{G}_X^\varkappa = \mathscr{G}_Y^\varkappa$), then the body is *homogeneous*, and $\varkappa$ is an *undistorted* configuration of the homogeneous body. Every configuration $\varkappa'$ related to an undistorted configuration $\varkappa$ of a homogeneous body by an affine transformation of $\mathscr{E}$ is also undistorted. If we think of $\overline{\mathscr{B}}$ as an elastic body, it is then clear that an undistorted configuration need not be an "equilibrium" or stress free, "natural" configuration of $\overline{\mathscr{B}}$. A homogeneous body may never have existed in an undistorted configuration.

Consider a uniform body, so that, in any configuration $\varkappa$, there exists an $L_{XY}^\varkappa$ such that

$$\mathfrak{G}_X^\varkappa(S) = \mathfrak{G}_Y^\varkappa(S \circ \overset{-1}{L_{YX}^\varkappa}) \tag{2.24}$$

The $L_{XY}^\varkappa$ are determined by the constitutive relations $\mathfrak{G}_X^\varkappa$ only up to a transformation $L_{XY}^\varkappa \to P_X^{\varkappa-1} \circ L_{XY} \circ P_Y^\varkappa$, where $P_X^\varkappa$ is an element of the representation $\mathscr{G}_X^\varkappa$ of the symmetry group $\mathscr{G}_X$. Clearly, we can choose the $L_{XY}^\varkappa$ in a uniform body such that the three conditions analogous to (2.5) are satisfied. Now set

$$\gamma_{XY} = (\nabla_X \varkappa)^{-1} \circ P_X^\varkappa \circ L_{XY}^\varkappa \circ P_Y^{\varkappa-1} \circ \nabla_Y \varkappa = P_X^{-1} \circ (\nabla_X \varkappa)^{-1} \circ L_{XY}^\varkappa \circ \nabla_Y \varkappa \circ P_Y. \tag{2.25}$$

We verify that $\{\gamma_{XY}\}$ satisfies the three conditions (2.5) and therefore defines a material parallelism. There is one such material parallelism for each choice of a field $\{P_X\}$ of symmetry transformations.

3. Crystalline Media

The definition of crystalline media rests on the concept of a periodic function and its invariance group. Let $f: \mathscr{E} \to \mathbb{R}$ be a function of points in Euclidean space. It is sufficient for present purposes that the range of f be at least two valued; say, black and white, or 0 and 1. Beyond this, the nature of the set $\mathbb{R}$ is unimportant. Let

$$T_{\boldsymbol{v}}: \mathscr{E} \to \mathscr{E} \tag{3.1}$$

denote the translation of $\mathscr{E}$ by $\boldsymbol{v} \in \mathscr{V}$ so that

$$T_{\boldsymbol{v}}(x) = x + \boldsymbol{v}. \tag{3.2}$$

Say that f is *periodic* in the direction $\boldsymbol{v}$ with period $|v| = \sqrt{\boldsymbol{v} \cdot \boldsymbol{v}}$ if

$$f\big(T_{n\boldsymbol{v}}(x)\big) = f(x) \tag{3.3}$$

for every integer n and $x \in \mathscr{E}$. Suppose next that f is periodic in at least three linearly independent directions $\underset{a}{\boldsymbol{D}}$, $a = 1, 2, 3$, and consider the set $\mathscr{D}$ of all such sets of three linearly independent directions in which f is periodic. Put vol $\{\underset{a}{\boldsymbol{D}}\} = |\underset{1}{\boldsymbol{D}} \cdot (\underset{2}{\boldsymbol{D}} \times \underset{3}{\boldsymbol{D}})|$. Call f a *pattern* if $V_m = \inf\limits_{\{\underset{a}{D}\} \in \mathscr{D}} \text{vol}\,\{\underset{a}{\boldsymbol{D}}\} > 0$ and if there exists at least one set of directions $\{\underset{a}{D}\} \in \mathscr{D}$ such that vol $\{\underset{a}{\boldsymbol{D}}\} = V_m$. Any set $\{\underset{a}{\boldsymbol{D}}\} \in \mathscr{D}$ satisfying this condition is a *basis* for the *lattice vectors* of the pattern f. A lattice vector of f is any vectorof the form

$$\boldsymbol{v}_n = \overset{a}{n}\, \underset{a}{\boldsymbol{D}} \tag{3.4}$$

with integer coefficients $\overset{a}{n}$, $a = 1, 2, 3$. We use the summation convention. The set of all lattice vectors is independent of the basis. If f is periodic in any direction $\boldsymbol{v}$, then $\boldsymbol{v}$ is a lattice vector of f. The point set

$$\Lambda_p = \bigcup_{\boldsymbol{n}} (p + \boldsymbol{v}_{\boldsymbol{n}}) = \bigcup_{n} T_{\boldsymbol{v}_{\boldsymbol{n}}}(p) \tag{3.5}$$

is a *lattice* in $\mathscr{E}$. Clearly, $\Lambda_{p+\boldsymbol{v}_{\boldsymbol{n}}} = \Lambda_p$.

Consider next the automorphisms of $\mathscr{E}$ which we call *rigid transformations*. Every rigid transformation $l: \mathscr{E} \to \mathscr{E}$ can be represented as a composition

$$l = T_{\boldsymbol{v}} \cdot O_p \tag{3.6}$$

of a *rotation* about a point p given by

$$O_p(x) = p + O \cdot (x - p) \tag{3.7}$$

where $O : \mathscr{V} \to \mathscr{V}$ is an orthogonal transformation of the translation space $\mathscr{V}$ of $\mathscr{E}$, followed by a translation $T_{\boldsymbol{v}}$. O_p is a *proper rotation* if $\det O = +1$, and an *improper rotation* if $\det O = -1$. Every l has also a representation

$$l = O_p \cdot T_{\boldsymbol{v}} \tag{3.8}$$

as a translation followed by a rotation. The representations (3.6) and (3.7) are not unique. In fact, the following identities are true.

$$l = T_{\boldsymbol{v}} \cdot O_p = T_{\boldsymbol{v}+O\cdot\boldsymbol{u}-\boldsymbol{u}}\, O_{p+\boldsymbol{u}} = O_p \circ T_{O^{-1}\cdot\boldsymbol{v}} = O_{p+\boldsymbol{u}} \circ T_{\boldsymbol{v}+\boldsymbol{u}-O^{-1}\cdot\boldsymbol{u}}. \tag{3.9}$$

Also,

$$O_p \circ T_{\boldsymbol{v}} \cdot O_p^{-1} = T_{O\cdot\boldsymbol{v}}. \tag{3.10}$$

A *space group* $\mathscr{G}_f$ is the group of all rigid transformations l such that

$$f(l(x)) = f(x), \tag{3.11}$$

for all x, where f is a pattern. In other words, a space group is the invariance group of rigid transformations of a pattern.

Our discussion of dislocations in crystalline media will require some elementary facts about all space groups which will be recorded here without proof since they are all well known. It has been noted already that any *pure translation* $T_{\boldsymbol{v}} \in \mathscr{G}_f$ is of the form $T_{\boldsymbol{v}_n}$, where $\boldsymbol{v}_n$ is a lattice vector of f. The subgroup $\mathscr{G}_f^p$ of $\mathscr{G}_f$ of all pure rotations about the point p such that

$$f(O_p(x)) = f(x) \tag{3.12}$$

is the *group of the point* p. Each $O_p \in \mathscr{G}_f^p$ determines an orthogonal transformation O of $\mathscr{V}$, and the set G_f^p of all these orthogonal transformations, $O_p \in \mathscr{G}_f^p$, is a faithful representation of $\mathscr{G}_f^p$. If $O \in G_f^p$, then it can be shown that the order of O cannot be five nor greater than six. Also, if $O \in G_f^p$, then $O(\boldsymbol{\Lambda}) = \boldsymbol{\Lambda}$, where $\boldsymbol{\Lambda}$ is the set of all lattice vectors of f. Also if $O \neq I$ (the identity transformation) and $O \in G_f^p$, then any proper vector of O must be proportional to a lattice vector. To put it otherwise, if $O\cdot\boldsymbol{v} = \pm\boldsymbol{v}$ and $O \neq I \in G_f^p$, the straight line $\mathscr{L}(O_p) = \{x; x = p + \lambda\,\boldsymbol{v}\}$ or *axis* of O_p passes through p and at least one other point of the lattice Λ_p. Of course this means it passes through infinitely many points of Λ_p.

Two points p and p' are called *equivalent points* of a pattern f if there exists an element $l \in \mathscr{G}_f$ such that $l(p) = p'$. The groups $\mathscr{G}_f^p$ and $\mathscr{G}_f^{p'}$ of equivalent points p and p' are conjugate subgroups of $\mathscr{G}_f$; i.e., there exists an element $l \in \mathscr{G}_f$ such that $\mathscr{G}_f^{p'} = l\,\mathscr{G}_f^p\, l^{-1}$. Clearly, the latter property of $\mathscr{G}_f^{p'}$ and $\mathscr{G}_f^p$ is not sufficient that p and p' be equivalent points of f.

The set of points τ_p which lie closer to the point p than to any other point p' in the lattice Λ_p is called the *symmetric unit cell about* p. Let $\boldsymbol{\tau}$ denote the set of all vectors in $\mathscr{V}$ such that $|\boldsymbol{u}| < |\boldsymbol{u} - \boldsymbol{v}_n|$ for all lattice vectors $\boldsymbol{v}_{\boldsymbol{n}} \neq 0$. (Then τ_p is the point set

$$\boldsymbol{\tau}_p = \bigcup_{\boldsymbol{v} \in \boldsymbol{\tau}} (p + \boldsymbol{v}), \qquad (3.13)$$

$$= \bigcup_{\boldsymbol{v} \in \boldsymbol{\tau}} T_{\boldsymbol{v}}(p),$$

and $\bar{\tau}_p$ is a convex polyhedral region in $\mathscr{E}$.

If $l \in \mathscr{G}_f$ has a representation $T_{\boldsymbol{v}} \circ O_p$ then either $\boldsymbol{v}$ is a lattice vector or O_p is not an element of $\mathscr{G}_f^p$. In either case, O_p must commute with every lattice translation $T_{\boldsymbol{v}_{\boldsymbol{n}}}$ so that $O_p(\Lambda_p) = \Lambda_p$, and the order of O_p must be finite; not five and not greater than six. It follows that for any $l \in \mathscr{G}_f$, $l(\tau_p) = T_{\boldsymbol{v}}(\tau_p)$. That is, l applied to τ_p is equivalent to a translation of τ_p. Every $l \in \mathscr{G}_f$ is the composition of a lattice translation and a transformation of the form $l' = T_{\frac{\boldsymbol{v}_{\boldsymbol{n}}^o}{r} + \boldsymbol{w}} \circ O_p$, where $O\,\boldsymbol{v}_{\boldsymbol{n}}^o = \pm \boldsymbol{v}_{\boldsymbol{n}}^o$, $\boldsymbol{v}_{\boldsymbol{n}}^o$ is a lattice vector, O has order r, $\boldsymbol{v}_{\boldsymbol{n}}^o \cdot \boldsymbol{w} = 0$, and $\boldsymbol{w} \in \tau$.

Consider now a simple body $\bar{\mathscr{B}}$ with constitutive relations $\mathfrak{G}_X$. We say that $\bar{\mathscr{B}}$ is a *crystalline medium*, or simply a *crystal* if for every point $X \in \mathscr{B}$ there is a neighborhood n_X of X and a faithful configuration $\varkappa$ of $\mathscr{B}$ such that

$$\mathfrak{G}_X^{\varkappa'} = \mathfrak{G}_Y^{\varkappa} \qquad (3.14)$$

whenever $\varkappa' = l \circ \varkappa$, X, $Y \in n_X$, $x_{\varkappa'} = y_\varkappa$, and l is any element of a space group $\mathscr{G}_f$. The constitutive relations $\mathfrak{G}_X^\varkappa$ may be viewed as a set of functions $\mathfrak{G}^\varkappa$ with domain $\mathscr{L}(\mathscr{V}, \mathscr{V}) \times \mathscr{B}_\varkappa$ defined by

$$\mathfrak{G}^\varkappa(S, x_\varkappa) = \mathfrak{G}_X^\varkappa(S). \qquad (3.15)$$

Expressed in terms of the $\mathfrak{G}^\varkappa$, the condition (3.14) reads

$$\mathfrak{G}^\varkappa(S \circ \nabla l^{-1}, l(x_\varkappa)) = \mathfrak{G}^\varkappa(S, x_\varkappa) \qquad (3.16)$$

for all $l \in \mathscr{G}_f$ and $X = \varkappa^{-1}(x_\varkappa)$, $Y = \varkappa^{-1}(y_\varkappa) \in n_X$, $y_\varkappa = l(x_\varkappa)$. It follows from the definitions of § 2, that if X and Y are two material points in n_X with positions $x_\varkappa$ and $y_\varkappa$ in the configuration $\varkappa$ such that $l(x_\varkappa) = y_\varkappa$, $l \in \mathscr{G}_f$, then X and Y are equivalent material points. As a further condition defining a crystal, we shall assume that the set of all transformations l satisfying (3.16) is contained in $\mathscr{G}_f$, and we call $\varkappa$ an *undistorted* configuration of n_X.

If $l \in \mathscr{G}_f$ and l leaves the point $x_\varkappa$ invariant $(l(x_\varkappa) = x_\varkappa)$, then (3.16) requires that

$$\mathfrak{G}^\varkappa(S\,\nabla l^{-1}, x_\varkappa) = \mathfrak{G}^\varkappa(S, x_\varkappa) \qquad (3.17)$$

or, equivalently, that

$$\mathfrak{G}_X^\varkappa(S \circ \nabla l^{-1}) = \mathfrak{G}_X^\varkappa(S). \qquad (3.18)$$

This asserts that ∇l^{-1} is an element $P_X^{\varkappa}$ of the representation $\mathscr{G}_X^{\varkappa}$ of the symmetry group $\mathscr{G}_X$ of X. Thus if $l = T_v \cdot O_{x\varkappa}$, then $O^{-1} = \nabla l^{-1} \in \mathscr{G}_X^{\varkappa}$.

If $\mathscr{G}_f$ is a space group with elements l, the *point group* $\mathscr{P}_f$ is the factor group $\mathscr{G}_f/\mathscr{G}_f^T$, where $\mathscr{G}_f^T$ are the pure translations. The point group is isomorphic to the group of orthogonal transformations $P_f = \{\nabla l;\, l \in \mathscr{G}_f\}$. In the classical theory of elastic crystals, one consides simple media $\mathscr{B}$ which are uniform and homogeneous for which the common symmetry group $\mathscr{G}_X^{\varkappa}$ of each point in an undistorted configuration $\varkappa$ is a point group representation $P_f^{\varkappa}$. Every such group is a subgroup of the group of all orthogonal transformations which leave some set of lattice vectors invariant; i.e., if $O \in P_f$, then there exists a set of lattice vectors Λ such that $O(\Lambda) = \Lambda$. In general, the crystalline media considered here are neither uniform nor homogeneous. The effects of non-uniformity are unimportant in many applications; however, in others they are of essence.

4. Dislocated Crystalline Media

The definition of a crystalline medium in § 3 is based on local properties of $\overline{\mathscr{B}}$. The body and its image $\overline{\mathscr{B}}_\varkappa$ under a faithful configuration $\varkappa$ is connected, but it need not be simply connected. There may not exist a single neighborhood n_X which covers $\mathscr{B}$ and a configuration $\varkappa$ for which the condition (3.13) holds. Let $x_\varkappa$ be a point of $\mathscr{B}_\varkappa$ and suppose that $\varkappa$ is an undistorted configuration of $n_{x_\varkappa} = \varkappa(n_X)$. Consider the lattice $\Lambda_{x_\varkappa}$ and the symmetric unit cells $\bar{\tau}_{y_n}$, $y_n \in \Lambda_{x_\varkappa}$. Call two closed polyhedral n-cells *non-overlapping* if their intersection is of lower dimension. Then $\bar{\tau}_{y_n}$ and $\bar{\tau}_{y_{n'}}$, are non-overlapping if $y_n \neq y_{n'}$, and $\mathscr{E} = \bigcup\limits_{y_n \in \Lambda_{x_\varkappa}} \bar{\tau}_{y_n}$ is a subdivision of space into symmetric unit cells of a lattice. It follows that $n_{x_\varkappa}$ is a union of non-overlapping symmetric unit cells or pieces of cells, and that $n_X = \varkappa^{-1}(n_{x_\varkappa})$ is a union of non-overlapping "*curvilinear*" symmetric unit cells, or pieces of curvilinear cells, $\tilde{\tau}_X = \varkappa^{-1}(\bar{\tau}_{x_\varkappa})$. The whole of $\mathscr{B}$ may be subdivided in this way and for simplicity and definiteness, let us suppose that $\overline{\mathscr{B}}$ is a union of whole curvilinear symmetric unit cells. Each point $X \in \overline{\mathscr{B}}$ is contained in a subset $\bar{n}_X$ which can be mapped faithfully by some $\varkappa$ onto a subset of the cells $\bar{\tau}_{y_n}$, $y_n \in \Lambda_x$ in $\mathscr{E}$. If $\overline{\mathscr{B}}_\varkappa$ is simply connected and for n_X we can choose $\mathscr{B}$, let us say that $\overline{\mathscr{B}}$ is a *perfect crystal*. If for n_X we can choose $\mathscr{B}$, but $\mathscr{B}$ is *not* simply connected, let us say that $\overline{\mathscr{B}}$ is a perfect crystal with *vacancies*. If $\overline{\mathscr{B}}$ is a perfect crystal with vacancies, there exists a faithful configuration $\overline{\mathscr{B}}_\varkappa$ of $\overline{\mathscr{B}}$ and a simply connected region $\overline{\mathscr{B}}_\varkappa^*$ in $\mathscr{E}$ such that

$\overline{\mathscr{B}}_\varkappa^* - \overline{\mathscr{B}}_\varkappa$ is a union of non-overlapping symmetric unit cells τ_{y_n}, $y_n \in \Lambda_{\dot{x}_\varkappa}$ for some $x_\varkappa \in \overline{\mathscr{B}}_\varkappa$.

A crystalline medium $\overline{\mathscr{B}}$ can fail to be a perfect crystal or a perfect crystal with vacancies if $\overline{\mathscr{B}}$ is not simply connected, in which case we say that it is *dislocated*, or *contains dislocations*. We give an example of a dislocated crystalline medium. For this purpose it is sufficiently general to consider a crystal $\overline{\mathscr{B}}$ of the cubic system for which there exists a

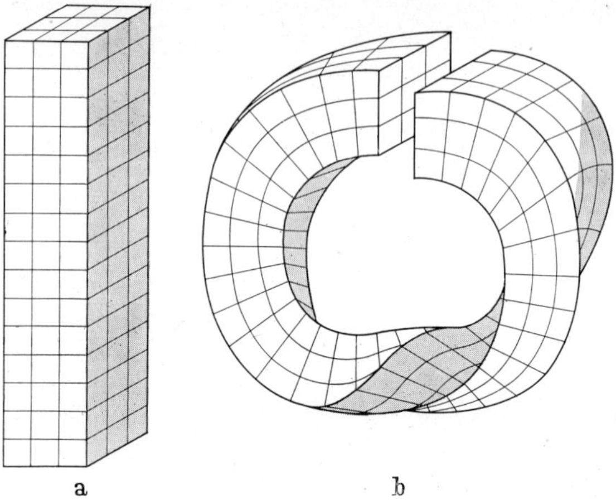

Fig. 1. a) Reference crystal; b) Dislocated crystal with 180° twist.

basis $\{\underset{a}{\boldsymbol{D}}\}$ of mutually orthogonal vectors of equal length. We shall suppose further that there exists a point p which has the full symmetry of the cubic lattice; i.e., for some p, $f(O_p\, x) = f(x)$ for *all* O for which $O(\boldsymbol{\Lambda}) = \boldsymbol{\Lambda}$. Consider a perfect crystal $\overline{\mathscr{C}}$ with this symmetry group $\mathscr{G}_f$ and suppose that $\overline{\mathscr{C}}_\varkappa = \varkappa(\overline{\mathscr{C}})$ is a faithful configuration of $\overline{\mathscr{C}}$. Let $\overline{\mathscr{C}}_\varkappa$ be a prism of square cross section (see Fig. 1). Then its length and two sides are integer multiples of a common length, the side of a symmetric unit cell $\tau_{x_\varkappa}$, which in this case is a perfect cube. Let $\overline{\mathscr{E}}_\varkappa^1$ and $\overline{\mathscr{E}}_\varkappa^2$ denote the ends of the rectangular region $\overline{\mathscr{C}}_\varkappa$. Now let $\lambda\colon \overline{\mathscr{C}}_\varkappa \to \mathscr{E}$ be a map of $\overline{\mathscr{C}}_\varkappa$ which is smooth in $\mathscr{C}_\varkappa$ and in every $\partial\mathscr{B}_\varkappa - \partial_0\mathscr{B}_\varkappa$, but fails to be faithful only because a subset $Q_\varkappa^1$ of $\overline{\mathscr{E}}_\varkappa^1$ and a subset $Q_\varkappa^2$ of $\overline{\mathscr{E}}_\varkappa^2$ are mapped by λ onto a common set of points in $\mathscr{E}$; i.e., $\lambda(Q_\varkappa^1) = \lambda(Q_\varkappa^2)$. Let λ_i denote the restriction of λ to $Q_\varkappa^i$. Then $\lambda_2^{-1} \circ \lambda_1, \colon Q_\varkappa^1 \to Q_\varkappa^2$ and if $\lambda_2^{-1} \circ \lambda_1$, is the restriction of an $l \in \mathscr{G}_f$ to $Q_\varkappa^1$, then equivalent points of $Q_\varkappa^1$ are mapped onto equivalent points of $Q_\varkappa^2$. But if this be true, then the unfaithful configuration $\overline{\mathscr{C}}_{\lambda\circ\varkappa}$ of the perfect crystal $\overline{\mathscr{C}}$ may be viewed as a faithful configuration, or identified with a crystal $\overline{\mathscr{B}}$ which contains a dislocation.

No single faithful configuration $\varkappa'$ of $\overline{\mathscr{C}}_{\lambda\circ\varkappa} = \overline{\mathscr{B}}$ satisfies the conditions of an undistorted configuration of a perfect crystal, or of a perfect crystal with vacancies. Yet, $\overline{\mathscr{C}}_{\lambda\circ\varkappa}$ satisfies all the conditions of a crystalline medium.

5. The Director Fields of a Crystal

Every material covector field $\boldsymbol{A}_X$ in $\mathscr{B}$ determines a corresponding spatial covector field $\boldsymbol{a}_{x_\varkappa}$ in $\mathscr{B}_\varkappa$ such that

$$\int_{\mathscr{C}} \boldsymbol{A}_X \cdot d\boldsymbol{X} = \int_{\mathscr{C}_\varkappa} \boldsymbol{a}_{x_\varkappa} \cdot d\boldsymbol{x}_\varkappa, \tag{5.1}$$

where $\mathscr{C}$ is any smooth curve in $\mathscr{B}$ and $\mathscr{C}_\varkappa = \varkappa(\mathscr{C})$ is its image under $\varkappa$, a smooth regular transformation of $\mathscr{B}$. The field $\boldsymbol{A}_X$ determines the exterior differential 1-form $A = \boldsymbol{A}_X \cdot d\boldsymbol{X}$, and $\boldsymbol{a}_{x_\varkappa}$ the 1-form $a = \boldsymbol{a}_{x_\varkappa} \cdot d\boldsymbol{x}_\varkappa$. In much the same way, material vector fields $\boldsymbol{V}_X$ and spatial vector fields $\boldsymbol{v}_{x_\varkappa}$ are set in correspondence by $\varkappa$ according to the rule

$$\boldsymbol{v}_{x_\varkappa} = \nabla_X \varkappa \cdot \boldsymbol{V}_X. \tag{5.2}$$

If $\overline{\mathscr{B}}$ is a perfect crystal, or a perfect crystal with vacancies, then there exists a faithful configuration $\varkappa$ of $\overline{\mathscr{B}}$ mapping $\overline{\mathscr{B}}$ onto a connected set of non-overlapping symmetric unit cells $\{\overline{\tau}_{y_n}\}$ where each y_n is a point of a space lattice $\Lambda_{x^\circ_\varkappa}$. Now a symmetric unit cell has a certain number k of pairs of parallel opposite faces separated by a lattice vector $\boldsymbol{d}$ which is perpendicular to these faces. Therefore, with each symmetric unit cell in $\mathscr{B}_\varkappa$, there is associated a finite set $\boldsymbol{\Omega}_{y_n} = \{\pm \underset{1}{\boldsymbol{d}}, \pm \underset{2}{\boldsymbol{d}}, \ldots, \pm \underset{k}{\boldsymbol{d}}\}$ of lattice vectors and $k \geqq 3$. Moreover, one sees that any three linearly independent vectors in $\boldsymbol{\Omega}_{y_n}$ is a basis for the set $\boldsymbol{\Lambda}$ of all lattice vectors. In each cell of $\mathscr{B}_\varkappa$, we may consider any of the *cell vectors* $\boldsymbol{d} \in \boldsymbol{\Omega}_{y_n}$ as a constant vector field in that cell. That is, we define $\boldsymbol{d}_x = \boldsymbol{d}$ for $x \in \overline{\tau}_{y_n}$. Now any symmetry element l of the space group $\mathscr{G}_f$ with gradient O has the property $O(\boldsymbol{\Omega}_{y_n}) = \boldsymbol{\Omega}_{y_n}$. Moreover, if we choose any basis $\{\underset{a}{\boldsymbol{d}}\}$, $a = 1, 2, 3$, from the set $\boldsymbol{\Omega}_{y_n}$, we can construct the reciprocal set $\{\overset{a}{\boldsymbol{d}}\}$ of covectors $(\overset{a}{\boldsymbol{d}} \cdot \underset{b}{\boldsymbol{d}} = \delta^a_b)$ and $O(\{\overset{a}{\boldsymbol{d}}\}) = \{\overset{a}{\boldsymbol{d}}\}$. In a perfect crystal with or without vacancies, there exists at least one way of choosing a basis $\underset{a}{\boldsymbol{d}}{}_{y_n} \in \boldsymbol{\Omega}_{y_n}$ in each cell of $\mathscr{B}_\varkappa$ such that the fields defined in $\mathscr{B}$ by $\underset{a}{\boldsymbol{d}}{}_{x_\varkappa} = \underset{a}{\boldsymbol{d}}{}_{y_n}$, $x \in \overline{\boldsymbol{\tau}}_{y_n}$, $y_n \in \Lambda_{x^\circ_\varkappa}$, are continuous, and the corresponding covector fields $\overset{a}{\boldsymbol{d}}{}_{x_\varkappa}$ reciprocal to the $\underset{a}{\boldsymbol{d}}_{x_\varkappa}$ will be continuous, and the 1-forms $\overset{a}{d}_\varkappa = \overset{a}{\boldsymbol{d}}_{x_\varkappa} \cdot \boldsymbol{dx}_\varkappa$ will be continuous throughout $\mathscr{B}_\varkappa$. Thus the material covector fields, the material vector fields, and the material 1-forms, $\overset{a}{\boldsymbol{D}}_X$, $\underset{a}{\boldsymbol{D}}_X$, and $\overset{a}{D}$ set in correspondence with $\overset{a}{\boldsymbol{d}}_{x_\varkappa}$, $\underset{a}{\boldsymbol{d}}_{x_\varkappa}$, and $\overset{a}{d}_\varkappa$ by $\varkappa$ will be smooth throughout $\mathscr{B}$. In an imperfect crystal containing dislocations, we *cannot* make this same construction of smooth fields in $\mathscr{B}$. The same construction can be carried out locally in each n_X and we may cover $\mathscr{B}$ by a finite set of such neighborhoods so as to obtain fields $\overset{a}{\boldsymbol{D}}_X$, $\underset{a}{\boldsymbol{D}}_X$, and forms $\overset{a}{D} = \overset{a}{\boldsymbol{D}} \cdot \boldsymbol{dX}$ defined throughout $\mathscr{B}$; but, in general, there is no way to choose the basis $\underset{a}{\boldsymbol{d}}_{y_{\boldsymbol{n}}}$ in each cell of a $\varkappa(n_X) = n_{x_\varkappa}$ such that the resulting fields $\underset{a}{\boldsymbol{D}}_X$ are all continuous throughout $\mathscr{B}$. It can be required that the $\underset{a}{\boldsymbol{d}}_{y_{\boldsymbol{n}}}$ in adjacent cells be equivalent in the sense that $\{\underset{a}{\boldsymbol{d}}_{y_{\boldsymbol{n}}}\} = O\{\pm \underset{a}{\boldsymbol{d}}_{y_{\boldsymbol{n}}}\}$ for some $O = \nabla l$, $l \in \mathscr{G}_f$.

To illustrate these ideas consider again the example of the cubic crystal of § 4. If the mapping $\lambda_2^{-1} \circ \lambda_1 : Q_\varkappa^1 \to Q_\varkappa^2$ that we considered in § 3 is a pure translation composed with a rotation of 180° about a symmetry axis in the plane of $Q_\varkappa^1$, then any constant field of cell vectors $\underset{a}{\boldsymbol{d}}_{x_\varkappa}$ in the perfect rectangular crystal is mapped into a continuous field $\underset{a}{\boldsymbol{D}}_{X_{(\lambda\circ\varkappa)^{-1}}} = \underset{a}{\boldsymbol{D}}_X$ in the dislocated crystal. In Addition, $\lambda_2^{-1} \circ \lambda_1$, may contain as a factor any number of complete revolutions about an axis perpendicular to $Q_\varkappa^1$. But if $\lambda_2^{-1} \circ \lambda_1$, contains a factor $l \in \mathscr{G}_f$ which is a rotation about an axis perpendicular to $Q_\varkappa^1$ which is not a complete revolution, then at least one of the fields $\underset{a}{\boldsymbol{D}}_X$ is not continuous in $\mathscr{B}_{\lambda\circ\varkappa}$, but suffers a jump discontinuity across the surface $\lambda(Q_\varkappa^1) = \lambda(Q_\varkappa^2)$ in the dislocated crystal.

Consider, in this same example, a closed curve $C_{\lambda\circ\varkappa}$ in the dislocated crystal which is the image $\lambda(\mathscr{C}_\varkappa)$ of a straight line $\mathscr{C}_\varkappa$ in $\mathscr{B}_\varkappa$ joining points $p_1 \in Q_\varkappa^1$ and $p_2 \in Q_\varkappa^2$ brought into conjunction by λ. Then we always have

$$\oint_{\mathscr{C}_{\lambda\circ\varkappa}} \overset{a}{\boldsymbol{D}}_X \cdot \boldsymbol{dX} = \int_{\mathscr{C}_\varkappa} \overset{a}{\boldsymbol{d}}_{x_\varkappa} \cdot \boldsymbol{dx}_\varkappa. \tag{5.2}$$

In the right-hand integral, the $\overset{a}{\boldsymbol{d}}_{x_\varkappa}$ are constant vector fields and $\mathscr{C}_\varkappa = \int_{\mathscr{C}_\varkappa} \boldsymbol{dx}_\varkappa$ is the vector joining the end points of the straight line $\mathscr{C}_\varkappa$. Let p be a reference point in $Q_\varkappa^1$, and set $p_1 - p = \boldsymbol{w}$. Then $\mathscr{C}_\varkappa$ has the form $\mathscr{C}_\varkappa = \boldsymbol{v_n} + \boldsymbol{w} - O\,\boldsymbol{w}$, and the integral on the right of (5.2) is seen to have the value

$$\overset{a}{\nu}(\mathscr{C}_{\lambda\circ\varkappa}) = \overset{a}{\boldsymbol{d}} \cdot (\boldsymbol{v}_n + \boldsymbol{w} - O \cdot \boldsymbol{w}), \tag{5.3}$$

where $\boldsymbol{v}_n$ is a lattice vector. Thus $\overset{a}{\boldsymbol{d}} \cdot \boldsymbol{v}_n = \overset{a}{n}$ is an integer equal to the number of cells along the length of the rectangle $\mathscr{B}_\varkappa$, and is independent of $\boldsymbol{w} = p_1 - p_0$. The *Burgers vector* for the circuit $\mathscr{C}_{\lambda\circ\varkappa}$ in the dislocated crystal is defined by

$$\boldsymbol{\nu}(\mathscr{C}_{\lambda\circ\varkappa}) = \overset{a}{\boldsymbol{\nu}}(\mathscr{C}_{\lambda\circ\varkappa})\, \underset{a}{\boldsymbol{d}} \tag{5.4}$$

which is independent of $\boldsymbol{w}$ and the same for every circuit $\mathscr{C}_{\lambda\circ\varkappa}$ *if and only if* $O = 1$ in (5.3). When $O = 1$, the forms $\overset{a}{D}$ in the dislocated crystal are smooth and continuous throughout $\mathscr{B}_{\lambda\circ\varkappa}$ the dislocated crystal; moreover, they are irrotational, *rot* $\overset{a}{D}$, the exterior derivative of $\overset{a}{D}$ is zero. The value of $\oint_{\mathscr{C}} \overset{a}{D}$ over all homologous cycles in the dislocated crystal is the same, and the Burgers vector

$$\boldsymbol{\nu}(\mathscr{D}) = \left(\oint_{\mathscr{C}} \overset{a}{D}\right) \underset{a}{\boldsymbol{d}} \tag{5.5}$$

is a function on the homology classes of 1-cycles in the dislocated crystal. But this property of the Burgers vector does not hold generally in a dislocated crystal as defined here and represents a restriction on the kind of dislocations to be considered. If the Burgers vector does not have the same value for every homologous cycle, then the dislocation is known to have *twist* (i.e., $O \neq 1$). On the other hand, one sees from the example that knowledge only of the Burgers vector for every cycle in a dislocated crystal does not permit one to distinguish between the cases where $\lambda\colon \mathscr{C}_\varkappa \to \mathscr{C}_{\lambda\circ\varkappa}$ entails different numbers of whole relative rotations of the ends $Q^1_\varkappa$ and $Q^2_\varkappa$.

The construction of the vector and covector fields $\underset{a}{\boldsymbol{D}}_X$ and $\overset{a}{\boldsymbol{D}}_X$ in a perfect or dislocated crystal provides a corresponding law of material parallelism defined by setting

$$\gamma_{X\,Y} = \underset{a}{\boldsymbol{D}}_X \otimes \overset{a}{\boldsymbol{D}}_Y. \tag{5.6}$$

We have seen that in perfect crystals with or without vacancies, this law of parallelism is irrotational and torsionless. In a dislocated crystal it is, in general, not continuous, but if dislocations without twist only are considered, it is irrotational but not torsionsless. In all cases, we see how one may view a crystalline medium in a fairly definite way as an oriented medium with directors $\underset{a}{\boldsymbol{D}}_X$ and reciprocal directors $\overset{a}{\boldsymbol{D}}_X$ interpreted as "*crystal axes*". The theory of exterior differential forms can then be used to assist in the analysis and description of the dislocated crystalline medium, and a connection can be made with the geometrical idea of distant parallelism. Defining a crystalline medium as we have done here allows also the possibility of a definite subdivision of a crystal into curvilinear ("deformed") symmetrical unit cells and offers the possibility of describing the global topological structure of the crystal using the methods of algebraic topology. Certain aspects of the dynamics of a dislocated crystal can also be inferred from established results in the theory of oriented media.

In summary then, I believe that even the incomplete and cursory results treated here are sufficient indication of how the methods of continuum mechanics and differential geometry can be applied to the problem of crystal structures and crystal dynamics. It is also conceivable that the model of a crystal considered here might be superior to one in which the body $\bar{\mathscr{B}}$ is a finite or countable set of mass; charge, and spin bearing points or small rigid spheres.

References

[1] Cf. Love, A. E. H.: The Mathematical Theory of Elasticity, Fourth Edition, Appendix to Chaps. VIII and IX. New York: Dover Publications 1927.

[2] Cosserat, E., and F. Cosserat: Théorie des Corps Déformable. Paris: Hermann 1909.

[3] Günther, W.: Zur Statik und Kinematik des Cosseratschen Kontinuums. Abh. d. Braunschweigischen Wiss. Ges. **10**, 195 (1958).

[4] Fox, N.: A Continuum Theory of Dislocations for Single Crystals. J. Inst. Maths. Appl. **2**, 285 (1966).

[5] Noll, W.: Materially Uniform Simple Bodies with Inhomogenities, Report 67-26, Dept. of Math., Carnegie Inst. of Tech. 1967.

[6] Whitney, H.: Geometric Integration Theory. Princeton University Press 1957.

The Generalized Dual Continuum in Elasticity and Dislocation Theory

By

M. Mişicu

Center of Mechanics of Solids
Academy of Roumanian Socialist Republic, Bucharest

The dual continuum can be defined with the aid of the principles of mechanics formulated in terms including the internal energy and coenergy as functions depending respectively on the product of internal forces and displacement variations or on the product of displacements and forces variations. In a more general sense, the internal energy ε can be developed by expanding the forces and velocities into series with terms rapidly increasing in the neighborhood of the contact areas of the structural elements, or with terms slowly varying at a microscale. Hence, after suitable integrations over the microvolumes ε splits into parts including stress or couple-stress effects or including terms of dual nature called, according to the adopted terminology, locations and couples (poles) of locations. The locations, according to the definitions, stand for the relative displacement between elements belonging to different material phases. Thus, in order to define these functions, a heterogeneous body must be considered. However, in the limit case of a homogeneous body, the theory shows the relevant fact that the location effect reduces to a non-vanishing effect, which we call self-coupling effect.

The aim of the present analysis is to elaborate a consequent dual model of non elastic processes including the dislocation theory according to the non-dual theory of K. Kondo [1], E. Kröner [2—6] and others [7]. Meanwhile we include both dual incompatibility effects of the strain and of the symmetrized gradients of the interaction forces between the structural elements. The respective effects shall be called dislocations and distensions and, in order to distinguish the corresponding synthetic treatment, a new term is required, for instance clasticity (since phase separation is involved). We note that a more complete asymmetric theory involves the consideration of dual compatible parts of the stress and location fields based on extensions of the representations given by W. Günther [8] and H. Schaefer [9]. However, we restrict ourselves to the above mentioned incompatibility fields, of a more pertinent significance, the additional representations being also useful, but in an usual sense [10, 11]. Dual theories of higher order of multi-

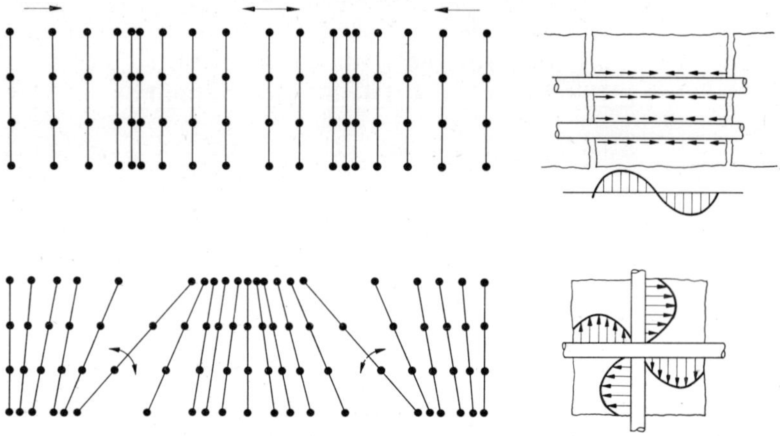

Fig. 1.

polar type can be assessed on the same basis, but resorting to generalized geometrical spaces [12]. Other extensions appear as significant in the case of asymmetric plastic models [13, 14].

The dual continuum constitutes an unitary modelation of asymmetric, heterogeneous bodies with microstructural effects [15—29] and more especially of the coupled continuum [30]. The qualitative new features appear as useful in the analysis of equivalent reinforced materials (with adherent and prestressed armatures) or of technical lattices in the sense of previously formulated equivalences [31]. Meanwhile, the dual continuum appears more suitable in the description of the phase-separation or limit interactions between structural parts of the metallic alloys.

1. The Dual Fields

The dual fields are defined below by using a particular structure. However the results lead to a general description.

The actual point configuration of a single material component of a heterogeneous body includes the macrostructure (x^i). The macrostructure includes the points of the microstructure $(y^i = x^i + \eta^i)$, identical with the barycenters of the understructural elements including the points $(z^i = x^i + \eta^i + \xi^i)$. Let then be $v^i(x, \eta, \xi)$ the velocity of an understructural particle. If in the Taylor development

$$v^i(x, \eta, \xi) = v^i(x, \eta) + \xi^j v^i_j(x, \eta) \tag{1}$$

it is possible to retain the first terms including velocities and their structural gradients v^i_j, then the volume integral over an understructural element

$$I_{x,\eta} = \int_{v_{x,\eta}} (f^i\, v_i)_{x,\eta,\xi}\, dv_\xi \tag{2}$$

can be equivalently written, under suitable assumptions, as follows

$$I = \int_{v_x} (f^i\, v_i + f^{ij}\, v_{ij})_{x,\eta}\, dv_\eta. \tag{3}$$

Here

$$f^i(x,\eta)\, dv_\eta = \int_{v_{x,\eta}} f^i(x,\eta,\xi)\, dv_\xi, \qquad f^{ij}(x,\eta)\, dv_\eta = \int_{v_{x,\eta}} (f^i\, \xi^j)_{x,\eta,\xi}\, dv_\xi \tag{4}$$

stand for interaction forces and couples. If f^i in (2) is expressed as v^i in (3), we shall consider structural gradients instead polar terms. Further, using the small parameter ε, subsequently included in the respective multiplied factors, we assume the existence of the developments

$$v^i(x,\eta) = v^i(x) + \varepsilon\, \bar{v}^i(x,\eta), \qquad f^i(x,\eta) = f^i(x) + \varepsilon\, \bar{f}^i(x,\eta) \tag{5}$$

where the second order terms are rapidly decreasing functions of η. (5) shall be called the assumption of the areolar localization of the dual interactions.

The above defined functions intervene in the balance equation [19, 27, 31] defined at a macroscale

$$\left((D/Dt)\int_V \varrho\, Q\, dv - \int_V \varrho\, S\, dv\right)_{/x} = I \tag{6}$$

where

$$Q = \varepsilon + (v^i\, v_i + v^{ij}\, v_{ij})/2,$$
$$\varrho\, S = \varrho(\Phi^i\, v_i + \Phi^{ij}\, v_{ij}) + f^i(\bar{v} - v)_i + f^{ij}(\bar{v} - v)_{ij}. \tag{7}$$

Q represents the internal and kinetic energy referred to the unit mass, S stands for the energy source and I for the interaction force. ϱ represents the density function assumed, for the simplicity sake, identical in macro- and microstructure. (7), after using (5) can be reformulated with the aid of the location velocities and stress functions

$$r^i = (1/ds)\int_{dv} \bar{v}^i(x,\eta)\, dv_\eta = \varrho^{ij}\, n_j, \qquad t^i = (1/ds)\int_{dv} \bar{f}^i(x,\eta)\, dv_\eta = \sigma^{ij}\, n_j. \tag{8}$$

dv are microvolumes located in the neighborhood of the areas $ds = ds^i n_i$ of the structural elements, regions where, according to the assumptions, the integrands are non-vanishing functions. Analogously, shall be considered the velocities of the location couples $r^{ij} = \varrho^{ijk}\, n_k/2$ or the couple-stresses $t^{ij} = \sigma^{ijk}\, n_k/2$. (6) can be reformulated under the form

$$\varrho\, \dot{\varepsilon} = P_v + P_f, \quad P_v = \pi^i\, v_i + \sigma^{ij}\, v_{i;j} + \pi^{ij}\, v_{ij} + \sigma^{ijk}\, v_{ij;k}/2, \tag{9}$$
$$P_f = \nu^i\, f_i + \varrho^{ij}\, f_{i;j} + \nu^{ij}\, f_{ij} + \varrho^{ijk}\, f_{ij;k}/2.$$

The result is obtained on the basis of thermodynamical principles and divergence theorems as in the non-dual case [31]. The covariant derivatives are relative to the connexion defined in (17). We have used the notations

$$\pi^i = \sigma^{ij}_{;j} + \varrho(\Phi^i - \dot v^i) + f^i, \quad \pi^{ij} = \sigma^{ijk}_{;k}/2 + \varrho(\Phi^{ij} - \dot v^{ij}) + f^{ij},$$
$$\nu^i = \varrho^{ij}_{;j} + \tilde v^i - v^i, \quad \nu^{ij} = \varrho^{ijk}_{;k}/2 + \tilde v^{ij} - v^{ij}, \tag{10}$$

$\tilde v^i$ and $\tilde v^{ij}$ can be considered as the velocities of the link points between the considered material component and the rest of the body, since the respective energetic factors in (3) constitute interaction forces or couples.

The following restrictions shall be made: a) the constraints $\tilde v^i = v^i$, $\tilde v^{ij} = v^{ij}$, b) the number of the material components is 2, c) the interaction is characterized by the action-reaction conditions

$$f^i_{(1)} = -f^i_{(2)} = f^i, \quad f^{ij}_{(1)} = -f^{ij}_{(2)} = f^{ij}. \tag{11}$$

Indices in brackets reffer to the material components. It is then suitable to reformulate P_f using coenergetic terms, as for instance replacing $v \cdot f$ by $-p \cdot v + (f \cdot u)^{\cdot}$, where $v = \dot u$, $p = \dot f$, so that, after some operations, we obtain

$$\varrho(\varepsilon - \varepsilon')^{\cdot} = j M = P_v + co P_v. \tag{12}$$

g_{ij} is defined in (16). We have used the notations

$$j = \varrho/\varrho_0 = |g_{ij}|, \quad co P_v = \chi^i p_i + \lambda^{ij} p_{i;j} + \chi^{ij} p_{ij} + \lambda^{ijk} p_{ij;k}/2. \tag{13}$$

$p = p_{(1)} = p_{(2)}$ stands for the absolute velocities of the interaction forces or couples and the respective structural gradients. We also denote

$$\pi^i = \pi^i_{(1)} + \pi^i_{(2)} = \sigma^{ij}_{;j} + \varrho(\Phi^i - \dot v^i), \quad \sigma^{ij}_{(1)} + \sigma^{ij}_{(2)} = \sigma^{ij},$$
$$\pi^{ij} = \pi^{ij}_{(1)} + \pi^{ij}_{(2)} = \sigma^{ijk}_{;k} + \varrho(\Phi^{ij} - \dot v^{ij}), \quad \sigma^{ijk}_{(1)} + \sigma^{ijk}_{(2)} = \sigma^{ijk}, \tag{14}$$
$$\chi^i = \chi^i_{(1)} - \chi^i_{(2)} = \lambda^{ij}_{;j}, \quad \chi^{ij} = \chi^{ij}_{(1)} - \chi^{ij}_{(2)} = \lambda^{ijk}_{;k}/2, \quad \dot\lambda_{(i)} = \varrho_{(i)},$$
$$(\lambda = \lambda_{(1)} + \lambda_{(2)}).$$

The self-coupled material defined above behaves as a simple body, including additional effects. Variational formulations can be established on the basis of the relations (9) and (13). The boundary conditions can be deduced as in [31], i.e. concerning any kinematic or dynamic unknown function r^i, t^i, f_i, v_i, etc.

2. Internal Dual States

Let be i, I, α indices concerning the actual, material and diacoptic states. The diacoptic state is defined as resulting by a double process of uncoupling the material components and cutting of the infinitesimal volume elements (we use Kondo's terminology of the non-dual considerations). The differential relations in macro- and microstructure

$$dx^i = A^i_\alpha\, dx^\alpha, \quad df^i(x) = F^i_\alpha\, dx^\alpha; \quad d\xi^i = a^i_\alpha\, d\xi^\alpha, \quad df^i(x, \xi) = f^i_\alpha\, d\xi^\alpha \tag{15}$$

shall characterize the kinematic and dynamic distortions at a macro- (A, F) or microscale (a, f). We assume that x^α, ξ^α are cartesian and that dx^i, ξ^i belong to the same local coordinate system, so that $a^i_\alpha - A^i_\alpha$ shall represent the relative distortion between macro- and microstructure. We admit the existence of inverse elements ($A^i_\alpha A^\alpha_j = \delta^i_j$, etc.). Let then be

$$d\sigma^2 = \delta_{\alpha\beta}\, d\xi^\alpha\, d\xi^\beta, \quad d\Sigma^2 = \delta_{\alpha\beta}\, dx^\alpha\, dx^\beta, \quad ds^2 = g_{ij}\, d\xi^i\, d\xi^j,$$
$$dS^2 = g_{ij}\, dx^i\, dx^j \tag{16}$$

the associated metric forms. According to (19) we consider the Euclidean connexion characterized by

$$g^i_{jk} = g^{il}\, g_{jkl}, \quad g_{jkl} = (g_{lj,k} + g_{kl,j} - g_{jk,l})/2 \tag{17}$$

with vanishing curvature and torsion tensors

$$r^i_{jkl} = 2\,(g^i_{[kl,j]} + g_{[jr}\, g^r_{k]l}) = 0, \quad t^i_{jk} = 2 g^i_{[jk]} = 0, \tag{18}$$

as a consequence of the continuous motion

$$x^i = x^i(x^I, t). \tag{19}$$

3. Constitutive Equations

(13) can be transformed into a total exact differential. For this purpose we observe that the total derivative in microstructure is

$$\mathfrak{d}/\mathfrak{d}t = \partial/\partial t + v^i\, \partial/\partial x^j + \dot\xi^j\, \partial/\partial \xi^j, \ (v^i = \partial x^i/\partial t, \ \dot\xi^i = \partial \xi^i/\partial t + \dot x^j\, \partial \xi^i/\partial x^j) \tag{20}$$

so that the derivation rule

$$\mathfrak{d}\boldsymbol{F}/\mathfrak{d}t = (\mathfrak{D} F^i/\mathfrak{D}t) \cdot \boldsymbol{g}_i, \quad (\boldsymbol{F} = (F^i\, \boldsymbol{g}_i)_{x,\xi}, \quad \gamma^i_{kj} = \boldsymbol{g}^i\, \partial \boldsymbol{g}_j/\partial \xi^k) \tag{21}$$

furnishes the expression of the material derivative

$$\mathfrak{D} F^i/\mathfrak{D}t = \mathfrak{d} F^i/\mathfrak{d}t + (g^i_{kj}\, v^k + \gamma^i_{kj}\, \dot\xi^k)\, F^j. \tag{22}$$

We choose $\boldsymbol{g}_i = \boldsymbol{g}_i(x)$ so that $\gamma \ldots = 0$. Hence can be proofed that, if $dh^i = h^i_\alpha(x)\, d\xi^\alpha$ for $x^i = \text{const.}$, and if $dH^i = H^i_\alpha\, dx^\alpha$, then

$$k^i_\alpha = Dh^i_\alpha/Dt = k^i_j\, a^j_\alpha, \quad K^i_\alpha = DH^i_\alpha/Dt = H^i_j\, A^j_\alpha,$$
$$k^i_j = (dh^i_\alpha/dt + g^i_{kl}\, v^k\, h^l_\alpha)\, a^\alpha_j, \tag{23}$$

where $D/Dt = \mathfrak{D}/\mathfrak{D}t$, $d/dt = \mathfrak{d}/\mathfrak{d}t$, for $\dot\xi^i = 0$. For instance, if in (23) we take $h^i_\alpha = \xi^i_{;\alpha}$, $H^i_\alpha = x^i_{;\alpha}$ then $\dot v^i_\alpha = v^i_j\, a^j_\alpha$, $v^i_{;\alpha} = v^i_{;j}\, A^j_\alpha$. Other property is formulated below

$$k^i_{j;k} = a^\alpha_j\, A^\beta_k\, Dh^i_{\alpha:\beta}/Dt + a^\alpha_{j;k}\, Dh^i_\alpha/Dt + d^i_{jk} \tag{24}$$

where

$$(\ )_{\alpha:\beta} = (\ )_{\alpha;j}\, A^j_\alpha, \qquad d^i_{jk} = a^\alpha_j\,(r^i_{kmp}\, h^p_\alpha - t^p_{km}\, h^i_{\alpha:p})\, v^m. \tag{25}$$

(18) implies that $d \ldots = 0$.

(13), (23) and (24) together with the assumptions that M is a total exact differential and that the mass is conserved, lead to the constitutive equations

$$\left.\begin{aligned}
&\pi_i = j\partial M/\partial x^i, \quad \sigma^{jk}_{.i.} = 2a^j_\beta\, A^k_\gamma\, j\, \partial M/\partial a^i_{\beta:\gamma},\\
&\sigma^j_{i.} = j\, A^j_\alpha(\partial M/\partial A^i_\alpha - \partial M/\partial (a^i_\alpha - A^i_\alpha)),\\
&\pi^j_{.i} = j(a^j_\alpha\, \partial M/\partial (a^i_\alpha - A^i_\alpha) + a^j_{\beta:\gamma}\, \partial M/\partial a^i_{\beta:\gamma}), \qquad \chi_i = j\, \partial M/\partial f^i,\\
&\lambda^{jk}_{.i.} = 2a^j_\beta\, A^k_\gamma\, j\, \partial M/\partial f^i_{\beta:\gamma}, \quad \lambda^j_{i.} = j\, A^j_\alpha(\partial M/\partial F^i_\alpha - \partial M/\partial (f^i_\alpha - F^i_\alpha)),\\
&\chi^j_{.i} = j(a^j_\alpha\, \partial M/\partial (f^i_\alpha - F^i_\alpha) + a^j_{\beta:\gamma}\, \partial M/\partial f^i_{\beta:\gamma}).
\end{aligned}\right\} \tag{26}$$

Let also be the additional rigid motions

$$x'^i = x^i + c^i, \qquad \xi'^i = \xi^i + c^i, \qquad dx'^i = Q^i_{.j}(t)\, dx^j, \qquad d\xi'^i = Q^i_{.j}(t)\, d\xi_j,$$
$$(Q^i_{.k}\, Q^k_{j.} = \delta^i_j, \ g'_{ij}\, Q^i_{.k}\, Q^j_{.l} = g_{kl}), \tag{27}$$

so that the metric forms ds, dS in (16) remain invariant and so that the velocity gradients transform according to the relations

$$v'_{ij} = R_{ij} + g_{il}\, Q^l_{.k}\, v^k_{.m}\, Q^m_{j.}, \qquad R_{ij} = -R_{ji} = g_{il}\, Q^l_{.k}\, \dot{Q}^k_{j.}, \quad \text{etc.} \tag{28}$$

Then (12) remains invariant if

$$\pi^i = 0, \quad \sigma^{[ij]} - \pi^{[ij]} + \chi^{[i}\, f^{j]} + \lambda^{[ik}\, F^{j]}_\alpha\, A^\alpha_k + \chi^{k[i}\, f^{j]}_\alpha\, a^\alpha_k +$$
$$+ \lambda^{m[ik}(a^\beta_m\, f^{j]}_\beta);\ \gamma\, A^\gamma_k/2 = 0. \tag{29}$$

Other restrictions can also be formulated as for instance: the internal energy is independent of additional uniform interaction forces $f'^i - f^i = c^i$ or additional force gradients $f'_{i;j} - f_{i;j} = f'_{ij} - f_{ij} = c_{[ij]}$. Thus (12) leads to a set of conditions characterizing the rigid behavior of the material links acted by f_i and by the couples resulting by summation over a structural element of the macro- and microstructural forces interactions $f_{[ij]}$, $f_{[i;j]}$. These conditions are

$$\chi^i = 0, \qquad \lambda^{[ij]} - \chi^{[ij]} = 0. \tag{30), (31}$$

The additional restrictions can be justified by the fact that generally no important relative uniform translations or rotations produce in simple materials, except some failure processes, not analysed here.

Resuming ourselves, for the time being, to the functions invariant under the additional motions (27)

$$\left.\begin{aligned}
&(A_\alpha = g_{ij}\, A^i_\alpha\, x^j), \qquad A_{\alpha\beta} = g_{ij}\, A^i_\alpha\, A^j_\beta = 2e_{\alpha\beta} + \delta_{\alpha\beta},\\
&\alpha_{\alpha\beta} = g_{ij}\, A^i_\alpha(a^j_\beta - A^j_\beta), \qquad \alpha_{\beta\alpha\gamma} = g_{ij}\, A^i_\alpha\, a^j_{\beta:\gamma},\\
&B_\alpha = g_{ij}\, A^i_\alpha\, f^j, \qquad B_{\alpha\beta} = g^{ij}\, A^i_\alpha\, F^j_\beta = f_{ij}\, A^i_\alpha\, A^j_\beta,\\
&\beta_{\alpha\beta} = g_{ij}\, A^i_\alpha(f^j_\beta - F^j_\beta), \qquad \beta_{\beta\alpha\gamma} = g_{ij}\, A^i_\alpha\, f^j_{\beta:\gamma},
\end{aligned}\right\} \tag{32}$$

it follows the new form of the Eqs. (26)

$$\left.\begin{aligned}
&\pi^i = j\, A^i_\alpha\, \partial M/\partial A_\alpha = 0, \quad \sigma^{ij} = \sigma^j_{k.}\, g^{ik} = A^j_\beta\, j\,(A^i_\alpha\, \partial M/\partial e_{\alpha\beta} + \\
&\quad + a^i_{\alpha:\gamma}\, \partial M/\partial \alpha_{\alpha\beta\gamma} + (a^i_\alpha - A^i_\alpha)\, \partial M/\partial \alpha_{\beta\alpha} - A^i_\alpha\, \partial M/\partial \alpha_{\alpha\beta} + \\
&\quad + f^i\, \partial M/\partial B_\beta + F^i_\alpha\, \partial M/\partial B_{\beta\alpha} + f^i_{\alpha:\gamma}\, \partial M/\partial \beta_{\alpha\beta\gamma} + \\
&\quad + (f^i_\alpha - F^i_\alpha)\, \partial M/\partial \beta_{\beta\alpha}), \quad \sigma^{jik} = 2 a^j_\beta\, A^i_\alpha\, A^k_\gamma\, j\, \partial M/\partial \alpha_{\beta\alpha\gamma}, \\
&\pi^{ji} = A^i_\alpha\, j\,(a^j_\beta\, \partial M/\partial \alpha_{\alpha\beta} + a^j_{\beta:\gamma}\, \partial M/\partial \alpha_{\beta\alpha\gamma}), \quad \chi^i = j\, A^i_\alpha\, \partial M/\partial B_\alpha, \\
&\lambda^{ij} = A^i_\alpha\, A^j_\beta\, j\,(\partial M/\partial B_{\alpha\beta} - \partial M/\partial \beta_{\alpha\beta}), \\
&\lambda^{jik} = 2 a^j_\beta\, A^i_\alpha\, A^k_\gamma\, j\, \partial M/\partial \beta_{\beta\alpha\gamma}, \\
&\chi^{ji} = A^i_\alpha\, j\,(a^j_\beta\, \partial M/\partial \beta_{\alpha\beta} + a^j_{\beta:\gamma}\, \partial M/\partial \beta_{\beta\alpha\gamma}).
\end{aligned}\right\} \quad (33)$$

(29) is implicitly satisfied since (32) are invariant functions.

(32) can be referred to an actual frame of reference

$$\left.\begin{aligned}
&A_{ij} = \delta_{\alpha\beta}\, A^\alpha_i\, A^\beta_j = 2 e_{ij} + g_{ij}, \quad \alpha_{ij} = \delta_{\alpha\beta}\, A^\alpha_i\,(a^\beta_j - A^\beta_j), \\
&\alpha_{j\,ik} = \delta_{\alpha\beta}\, A^\alpha_i\, A^\beta_{j:k}, \quad B_{ij} = F_{ij} = g_{ik}\, F^k_\alpha\, A^\alpha_j, \\
&\beta_{ij} = f_{ij} - F_{ij}, \quad \beta_{j\,ik} = \delta_{\alpha\beta}\, A^\alpha_i\, f^\beta_{j;k}.
\end{aligned}\right\} \quad (34)$$

We denote

$$F^k_\alpha = A^k_\beta\, F^\beta_j\, A^j_\alpha, \quad f^k_\alpha = a^k_\beta\, f^\beta_j\, a^j_\alpha. \tag{35}$$

Consequently, the following relations are satisfied

$$\left.\begin{aligned}
&e_{\alpha\beta} = A^i_\alpha\, A^j_\beta\, e_{ij}, \quad \alpha_{\alpha\beta} = 2 A^i_\alpha\,(a^j_\beta - A^j_\beta)\, e_{ij} - A^i_\alpha\, a^j_\beta\, \alpha_{ij}, \\
&\alpha_{\beta\alpha\gamma} = 2 A^i_\alpha\, a^j_{\beta:\gamma}\, e_{ij} - \delta_{\alpha\delta}\, A^\delta_m\, a^m_\varepsilon\, \delta^{\varepsilon\varphi}\, a^j_\beta\, A^i_\varphi\, A^k_\gamma\, \alpha_{j\,ik}, \\
&B_{\alpha\beta} = A^i_\alpha\, A^j_\beta\, B_{ij}, \quad \beta_{\alpha\beta} = A^i_\alpha\, A^j_\beta\, \beta_{ij} + A^i_\alpha\, A^j_\delta\, \delta^{\delta\gamma}\,(B_{ij} + \beta_{ij})\, \alpha_{\gamma\beta}, \\
&\beta_{\beta\,\alpha\gamma} = 2 A^i_\alpha\, f^j_{\beta:\gamma}\, e_{ij} + \delta_{\alpha\mu}\, A^\mu_m\,(a^m_\delta\, \delta^{\delta\lambda}\, A^i_\lambda\, a^j_\beta\, A^k_\gamma\, \beta_{j\,ik} + \\
&\qquad + \delta^{\nu\delta}\, A^n_\nu\,(a^m_{\delta:\gamma}\, a^k_\beta + a^m_\delta\, a^k_{\beta:\gamma})\,(B_{nk} + \beta_{nk})).
\end{aligned}\right\} \quad (36)$$

Inverse relations can be easily deduced after inversion of the indices (i, α) and the tensors (g_{ij}, δ_{ij}). In the frame of a first order theory, (37) leads to the approximation $\beta_{\alpha\beta} = A^i_\alpha\, A^j_\alpha\, \beta_{ij}$ so that, A_α and B_β being omitted, (33) remain still valid under the assumptions (30) and (31.) Indeed, these assumptions require the independence of M on B_α and the replacements

$$B_{\alpha\beta} \longleftrightarrow \varphi_{\alpha\beta} = B_{(\alpha\beta)}, \quad \beta_{\alpha\gamma} \longleftrightarrow \beta'_{\alpha\beta} = A^i_\alpha\, A^j_\beta\, \beta_{ij}. \tag{37}$$

The linearized form of (36) results by the replacements

$$\left.\begin{aligned}
&e_{\alpha\beta} = \Delta^i_\alpha\, \Delta^j_\beta\, e_{ij}, \quad \alpha_{\beta\alpha\gamma} = -\Delta^i_\alpha\, \Delta^j_\beta\, \Delta^k_\gamma \alpha_{j\,ik}, \quad \alpha_{\alpha\beta} = -\Delta^i_\alpha\, \Delta^j_\beta \alpha_{ij}, \\
&\beta_{\alpha\beta} = \Delta^i_\alpha\, \Delta^j_\beta B_{ij}, \quad \beta_{\alpha\beta} = \Delta^i_\alpha\, \Delta^j_\beta\, \beta_{ij}, \quad \beta_{\beta\alpha\gamma} = \Delta^i_\alpha\, \Delta^j_\beta\, \Delta^k_\gamma \beta_{j\,ik},
\end{aligned}\right\} \quad (38)$$

where $\Delta^i_\alpha = A^i_\alpha = a^i_\alpha$, for $t = 0$. If M is expressed under the form of a second order polynomial, the linearized equations can be deduced as in the previously studied cases [31]. If M is independent on products of dual kinematic and dynamic functions (32) except e_{ij} and φ_{ij} we obtain the equations

$$\sigma^{ij} = E_1^{ijkl}\, e_{kl} + C_1^{ijkl}\, \alpha_{kl} + G^{ijkl}\, \varphi_{kl} - \pi^{ji}, \qquad \lambda^{jik} = 2A_1^{jikmln}\, \alpha_{mln},$$

$$\pi^{ji} = 2A_1^{ijkl}\, \alpha_{kl} + C_1^{lkij}\, e_{kl}, \tag{39}$$

$$\begin{gathered} \lambda^{ij} = E_2^{ijkl}\, \varphi_{kl} + C_2^{ijkl}\, \beta_{kl} + C_2^{lkij}\, e_{kl} - \chi^{ji}, \\ \lambda^{jik} = A_2^{jikmln}\, \beta_{mln}, \qquad \chi^{ji} = A_2^{ijkl}\, \beta_{kl} + C_2^{lkij}\, \varphi_{kl}. \end{gathered} \tag{40}$$

4. The Compatibility Conditions

According to the structure of the functions (32) it is suitable to consider the asymmetric forms

$$\overset{pq}{ds^2} = g_{ij}\, \overset{p}{dx^i}\, \overset{q}{dx^j}, \qquad \big(\overset{1}{x^i} = x^i,\ \overset{2}{x^i} = \xi^i,\ \overset{3}{x^i} = f^i(x),\ \overset{4}{x^i} = f^i(x, \xi)\big) \tag{41}$$

as well as the symbols concerning the distortions

$$\begin{gathered} \overset{1}{A^i_\alpha} = A^i_\alpha, \qquad \overset{2}{A^i_\alpha} = a^i_\alpha, \qquad \overset{3}{A^i_\alpha} = F^i_\alpha, \qquad \overset{4}{A^i_\alpha} = f^i_\alpha, \\ \overset{pq}{A}_{ij} = \delta_{\alpha\beta}\, \overset{p}{A}{}^\alpha_i\, \overset{q}{A}{}^\beta_j, \qquad \overset{pq}{\Gamma}{}^{kj}_i = \overset{p}{A}{}^i_\alpha\, \overset{q}{A}{}^\alpha_{j;k}, \qquad \overset{pq}{A}_{ijk} = \delta_{\alpha\beta}\, \overset{p}{A}{}^\alpha_j\, \overset{q}{A}{}^\beta_{i;k}. \end{gathered} \tag{42}$$

The following relations are valid (the p, q-indices being omitted)

$$\Gamma_{ijk} = (C + T + g)_{ijk}, \qquad A_{ijk} = (\Gamma - g)_{kij} = (C + T)_{kij} \qquad (43),\ (44)$$

where

$$\left.\begin{aligned} \overset{pq}{C}_{kjl} &= \overset{pq}{A}_{lj,k} - \overset{pp}{A}{}^{nm}\, \overset{qp}{A}_{jn}(A_{lm,k} + \overset{pp}{A}_{mk,l} - A_{kl,m})/2, \\ \overset{pq}{T}_{kjl} &= -\overset{pp}{A}{}^{nm}\, \overset{qp}{A}_{jn}\, \overset{pp}{t}_{klm}, \qquad t_{klm} = T_{kl,m} + T_{mk,l} - T_{lm,k}, \\ T_{kl,m} &= 2\overset{pq}{\Gamma}_{[kl]m}, \qquad g_{kij} = \delta_{\alpha\beta}\, \overset{p}{A}{}^\alpha_j\, \overset{q}{A}{}^\beta_m\, g^m_{ki} = \overset{pq}{A}_{jm}\, g^m_{ki}. \end{aligned}\right\} \tag{45}$$

The coefficients g^i_{kj} are introduced in (17). The symbols (43) appear in the expressions of the derivatives of the invariant

$$\overset{pq}{\boldsymbol{T}} = \boldsymbol{e}_\alpha\, \boldsymbol{e}_\beta\, \overset{p}{A}{}^\alpha_i\, \overset{q}{A}{}^\beta_j\, \overset{pq}{T}{}^{ij}. \tag{46}$$

The derivation is not commutative ($T_{klm} \neq 0$). The torsion (43) appears in (45) and the curvature tensor is expressible under the dual form

$$\overset{pq}{R}{}^{ij} = \varepsilon^{jnm}\, \varepsilon^{jlk}(\overset{pq}{A}_{klm;n} - \overset{ab}{A}{}^{rs}\, \overset{ap}{A}_{rkn}\, \overset{bq}{A}_{slm})/2 \tag{47}$$

as in [4], for $p = q = 1$. We assume that (47) vanishes identically if $p = 1$, $q = 1$ or 3. Hence, if we put $t \ldots$ in (45) under the dual form

$$\overset{rr}{t}_{kml} = -\varepsilon_{mlp}\, \overset{rr}{\alpha}{}^p_{.k}/2. \tag{48}$$

$\alpha^{p}_{.k}$ corresponds, in generalized sense, to the density of dislocations [3]. According to the additional notations (see [2, 5])

$$\begin{aligned} \overset{pq}{\eta}{}^{ij} &= -\varepsilon^{(jnm}\,\delta^{i)}_{[a}\,\delta^{l}_{c]}\,\overset{pp}{A}{}^{ba}\,\overset{qp}{A}_{lb}\,\overset{pp}{\alpha}{}^{c}_{.m;n}/2, \\ \overset{pq}{\xi}{}^{ij} &= -\varepsilon^{(jnm}\,\varepsilon^{i)lk}\,\overset{ab}{A}{}^{rs}\,\overset{pa}{A}_{rkn}\,\overset{bq}{A}_{slm}/2 \end{aligned} \tag{49}$$

the vanishing of (47) implies the 3 independent compatibility conditions (since $I^{ij}_{;j} = 0$)

$$\overset{pq}{I}{}^{(ij)} = (\overset{pq}{\eta} + \overset{pq}{\xi})^{ij} = -\varepsilon^{(jmn}\,\varepsilon^{i)kl}\,\overset{pq}{C}_{mlk;n}/2. \tag{50}$$

Finally we restrict our attention to the symmetric conditions

$$\overset{11}{I}_{(ij)} = \cdots, \qquad \overset{13}{I}_{(ij)} = \cdots \tag{51}$$

(50) can be used in connection with the previously introduced elements (32) since

$$\begin{aligned} &\overset{11}{A}_{ij} = -2e_{ij} + g_{ij}, \quad \overset{12}{A}_{ij} = \alpha_{ij} + \overset{11}{A}_{ij}, \quad \overset{12}{A}_{ijk} = \alpha_{ijk}, \\ &\overset{13}{A}_{(ij)} = \varphi_{ij}, \quad \overset{14}{A}_{ij} = \beta_{ij} + \overset{13}{A}_{ij}, \quad \overset{13}{A}_{ijk} = \beta_{ijk}. \end{aligned} \tag{52}$$

In the linearized theory both conditions (51) take similar forms so that $I^{(ij)} = -\varepsilon^{(ikl}\,\varepsilon^{j)mn}(e \text{ or } \varphi)_{ln,km}$ for $p = 1$, $q = 1$ or 3. The functions $\overset{22}{A}_{ij}$, $\overset{44}{A}_{ij}$ are incompatible in the above considered sense. The functions π_{ij}, χ_{ij} play an auxiliary role, according to (29)—(31).

The 174 unknown functions (34), (39) can be determined with the aid of the Eqs. (29); (30), (31), $(14)_{1,2}$, (44); (51), (33), (37), their number being $12 + 12 + 60 + 90$.

The linearized functions (34) take the form

$$\begin{aligned} &e'_{ij} = u_{(i,j)}, \quad \alpha'_{ij} = u_{ji} - u_{i,j}, \quad \alpha'_{jik} = u_{ji,k}, \quad (u_{ji} = u^{i}_{i.}), \\ &\varphi_{ij} = f_{(i,j)}, \quad \beta'_{ij} = f_{ji} - f_{i,j}, \quad \alpha'_{jik} = f_{ji,k}. \end{aligned} \tag{53}$$

A more general situation occurs if $dA^{\alpha}_{i} = A^{\alpha}_{ij}\,dx^{j}$, $da^{\alpha}_{i} = a^{\alpha}_{ij}\,d\xi^{j}$. It is then necessary to introduce the new elements

$$\overset{pq}{\Gamma}{}^{i}_{lkj} = \overset{p}{A}{}^{\alpha}_{i}\,\overset{q}{A}{}^{\alpha}_{jk;l}, \qquad \overset{q}{\boldsymbol{T}} = \boldsymbol{e}_{\alpha}\,\overset{q}{A}{}^{\alpha}_{jk}\,T^{jk}, \qquad (\overset{1}{A}_{ij} = A_{ij},\ \overset{2}{A}_{ij} = A_{ij}) \tag{54}$$

so that

$$\overset{q}{\boldsymbol{T}}_{,l} = \boldsymbol{e}_{\alpha}\,\overset{p}{A}{}^{\alpha}_{i}(\overset{pq}{\Gamma}{}^{i}_{kj}\,\overset{q}{T}{}^{jk}_{,l} + \overset{pq}{\Gamma}{}^{i}_{lkj}\,\overset{q}{T}{}^{jk}), \qquad \overset{q}{\boldsymbol{T}}_{,[lm]} = -\boldsymbol{e}_{\alpha}\,\overset{p}{A}{}^{\alpha}_{i}\,\overset{pq}{R}{}^{i}_{mlkj}\,\overset{q}{T}{}^{jk}/2, \tag{55), (56}$$

where

$$\overset{pq}{R}{}^{i}_{mlkj} = 2(\overset{pq}{\Gamma}{}^{i}_{[lkj.m]} - \overset{rp}{\Gamma}{}^{i}_{[mn}\,\overset{rp}{\Gamma}{}^{n}_{l]kj}). \tag{57}$$

Thus, an analogous development leads to the substitution of the conditions (44) by

$$R^{ijk} = \varepsilon^{jnm} \varepsilon^{ilk} R_{nmlk}/4 = 0, \quad (p = 1, \; q = 1, 3), \quad \text{etc.} \tag{58}$$

5. Conclusions

The solving methods adopted in some subsequent applications are based on the following considerations. We assume that the singular surfaces can be isolated by continous closed boundaries. Thus, the body becomes a multiconnected elastic continuum and the kinematic and dynamic incompatibility conditions are replaced by special additional boundary conditions. Since, usually, the asymptotic behavior of the solutions call important specific boundary effects, it seems suitable to elaborate general asymptotic solving methods. For instance, we decompose the solutions S in parts $S' + S''$ so that S' vanish on a boundary in the neighborhood of a singular region. Then S'' expresses the compatibility conditions. S' constitutes a correction which is determined according to the conditions on the remaining boundaries. Such method, reductible to other recent solving methods, transposed in our case leads to the decomposition $e = e' + e''$, etc., e' being a compatible solution and e'' a particular solution of (51). As a matter of fact, the compatible solutions of the linear theory of self-coupled bodies is of biharmonic type. Some results concerning the elastic phenomena are included in recent studies[1].

References

[1] Kondo, K.: Japan Soc. Appl. Mech. **3**, 107—110 (1950).
[2] Kröner, E.: Kontinuumstheorie der Versetzungen und Eigenspannungen. Berlin/Göttingen/Heidelberg: Springer 1958.
[3] Kröner, E.: Arch. Rat. Mech. Anal. **4**, 1, 273—334 (1959).
[4] Kröner, E., and A. Seeger: Arch. Rat. Mech. Anal. **3**, 1, 17—119 (1959).
[5] Kröner, E.: Int. J. Engng. Sci. **1**, 261—278 (1963); Proc. 11th Int. Congr. Appl. Mech. München 1964. Berlin/Heidelberg/New York: Springer 1966.
[6] Hehl, F., and E. Kröner: Z. Naturf. **20a**, 3, 336—350 (1965).
[7] Bilby, B. A., R. Bullough and E. Smith: Proc. Roy. Soc. London, Ser. A, **236**, 263—273 (1955).
[8] Günther, W.: Abh. d. Braunschweigischen Wiss. Ges. **10**, 195—213 (1958).

---

[1] A more complete analysis concerning the non-dual theory of elastic coupled bodies is included in [32]. The methods using dynamic and kinematic potentials, conformal mapping, Fourier transforms and complex functions are adapted for the case of asymmetric elastic stress concentration effects in the presence of coupling reactions. The obtained solutions concern the plane, antiplane, axisymmetric and bending problems. The theory of coupled shells developed recently corrects the Love-Kirchhoff theory in order to consider the distortional effects on the basis of complex formulations and field-correspondency principles.

[9] SCHAEFER, H.: Abh. d. Braunschweigischen Wiss. Ges. **7**, 107–112 (1955).
[10] COHEN, H.: J. Math. Phys. **45**, 1, 35–44 (1966).
[11] MIŞICU, M.: Rev. Roum. Techn. Mec. Appl. **10**, 1, 35–46 (1965).
[12] MIŞICU, M.: Rev. Roum. Sci. Techn. Mec. Appl. **11**, 1, 109–123 (1966).
[13] MIŞICU, M.: Rev. Roum. Sci. Techn. Mec. Appl. **9**, 3, 477–495 (1964).
[14] GREEN, A. E., and P. M. NAGHDI: Matematika **12**, 23, 21–26 (1965).
[15] VOIGT, W.: Abh. Ges. Wiss. Göttingen **34** (1887).
[16] COSSERAT, E., and F. COSSERAT: Théorie des corps déformables, in O. D. CHWOLSON: Traité de Physique. Paris: 1909 pp. 953–1173.
[17] TRUESDELL, D., and R. A. TOUPIN: The Classical Field Theories. Handbuch der Physik III/1. Berlin/Göttingen/Heidelberg: Springer 1960, pp. 226–790.
[18] AERO, E. L., and H. F. KUVSHINSKI: Fiz. Tverdogo Tela **2**, 7, 1399–1409 (1960).
[19] TOUPIN, R. A.: Arch. Rat. Mech. Anal. **1**, 5, 385–414 (1962).
[20] MINDLIN, R. D., and H. F. TIERSTEN: Arch. Rat. Mech. Anal. **1**, 5, 415–448. (1962).
[21] GRIOLI, G.: Ann. Mat. pura e Appl. **30**, 389–417 (1960).
[22] MIŞICU, M.: Rev. Roum. Sci. Techn. **6**, 9, 1351–1359 (1964).
[23] GREEN, A. E., and R. S. RIVLIN: Arch. Rat. Mech. Anal. **17**, 2, 113–148 (1964).
[24] MINDLIN, R. D.: Arch. Rat. Mech. Anal. **16**, 1, 51–78 (1964).
[25] ERINGEN, A. C., and E. S. SUHUBI: J. Engng. Sci. **2**, 2, 189–205 (1965); **2**, 4, 389–455 (1965).
[26] ESHELBY, J. D.: Solid state physics. New York: Acad. Press 3/2, 1965, p. 79–114.
[27] ERINGEN, A. C., and D. INGRAM: Int. J. Engng. Sci. **3**, 2, 197–213 (1965).
[28] GREEN, A. E.: Int. J. Engng. Sci. **3**, 2, 231–241 (1965).
[29] MIŞICU, M.: Rev. Roum. Sci. Techn. Mec. Appl. **10**, 4, 843–891 (1965).
[30] MIŞICU, M.: Rev. Roum. Sci. Techn. Mec. Appl. **12**, 1, 177–199 (1967).
[31] MIŞICU, M.: Mechanics of Deformable Media. Bucarest: Academia R.S.R. 1967.
[32] MIŞICU, M.: Asymmetric heteroelasticity. Bucarest: Academia R.S.R. (in press).
[33] MAWARDI, O. K.: J. Franklin Inst. **264**, 4, 313–336 (1957).

Dislocations in the Generalized Elastic Cosserat Continuum

By

R. Stojanović

Department of Mechanics
University of Belgrade

List of Notations

ϱ	Density of matter	$\boldsymbol{t}$	Stress tensor
ε	Internal energy density	$\boldsymbol{m}$	Couple-stress tensor
$\boldsymbol{d}_{(\lambda)}$	Directors in the deformed state	$\boldsymbol{h}^{(\lambda)}$	Stresses of rotation
$\boldsymbol{d}^{(\lambda)}$	Reciprocal directors	$\boldsymbol{h}$	Hyperstress tensor
$\boldsymbol{D}_{(\lambda)}$	Directors in the initial state	$\Delta \boldsymbol{F}$	Surface element
X^K	Material (Lagrangean) coordinates	$\boldsymbol{b}$	Burgers vector
x^k	Spatial (Eulerian) coordinates	$(\cdots)_{;K}$	Total covariant derivative
$\boldsymbol{g}$	Metric tensor of the deformed configuration	$(\cdots)_{,k}$	Partial covariant derivative
$\boldsymbol{f}$	Body forces	$(\dot{\cdots})$	Material time-derivative
$\boldsymbol{L}$	Body couples	$(\cdots)_{(ij)} \equiv \frac{1}{2}[(\cdots)_{ij} + (\cdots)_{ji}]$	
$\boldsymbol{k}^{(\lambda)}$	Director forces	$(\cdots)_{[ij]} \equiv \frac{1}{2}[(\cdots)_{ij} - (\cdots)_{ji}]$	
		$(\cdots)_{[i\|j\|k]} \equiv \frac{1}{2}[(\cdots)_{ijk} - (\cdots)_{kji}]$	

Greek indices in round brackets are non-tensorial indices.

The aim of this communication is to present some results in the continuum theory of dislocations and internal stresses which follow from the assumption that the elastic generalized Cosserat continuum is a continuum-theoretical model for crystals.

In the elasticity theory of the generalized Cosserat continuum developed by Stojanović, Djurić and Vujošević [1, 5] it is assumed that the strain energy ε of a body suffering simultaneous deformations of position $x^k = x^k(X^1, X^2, X^3; t)$ and of orientation $d^k_{(\lambda)} = d^k_{(\lambda)}(D^K_{(\lambda)}; t) = d^k_{(\lambda)}(X^k; t)$ is a function of the position gradients $x^k_{;K}$, of the second-order position gradients $x^k_{;KL}$ and of the director-gradients $d^k_{(\lambda);K}$. A slight generalization of the conservation laws used by Ericksen [6] in the theory of liquid crystals leads to equations of motion in the form

$$\varrho\, \ddot{x}^i = t^{ij}{}_{,j} + f^i, \tag{1}$$

$$\varrho\, i^{\lambda\mu}\, \ddot{d}^i_{(\mu)} = h^{(\lambda)\,ij}{}_{,j} + k^{(\lambda)\,i}, \tag{2}$$

$$t^{[ij]} = m^{ijk}{}_{,k} + d^{[i}_{(\lambda),k}\, h^{(\lambda)j]k} + L^{ij}. \tag{3}$$

The symbols ϱ, $\boldsymbol{t}$, $\boldsymbol{m}$, $\boldsymbol{f}$ and $\boldsymbol{L}$ have the usual meaning: density of matter, stress tensor, couple-stress tensor, volume force and volume couple, respectively. The newly introduced quantities are the *density of inertia* $i^{\lambda\mu}$, *stress of rotation* $h^{(\lambda)ij}$, and the $k^{(\lambda)i}$ are certain extrinsic forces acting on the directors, but their nature remains without an interpretation[1].

From the conservation law for the total energy and from the principle of material frame indifference under rigid motions follow the constitutive relations for the elastic generalized Cosserat continuum,

$$t^{(ij)} = \varrho \left(\frac{\partial \varepsilon}{\partial E_{AB}} x^i_{;A} x^j_{;B} + 2 \frac{\partial \varepsilon}{\partial D_{ABC}} x^i_{;A} x^j_{;BC} + \frac{\partial \varepsilon}{\partial F_{\alpha AB}} d^i_{(\alpha);A} x^j_{;B} \right), \quad (4)$$

$$m^{i(jk)} = -\varrho \frac{\partial \varepsilon}{\partial D_{ABC}} x^i_{;A} x^{(j}_{;B} x^{k)}_{;C}, \quad (5)$$

$$h^{(\alpha)ij} = \varrho \frac{\partial \varepsilon}{\partial F_{\alpha AB}} x^i_{;B} x^j_{;A}. \quad (6)$$

Here $\boldsymbol{E}$ and $\boldsymbol{D}$ are the material measures of strain already appearing in the theory of elastic materials of grade two[2] [8],

$$E_{AB} \equiv \tfrac{1}{2}(C_{AB} - g_{AB}), \qquad C_{AB} \equiv g_{ij} x^i_{;A} x^j_{;B}, \quad (7)$$

$$D_{ABC} \equiv C_{C[A,B]}, \quad (8)$$

and $F_{\alpha AB}$ are three tensors ($\alpha = 1, 2, 3$) of the *strain of orientation*,

$$F_{\alpha AB} \equiv g_{ij} d^i_{(\alpha);A} x^j_{;B}. \quad (9)$$

If the directors $\boldsymbol{D}_{(\alpha)}$ were material vectors, the directors $\boldsymbol{d}_{(\alpha)}$ in the deformed configuration would be $\overset{*}{d}{}^i_{(\alpha)} = D^A_{(\alpha)} x^i_{;A}$. In the Cosserat continuum this is not the case and the vectors

$$\Delta_{(\alpha)i} \equiv d_{(\alpha)i} - D_{(\alpha)A} X^A_{;i} \quad (10)$$

do not vanish.

If the vectors $\boldsymbol{D}_{(\alpha)}$ coincide with the lattice vectors of a perfect crystal and the directors $\boldsymbol{d}_{(\alpha)}$ coincide with the lattice vectors of a dislocated crystal, the directors $\boldsymbol{D}_{(\alpha)}$ represent three fields of absolutely parallel vectors in Euclidean space, but this does not hold true for the directors $\boldsymbol{d}_{(\alpha)}$.

Let l be a closed contour connecting lattice points in a dislocated crystal, passing through a "good" region. The integration of $d_{(\alpha)i}\, dx^i$

---

[1] If we put $i^{\lambda\mu} \dot{d}^{[i}_{(\lambda)} d^{j]}_{(\mu)} \equiv \dot{s}^{ij}$ and $L^{ij} + d^{[i}_{(\lambda)} k^{(\lambda)j]} = l^{ij}$, the relations (2) and (3) will yield the differential equations of motion which coincide with the equations of motion derived by Toupin [7]. The quantity $\dot{s}^{ij}$ may be interpreted as Toupin's spin angular momentum.

[2] Toupin [8] uses the tensor D_{ABC} with the value $D_{ABC} = \frac{2}{3} C_{C[A,B]}$.

along l gives the components of the Burgers vector in the directions of the directors $d_{(\alpha)}$,

$$\Delta b_\alpha = \oint d_{(\alpha)\,i}\, dx^i, \quad \Delta b_k = \Delta b_\alpha\, d_k^{(\alpha)}. \qquad (11)$$
$$(d_k^{(\alpha)}\, d_{(\beta)}^k = \delta_\beta^\alpha)$$

Substituting $d_{(\alpha)\,i}$ from (10) we obtain

$$\Delta b_\alpha = \iint\limits_{\Delta F} \Delta_{(\alpha)\,[i,j]}\, dF^{j\,i}.$$

The mean-value theorem for integrals gives $\Delta b_{(\alpha)} = \Delta_{(\alpha)[i,j]}\, \Delta F^{j\,i}$ and for the dislocation density tensor we may write the formula

$$\alpha_{ijk} = \lim_{\Delta F \to 0} \frac{\Delta b_k}{\Delta F^{ij}} = d_k^{(\alpha)}\, \Delta_{(\alpha)\,[j,\,i]}\,. \qquad (12)$$

Hence, the antisymmetric part of the strain of orientation is directly connected with the distribution of dislocations,

$$F_{\alpha\,[A\,B]} = d_{(\alpha)\,[i,j]}\, x^i_{;\,B}\, x^j_{;\,A} = \alpha_{ijk}\, x^i_{;\,A}\, x^j_{;\,B}\, d^k_{(\alpha)}\,. \qquad (13)$$

For infinitesimal deformations this formula reduces to a formula given by GÜNTHER [9] which represents the connection of the infinitesimal strain of orientation with NYE's [10] structural curvature.

Since the distribution of dislocations influences only the antisymmetric part of the strain of orientation tensor, we assume that the strain energy does not depend on the symmetric part $F_{\alpha\,(A\,B)}$. From (6) we now have

$$h^{(\alpha)\,ij} = \varrho\, \frac{\partial \varepsilon}{\partial F_{\alpha\,[A\,B]}}\, x^i_{;\,B}\, x^j_{;\,A} \qquad (14)$$

and it follows that the stress of rotation is represented by an antisymmetric tensor.

Denoting by $\overset{*}{t}{}^{(ij)}$ and $\overset{*}{m}{}^{ij\,k}$ those parts of the right-hand side of the constitutive relations (4) and (5) which coincide in form with the constitutive relations of the elasticity theory of materials of grade two [8], from the constitutive relations (4)—(6) and from the equations of motion (1)—(3) in the static case, and in the absence of the volume forces and couples there follows the expression for the total stress,

$$t^{ij} = \overset{*}{t}{}^{(ij)} + \overset{*}{m}{}^{ij\,k}{}_{,\,k} + (d^i_{(\alpha)}\, h^{(\alpha)\,j\,n})_{,\,n}\,. \qquad (15)$$

The third-order antisymmetric tensor

$$h^{j\,n\,i} \equiv d^i_{(\alpha)}\, h^{(\alpha)\,j\,n} = -h^{n\,j\,i} \qquad (16)$$

corresponds to the *hyperstress* obtained by TOUPIN [7]. The antisymmetric part of the stress tensor is now

$$t^{[ij]} = (\overset{*}{m}{}^{ij\,k} + h^{[j\,|k|\,i]})_{,\,k}\,. \qquad (17)$$

Owing to the antisymmetry of the hyperstress tensor, the equilibrium conditions for the total stress reduce in form to the equilibrium conditions for ordinary continua,

$$t^{ij}{}_{,j} = \overset{*}{t}{}^{(ij)}{}_{,j} + \overset{*}{m}{}^{i\,(jk)}{}_{,jk} = 0. \tag{18}$$

From the expression (17) for the antisymmetric part of the stress tensor it follows that the hyperstress tensor $\boldsymbol{h}$ and the couple-stress tensor $\boldsymbol{m}$ play the same role. However, HEHL and KRÖNER [11] give an analysis from which it follows that the couple-stress $\overset{*}{\boldsymbol{m}}$, connected with the second-order position gradients, is negligible. Accordingly, it may be assumed that the strain energy does not depend on the second-order position gradients and we have

$$\varepsilon = \varepsilon(x^k_{;K}, d^k_{(\lambda);K}). \tag{19}$$

The explicit form of the stress-strain relations reduces now to

$$t^{(ij)} = \varrho \frac{\partial \varepsilon}{\partial E_{AB}} x^i_{;A} x^j_{;B} + h^{(i|k|j)}{}_{,k}, \tag{20}$$

$$h^{[ij]k} = -\varrho\, d^k_{(\alpha)} \frac{\partial \varepsilon}{\partial F_{\alpha[AB]}} x^i_{;A} x^j_{;B}. \tag{21}$$

The linear approximation of (21) in the isotropic case reduces to the expression for the couple-stress obtained by HEHL and KRÖNER. Since the antisymmetric part of the stress tensor, in virtue of (13), (17) and (21), is determined directly by the given distribution of dislocations, and since the equilibrium conditions (18) reduce to the usual form

$$t^{ij}{}_{,j} = \overset{*}{t}{}^{(ij)}{}_{,j} = 0, \tag{22}$$

only the symmetric part of the stress tensor remains to be determined. This may be achieved following the procedure proposed by KRÖNER and SEEGER [12] as though the continuum were an ordinary, non-oriented medium.

References

[1] STOJANOVIĆ, R., S. DJURIĆ and L. VUJOŠEVIĆ: Mat. Vesnik **1** (**16**) 127 (1964).
[2] DJURIĆ, S.: Dynamics and small vibrations of the Cosserat continuum. Belgrade: Thesis 1964.
[3] DJURIĆ, S.: Proc. VIII Yugoslav Congress on Mechanics 1966 (in print).
[4] STOJANOVIĆ, R., and S. DJURIĆ: Proc. VIII Yugoslav Congress on Mechanics 1966 (in print).
[5] STOJANOVIĆ, R., S. DJURIĆ and L. VUJOŠEVIĆ: Conference on Mechanics of Continua. Zakopane 1965.
[6] ERICKSEN, J. L.: Trans. Soc. Rheol. **5**, 23 (1961).
[7] TOUPIN, R.: Arch. Rat. Mech. Anal. **17**, 85 (1964).
[8] TOUPIN, R.: Arch. Rat. Mech. Anal. **11**, 385 (1962).
[9] GÜNTHER, W.: Abh. d. Braunschweigischen Wiss. Ges. **10**, 195 (1958).
[10] NYE, J. F.: Acta Metallurgica **1**, 153 (1953).
[11] HEHL, F., and E. KRÖNER: Z. Naturf. **20a**, 336 (1965).
[12] KRÖNER, E., and A. SEEGER: Arch. Rat. Mech. Anal. **3**, 97 (1959).

Some Considerations on the Mechanics of Granular Materials

By

M. Satake

Faculty of Engineering
Tohoku University, Sendai

1. Introduction

Some approaches to the mechanics of granular materials have been presented from various viewpoints [1—4]. This paper will deal with some basic properties of granular materials which seem useful for the formation of the mechanics of the materials.

2. Deformation of Granular Materials

Assuming that each grain is completely rigid and that deformation of the medium arises only from relative displacements and rotations of inner grains, we can write for a grain-series

$$[\boldsymbol{u}]_A^B = \sum (\Delta \boldsymbol{u}_i' - \Delta \boldsymbol{r}_i \times \boldsymbol{w}_{i-1}), \tag{2.1}$$

where $[\boldsymbol{u}]_A^B$ denotes difference of absolute displacements of two end grains, and $\Delta \boldsymbol{u}_i'$, $\Delta \boldsymbol{r}_i$ and $\boldsymbol{w}_i$ are relative displacement, difference of position vectors and rotation vector, of the i-th grain respectively. It should be noted that $\boldsymbol{w}$ is considered as independent on $\boldsymbol{u}$ in granular materials.

Next, we proceed to the macroscopic consideration. We introduce relative rotation (tensor) α and relative displacement (tensor) γ which may be regarded as distributed along a curve in the medium. If we can assume that resources of deformation of the curve are distributed only on the inner points and also on the two end points of the curve, we can introduce the following equilibrium equations of deformation just as in the equilibrium of forces and couples.

$$\frac{d\boldsymbol{w}}{d\boldsymbol{r}} = -\alpha, \tag{2.2}$$

$$\frac{d\boldsymbol{u}}{d\boldsymbol{r}} = -\gamma - \boldsymbol{I} \times \boldsymbol{w}. \tag{2.3}$$

Integrating these equations, we have

$$\int_A^B d\boldsymbol{r}\cdot\alpha + [\boldsymbol{w}]_A^B = 0, \tag{2.4}$$

$$\int_A^B d\boldsymbol{r}\cdot\gamma + \int_A^B \boldsymbol{r}\times(d\boldsymbol{r}\cdot\alpha) + [\boldsymbol{u}]_A^B + [\boldsymbol{r}\times\boldsymbol{w}]_A^B = 0. \tag{2.5}$$

Eq. (2.5) corresponds to Eq. (2.1). Using *dislocations*[1], we can write as

$$\int_A^B d\boldsymbol{r}\cdot\alpha = -\tilde{\boldsymbol{w}}_A^B, \tag{2.6}$$

$$\int_A^B d\boldsymbol{r}\cdot\gamma + \int_A^B \boldsymbol{r}\times(d\boldsymbol{r}\cdot\alpha) = -\tilde{\boldsymbol{u}}_A^B. \tag{2.7}$$

It follows, from Eqs. (2.4) and (2.5), that

$$\alpha + \nabla\boldsymbol{w} = 0, \tag{2.8}$$

$$\gamma + \nabla\boldsymbol{u} = -\boldsymbol{I}\times\boldsymbol{w}, \tag{2.9}$$

and

$$-\gamma_s = \tfrac{1}{2}(\nabla\boldsymbol{u} + \boldsymbol{u}\nabla), \tag{2.10}$$

$$-\boldsymbol{I}\cdot\times\gamma = \nabla\times\boldsymbol{u} - 2\boldsymbol{w}, \tag{2.11}$$

where subscript s denotes the symmetric part of the tensor. If γ is symmetrical, $\boldsymbol{I}\cdot\times\gamma$ vanishes as in the case of ordinary continua, and we obtain

$$-\alpha = \gamma_s\times\nabla. \tag{2.12}$$

---

[1] Rotation-dislocation $\tilde{\boldsymbol{w}}_A^B$ and displacement-dislocation $\tilde{\boldsymbol{u}}_A^B$ are defined by the form:

$$\tilde{\boldsymbol{w}}_A^B = [\boldsymbol{w}]_A^B, \tag{1}$$

$$\tilde{\boldsymbol{u}}_A^B = [\boldsymbol{u}]_A^B + [\boldsymbol{r}\times\boldsymbol{w}]_A^B. \tag{2}$$

The meanings of these quantities are illustrated in Fig. 1, in which $\underset{0}{\boldsymbol{r}}_A, \underset{0}{\boldsymbol{r}}_B$ indicate position vectors in the initial state obtained by regarding that $\boldsymbol{r}_A, \boldsymbol{r}_B$ (position vectors in the deformed state) are fixed rigidly to the points A, B. From the consideration on this figure, we can write

$$\underset{k}{\boldsymbol{i}}_B - \underset{k}{\boldsymbol{i}}_A = \underset{k}{\boldsymbol{i}}\times\tilde{\boldsymbol{w}}_A^B, \tag{3}$$

$$\overrightarrow{0_A 0_B} = -\tilde{\boldsymbol{u}}_A^B. \tag{4}$$

The concept of these dislocations originates with Moriguchi [5].

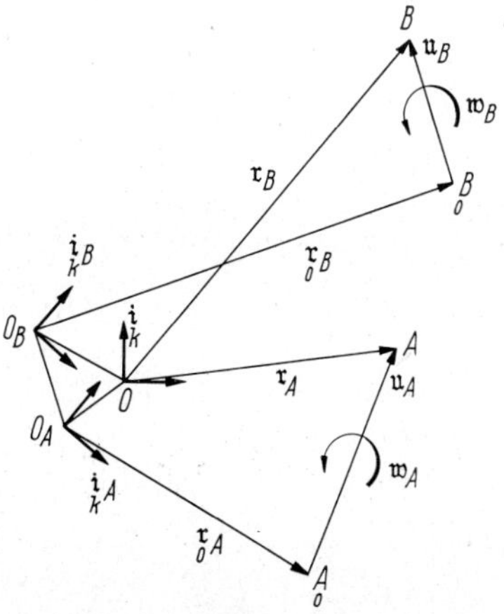

Fig. 1.

It is seen that $-\underset{s}{\gamma}$ is the ordinary strain tensor, and $-\alpha$ is the secondary strain tensor introduced by OSHIMA [2].

When the curve is closed, the right-hand sides of (2.6) and (2.7) must vanish. There is a case, however, where these do not vanish but have finite values in spite of simply connected region. We must consider, in this case, that resources of deformation are distributed not only along the curve but also on the interior region bounded by the curve. Then, the equilibrium equations may become

$$\int d\boldsymbol{S}\cdot\boldsymbol{J} + \oint d\boldsymbol{r}\cdot\alpha = 0, \tag{2.13}$$

$$\int d\boldsymbol{S}\cdot\boldsymbol{K} + \int \boldsymbol{r}\times(d\boldsymbol{S}\cdot\boldsymbol{J}) + \oint d\boldsymbol{r}\cdot\gamma + \oint \boldsymbol{r}\times(d\boldsymbol{r}\cdot\alpha) = 0, \tag{2.14}$$

where, $\boldsymbol{J}$ and $\boldsymbol{K}$ represent relative rotation and relative displacement distributed in the area respectively. It follows that

$$\boldsymbol{J} + \nabla\times\alpha = 0, \tag{2.15}$$

$$\boldsymbol{K} + \nabla\times\gamma = -\boldsymbol{I}\times\times\alpha. \tag{2.16}$$

Eliminating α from these equations, we have

$$\underset{s}{\boldsymbol{J}} = \nabla\times\underset{s}{\gamma}\times\nabla + (\boldsymbol{K}\times\nabla)_s, \tag{2.17}$$

$$I\cdot\times\boldsymbol{J} = -\nabla\cdot\boldsymbol{K}. \tag{2.18}$$

In the case where Eq. (2.12) holds, $\boldsymbol{K}$ vanishes. $\underset{s}{\boldsymbol{J}}$ is a tensor usually called the *incompatibility tensor*.

It has been seen that equilibrium equations of deformation are quite similar to those of forces and couples. Thus, equations in the case where body forces and couples are absent may be obtained by replacing $\boldsymbol{J}$, $\boldsymbol{K}$, α and γ with σ, μ, φ and χ respectively in Eqs. (2.13) to (2.18), where σ and μ denote stress and couple stress tensors, and φ and χ distributed force and couple along the boundary respectively. $\underset{s}{\chi}$ is a tensor usually called the *stress function tensor*.

3. Modification of Mohr's Circle

Here we shall present the modification of Mohr's circle in the case where σ is not symmetrical. For the simplicity, we treat the case where $\boldsymbol{I}\cdot\times\sigma$ takes a form of $(-2q, 0, 0)$ with respect to the principal axes of $\underset{s}{\sigma}$. Letting σ_n and τ_n be normal and tangential stress components on a surface element with normal $\boldsymbol{n} = (l, m, n)$ and σ_1, σ_2 and σ_3 three principal stresses, we obtain

$$\left(\sigma_n - \frac{\sigma_2+\sigma_3}{2}\right)^2 + (\tau_n \pm q)^2 = \left(\frac{\sigma_2-\sigma_3}{2}\right)^2 \tag{3.1}$$

for surfaces with $l = 0$,

$$\left\{\sigma_n - \frac{\sigma_3 + \sigma_1}{2} - \frac{q^2}{2(\sigma_3 - \sigma_1)}\right\} + \tau_n^2 = \left\{\frac{\sigma_3 - \sigma_1}{2} + \frac{q^2}{2(\sigma_3 - \sigma_1)}\right\}^2 \tag{3.2}$$

for $m = 0$ and a similar expression for $n = 0$. These equations represent three circles in the $\sigma_n - \tau_n$ plane (Fig. 2), which may be useful for investigations of shear strength of granular materials.

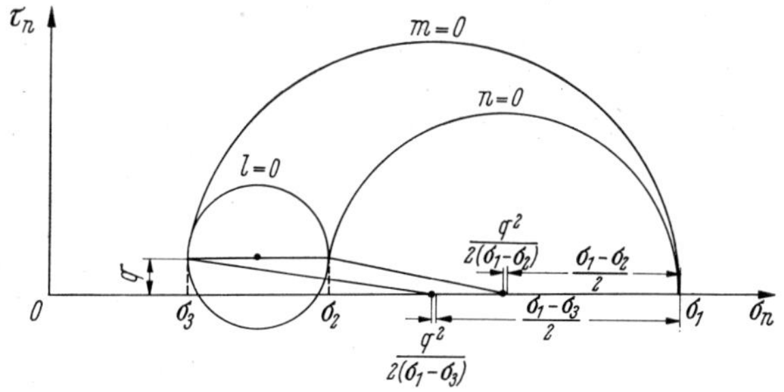

Fig. 2. Modification of Mohr's circle.

References

[1] OSHIMA, N.: Proc. 2nd Jap. Nat. Cong. Appl. Mech., 1952, p. 5—8.
[2] OSHIMA, N.: Memoirs of RAAG, 1 Div. D, (1955), p. 563—572.
[3] MOGAMI, T.: Soil and Foundation, **5**, No. 2, 26—36 (1965).
[4] SOKOLOVSKII, V. V.: Statics of Granular Media. Oxford: Pergamon Press 1965.
[5] MORIGUCHI, S.: RAAG Research Notes, 3rd Series No. 65, 1963.

A comment on the communication by Dr. M. Satake

On the Influence of Couple-Stresses on the Distribution of Velocities in the Flow of Polar Fluids

By

M. Plavšić

Department of Mechanics
University of Belgrade

Dr. M. Satake pointed out in his lecture the influence of shear stress on the flow of granular materials, considered as continuous media with couple-stress.

In the theory of elasticity the influence of couple-stresses is theoretically predicted and investigated in the great number of papers, but there are no indications about the order of magnitude of this influence (cf. the lecture by Prof. E. Sternberg).

It seems that it is much easier to detect the influence of couple-stresses in fluids than in solids. For that purpose we considered theoretically a fluid for which we assumed that the dissipation function ϕ is a function of the first and second gradients of velocity of the particles [1]. From Ziegler's principle of the least irreversible force [2] we obtained the constitutive relations of the form

$$
\begin{aligned}
t^{(ij)} &= -p\,g^{ij} + \varrho\left(\frac{\partial\phi}{\partial d_{pq}}\,d_{pq} + \frac{\partial\phi}{\partial\omega_{pq,r}}\,\omega_{pq,r}\right)^{-1}\phi\,\frac{\partial\phi}{\partial d_{ij}}\,,\\
\mu^{ijk} &= -\varrho\left(\frac{\partial\phi}{\partial d_{pq}}\,d_{pq} + \frac{\partial\phi}{\partial\omega_{pq,r}}\,\omega_{pq,r}\right)^{-1}\phi\left(\frac{\partial\phi}{\partial\omega_{ij,k}} - \frac{1}{3!}\,\delta^{ijk}_{pqr}\,\frac{\partial\phi}{\partial\omega_{pq,r}}\right),
\end{aligned} \tag{1}
$$

where $t^{(ij)}$ is the symmetric part of the stress tensor, g^{ij} is the contravariant fundamental tensor, d_{ij} is the rate of strain tensor, $d_I = d^{k}_{.k}$, $\mu^{lk} = \frac{1}{2}\varepsilon^{lij}\,{}^{s}u_{ij}^{..k}$ is the deviatoric part of the couple-stress tensor. $w_{ij} = \frac{1}{2}\varepsilon_{ilm}\,\omega^{lm}_{,j}$ is the gradient of the vorticity tensor ω^{lm}, and $\eta_1, \eta_2, \eta_3, \eta_4$ are material constants which appear in the linearized constitutive equations,

$$
\begin{aligned}
t^{(ij)} &= -p\,g^{ij} + \eta_1\,d_I g^{ij} + 2\eta_2\,d^{ij},\\
\mu^{ij} &= -(2\eta_3\,w^{ij} + 2\eta_4\,w^{ji}).
\end{aligned} \tag{2}
$$

The differential equations of motion read

$$\varrho \frac{dV}{dt} = -\operatorname{grad} p + \left(\eta_v + \frac{1}{3}\eta_2\right) \operatorname{grad}(\operatorname{div} V) + \eta_2 \Delta V - \\ - \eta_3 \Delta\Delta V + \eta_3 \Delta[\operatorname{grad}(\operatorname{div} V)], \tag{3}$$

where η_V is the volume viscosity, η_2 is the shear viscosity and η_3 the "rotational viscosity" of the fluid.

We integrated [3] the equations of motion (3) in some cases of the viscometric flow. The obtained results show us that the velocities are smaller than in the case of non-polar fluids, which case is obtained when the rotational viscosity is neglected. For instance, in the case of the stationary flow through a circular pipe of radius R with the constant gradient of pressure k the distribution of velocities is given by the expression

$$V = \frac{k}{4\eta_2}(-r^2 + R^2) + \frac{k\eta_3}{\eta_2^2} \frac{I_0\left(\sqrt{\frac{\eta_2}{\eta_3}}R\right)}{I_0\left(\sqrt{\frac{\eta_2}{\eta_3}}R\right) + I_2\left(\sqrt{\frac{\eta_2}{\eta_3}}R\right)} \left[\frac{I_0\left(\sqrt{\frac{\eta_2}{\eta_3}}r\right)}{I_0\left(\sqrt{\frac{\eta_2}{\eta_3}}R\right)} - 1\right], \tag{4}$$

where I_0 and I_2 are the modified Bessel's functions. A similar result is obtained by Bleustein and Green [4]. The volume discharge Q per unit time through a cross-section of the pipe is given by the expression

$$Q = \frac{R^4 \bar{u} k}{8\eta_2} - \frac{R^2 \bar{u} k \eta_3}{\eta_2^2} \frac{I_2\left(\sqrt{\frac{\eta_2}{\eta_3}}R\right)}{I_0\left(\sqrt{\frac{\eta_2}{\eta_3}}R\right) + I_2\left(\sqrt{\frac{\eta_2}{\eta_3}}R\right)}. \tag{5}$$

The expression (5) may serve for the experimental determination of the coefficient of rotational viscosity η_3 and for the detection of the existence of couple-stresses in viscous fluids and of their influence on the velocities.

Granular materials are considered also by Dr. Cowin [5], but it seems that for experimental verifications it would be better to consider first the ordinary fluids, since in the case of granular materials probably the simultaneous effects of independent rotations of grains and of the second-order gradients of velocities of the grains can not be observed separately.

References

[1] Plavšić, M., and R. Stojanović: Naučno-tehnički pregled **7**, 3 (1966) (in Serbian).

[2] Ziegler, H.: Progress in Solid Mechanics, Vol. IV (Eds. Sneddon and Hill). Amsterdam: North-Holland 1963.

[3] Plavšić, M.: Thesis. Belgrade: 1966 (in Serbian).

[4] Bleustein, J. L., and A. E. Green: Int. J. Engng. Sci. **5**, 323 (1967).

[5] Cowin, S. C.: Thesis. Pennsylvania State Univ. 1962.

Answer to the comment by Dr. M. Plavšić

Masao Satake (Sendai, Japan): The author wishes to express his appreciation to Dr. M. Plavšić for his suggestive comment on the author's communication. It would be remarkable that Dr. M. Plavšić turns his attention to detect the influence of couple-stresses by experiments and that he points out the possibility of detection by using "rotational viscosity" η_3, a quantity indicating the influence of couple-stresses, in viscous fluids.

In granular materials, to the author's thinking, the influence of couple-stresses may appear in measurements of shear strength. To measure the shear strength, we usually employ the triaxial compression test and determine the internal friction angle using Mohr's circles. In the case where couple-stresses exist, however, the Mohr's circles are to be modified, as the author pointed out, and consequently the value of internal friction angle obtained by using unmodified Mohr's circles in usual manner may become slightly smaller than the real one. Thus, the author expects that this difference may be verified in two different shear tests, the triaxial compression test and the direct shear test, and that it may also detect the influence of couple-stresses in granular materials.

On Plastic Strain

By

N. Fox

Department of Applied Mathematics and Computing Science
University of Sheffield

1. Introduction

Truesdell and Noll [1] have given a continuum theory of dislocations for materially uniform, inhomogeneous, simple bodies. If such a body is also elastic, the current stress at each point X is a function of the deformation gradients, calculated relative to a fixed local reference configuration of a neighbourhood of X. In general, these deformation gradients do not coincide with the deformation gradients calculated from any global reference configuration of the whole body. The purpose of this communication is to show how certain theories may be interpreted in terms of moving local reference configurations and associated time dependent dislocation densities.

In other words, we suppose that the current stress at X is a function of the deformation gradients at X calculated relative to a moving local reference configuration. A further constitutive equation must then be specified to determine the gradient of the motion of the local reference configurations. This last gradient may be identified as the plastic strain and if a suitable equation is postulated for the plastic strain rate the theory can be shown (see Fox [2]) to be consistent with the plasticity theory of Green and Naghdi [3]. The essential difference is that instead of taking the total strain equal to the elastic strain plus the plastic strain we take the gradient of the total deformation to be the product of two other tensors from which we define the plastic and elastic strains. This makes possible an immediate interpretation of large deformation plasticity theory in terms of dislocation density. Another choice of the constitutive relations makes possible a similar interpretation of hypoelasticity.

2. Constitutive Equations

We use fixed cartesian axes. Let $\mathscr{C}_0$ be a global reference configuration for the whole body and let $\mathscr{C}_t$ be the configuration at the current time t. We denote by X_A the position of a typical particle in the configuration $\mathscr{C}_0$ and by x_i its position in $\mathscr{C}_t$.

The motion of the body is given by functions

$$x_i = x_i(X_A, t) \tag{2.1}$$

which we assume to be differentiable as many times as required. And we suppose that

$$\det(x_{i,A}) > 0. \tag{2.2}$$

The comma followed by an upper case latin suffix denotes differentiation with respect to X_A.

In order to consider the requirements of the principle of material indifference we need to consider a second motion obtained from (2.1) by superposing an arbitrary rigid body motion. That is, we consider the motion

$$x_i^* = Q_{ij}(t)\, x_j(X_A, t) + c_i(t) \tag{2.3}$$

where $Q_{ij}(t)$ is an arbitrary rotation tensor and $c_i(t)$ an arbitrary vector. The asterisk is used throughout to denote quantities associated with the motion (2.3).

Now suppose that for every motion of the body there exists a tensor field $p_{AB}(X_A, t)$ defined over $\mathscr{C}_0$ for all time such that

$$\det(p_{AB}) > 0 \tag{2.4}$$

and

$$p_{AB}^* = p_{AB}. \tag{2.5}$$

Then there also exists the inverse tensor P_{AB};

$$p_{AK}\, P_{KB} = \delta_{AB} = P_{AK}\, p_{KB} \tag{2.6}$$

and

$$P_{AB}^* = P_{AB}. \tag{2.7}$$

The tensor P_{AB} may be regarded as defining at time t a uniform reference for the body. That is, there exists a class of local reference configurations $\varkappa$ associated with each particle such that P_{AB} is the deformation gradient tensor from $\varkappa$ to $\mathscr{C}_0$ at that particle at time t. Then the gradient at time t of the deformation from $\varkappa$ to $\mathscr{C}_t$ is given by

$$x_{iA} = x_{i,K}\, P_{KA}. \tag{2.8}$$

And in virtue of (2.3), (2.7)

$$x_{iA}^* = Q_{ij}\, x_{jA}. \tag{2.9}$$

Hence, x_{iA} is a particular dipolar field as defined by GREEN and RIVLIN [4].

If we now suppose that the current stress σ_{ij} is a function of the gradients x_{iA} then, by the usual arguments of material indifference, this relationship must take a form equivalent to

$$\sigma_{ij} = x_{iA}\, x_{jB}\, F_{AB}(l_{KL}), \tag{2.10}$$

where

$$l_{KL} = x_{iK} x_{iL}. \tag{2.11}$$

To this relation must be added the usual equations of motion and continuity equation. Then to complete the theory, a further constitutive relation is required to determine P_{AB}. If P_{AB} is a given function of X_A independent of the time, the theory reduces to that of an inhomogeneous elastic material. If however we wish to consider any form of stress relaxation theory it would seem appropriate to take rate-type equations of the form

$$g_{AB}(p_{KL}, \dot{p}_{KL}, l_{KL}, \dot{l}_{KL}) = 0. \tag{2.12}$$

Equations of this type describe how the material shares out an increment of total deformation between the tensor l_{KL}, which changes the stress, and the tensor p_{KL} which alters the reference configurations.

If a special form of this relation is taken which is homogeneous of degree one in $\dot{l}_{KL}$ and $\dot{p}_{KL}$ and involves an independently specified yield function, the theory has been shown by Fox [2] to be consistent with the plasticity theory of Green and Naghdi [3]. If on the other hand, the stress relation (2.10) is taken to be isotropic with respect to the configurations $\varkappa$ and a simple relation is taken for $\dot{p}_{KL}$ in a form involving only quantities independent of the choice of $\mathscr{C}_0$ the theory has been shown by Fox [5] to reduce to that of hypoelasticity. The time dependent dislocation density field associated with each of these theories is then easily determined. For example, one measure of the dislocation density may be taken as

$$\omega_{Ai} = \varepsilon_{ijk} X_{Ak,j}. \tag{2.13}$$

It seems possible that more general forms of equation (2.12), perhaps homogeneous of degree two in $\dot{l}_{KL}, \dot{p}_{KL}$ would show further light on yield type phenomena. Moreover on physical grounds, one would expect that for materials in which stress relaxation takes place through the motion of dislocations, Eq. (2.12) should also depend on the dislocation density and possibly its gradients.

References

[1] Truesdell, C. A., and W. Noll: The Non-linear Field Theories of Mechanics. Handbuch der Physik (Ed. S. Flügge) III/3. Berlin/Heidelberg/New York: Springer 1965.
[2] Fox, N.: Quart. J. Mech. App. Math. **21**, 67 (1968).
[3] Green, A. E., and P. M. Naghdi: Arch. Rat. Mech. Anal. **18**, 251 (1965).
[4] Green, A. E., and R. S. Rivlin: Arch. Rat. Mech. Anal. **17**, 113 (1964).
[5] Fox, N. (Submitted for publication).

Physical Foundations of Dislocation Theory

By

V. L. Indenbom and **A. N. Orlov**

Institute of Crystallography
Academy of Sciences, Moscow

A. F. Yoffe Physico-Technical Institute
Academy of Sciences, Leningrad

1. Introduction

From the physical point of view the mechanical properties of crystalline solids are determined by their atomic structure and its transformations during the deformation process. The phenomenae of elasticity and plasticity hardly distinguishable in macroscopic mechanics differ principally in the atomic language [1, 2]. The elastic distortion u_{ij} corresponds to atomic displacements *without* replacements of nearest neighbour atoms (Fig. 1a), the interatomic bonds being distorted and

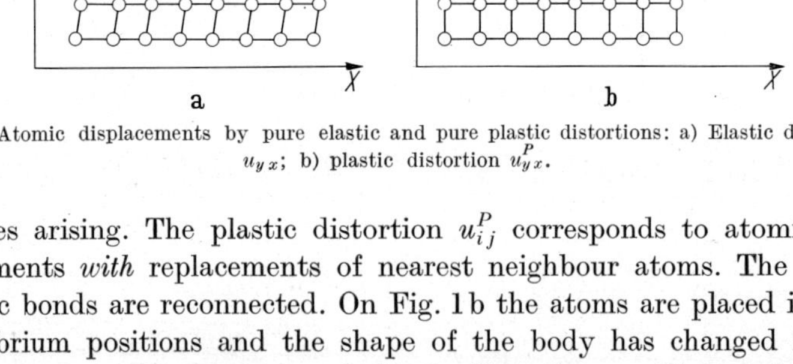

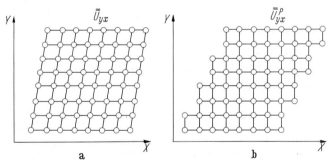

Fig. 1. Atomic displacements by pure elastic and pure plastic distortions: a) Elastic distortion u_{yx}; b) plastic distortion u^P_{yx}.

stresses arising. The plastic distortion u^P_{ij} corresponds to atomic displacements *with* replacements of nearest neighbour atoms. The interatomic bonds are reconnected. On Fig. 1b the atoms are placed in new equilibrium positions and the shape of the body has changed in the same manner as in the case of elastic distortion on Fig. 1a, but the interatomic bonds are not stressed and lattice twists are absent.

The successive reconnections of atomic bonds correspond to the movements in the crystal of some characteristic atomic configurations called lattice defects. Some typical exemples are shown on Fig. 2: point defects (vacancy and interstitial atom) and a line defect — the dislocation.

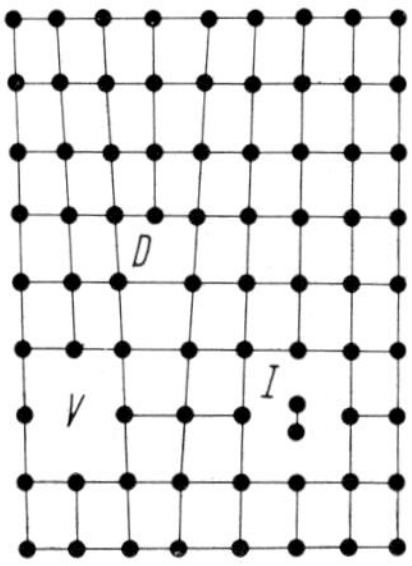

Fig. 2. Lattice defects in f.c.c. lattice: *V* Vacancy, *I* interstitial atom (dumb-bell configuration), *D* dislocation. The plane of the drawing is (110), the horizontal direction is [001].

The movement of dislocations plays the main role in plastic deformation. That is why the physical theory of plasticity of crystalline solids is based on the study of their dislocation structure. The comparison

a b c d

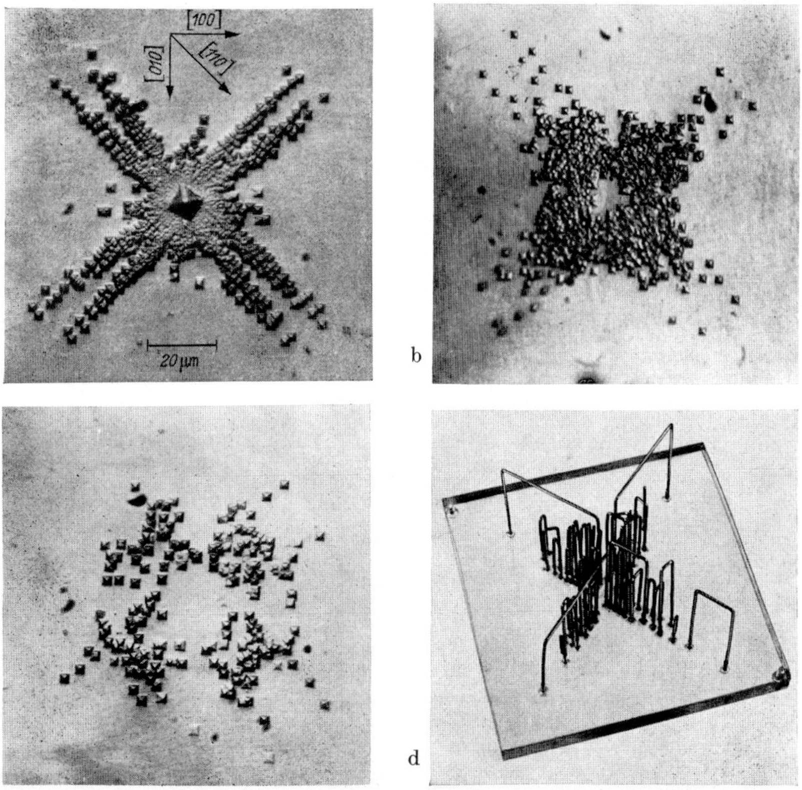

Fig. 3. Dislocation rosettes on a NaCl crystal arising by local loading [3]: a) Dislocation etch pits on the surface (001); b), c) the same after polishing away a layer 20 μ and 30 μ thick; d) reconstruction of the dislocation configuration of a rosette (wire model).

between the bodies' macroscopic properties and their dislocation structure becomes the foundation of modern material science. Instead of

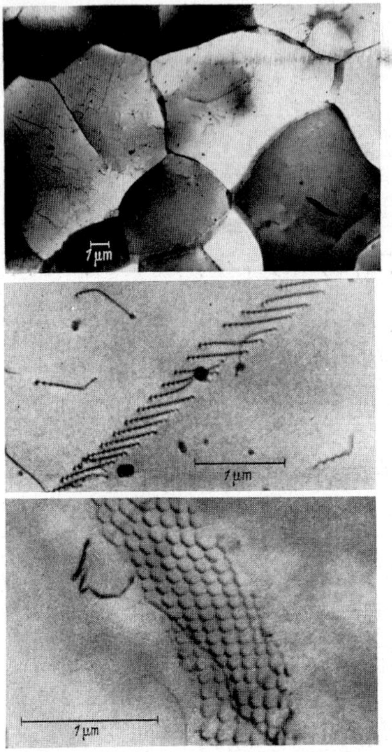

a

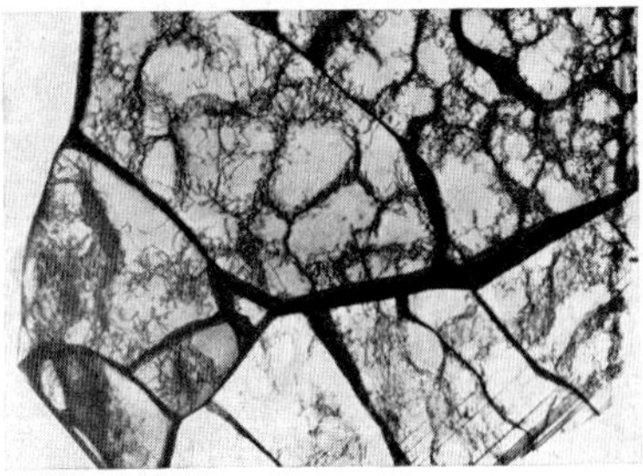

b

Fig. 4. a) Subgrain structure and dislocation structure of different subboundaries (dislocation networks) in Al after creep test, electron microscopy (MYSHLAJEV [4]); b) dislocation structure of subboundaries in MgO, x-ray topography (MIUSKOV [4]).

measuring the microhardness one explores the dislocation rosettes arising during indentation (Fig. 3). Instead of x-ray analysis of subgrain structure the dislocation networks forming thes ubboundaries are investigated by direct visualization (Fig. 4). The results of mechanical tests are compared not only with metallographic data but also with direct observations of the behaviour of dislocations in the material under consideration (Fig. 5).

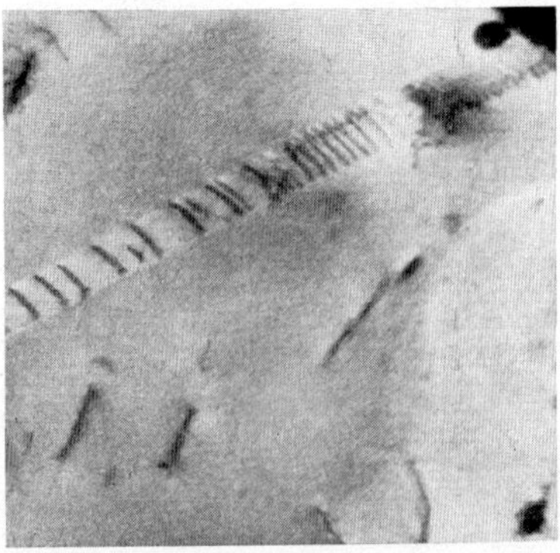

Fig. 5. The dislocations travel successively along a single slip plane and form a pile-up pressed against a grain boundary (alloy Ni + 20% Cr, deformation $\varepsilon = 0.7\%$) [5]. $\times 30000$.

2. Theory of Dislocations at Rest

In the approach of continuum elasticity dislocations are the vortex lines of the elastic field. As is well known, the movement of a fluid with vortices is no more potential and the circulation of the velocity on a closed contour is not zero but is proportional to the total intensity of all vortices encircled by the contour

$$\oint \mathfrak{v}\, d\mathfrak{l} = Q. \tag{1}$$

In the presence of vortex lines the potential φ of the velocity field may be constructed only by non-unique multifolded functions branching when encircling a vortex

$$\oint d\varphi = Q. \tag{2}$$

In the case of dislocations the role of the potential is played by the displacement vector $\mathfrak{u}$ branching when encircling a dislocation line

$$\oint_{\Gamma} d\mathfrak{u} = -\mathfrak{b}. \tag{3}$$

The vector $\mathfrak{b}$ is termed the total Burgers vector of the dislocations encircled by the contour Γ. Dislocations are the vortex lines of the tensor field of the elastic distortion. The Burgers vector determines the circulation of the elastic distortion around the dislocation

$$\mathfrak{b}_j = -\oint u_{ij}\, d\mathfrak{x}_i \quad \text{or} \quad \operatorname{Rot} \boldsymbol{u} = -\boldsymbol{\alpha}. \tag{4}$$

The tensor α_{ij} is named the dislocation density (Burgers vector flux density) and determines the sum of the Burgers vectors of all dislocations penetrating a unit area of the corresponding orientation. In the case of dislocations in crystals the Burgers vector must be one of the lattice vectors. In this respect dislocations resemble quantized vortices in superfluid helium where the circulation of the velocity is a multiple of PLANCK's constant.

Eq. (4) together with the equilibrium condition for the stresses

$$\operatorname{Div} \boldsymbol{\sigma} = 0 \tag{5}$$

and Hooke's law

$$\sigma_{ij} = c_{ijkl}\, u_{kl} \tag{6}$$

form a closed system of equations determining the internal stresses due to dislocations. Thus the distribution of the dislocations determines completely the structure of the internal stress field including macrostresses as well as microstresses [6] (and all the moments of the stress field too!).

The distortion $u_{mn}(\mathfrak{x}, \mathfrak{l}, \mathfrak{b}) = \mathfrak{b}_k\, u_{mnk}(\mathfrak{x}, \mathfrak{l})$ of a straight dislocation directed along the vector $\mathfrak{l}$ depends only on the component of the vector $\mathfrak{x}$ normal to $\mathfrak{l}$. The corresponding plane problem is solved by the methods of the theory of analytical functions. For any curvilinear dislocation loop Γ in an arbitrary anisotropic medium the solution may be represented in the following manner [7]:

$$\boldsymbol{u}(\mathfrak{x}) = -\frac{1}{2} \frac{\delta^2}{\delta \mathfrak{x}_i \delta \mathfrak{x}_j} \oint\limits_{\Gamma} \mathfrak{l}_i \, \boldsymbol{u}(\mathfrak{l}, \mathfrak{x} - \mathfrak{x}', \mathfrak{b}) \, d\mathfrak{x}_j'. \tag{7}$$

$\mathfrak{l}(\mathfrak{x})$ being the tangent vector of the loop Γ. If the shape of the loop is a polygon

$$\boldsymbol{u}(\mathfrak{x}) = \frac{1}{2} \sum_{k=1}^{n} (\mathfrak{x}_\alpha^{(k+1)} - \mathfrak{x}_\alpha^{(k)}) \frac{\delta}{\delta \mathfrak{x}_\alpha} [\boldsymbol{u}(\mathfrak{x}^{(k+1)} - \mathfrak{x}^{(k)}, \mathfrak{x} - \mathfrak{x}^{(k+1)}, \mathfrak{b}) - \\ - \boldsymbol{u}(\mathfrak{x}^{(k+1)} - \mathfrak{x}^{(k)}, \mathfrak{x} - \mathfrak{x}^{(k)}, \mathfrak{b})]. \tag{8}$$

$\mathfrak{x}^{(k)}$ being the vector-radius of the k-th angle of the polygon consisting of n sides ($\mathfrak{x}^{(n+1)} = \mathfrak{x}^{(1)}$). In the general case

$$u_{mn}(\mathfrak{x}) = -\frac{1}{2} \frac{\delta^2}{\delta \mathfrak{x}_i \delta \mathfrak{x}_j} \int \Big(\delta_{ik} \mathfrak{l}_j + \delta_{ij} \mathfrak{l}_i + \\ + \mathfrak{l}_i \mathfrak{l}_j \frac{\delta}{\delta \mathfrak{l}_k} \Big) u_{mnl}(\mathfrak{l}, \mathfrak{x} - \mathfrak{x}') \, \alpha_{kl}(\mathfrak{x}') \, (d\mathfrak{x}'). \tag{9}$$

Modern experimental techniques provide direct observations of stress and displacement fields of single dislocations. As an example

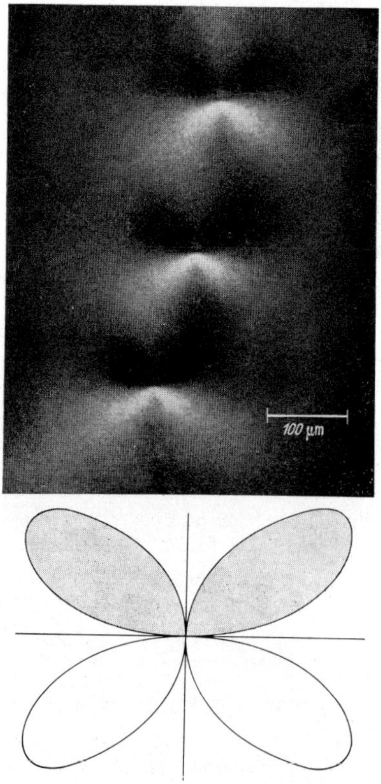

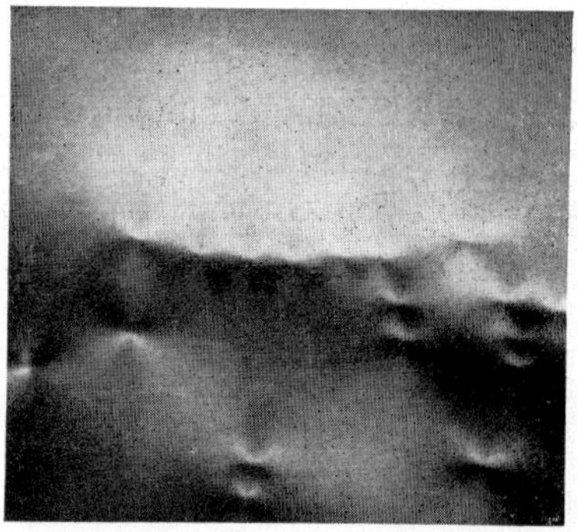

Fig. 7. Macro- and microstresses due to a set of dislocations in Si (photoelastic method [8]).

Fig. 6. Stress fields of single dislocations in Si revealed by photoelastic method. × 190. The calculated picture is shown at the bottom [8].

the stress field of single dislocations in a silicon crystal is represented on Fig. 6 (infra-red photoelasticity). The stress decreases as r^{-1} and may be detected as far as $r = 10^{-2}$ cm away from the dislocation whereas the Burgers vector amounts only to $b = 4 \cdot 10^{-8}$ cm. By superposition of the stresses of neighbour dislocations either long-range stresses arise (macrostresses) or the stresses are eliminated at distances of the order of the distance between the dislocations (microstresses). A typical example is represented on Fig. 7.

The system of Eqs. (4)—(6) determines the stresses provided the shapes of the dislocation lines are known. In this stress field a line element of the dislocation is acted upon by a configurational force

$$\mathfrak{f} = \mathfrak{l} \times \boldsymbol{\sigma}\, \mathfrak{b}\,.$$

In a general case $\boldsymbol{\sigma}$ means the sum of internal stresses $\boldsymbol{\sigma}_i$, stresses $\boldsymbol{\sigma}_e$ due to external loads and stresses $\boldsymbol{\sigma}_f$ corresponding to "frictional" forces (interactions of the dislocation with the crystal matrix). Completing the system (4)—(6) by the equilibrium condition of the dislocations

$$\mathfrak{f} = 0\,, \tag{10}$$

we get a closed system describing equilibrium dislocation configurations.

In such a framework flat dislocation arrays are thoroughly in-

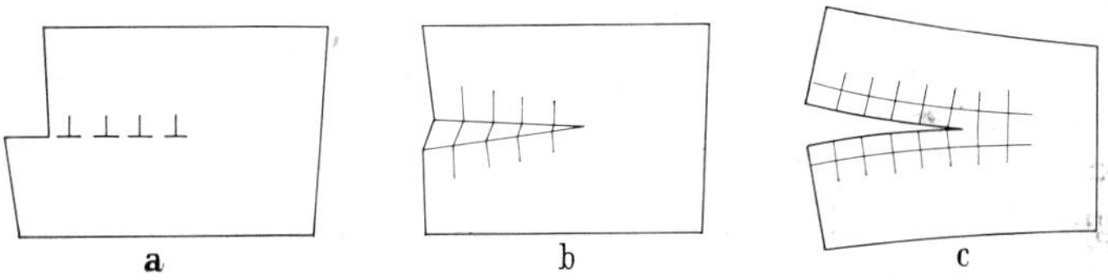

Fig. 8. Schema of the main applications of the theory of plane dislocation arrays: a) Glide bands; b) thin twins; c) plane cracks.

vestigated (glide bands, thin twins, plane cracks — Fig. 8). In these cases the system (4)—(6), (10) is reduced to the system of equations of equilibrium

$$\sigma_e(\mathfrak{x}_i) + \sigma_f(\mathfrak{x}_i) + \sum_j K(\mathfrak{x}_i, \mathfrak{x}_j) = 0\,. \tag{11}$$

$\mathfrak{x}_i$ is the coordinate of the i-th dislocation and K describes the interaction between the dislocations. In the case of an infinite body the kernel K is simply

$$K = K_0 (\mathfrak{x}_i - \mathfrak{x}_j)^{-1}. \tag{12}$$

Usually the dislocation distribution is supposed to be smeared out and the sum in (11) is replaced by an integral. The corresponding integral equation is the basis of the theories of glide bands [9], thin twins [10, 11] and plane cracks [12—14].

Curvilinear configurations are the most important ones in the applications of the dislocation theory. These configurations were usually investigated in the line tension approximation. To each line element of the dislocation a line energy $E(\mathfrak{l}, \mathfrak{b})$ is ascribed depending on the orientation of the element, and instead of searching for the minimum of the elastic energy [this search resulting in Eq. (10)] one searches for the minimum of the net line energy $\int E\,dl$.

Curvilinear configurations were investigated in the framework of an exact theory only recently. It turned out particularily that the

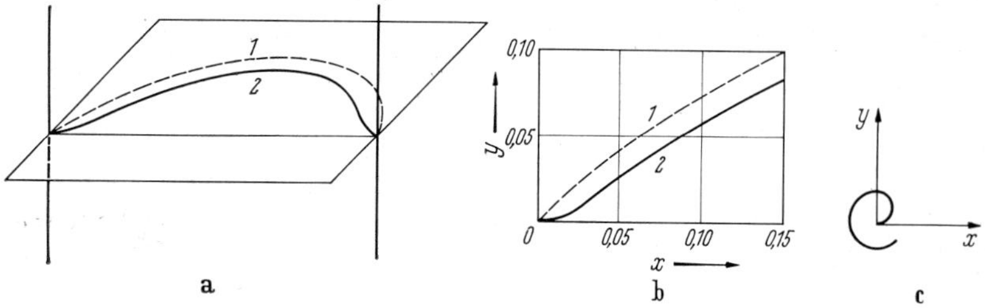

Fig. 9. Dislocation configuration of a Frank-Read source. *1* Line tension approximation, *2* exact calculation. a) General view; b) configuration near the node, x — distance along the segment, y deflection of the segment, both in units of the segment's length AB; c) expected configuration after the emission of the first loop.

elastic interactions of dislocations at the nodes of the dislocation network uniquely determine the orientations of the dislocation lines near the nodes, the nodes become rigid and strictly oriented [15]. As a result for example, the well known Frank-Read source has a more complicated configuration than is usually assumed (Fig. 9).

3. Theory of Mobile Dislocations

Every kind of plastic deformation consists of movements of point defects and dislocations in various proportions [16, 17]. In the case of diffusional plasticity the fluxes of point defects $\mathfrak{j}$ determine directly the transfer of matter (Fig. 10a). The gradient of the flux gives the rate of change of the plastic distortion

$$\dot{u}_{ik}^{P} = \mathfrak{j}_{k,i}. \tag{13}$$

The trace of the tensor Eq. (13) determines the local change of the density of the body

$$\frac{d}{dt}\frac{\Delta v}{v} = \dot{u}_{kk}^{P} = \operatorname{div}\mathfrak{j}. \tag{14}$$

In the case of dislocational plasticity the rate of plastic distortion is directly determined by the dislocation flux (Fig. 10b)

$$\dot{u}_{ij}^{P} = I_{ij}, \tag{15}$$

where I_{ij} is the j-th conponent of the net Burgers vector of all dislocations crossing per unit time a unit segment along the i-th axis. In the

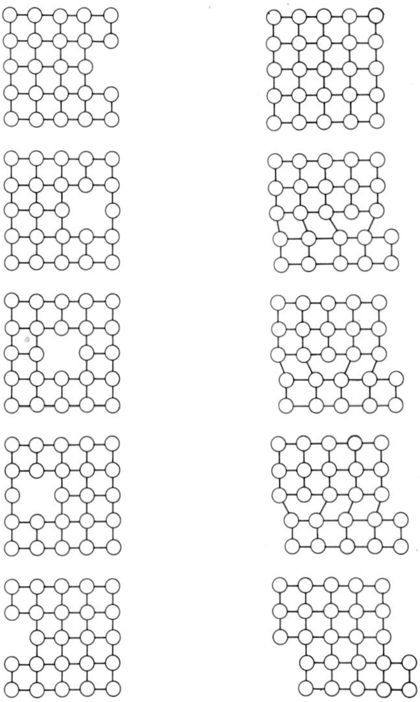
a b

Fig. 10. Plastic distortion as a result of migration of lattice defects. a) Diffusional transfer of matter by the movement of a vacancy through the crystal; b) translational slip by the movement of a dislocation through the crystal.

simplest case of parallel dislocations with equal Burgers vectors $b = \mathfrak{b}_j$ moving along the i-th axis with a velocity $v = \mathfrak{v}_i$ one has

$$\dot{u}_{ij}^{P} = \mathfrak{b}_j \, N \, \mathfrak{v}_i, \tag{16}$$

N being the dislocation density (the total line length per unit volume or the number of lines penetrating a unit area normal to the dislocation lines). For practical purposes the rate of plastic distortion may be estimated by the product of the density of mobile dislocations and their average velocity

$$\dot{u}^{P} = b \, N \, v. \tag{17}$$

If the mobility of single dislocations is known in terms of the local stress $\boldsymbol{\sigma}$ and the temperature T and the law governing the changes of the density of mobile dislocations is also known, then Eq. (17) enables one to predict the macroscopic plastic properties of the crystal. During the initial stages of plastic deformation of crystals with few dislocations the dislocation density depends on the amount of deformation, and the difference between the local and the average stresses may be estimated by the magnitude of N. Thus, the velocity of dislocations may be expressed through the external stress and the accumulated deformation. In this approximation one gets a satisfactory quantitative description of such phenomenae as the dynamical yield point, creep and stress relaxation [18—20], the macroscopic temperature, rate and time characteristics of these phenomenae being directly expressed in

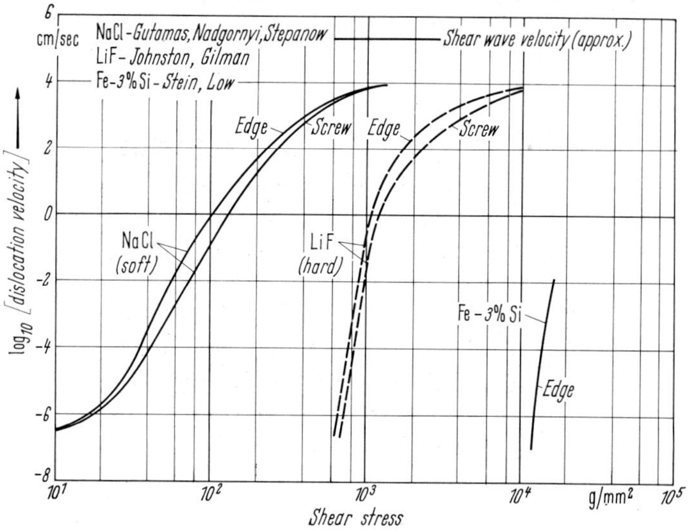

Fig. 11. Dislocation velocity vs. applied stress [21].

terms of the functions $v = v(\sigma, T)$ and $N = N(u^P)$, that were established by experimental investigations of the mobility of single dislocations and the peculiarities of their multiplication.

As a rule, the mobility of dislocations depends on the stress in a very drastic manner (Fig. 11). This enables one to introduce the concept of a start stress σ_0 necessary to force the dislocation to move. By equating σ_0 to the yield stress for a given slip system one can draw up a yield surface in the configurational space of stresses, this surface turning out to be a polyhedron [22]. Its shape corresponds to the symmetry of the crystal. The edges and corners of the polyhedron correspond to plastic instability resulting in the appearance of regions with deformations on different slip systems (deformation bands, irrational twins [23]). A simple example is illustrated on Fig. 12.

The theoretical investigation of the mobility of dislocations is a very hard problem complicated in all its aspects, starting from core structure up to the account of the influence of impurities. The most detailed theoretical investigations concern those mechanisms of dissipation that result in a linear stress dependence of the dislocation velocity. These mechanisms are essential only at very high velocities (thermo-

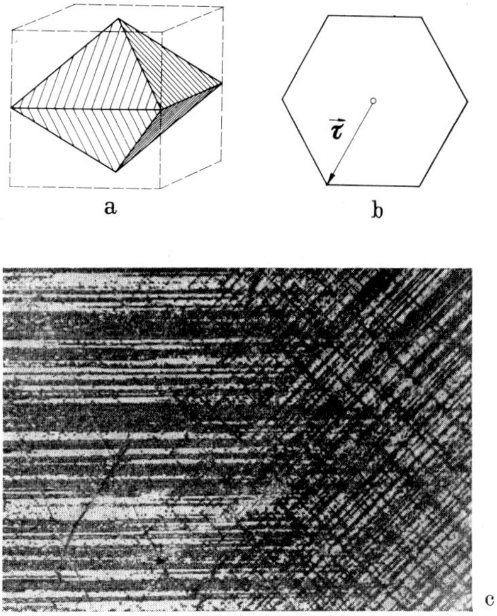

Fig. 12. An example of yield surface for an anisotropic medium [22]. a) Slip systems $\langle 110 \rangle$ $\{1\bar{1}0\}$ in a crystal of NaCl type; b) the corresponding yield surface in the configurational space of stresses; c) if the stress vector $\boldsymbol{\tau}$ points to a corner plastic instability arises and irrational twins [23] are formed.

elastic effect, scattering of phonons, forced scattering of thermal vibrations, dispersion of elastic moduli, cf. [24—26]).

In this case in Eq. (5) the inertial forces must be taken into account by a term $\varrho\, \ddot{\mathfrak{u}}$, ϱ is the density of the body, $\mathfrak{u}$ the total displacement, the gradient of $\mathfrak{u}$ being the total distortion, the sum of the elastic and the plastic distortions

$$\frac{\delta}{\delta \mathfrak{x}_i}\, \mathfrak{u}_j = u_{ij} + u^P_{ij}\,.$$

The establishment of the mechanisms controlling the dislocation mobility in real conditions is up to now a matter of discussion. Thus we are not yet able to write down instead of Eq. (10) a reasonable equation of motion of dislocations in a given stress field, thus setting up a closed system of equations governing moving dislocations. Various attempts to set up such a system that may be found in existing literature seem to us too formal.

The only exception to this is the movement of dislocations during diffusional creep. In this case the dissipative mechanism is well known, it is due to the diffusion of vacancies to and from the dislocations.

4. Kinetics of Dislocation Ensembles

The decisive problem of the dislocation theory of plasticity is to-day the kinetics of ensembles of a large number of strongly interacting dislocations. In this direction very many attempts are known that were undertaken in connection with searches for physical mechanisms of work hardening. In a first approximation the work hardening is due to the difference between the average and the local stress acting upon the dislocations. This difference depends on the average distance d between the dislocations as follows

$$\bar{\sigma} - \sigma_i \propto G\, b/d \simeq G\, b\, \sqrt{N}, \tag{18}$$

G being the shear modulus. It turns out, however, that essential points are the distribution of dislocations among the slip systems, the spatial inhomogenity of the dislocation density and various fine details of the dislocation structure of the crystals (cf. [27] and the reports of the recent Ottawa Conference [28]). One may attempt to describe the dislocation arrangement by a distribution function $\Phi_{\mathfrak{b}}(\mathfrak{l}, \mathfrak{r})$, $d\,\Phi_{\mathfrak{b}}(\mathfrak{l}, \mathfrak{r}) = \Phi_{\mathfrak{b}}\, dO\, dV$ being the total length of dislocations with Burgers vector $\mathfrak{b}$ penetrating the volume dV around the point $\mathfrak{r}$ and situated inside the solid angle dO near the direction $\mathfrak{l}$ [16].

Integrating the distribution function $d\,\Phi_{\mathfrak{b}}$ over all directions and summing up over all possible Burgers vectors one gets the total length of dislocations in a unit volume (the scalar density of dislocations)

$$N = \sum_{\mathfrak{b}} \int dO\, \Phi_{\mathfrak{b}}. \tag{19}$$

The total Burgers vector of all dislocations crossing a unit area of given direction determines the above mentioned tensor density of dislocations

$$\alpha_{ij} = \sum_{\mathfrak{b}} \int dO\, \mathfrak{l}_i\, \mathfrak{b}_j\, \Phi_{\mathfrak{b}}. \tag{20}$$

To describe the movement of dislocations one needs a third rank tensor

$$N_{ijk} = \sum_{\mathfrak{b}} \int dO\, \mathfrak{v}_i\, \mathfrak{b}_j\, \mathfrak{l}_k\, \Phi_{\mathfrak{b}}, \tag{21}$$

$\mathfrak{v} = \mathfrak{v}(\mathfrak{l}, \mathfrak{b})$ being the velocity of dislocations directed along $\mathfrak{l}$ with Burgers vector $\mathfrak{b}$. Of course, $\mathfrak{v}$ is normal to $\mathfrak{l}$. In the general case the macroscopic plastic distortion and the dislocation movement tensor are connected by the formula

$$\dot{u}^p_{ij} = \varepsilon_{mni}\, N_{mnj}. \tag{22}$$

Unfortunately up to now we have not succeeded in setting up a kinetic equation for the function $\Phi_{\mathfrak{b}}$, because the dislocation interactions display essential peculiarities at short distances and depend sensitively on the fine structure of the dislocation lines. When investigating the moments of the dislocation distribution one reveals important details of the dislocation structure only in the moments of very high order.

In this connection a general question arises concerning the choice of kinetic units (internal parameters) necessary to describe the evolution of the dislocation structure. It seems promising to choose as those units the most important elements of the dislocation structure (nodes, bends, different kinds of jogs and kinks on the dislocation lines, Fig. 13). Along this line we succeeded in setting up the kinetic equations in some simple cases [29], when the forces acting upon the dislocations were supposed to be known. But in the general case the connection between the fluxes of the elements of dislocation structure and the corresponding generalized forces is up to now a matter of exploration.

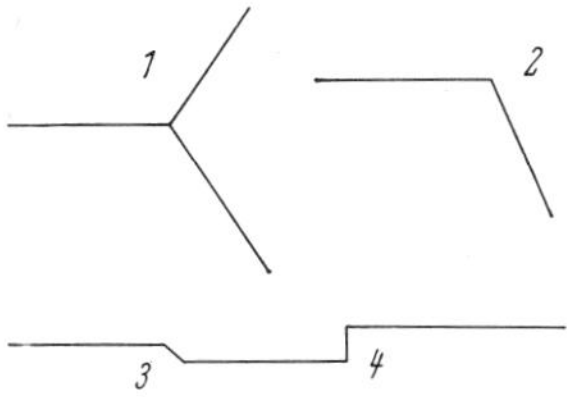

Fig. 13. Elements of dislocation structure: *1* Node; *2* bend; *3* kink; *4* jog.

If we look back to the analogy between dislocations and vortex lines it is possible to put forward a more exotic picture concerning the future progress of the dislocation theory. Our present attempts of building a dislocation theory of plasticity are similar to the attempts of constructing a theory of turbulence by the study of singular acts of motion intersection and interaction of vortex lines, loops and chains. In the theory of turbulence it was, however, found advantageous to utilize another way: the statistical study of the gradual transfer of energy from large pulses to small ones. In connection with this fact the question arises of applying this statistical procedure to the description of dislocation plasticity and other effects determined by the dislocation structure of the crystal. We mention one of these attempts [30].

5. Conclusions

The problems already solved in the dislocation theory make up the mechanics of dislocations rather than the mechanics of bodies with dislocations.

Only recently investigations of the mechanical properties of crystalline solids on single dislocation level have started to result in the necessary quantitative characteristics of the behaviour of dislocation ensembles. We mention for instance the refined experiments of Pred-

VODITELEV and co-workers [31], who measured directly the stresses necessary for pressing the dislocations through opposing dislocation arrays of various nature (Fig. 14). It seems that by proceeding in this

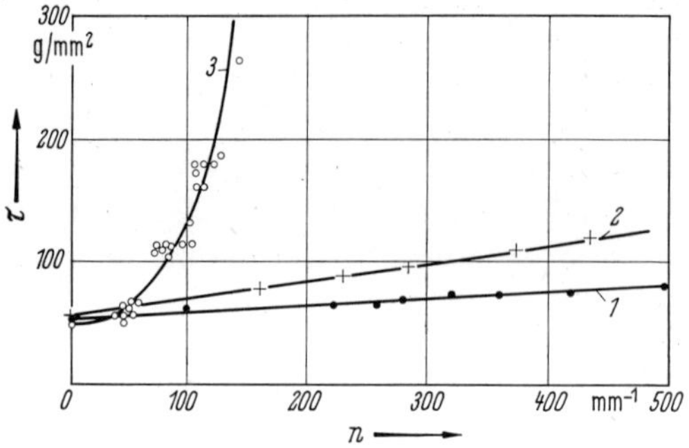

Fig. 14. Stress necessary to press a dislocation through arrays of opposite dislocations vs. the linear density of the dislocations in the subboundary (*1*) and slip bands without (*2*) and with (*3*) dislocation reactions [31].

way and prooving by experiment step by step each assumption of theory one should reach the desired goal sooner than by attempting to guess the mechanisms controlling dislocation movement and work hardening.

For different approaches see also [27, 32, 33].

References

[1] KRÖNER, E.: Kontinuumstheorie der Versetzungen und Eigenspannungen. Berlin/Göttingen/Heidelberg: Springer 1958.
[2] INDENBOM, V. L.: in: Plasticity of Crystals, Ed. by M. V. KLASSEN-NEKLYUDOVA. New York: Consultant Bureau 1962, p. 105.
[3] PREDVODITELEV, A. A., V. N. ROZHANSKI and V. M. STEPANOVA: Kristallografia **7**, 418 (1962).
[4] MIUSKOV, V. F., and A. R. LANG: Kristallografia **8**, 652 (1964). — MYSHLAJEV, M. M.: Fiz. tv. tela **9**, 1203 (1967).
[5] USSIKOV, M. P., and L. M. UTEVSKI: Fiz. metallov i met. **13**, 701 (1962).
[6] INDENBOM, V. L.: Theory of Crystal defects. Prague: Academia 1966, p. 257.
[7] INDENBOM, V. L., and S. S. ORLOV: Kristallografia **12**, 971 (1967).
[8] NIKITENKO, V. I.: in: Photoelastic method of stress analysis (in Russian). University of Leningrad 1966, p. 145.
[9] LEIBFRIED, G.: Z. Phys. **130**, 214 (1951).
[10] VLADIMIRSKI, K. V.: J. exp. theor. Phys. (USSR) **17**, 530 (1947).
[11] KOSEVICH, A. M., and L. A. PASTUR: Fiz. tv. tela **3**, 1291 (1961).
[12] STROH, A. N.: Adv. Phys. **6**, 418 (1957).
[13] BARENBLATT, G. I.: Prikl. mech. i techn. fiz. **4**, 3 (1961).
[14] INDENBOM, V. L.: Fiz. tv. tela **3**, 2071 (1961).
[15] INDENBOM, V. L., and G. I. DUBNOVA: Fiz. tv. tela **9**, 1171 (1967).
[16] INDENBOM, V. L., and A. N. ORLOV: Uspekhi fiz. nauk **76**, 557 (1962).
[17] SEEGER, A.: Handbuch der Physik VII/2. Berlin/Göttingen/Heidelberg: Springer 1958.

[18] JOHNSTON, W. G.: J. Appl. Phys. **33**, 2716 (1962). — GILLIS, P. P., and J. J. GILMAN: J. Appl. Phys. **36**, 3370, 3380 (1965).
[19] HAASEN, P.: Disc. Faraday Soc. **38**, 191 (1964).
[20] GOVORKOV, V. G., V. L. INDENBOM, V. S. PAPKOV and V. R. REGEL: Fiz. tv. tela **6**, 1039 (1964).
[21] GILMAN, J. J.: J. Appl. Phys. **36**, 3195 (1965).
[22] ORLOV, S. S., and YU. I. SIROTIN: Kristallografia **13**, 10 (1968).
[23] INDENBOM, V. L., and A. A. URUSOVSKAJA: Kristallografia **4**, 90 (1959).
[24] LOTHE, J.: J. Appl. Phys. **33**, 2116 (1962).
[25] SEEGER, A., and P. SCHILLER: in: Physical Acoustics, Ed. by P. MASON, Vol. 3A. New York: Academic Press 1966.
[26] INDENBOM, V. L., and A. N. ORLOV: Conference on Dynamic Behaviour of Dislocations. Kharkov 1967.
[27] NABARRO, F. R. N., Z. S. BASINSKI and D. B. HOLT: Adv. Phys. **13**, 193 (1964).
[28] Canad. J. Phys. **45**, No. 2 (1967), Part 2 and 3.
[29] ORLOV, A. N.: in: Theory of Crystal defects. Prague: Academia 1966, p. 317.
[30] BARENBLATT, G. I., and V. A. GORODTSOV: Prikl. mat. mech. **28**, 326 (1964).
[31] PREDVODITELEV, A. A., V. M. STEPANOVA and N. A. NOSOVA: Kristallografia **11**, 632 (1965).
[32] KROENER, E.: J. Math. Phys. **42**, N° 1, 27 (1963). — Chapt. 9 in A. SOMMERFELD: Mechanik der deformierbaren Medien, 5. Aufl. Leipzig: Akad. Verlagsges. Geest & Portig 1964.
[33] Moderne Probleme der Metallphysik (Ed. A. SEEGER), Bd. 1. Berlin/Heidelberg/New York: Springer 1965.

Geometry and Continuum Mechanics

By

B. A. Bilby

Department of the Theory of Materials
University of Sheffield

1. Introduction

"En somme jusqu'a present l'espace de RIEMANN est pour nous une collection de petits morceaux d'espace euclidien, mais il reste jusqu'a un certain point amorphe ..."

So said ÉLIE CARTAN [1] as a preface to a discussion of the idea of connexion. In 1922 he published two notes (CARTAN [2, 3]) introducing the geometrical idea of spaces with torsion[1], subsequently developing in further articles (CARTAN [4—6]) his resulting generalisations of varieties with affine connexion and their relation to the general theory of relativity; see also EINSTEIN [7]. CARTAN also noted in a later review on absolute parallelism (CARTAN [8]) the relevance of results of WEITZENBÖCK [9, 10] and of Ricci's coefficients of rotation (RICCI [11]).

CARTAN introduces at each point $\boldsymbol{m}$ of an X_n (SCHOUTEN [12]) an affine space $E_n(m)$ with basis $\boldsymbol{e}_i(m)$ and origin $\boldsymbol{m}$. If u_i' is a neighbouring point, he establishes a connexion by giving a $1:1$ mapping f: $E_n(m') \, \varepsilon \, \boldsymbol{m}' \rightarrow E_n(m) \, \varepsilon \, \boldsymbol{m}$, writing $f(\boldsymbol{m}', \boldsymbol{e}_i(m')) = \{\boldsymbol{m} + d\boldsymbol{m}, \boldsymbol{e}_i(u_i) + + d\boldsymbol{e}_i\}$ and

$$d\boldsymbol{m} = \omega^i \, \boldsymbol{e}_i(m), \quad d\boldsymbol{e}_i = \omega_i^j \, \boldsymbol{e}_j(m). \tag{1}$$

The $\omega^i \overset{\text{def}}{=} A_\varkappa^i \, d\xi^\varkappa$ and $\omega_i^j \overset{\text{def}}{=} \Gamma_i^j \overset{\text{def}}{=} d\xi^\varkappa \, \Gamma_{\varkappa i}^j$ are differential one-forms describing a formal *translation* and distortion ("affine rotation") in $E_n(m)$, the ξ^x being the coordinates of $\boldsymbol{m}$.[2] When the ω^i and Γ_i^j are not

---

Acknowledgment. The author is greatly indebted to Dr. J. D. ESHELBY for many valuable discussions.

[1] As Dr. HEHL remarked at the conference, CARTAN [2] refers in his paper to "les beaux travaux de MM. E. et F. COSSERAT", as well as to the work of H. WEYL. Great originators in mathematics have often been influenced by the stimulus of physical intuition; here perhaps is yet another example.

[2] CARTAN also uses this formulation when the $\boldsymbol{e}_i(m)$ depend on additional parameters.

perfect differentials, the X_n possesses *torsion* and *curvature* respectively. Evaluation of $\int d\boldsymbol{m}$ and $\int d\boldsymbol{e}_i$ for an infinitesimal closed circuit then gives (CARTAN [4], SCHOUTEN [12]) closure failures in a nearby E_n of

$$S^k \boldsymbol{e}_k \quad \text{and} \quad \Gamma_l^k \boldsymbol{e}_k \quad \text{where} \qquad [d\omega^k] \overset{\text{def}}{=} \Omega^k, \tag{2}$$

$$\Omega^k + \Gamma_s^k \wedge \omega^s \overset{\text{def}}{=} S^k \overset{\text{def}}{=} S_{\dot{m}\,l}^{\cdot\,\cdot\,k}\, \omega^m \wedge \omega^l, \tag{3}$$

$$[d\Gamma_l^k] + \Gamma_s^k \wedge \Gamma_l^s \overset{\text{def}}{=} \tfrac{1}{2} R_l^{\cdot k} \overset{\text{def}}{=} \tfrac{1}{2} R_{\dot{n}\,\dot{m}\,l}^{\cdot\,\cdot\,\cdot\,k}\, \omega^n \wedge \omega^m. \tag{4}$$

Here $[d\omega^k]$ and $[d\Gamma_l^k]$ denote the exterior derivative, the operator $\wedge$ is the exterior product of p-forms, $S_{\dot{m}\,l}^{\cdot\,\cdot\,k}$ and $R_{\dot{n}\,\dot{m}\,l}^{\cdot\,\cdot\,\cdot\,k}$ are the torsion and curvature tensors, and S^k and Γ_l^k are the torsion and curvature forms. Eqs. (3) and (4) are the "equations of structure" of the X_n. Exterior differentiation of these gives at once the second identity (SCHOUTEN [12]):

$$[dS^k] + S^s \wedge \Gamma_s^k = \tfrac{1}{2} \omega^s \wedge R_s^k \tag{5}$$

and the generalised Bianchi identity:

$$\tfrac{1}{2}[dR_l^{\cdot k}] + [d\Gamma_s^k] \wedge \Gamma_l^s - \Gamma_s^k \wedge [d\Gamma_l^s] = 0. \tag{6}$$

Cartan calls (5) and (6) the "théorème de conservation de la courbure et de la torsion"; they correspond (BILBY and SMITH [13], BILBY [14]) to node theorems in dislocation theory.

In the integration giving (3) we map the steps on the small circuit into one nearby E_n by using (1) and sum their images to obtain the resultant $S^i \boldsymbol{e}_i$. This corresponds exactly to the calculation of the local Burger's vector if we identify vectors parallel under (1) with those having the same crystallographic components at $\boldsymbol{m}'$ and $\boldsymbol{m}$ (see BILBY, BULLOUGH and SMITH [15] for $\Gamma_l^k = 0$, and GARDNER [16] and BILBY [14] for the general case). In a slightly different interpretation, Cartan regards (1) as specifying an infinitesimal "*affine displacement*"—that is, linear transformation—with components ω^i (translation) and Γ_i^j ("affine rotation") generating $E_n(m')$ from $E_n(m)$. There is a corresponding translation S^i [given by (3)] and "affine rotation" obtained by integrating the "affine displacement" round a small circuit (see e.g. [14]). In Cartan's words:

"La *translation* révèle la *torsion*, la *rotation* révèle la *courbure* de la variété donnée."

In some sense then the presence of torsion corresponds to an infinitesimal translation or sliding of basic vectors, of such a kind that a net translation results when a small closed circuit is executed about any point. We can say (BILBY and SMITH [13]) that the property of torsion arises when there is a resultant infinitesimal "local slip" in any small neighbourhood and that the resultant translation, identical when $\omega_i^k = 0$ with the infinitesimal affine displacement of CARTAN, corresponds to the resultant Burgers vector of the dislocations threading the small circuit considered. We also note (BILBY [14, 17, 18]) that dislocation density is an *affine concept*, independent of the notion of *metric*.

In the first part of my talk I shall describe briefly how we at Sheffield came to recognise this correspondence between dislocation density and torsion, which means that the dislocated crystal provides a truly remarkable realisation in Nature of a rather abstract geometrical concept originally conceived in a totally different context. It is especially interesting that this correspondence was discovered during the study of a very practical problem—the structure of steel.

When a suitable steel is quenched small lenticular regions within it change rapidly into another structure called martensite. The boundary between these regions and the matrix separates body-centred tetragonal crystal from face-centred cubic. We were interested at Sheffield in the theory of the crystallography of these martensitic transformations and required a quantitative description of such a boundary. So in July 1953 I came to consider the idea of a *surface dislocation* (BILBY [19])[1]—a two dimensional surface separating volumes between which there is a discontinuity both of lattice deformation and plastic deformation, and whose dislocation content is characterised by a dislocation tensor. Now NYE [20] refers to the relation between his theory of a crystal containing distributions of dislocations producing no far-reaching stress and FRANK's analysis [21] of the rotation boundary. NYE's work was stimulated by some then unpublished observations and ideas of WEST [22] on the bending of corundum crystals, which were used to make very fine dies. NYE rationalised the geometry of the lattice in a plastic deformation of this type by introducing a continuous distribution of dislocations, and a dislocation tensor α_{ij}. He incorporated the requirement that the dislocations should produce no far-reaching stress by supposing that, in the limit, the lattice basis was only *rotated* at each point, but not subject to any pure strain. By assuming that the dislocated lattice could be described by an arbitrary finite affine distortion varying smoothly from point to point we obtained[2] the general fundamental geometrical relation connecting the finite lattice distortion and the dislocation density, and the formulae for the surface dislocation tensor, see also BILBY, GARDNER and

---

[1] A preliminary report of work with R. BULLOUGH and E. SMITH; read at the Bristol Conference on Defects in Crystalline Solids July 1954, MS amended November 1954.

[2] READ [23] independently suggested a similar method for obtaining Nye's results, see [19].

SMITH [24]. This fundamental geometrical relation for small distortions was also reported by KRÖNER [25]. Subsequently the surface dislocation theory was applied to discuss the crystallography of martensitic transformations (BULLOUGH and BILBY [26], CROCKER and BILBY [27]).

The relation between dislocation density and torsion was recognised in May 1954 and reported in the following year (BILBY, BULLOUGH and SMITH [15][1]). In 1956 (see [13]) Dr. J. D. ESHELBY informed the present author of the work of K. KONDO, who had independently recognised the relation between torsion and dislocation density in 1952 (KONDO [28]). At this time also ESHELBY [29] obtained from the linear form of the fundamental geometrical relation, the formula giving the incompatibility tensor in terms of the dislocation tensor, showing it to be equivalent to the symmetrical formula of KRÖNER [25] because of the node theorem.

Once torsion is identified with dislocation density, of course, the whole apparatus of differential geometry and its supporting disciplines at any level of sophistication becomes on the one hand available for the development of dislocation theory, and on the other interpretable in its terms. We can see realisations of Lie groups and fibre bundles [18], and note how simple an interpretation is provided of many topological and geometrical theorems. If, for example, $S_{ml}^{\cdot\cdot k}$ is constant and $R_{nml}^{\cdot\cdot\cdot k} = 0$, the dislocated crystal is a direct representation of a simply transitive Lie group (EISENHART [30]). Associations with other branches of theoretical physics, are immediate, particularly with plasticity, electromagnetism, relativity and the theory of fundamental particles. There is thus a welcome unification, and a rich source of study for the interplay of ideas from widely different disciplines.

The state considered by NYE [20] provides simple and intriguing examples of these inter-relations. The presence of variable lattice *rotation* only, obviously incompatible without lattice strain, and the distinction between the behaviour of scribed lines and lattice lines shows the association with plasticity. The lattice connexion of distant parallelism is metric with respect to a Euclidean metric [13], and the lattice lines, which are the autoparallel curves or paths of the connexion, are congruences of curves in ordinary space.

---

[1] E. SMITH's most important contribution was to pose and solve the problem of following a constant crystallographic direction through the continuously rotated lattice of NYE. He used a kinematic argument, imagining a particular lattice direction carried round as the lattice turned infinitesimally about the instantaneous axis of rotation associated with a small spatial increment. R. BULLOUGH made a valuable contribution towards the derivation of the correct formula for the dislocation density when the lattice distortions are finite; the recognition of the relation between dislocation density and the torsion of a connexion making the lattice everywhere parallel was due to the present author.

Nye's lattice curvature relations correspond to the fact that the general expression for a linear connexion (see SCHOUTEN [12], Ch. III) now completely determines the connexion in terms of its torsion [13, 14]. In an anholonomic system (a) based on an orthonomal lattice basis, the lattice connexion Γ^a_{cb} vanishes identically and the dislocation tensor $S_{cb}^{\cdot\cdot a}$ is related to Ricci's coefficients of rotation γ^a_{cb} for the congruences by [13]

$$2 S_{cb}^{\cdot\cdot a} = -(\gamma^a_{cb} - \gamma^a_{bc}). \tag{7}$$

These coefficients appear in the Poisson (Lie) bracket $[A_c, A_b]$ of the differential operators (lattice tangent vectors) $A_a = A^x_a\, \partial_x$ on the congruences

$$[A_c, A_b] = A_c A_b - A_b A_c = -2 S_{cb}^{\cdot\cdot a} A_a. \tag{8}$$

The Jacobi identities

$$[A_a[A_b, A_c]] + [A_b[A_c, A_a]] + [A_c[A_a, A_b]] = 0 \tag{9}$$

now correspond to the second identity (5), that is to the node theorem [13]. A congruence, say $a = 3$, is normal (EISENHART [31]) if $\gamma^3_{cb} - \gamma^3_{bc} = -2 S_{cb}^{\cdot\cdot 3} = 0$, for $c, b = 1, 2$; this simply means that $\alpha^3_3 = 0$, that is, that no screw dislocations thread the surface normal to the $a = 3$ congruence [13]. There are thus no "steps" in this surface!

Our main interest at Sheffield has always been with the geometrical aspects of the theory, and with its application to the theory of anisotropic crystal plasticity. It was recognised from the outset (for the work of NYE and the surface dislocation theories of martensite [26] and BILBY and CHRISTIAN [32] show this most graphically) that the theory involved several separate types of shape change and that the total shape deformation could be resolved into a formal sequence of distortions ([13], KRÖNER and RIEDER [33], BILBY, GARDNER and STROH [34]). The resulting geometrical relations arising in various modes of deformation with lattice rotation only have been analysed in some detail (NYE [20], BILBY, BULLOUGH, GARDNER and SMITH [35], BILBY and GARDNER [36]). The (rather complicated) form of the glide surfaces and crystal shape predicted for finite twisting in conditions of single glide [36] are in reasonable agreement with observations of WHAPHAM and WILMAN [37]. In another application it was shown that the main equations of READ's theory of plastic-elastic bending [38] follow from the linear elastic stress-strain relations and the fundamental geometrical relation; GARDNER [16] treated plastic-elastic twisting by this method. KRÖNER [25, 39, 40, 41] treated extensively the calculation of internal elastic stresses from given dislocation distributions on the linear theory, while the above resolution of the total strain, together with the condition that the curvature tensor $R_{\nu\mu\lambda}^{\cdot\cdot\cdot\varkappa}$ of the lattice connexion must vanish, was made the basis of a non-linear theory of the stresses produced by a given distribution of dislocations and extra matter (Kröner and Seeger [42], Pfleiderer, Seeger and Kröner [43], Kröner [44], Stojanovitch [45], Bilby, Gardner, Grinberg and Zorawski [17]).

The principal accounts of all this work are in Kondo [46–49], Eshelby [29], Kröner [41, 44], Bilby [14, 18], Truesdell and Noll [50] and Noll [51]. Other relevant topics—the Cosserat continuum (see Günther [52], the "director" theories of A. E. Green and his co-workers (N. B. Green and Naghdi [53]), and the general questions of inhomogeneity and couple stresses in continuum mechanics—will be discussed by more appropriate members of this meeting. Clearly, the original description [13, 15] of the continuously dislocated crystal can be related to a three-director theory in which, for example, the directors give the lattice distortion, the total shape change is the deformation of the medium and the plastic distortion is their combination (Fox [54]). Dynamical theories may involve a fixed dislocation density ("inhomogeneity") or one which develops in time, giving an incremental theory appropriate to plasticity and the dislocation motion ([13, 14, 33, 34, 47, 48], Kröner [49], Holländer [55, 56], Ben Abraham [57], Simmonds [58], Lemprière [59], Indenbom and Orlov [60, 61], Mura [62–65], Bross [66], Fox [67], Sedov and Berditchevski [68], Günther [71]). With more than three directors we have the possibility of descriptions of inhomogeneous local distortions. Theories in which the lattice connexion has both curvature and torsion are also of interest ([16, 14, 47–49], Anthony, Essmann, Seeger and Träuble [69], Schaefer [70]). In these, surface dislocations are single entities [29, 14] and there are interesting interpretations of the generalised Bianchi identities which describe the geometrical interplay of dislocations, surface dislocations, holes and cracks [14, 46–48, 42]; they are related also to "terminating dislocations" (Eshelby and Laub [72]). However, we shall not discuss these questions further but confine ourselves firstly to some comments on Noll's work; and secondly to preliminary accounts of the idea of displacement in R_n, $n > 3$ and of a potential new application of these geometrical ideas—the theory of *folding*.

2. Materially Uniform Simple Bodies

The theory of continuous distributions of dislocations is associated with the idea of a continuous body in which at each point at least two independent structures can be defined.

Classically, these are the shape changes firstly of a crystal *lattice*, and secondly, of a "shape lattice" scribed on the crystal; plastic slip accounts for the difference between them. In the analysis we use auxiliary abstract continuous bodies which undergo these shape changes separately;

they can do so and remain continuous only in generalised spaces. We may identify a structure at each point of a current state of the body in real space [15] cut it up and specify the distortions needed to make the structures everywhere congruent (usually, that is, with corresponding vectors everywhere equal and parallel); from this knowledge we may construct a manifold in which the cut up elements fit continuously [46, 42, 44, 17]. Equivalently, we construct the cut-up elements (a "uniform reference") and use them to assign a geometric structure to a manifold $\mathscr{B}$ identified with the body ([50], Noll [51])[1]. A structure may be assigned by the mechanical response only [50]. Other physical responses are envisaged in a fuller account, which gives a valuable axiomatic treatment [51]. The original appeal to direct microscopic identification of a structure from its response to radiation also has merit, especially in dislocated crystals. Clearly from the affine character of the dislocation concept [17] and its relation to torsion [15] and curvature [14] any property defining a parallelism will define a dislocation density. The continuum theory of dislocations, e.g. [29], based on displacements and Euclidean parallelism without the notion of lattice, recognises this implicitly, and there have been other explicit applications, e.g. [73].

We now review [50] and [51], modifying their presentation and making clear its relation to the original work. They treat materially uniform bodies so that we do not need the notions of extra-matter or non-metric connexions [44], [17]. Also, the dislocation density is fixed in the motions considered so there is no account of the course of plastic flow. The particles P of the body $\mathscr{B}$ are identified with the points of a region of an X_3 [12] with coordinates $P^{\varkappa'}$, $(\varkappa')$ holonomic. A *global configuration* of $\mathscr{B}$ is a placing of the particles (Fig. 1) into positions $x^{k'}$, (k') holonomic, of Euclidean space R_x. This placing is defined by a one-one continuous map $\boldsymbol{\chi}$: $\mathscr{B} \to R_x$, $x^{k'} = \chi^{k'}(P^{\varkappa'})$; $P^{\varkappa'} = \overset{-1}{\chi}{}^{\varkappa'}(x^{k'})$ and represents the usual final or current state. A one-one linear mapping M_P: $T_P \to T_x$, $dx^{k'} = M^{k'}{}_{\varkappa'}(P^{\varkappa'})\, dP^{\varkappa'}$, of the tangent space T_P of $\mathscr{B}$ at P onto the tangent space T_x of R_x at x (T_x may be identified for all x with a centred R_3) is called a *local configuration.* The set of all such maps is denoted by $\mathscr{L}(T_P, T_x)$; it represents placings of small neighbourhoods $N(P)$ of $\mathscr{B}$ onto neighbourhoods $N(x)$ of R_x. For each P, $\boldsymbol{\chi}$ induces a local configuration

$$\boldsymbol{G}_P \in \mathscr{L}\,(T_P, T_x), \qquad dx^{k'} = G^{k'}_{\cdot\ \varkappa'}(P^{\varkappa'})\, dP^{\varkappa'}$$

---

[1] The MS of [51] was made available to the author at the conference, after he had presented the substance of the account here given of [50]. Accordingly [50] and [51] are discussed together here; equation numbers of these papers are prefixed by the numbers [50] and [51].

with

$$G^{k'}_{\cdot\ \varkappa'} = \partial \chi^{k'} / \partial P^{\varkappa'} \tag{10}$$

but not all $\boldsymbol{M}_P$ may be derived from a global configuration in this way. A *response descriptor* $\mathcal{G}: \mathcal{B} \to \mathcal{R}$ assigns to each P a physical response (e.g. a stress) $\mathcal{R}$, and a body is *simple* if $\mathcal{G}_P = \mathcal{G}(P)$ depends only on the local configuration $\boldsymbol{M}_P$ at P, $\mathcal{G}_P = \mathcal{G}_P(\boldsymbol{M}_P)$. Consider now a map

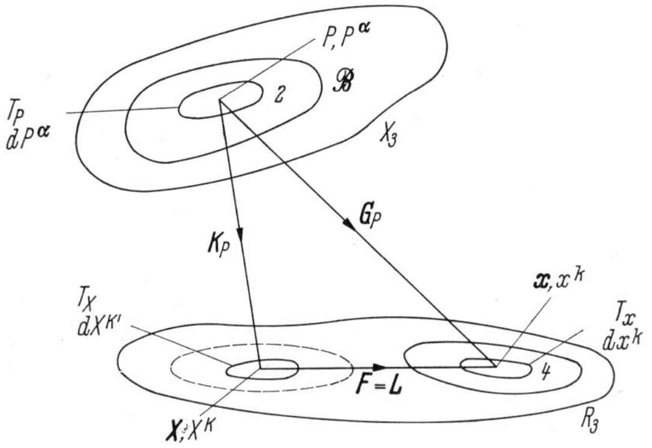

Fig. 1. The maps $\boldsymbol{K}_P$, $\boldsymbol{G}_P$ and $\boldsymbol{F}$.

$\boldsymbol{K}: \mathcal{B} \to \mathcal{L}(T_P, T_X)$ assigning to each $P \in \mathcal{B}$ a $\boldsymbol{K}_P = \boldsymbol{K}(P) \in \mathcal{L}(T_P, T_X)$. In general frames $(\varkappa)$ and (K) we write

$$d X^K = K^K_{\cdot\ \varkappa}(P^\varkappa)\, d P^\varkappa \tag{11}$$

and if $\boldsymbol{K}_P$ is derived from a global configuration we have

$$\overset{-1}{K}{}^{\varkappa}_{\cdot\ K}\, \partial_{[\mu} K^K_{\cdot\ \lambda]} + \Omega^{\varkappa}_{\mu\lambda} - \overset{-1}{K}{}^{\varkappa}_{\cdot\ K} K^M_{\cdot\ \mu} K^L_{\cdot\ \lambda} \Omega^K_{ML} = 0 \tag{12}$$

where $\Omega^{\varkappa}_{\mu\lambda}$ and Ω^K_{ML} are the anholonomic objects of $(\varkappa)$ and (K). $\boldsymbol{K}_P$ does not usually satisfy (12), and the $N(X)$ are disjoint or overlap, for only the placings of neighbourhoods are unique.

We ascribe the following property to $\boldsymbol{K}$ (cf. [17]):

In the placing of $\mathcal{B}$ into R_X defined by $\boldsymbol{K}$ all local distances are invariant and vectors parallel in R_X are called parallel in $\mathcal{B}$. (13)

As we now show this fixes the geometry of $\mathcal{B}$ when that of R_X is known. Let P be placed on X. Then $\boldsymbol{K}_P$ associates small differences of labels $dP^\varkappa$ with small differences of position dX^K. By (13) it thereby associates a distance with $dP^\varkappa$ and the equality $ds^2 = b_{KL}\, dX^K\, dX^L = g_{\varkappa\lambda}\, dP^\varkappa\, dP^\lambda$ defines a metric $g_{\varkappa\lambda}$ on $\mathcal{B}$ given by (cf. (TN 34.3) and (N 12.1))

$$g_{\varkappa\lambda} = b_{KL}\, K^K_{\cdot\ \varkappa} K^L_{\cdot\ \lambda}. \tag{14}$$

If $\mathcal{B}$ is simple, the response of P in the placing $\boldsymbol{K}$ is entirely fixed by the local placing, $\boldsymbol{K}_P$, and it is sufficient to make any three inde-

pendent assignments $\underset{a}{dP^\varkappa} = \underset{a}{e^\varkappa}$ of label differences in $\mathscr{B}$ to position differences
$\underset{A}{dX^K} = \underset{A}{E^K}$ in R_X, to fix all of them, where by (11)

$$\underset{a}{e^\varkappa} = \delta_a^A \overset{-1}{K^\varkappa_{\cdot\ K}} \underset{A}{E^K}. \tag{15}$$

Let us identify, in the placing of $N(P)$, the vectors $\underset{A}{E^K}(P)$ with the characteristic structure assigned by the descriptor $\mathscr{G}_P(\boldsymbol{K}_P)$; for example they could be a set of crystallographic directions in an anisotropic elastic crystal. Now consider the placing of another particle Q about which we have label differences $dQ^\varkappa$. The absolute position of Q is at the moment irrelevant; we have in fact only unique placings of neighbourhoods under $\boldsymbol{K}(P)$, but no global uniqueness. We now set down $N(Q)$ by $\boldsymbol{K}_Q$ and suppose that we can adjust $\boldsymbol{K}_Q$ (i.e. adjust the assignment of labels to positions) until the response of $N(Q)$ is the same as that of $N(P)$, that is, until $\mathscr{G}_P(\boldsymbol{K}_P) = \mathscr{G}_Q(\boldsymbol{K}_Q)$. We assume that the selected vectors $\underset{A}{E^K}(P)$ are then parallel and equal to the congruent structural vectors $\underset{A}{E^K}(Q)$ which arise from label differences $\underset{a}{e^\varkappa}(Q)$ at Q given by[1]

$$\underset{a}{\boldsymbol{e}}(Q) = \overset{-1}{\boldsymbol{K}}(Q)\,\underset{A}{\boldsymbol{E}}\,\delta_a^A = \boldsymbol{\Phi}_{QP}\,\underset{a}{\boldsymbol{e}}(P) \tag{16}$$

where

$$\boldsymbol{\Phi}_{QP} = \overset{-1}{\boldsymbol{K}}(Q)\,\boldsymbol{K}(P). \tag{17}$$

The $\underset{a}{e^\varkappa}(Q)$ so determined are structurally congruent to $\underset{a}{e^\varkappa}(P)$ and $\Phi_{QP}: T_P \to T_Q$ is called a *material isomorphism* with respect to $\mathscr{G}$. From (17), $\mathscr{G}_P(\boldsymbol{K}_P) = \mathscr{G}_Q(\boldsymbol{K}_P\,\Phi_{PQ})$. If $\boldsymbol{g}_{PQ}$ is the set of all $\boldsymbol{\Phi}_{PQ}$, $\boldsymbol{g}_{PP} \overset{\text{def}}{=} \boldsymbol{g}_P$ is the *intrinsic symmetry* (*isotropy*) *group* $(\boldsymbol{g}_P\,\boldsymbol{e}(P) = \boldsymbol{e}(P))$ of the material at P and obviously

$$\boldsymbol{g}_P = \boldsymbol{\Phi}_{PQ}\,\boldsymbol{g}_Q\,\boldsymbol{\Phi}_{QP}. \tag{18}$$

If a Φ_{PQ} exists for all $P, Q \in \mathscr{B}$, $\mathscr{B}$ is *materially uniform* and $\boldsymbol{K}$ is a *uniform reference*. From (17) and (18), $\boldsymbol{K}_{(P)}\,\boldsymbol{g}_P\,\overset{-1}{\boldsymbol{K}}_{(P)} = \boldsymbol{K}_{(Q)}\,\boldsymbol{g}_Q\,\overset{-1}{K}_{(Q)} = \boldsymbol{g}_{\boldsymbol{K}}$, say, where $\boldsymbol{g}_{\boldsymbol{K}}$ is independent of P and is called the *symmetry* (*isotropy*) *group of* $\mathscr{B}$ *relative to* $\boldsymbol{K}$ $(\boldsymbol{g}_{\boldsymbol{K}}\,\boldsymbol{K}(P)\,\boldsymbol{e}(P) = \boldsymbol{K}(P)\,\boldsymbol{e}(P))$. If $\boldsymbol{S}(P) \in \boldsymbol{g}_{\boldsymbol{K}}$ and $\boldsymbol{L}$ is constant, then clearly $\boldsymbol{L}\,\boldsymbol{S}(P)\,\boldsymbol{K}(P)\,\boldsymbol{e}(P)$ is constant for all P if $\boldsymbol{K}(P)\,\boldsymbol{e}(P)$ is, so that $\boldsymbol{K}'(P) = \boldsymbol{L}\,\boldsymbol{S}(P)\,\boldsymbol{K}(P)$ is also a uniform reference. If $\boldsymbol{g}_{\boldsymbol{K}}$ is discrete, $\boldsymbol{S}(P)$ must be constant for continuous $\boldsymbol{K}^K_{\cdot\varkappa}(P^\varkappa)$[2]; this corresponds to a fixed labelling of the lattice in dislocation theory

---

[1] It is sometimes desirable in dislocation theory *not* to require congruent vectors to have the *same length*; for example [17, 18], in order to have no dislocations in *elastic* thermal stress.

[2] See (N 14.1); Wang, this volume, discusses continuous $\boldsymbol{g}_{\boldsymbol{K}}$.

[15]. We assume the $\boldsymbol{K}^K_{.\varkappa}(P^\varkappa)$ to be finite, continuous and single-valued, thus excluding Möbius crystals [15] and other topological oddities[1]. Then they determine at each point P unique equivalent structural vectors $\underset{a}{e^\varkappa}$ which, with their reciprocals $\overset{a}{e}_\varkappa$, we may use to define on $\mathscr{B}$ a connexion $\Gamma^\varkappa_{\mu\lambda}$ making the $\underset{a}{e^\varkappa}$ everywhere parallel (cf. [15, 17])

$$\Gamma^\varkappa_{\mu\lambda} = \underset{a}{e^\varkappa}\, \partial_{[\mu} \overset{a}{e}_{\lambda]}\,. \tag{19}$$

If the $\underset{A}{E^K}$ are everywhere parallel in R_X, then $\partial_M \overset{A}{E}_L = b^K_{ML} \overset{A}{E}_K$ where $\overset{A}{E}_K$ are reciprocal to $\underset{A}{E^K}$ and b^K_{ML} are the Christoffel symbols in (K) of the Euclidean metric b_{ML}. Hence from (15) and (19) we get

$$\Gamma^\varkappa_{\mu\lambda} = \overset{-1}{K^\varkappa_{.}}{}_K\, \partial_\mu K^K_{.}{}_\lambda + \overset{-1}{K^\varkappa_{.}}{}_K\, K^M_{.}{}_\mu\, K^L_{.}{}_\lambda\, b^K_{ML}\,. \tag{20}$$

Thus (cf. [17]) $\Gamma^\varkappa_{\mu\lambda}$ is related to b^K_{ML} as it would be if $K^K_\varkappa$ specified a change of coordinates; we may say [17] that it is b^K_{ML} "dragged along" by $\overset{-1}{\boldsymbol{K}}$. If we have both $\boldsymbol{K}$ and a global map $\boldsymbol{\chi}$ then $\boldsymbol{F} = \boldsymbol{G}_P \overset{-1}{\boldsymbol{K}}(P)$ is a map $T_X \to T_x$ describing the local distortion $\boldsymbol{F}$ of a current state relative to the uniform reference, and a materially uniform body admits a $\boldsymbol{K}$ such that a *response function relative to* $\boldsymbol{K}$, $\mathscr{H}_{\boldsymbol{K}}(\boldsymbol{F}) = \mathscr{G}_P(\boldsymbol{F}\,\boldsymbol{K}(P))$ exists for all P and $\boldsymbol{F}$ which does not depend on P. Since $\boldsymbol{G}_P$ assembles $\mathscr{B}$ as a continuous body in R_x, it is evident that when the geometry of $\mathscr{B}$ has been assigned by [14] and [19], $\boldsymbol{G}_P$ also describes a distortion[2], in fact, the same distortion as $\boldsymbol{F}$. Unlike $\boldsymbol{K}$, $\boldsymbol{F}$ and $\boldsymbol{G}_P$ do not satisfy (13) and occur with a change of geometry. Thus they produce from b^K_{ML} and $\Gamma^\varkappa_{\mu\lambda}$ respectively not a^k_{ml} the Christoffel symbols of the Euclidean metric a_{ml} in the general frame (k), but $\overset{4}{\Gamma}{}^k_{ml}$, the connexion making parallel in the current state the vectors parallel in $\mathscr{B}$ under $\Gamma^\varkappa_{\mu\lambda}$ or in R_X under b^K_{ML} (see [15, 17])

$$\overset{4}{\Gamma}{}^k_{ml} = G^k_{.\varkappa}\, \partial_m \overset{-1}{G^\varkappa_{.l}} + G^k_{.\varkappa}\, \overset{-1}{G^\mu_{.m}}\, \overset{-1}{G^\lambda_{.l}}\, \Gamma^\varkappa_{\mu\lambda} \tag{21}$$

$$= F^k_{.K}\, \partial_m \overset{-1}{F^K_{.l}} + F^k_{.K}\, \overset{-1}{F^M_{.m}}\, \overset{-1}{F^L_{.l}}\, b^K_{ML}\,. \tag{22}$$

Since $\boldsymbol{K}$ relates vectors parallel under $\Gamma^\varkappa_{\mu\lambda}$ and b^K_{ML} we have if $\overset{K}{\nabla}$ is the total covariant derivative with respect to these connexions

$$\overset{-1}{K^\varkappa_{.}}{}_S\, \overset{K}{\nabla}_{[\mu} K^S_{.\lambda]} = 0 \tag{23}$$

---

[1] See Wang, this volume and [51], p. 41 for "local inhomogeneity".

[2] If, for a crystal, $\mathscr{G}_P(\boldsymbol{K}_P) = 0$, $\boldsymbol{G}_P$ is the lattice distortion from the natural to the final state [42, 44].

as is easily verified from (20). Similarly, from (21) and (22)

$$\overset{4}{\Gamma}{}^{k}_{ml} - a^{k}_{ml} = G^{k}_{.\varkappa} \overset{G}{\nabla}_{m} \overset{-1}{G}{}^{\varkappa}_{.l} = F^{k}_{.K} \overset{F}{\nabla}_{m} \overset{-1}{F}{}^{K}_{.l} \tag{24}$$

where $\overset{G}{\nabla}$ and $\overset{F}{\nabla}$ refer to a^{k}_{ML} and $\Gamma^{\varkappa}_{\mu\lambda}$ or b^{K}_{ML} respectively. Representations of the dislocation density in the general frames ($\varkappa$) and (k) are given by the corresponding torsions [12]

$$S_{\mu\lambda}^{\cdot\cdot\varkappa} = \Gamma^{\varkappa}_{[\mu\lambda]} + \Omega^{\varkappa}_{\mu\lambda}, \tag{25}$$

$$\overset{4}{S}{}_{ml}^{\cdot\cdot k} = \overset{4}{\Gamma}{}^{k}_{[ml]} + \Omega^{k}_{ml}. \tag{26}$$

The torsions of a^{k}_{ml} and b^{K}_{ML} are zero so that

$$a^{k}_{[ml]} = -\Omega^{k}_{ml}, \quad b^{K}_{[ML]} = -\Omega^{K}_{ML}. \tag{27}$$

However, a representation of the dislocation density in (K) is obtained by defining $\hat{S}_{ML}^{\cdot\cdot K}$ as

$$\hat{S}_{ML}^{\cdot\cdot K} = \overset{-1}{F}{}^{K}_{.k} F^{m}_{.M} F^{l}_{.L} \overset{4}{S}{}_{ml}^{\cdot\cdot k} = K^{K}_{.\varkappa} \overset{-1}{K}{}^{\mu}_{.M} \overset{-1}{K}{}^{\lambda}_{.L} S_{\mu\lambda}^{\cdot\cdot\varkappa}. \tag{28}$$

From (27) and (24) we also get

$$\overset{4}{S}{}_{ml}^{\cdot\cdot k} = G^{k}_{.\varkappa} \overset{G}{\nabla}_{[m} G^{\varkappa}_{.l]} = F^{k}_{.K} \overset{F}{\nabla}_{[m} \overset{-1}{F}{}^{K}_{.\,l]} \tag{29}$$

and so

$$\hat{S}_{ML}^{\cdot\cdot K} = F^{m}_{.\,M} F^{l}_{.\,L} \overset{F}{\overset{-1}{\nabla}}_{[m} F^{K}_{.\,l]} \tag{30}$$

$$= -\overset{-1}{F}{}^{K}_{.\,l} \overset{F}{\nabla}_{[M} F^{l}_{.\,L]}. \tag{31}$$

The above discussion has been given by the usual method of generating parallel fields in $\mathscr{B}$ and R_x from parallel fields in R_X [15, 17]. We now compare it with [50] and [51]. There are pitfalls for the unwary reader in [50] because it uses the same Greek letter (α) for a holonomic system $(\alpha) \in \mathscr{B}$ (i.e. for ($\varkappa'$)) and for the system $(K'') \in R_X$ obtained by dragging along under $\boldsymbol{K}$ and so given by[1]

so that

$$\begin{aligned} d X^{K''} &\overset{*}{=} \delta^{K''}_{\alpha} d P^{\alpha} \\ K^{K''}_{.\alpha} &\overset{*}{=} \delta^{K''}_{\alpha}. \end{aligned} \tag{32}$$

Since (32) establishes a one-one correspondence between labelling of particles and labelling of positions in R_X, (K'') must therefore be anholonomic if (12) is not true; the (α) of [50] must therefore not be taken as holonomic in R_X. From (20) with $(\varkappa) = (\varkappa') = (\alpha)$ and (32) we get the $\overset{TN}{\Gamma}{}^{\alpha}_{\gamma\beta}$ of ([50], 34.4)

$$\overset{TN}{\Gamma}{}^{\alpha}_{\gamma\beta} = \Gamma^{\alpha}_{\gamma\beta} \overset{*}{=} \delta^{\alpha}_{K''} \delta^{M''}_{\gamma} \delta^{L''}_{\beta} b^{K''}_{M''L''} \tag{33}$$

---

[1] They also use X^{α} for both $P^{\alpha} \equiv P^{\varkappa'}$ and for $X^{K''}$.

so that $\overset{TN}{\Gamma}{}^{\alpha}_{\gamma\beta}$ is numerically equal to the Christoffel symbols $b^{K''}_{M''L''}$ in the anholonomic (K''). Again if $\overset{N}{\Gamma}{}^{\varkappa}_{\mu\lambda}$ is the connexion Γ of [51] we have from (20), taking in ([51], 10.5) $\mathscr{H}^{\sigma} = \underset{\lambda}{\delta^{\sigma}}$, $\mathscr{K}^{\tau} = \underset{\mu}{\delta^{\tau}}$ and using ([51], 9.6), that

$$\overset{N}{\Gamma}{}^{\varkappa}_{\lambda\mu} \overset{*}{=} \overset{-1}{K}{}^{\varkappa}_{.K} \overset{N}{\nabla}_M (K^K_{.\sigma} \underset{\lambda}{\delta^{\sigma}})\, K^M_{.\mu}, \tag{34}$$

$$= \Gamma^{\varkappa}_{\mu\lambda} \tag{35}$$

where $\overset{N}{\nabla}$ refers to b^K_{ML} and $\overset{N}{\Gamma}{}^{\varkappa}_{\mu\lambda}$. From (25) and (20) we note that if (12) holds then $S_{\mu\lambda}^{\cdot\cdot\varkappa} = 0$ and hence $\Gamma^{\varkappa}_{[\mu\lambda]} = -\Omega^{\varkappa}_{\mu\lambda}$. Thus if (12) holds the expression ([51], 8.15) for the Lie bracket $[\boldsymbol{e}_\mu\, \boldsymbol{e}_\lambda]$ of two basic vectors of $(\varkappa)$ is the alternated form of (34) or ([51], 10.5) and

$$[\boldsymbol{e}_\mu, \boldsymbol{e}_\lambda] = -2\Omega^{\varkappa}_{\mu\lambda}\, \boldsymbol{e}_\varkappa. \tag{36}$$

Thus ([51], 9.11) is equivalent to (21) and for the torsion $\gamma(\boldsymbol{e}_\mu, \boldsymbol{e}_\lambda)$ there defined

$$\gamma(\boldsymbol{e}_\mu, \boldsymbol{e}_\lambda) = (2\overset{N}{\Gamma}{}^{\varkappa}_{[\lambda\mu]} + 2\Omega^{\varkappa}_{\mu\lambda})\, \boldsymbol{e}_\varkappa = 2S_{\mu\lambda}^{\cdot\cdot\varkappa}\, \boldsymbol{e}_\varkappa = 2\overset{TN}{S}{}_{\mu\lambda}^{\cdot\cdot\varkappa}\, \boldsymbol{e}_\varkappa. \tag{37}$$

On the other hand the "inhomogeneity" $\mathscr{S}_{ML}^{\cdot\cdot K}$ defined by ([51], 11.1) is either from (28) and (37) or from (30), which is ([51], 11.5), or from (31), which is ([51], 11.4)

$$\mathscr{S}_{ML}^{\cdot\cdot K} = -2\hat{S}_{ML}^{\cdot\cdot K}. \tag{38}$$

The essential correspondences with [50] and [51] are thus established and the further developments are straightforward. The fundamental geometrical relation is $R_{\nu\mu\lambda}^{\cdot\cdot\cdot\varkappa} = 0$, where $R_{\nu\mu\lambda}^{\cdot\cdot\cdot\varkappa}$ is the curvature tensor of $\Gamma^{\varkappa}_{\mu\lambda}$ [15, 13, 42, 44, 17, 14, 18]. The second identities for $R_{\nu\mu\lambda}^{\cdot\cdot\cdot\varkappa}$, $R_{[\nu\mu\lambda]}^{\cdot\cdot\cdot\varkappa} = 0$ correspond to the Jacobi identities [13, 18], or to the vanishing of the anti-symmetric part of the Einstein tensor $\Gamma^{\sigma\tau}$ of $R_{\nu\mu\lambda}^{\cdot\cdot\cdot\varkappa}$ [42, 44, 14], and represent the node theorem for translation dislocations. This is ([50], 34.18) or ([51], 9.18) and ([51], 11.11). The vanishing of the symmetric part of $\Gamma^{\sigma\tau}$ gives the relation between the Riemannian curvature tensor $K_{\nu\mu\lambda}^{\cdot\cdot\cdot\varkappa}$ of $g_{\varkappa\lambda}$ (or the incompatibility tensor) and the dislocation density [42, 44, 17, 14]. For no extra matter this is equivalent to ([50], 34.19) or ([51], 12.12). In [51], the Christoffel symbols $g^{\varkappa}_{\mu\lambda}$ of $g_{\varkappa\lambda}$ are written $\overset{*}{\Gamma}$ and $K_{\nu\mu\lambda}^{\cdot\cdot\cdot\varkappa}$ is denoted by $\overset{*}{R}$. $\Gamma^{\varkappa}_{\mu\lambda}$ is obviously metric with respect to $g_{\varkappa\lambda}$ (because $g_{\varkappa\lambda}\, \underset{a}{e^{\varkappa}}\, \underset{b}{e^{\lambda}} = \delta^A_a\, \delta^B_b\, b_{ML}\, \underset{A}{E^M}\, \underset{B}{E^L} =$ constant) and ([51], 12.4) and ([51], 12.10) or ([50], 34.11) thus correspond to the usual relation ([12], Chap. III) for a metric connexion [13, 17, 14].

To discuss the Cauchy equations we note first that if we use (32) we have $F^k{}_{K''} \overset{*}{=} G^k{}_{\alpha}\, \delta^{\alpha}_{K''} \overset{*}{=} \dfrac{\partial \chi^k}{\partial P^{\alpha}}\, \delta^{\alpha}_{K''} \overset{*}{=} \dfrac{\partial \chi^k}{\partial X^{K''}}$. The distortions, either from the uniform reference $\boldsymbol{K}$ or equivalently from the body $\mathscr{B}$ may be written as derivatives either with respect to holonomic coordinates (α) and an asymmetric connexion $\Gamma^{\alpha}_{\gamma\beta}$, or with respect to anholonomic coordinates (K'') with connexion $b^{K''}_{M''L''}$ numerically equal to $\Gamma^{\alpha}_{\gamma\beta}$. Let $t_{k'}^{m'}$ be the stress tensor[1], and introduce the first Piola-Kirchhoff stress tensor $T_{k'}^{\cdot K'}$ and the response function $\mathscr{H}_{k'}^{\cdot K''}$ so that

$$T_{k'}^{\cdot K''} = \mathscr{H}_{k'}^{\cdot K''}(F^{k'}_{\cdot K''}) \tag{39}$$

$$= J\, t_{k'}^{\cdot m'} \overset{-1}{F}{}^{M''}_{\cdot m'} \tag{40}$$

where

$$J = |a_{m'l'}|^{\frac{1}{2}} |b_{M''L''}|^{\frac{1}{2}} |F^{k'}_{\cdot K''}|. \tag{41}$$

Then the equations of motion become

$$\overset{F}{\nabla}_{K''} T_{k'}^{\cdot K''} - \overset{F}{\nabla}_{K''} (J \overset{-1}{F}{}^{K''}_{\cdot m'}) t_{k'}^{\cdot m'} + \varrho_R b_{k'} = \varrho_R \ddot{x}_{k'} \tag{42}$$

where ϱ_R is the density in the uniform reference (or in $\mathscr{B}$), $\ddot{x}_{k'}$ the double material time derivative, $b_{k'}$ the body force per unit mass and $\overset{F}{\nabla}$ refers to $a^{k'}_{m'l'}$ and $b^{K''}_{M''L''}$. The second term gives a contribution

$$2 T_{k'}^{\cdot K''} b^{S''}_{[K''S'']} = -2 T_{k'}^{\cdot K''} \Omega^{S''}_{K''S''} \tag{43}$$

$$= 2 T_{k'}^{\cdot K''} \delta^{\beta}_{K''} S_{\beta\sigma}^{\cdot\cdot\sigma} \tag{44}$$

which vanishes if (K'') is holonomic or $\Gamma^{\varkappa}_{\mu\lambda}$ is symmetric, that is, in the absence of dislocations or "inhomogeneity". Defining $A_{k'\cdot n'}^{\cdot K''\cdot L''} = \partial \mathscr{H}_{k'}^{\cdot K''} / \partial F^{n'}_{\cdot L''}$ (42) becomes[2]

$$A_{k'\cdot n'}^{\cdot K''\cdot N''} \overset{F}{\nabla}_{K''} F^{n'}_{\cdot N''} - 2 \mathscr{H}_{k'}^{\cdot K''} \Omega^{S''}_{K''\delta''} + \varrho_R b_{k'} = \varrho_R \ddot{x}_{k'}. \tag{45}$$

This is ([50], 44.7) if we multiply the second term by $-\frac{1}{2}$; the difference arises because there is a discrepancy between the torsion appearing in ([50], 34.8) or ([50], 34.11) and ([50], 44.7). Since the field $\mathbf{s}$ defined by ([51], 15.7) is given from (38), (28), (25), (27) and (32) by

$$s_{K''} = \mathscr{S}_{S''K''}^{\cdot\cdot\;S''} = 2 \hat{S}_{K''S''}^{\cdot\cdot\;S''} = -2 \Omega^{S''}_{K''S''} \tag{46}$$

we see that (45) agrees with ([51], 15.21) and ([51], 15.27). Moreover we see from (45) that this extra term is analogous to that arising when we take the divergence of a tensor density[3] with respect to an anholonomic system.

3. Displacements in N Dimensions

If $\boldsymbol{K}$ is a stress-free uniform reference for the lattice structure of a continuously dislocated crystal then F^{k}_{K} (or $G^{k}_{\varkappa}$) is the lattice distortion. The stress-free body can exist in R_3 only by occupying disjoint or

[1] A density of weight +1.

[2] We use the fact that $\overset{F}{\nabla}_{K''} \mathscr{H}_{k'}^{\cdot L''} = 0$ whenever $\overset{F}{\nabla}_{K''} F^{m'}_{\cdot L''} = 0$.

[3] cf. the example II 9, 1 (3) p. 101 of SCHOUTEN (1954).

overlapping neighbourhoods and we lose the concept of a unique *displacement* relating the *positions* x^k and X^K of a given particle in the current and reference states respectively. We now examine whether this concept can be preserved by considering formal displacements in a Euclidean space R_N with sufficiently large N[1]. We take $(k) = (K) = (a)$, a common cartesian system and augment it so that $(\alpha) = (a, \mathrm{x})$, $a = 1, 2, 3;\ \mathrm{x} = 4, 5 \ldots N,\ \alpha = 1, \ldots N$, is an N dimensional cartesian system spanning R_N, with the particles in the current state in positions x^a of a region $D_3(x)$ of the flat subspace R_3, $\xi^a = x^a$, $a = 1, 2, 3$; $\xi^X = 0$, $X = 4, \ldots N$. We can use either $\boldsymbol{F}$ or $\boldsymbol{G}$ to assess the relative distortions of neighbourhoods in $\boldsymbol{K}$ or $\mathscr{B}$; we choose $\boldsymbol{F}$ and investigate whether these distortions can be copied faithfully by a continuous deformation which is compatible in R_N. Under $\overset{-1}{\boldsymbol{F}}$ the cartesian basis $\underset{m}{e^a} = \delta^a_m$ of (a) becomes $\underset{\bar A}{E^a}(x) = \delta^b_{\bar A} \overset{-1}{F}{}^a_{.b}(x) \overset{\text{def}}{=} A^a_{\bar A}$, with reciprocals $\overset{\bar A}{E}_a(x) = \delta^{\bar A}_b F^b_{.a}(x) \overset{\text{def}}{=} A^{\bar A}_a$. In the anholonomic system $(\bar A)$ the distortion increment $d M_{\bar B \bar A}$ (that is, the change in $\underset{\bar A}{E^a}$) associated with a movement dx^a in $D_3(x)$ (or $G^a_k\, dP^\varkappa$ in $\mathscr{B}$) is easily seen to be seen to be[2] (cf. [17])

$$d M_{\bar B \bar A} = dx^c \Lambda_{c \bar B \bar A}, \qquad \Lambda_{c \bar B \bar A} = \partial_c A^b_{\bar B} \delta_{ba} A^a_{\bar A}. \tag{47}$$

Now let $D_N(\zeta)$ and $D_N(\mathrm{e})$ be regions of R_N with $D_N(\xi) \supset D_3(\mathrm{x})$, and let $\zeta^\alpha = \zeta^\alpha(\xi^\beta)$ define a one-one continuous deformation $D_N(\xi) \to D_N(\zeta)$, and define the displacement $u^\alpha = \xi^\alpha - \zeta^\alpha$. If (α') is the (α) system dragged along then $\zeta^{\alpha'} \overset{*}{=} \delta^{\alpha'}_\alpha \xi^\alpha$ and the cartesian basis $\underset{\alpha}{e^\beta} \overset{*}{=} \delta^\beta_\alpha$ of (α) becomes $\underset{\alpha'}{e^\alpha} = A^\alpha_{\alpha'} = \partial \zeta^\alpha / \partial \zeta^{\alpha'} \overset{*}{=} (\partial \zeta^\alpha / \partial \xi^\beta)\, \delta^\beta_{\alpha'}$ with reciprocals $\overset{\alpha'}{e}_\alpha = A^{\alpha'}_\alpha$ $= \partial \zeta^{\alpha'} / \partial \zeta^\alpha \overset{*}{=} (\partial \xi^\beta / \partial \zeta^\alpha)\, \delta^{\alpha'}_\beta$. As for $dM_{\bar B \bar A}$, the distortion increments $d\tilde M_{\beta' \alpha'}$ describing the variation of $\underset{\alpha'}{e^\alpha}$ are

$$d \tilde M_{\beta' \alpha'} = d\zeta^{\gamma'} \tilde \Lambda_{\gamma' \beta' \alpha'}, \quad \tilde \Lambda_{\gamma' \beta' \alpha'} = \partial_{\gamma'} A^\beta_{\beta'}\, \delta_{\beta\alpha} A^\alpha_{\alpha'}. \tag{48}$$

Under $-u^\alpha$, the subspace R_3, $\xi^\mathrm{x} = 0$, $x = 4, \ldots N$ becomes the subspace X_3, $\zeta^\mathrm{x} = 0$, $\mathrm{x}' = 4', \ldots N'$, and if at one point we make $\underset{0}{\zeta^a} = \underset{0}{\xi^a} = \underset{0}{x^a}$, $\underset{0}{\zeta^\mathrm{x}} = \underset{0}{\xi^\mathrm{x}} = 0$ and $\underset{\alpha'}{e^\alpha}(\zeta) = \underset{\beta}{e^a}(\xi)\, \delta^\beta_{\alpha'}$ (i.e. identity distortion at $\underset{0}{\zeta} = \underset{0}{\xi}$ then the relative distortions of neighbourhoods of X_3 will copy faithfully those $d M_{\bar B \bar A}$ of the reference (or of $\mathscr{B}$) if

$$\Lambda_{c \bar B \bar A} \overset{*}{=} \delta^{c'}_c\, \delta^{b'}_{\bar B}\, \delta^{a'}_{\bar A}\, \tilde \Lambda_{c' b' a'}. \tag{49}$$

[1] The discussion here is based partly on unpublished work by Dr. Zorawski and the author.

[2] We change the sign of Λ.

Since $A^{\beta}_{\beta'} = \partial \zeta^{\beta} / \partial \zeta^{\beta'}$, $\tilde{A}$ is symmetric in c' and b' and so, since $\overset{-1}{F}{}^{a}_{.b} = A^{a}_{\bar{A}}\, \delta^{\bar{A}}_{b}$,

$$\delta^{c}{}_{[c'}\, \delta^{b}{}_{b']}\, \partial_c \overset{-1}{F}{}^{a}_{b} = 0\,. \tag{50}$$

Thus, $\boldsymbol{F}$ is integrable, the dislocation density vanishes (no (local) inhomogeneity) and the lattice distortions are compatible. No displacement in R_N for any N will make a faithful copy of $\mathscr{B}$ (or the reference) in this way. A weaker requirement is that corresponding increments of pure strain only shall agree so that

$$\Lambda_{c(\bar{B}\bar{A})} \overset{*}{=} \delta^{c'}_{c}\, \delta^{b'}_{\bar{B}}\, \delta^{a'}_{\bar{A}}\, \tilde{\Lambda}_{c'(b'a')} \tag{51}$$

which may be written

$$\partial_c\, (A^{b}_{\bar{B}}\, \delta_{ba}\, A^{a}_{\bar{A}}) \overset{*}{=} \delta^{c'}_{c}\, \delta^{b'}_{B}\, \delta^{a'}_{A}\, \partial_{c'}\, (\partial_{b'} \zeta^{\beta}\, \delta_{\beta\alpha}\, \partial_{a'} \zeta^{\alpha}) \tag{52}$$

and may be obtained by requiring that

$$a_{\bar{B}\bar{A}} = A^{b}_{\bar{B}}\, \delta_{ba}\, A^{a}_{\bar{A}} \overset{*}{=} \delta^{b'}_{\bar{B}}\, \delta^{c'}_{\bar{A}}\, \partial_{b'} \zeta^{\beta}\, \delta_{\beta\alpha}\, \partial_{a'} \zeta^{\alpha}. \tag{53}$$

These are the familiar equations for the embedding of a Riemann space of metric numerically equal to $a_{\bar{B}\bar{A}}$ in a Euclidean space; as is well known for the n dimensional Riemann space V_n, $N \leqq n(n+1)/2$, and $N - n$ is the class of the V_n. Hence if we impose (51) only we can find a u^{α} in R_N, $N \leqq 6$ so that the pure strain increments in $\mathscr{B}$ are faithfully copied. Indeed there are many such u^{α}, corresponding to the class of V_3's applicable (isometric) to any one with the given metric. However, to represent the rotation increments also in V_3, we must introduce a system of congruences in V_3 representing the lattice lines [14], to describe fully the increments of rotation.

4. Folding

The physical significance of mappings into higher dimensional spaces is puzzling to the intuition. What meaning are we to attach to the existence of a displacement in six dimensions, while it does not exist in three? The operation of crushing flat a spherical cap (Fig. 2) throws some light on this problem. Here a two-dimensional Riemann surface is carried by a three dimensional displacement into a two-dimensional flat Euclidean space. Obviously the distortion in the (1, 2) plane corresponding to this map, and calculable from (53) if $\overset{-1}{F}{}^{a}_{.b}(x)$ is symmetric, is incompatible in the plane, and its incompatibility when it occurs at x corresponds to the foldings or overlappings which occur during the crushing. From Fig. 3 we see that a *pleat* or *fold* or flattened *ruck* is a kind of dislocation in which the extra "half-plane" is, as it were, "pinched out". Of course, the crushed curved surface is not really two-dimensional, for the folds do not make it strictly flat, but the formal compression into a smaller dimensional manifold provides a description of packing away by folding into a much smaller volume.

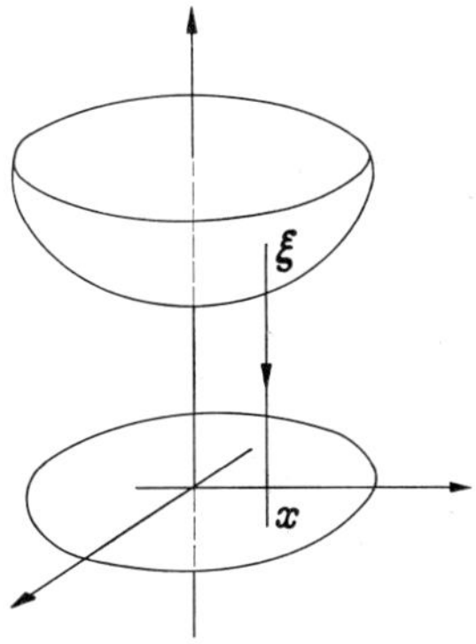

Fig. 2. Crushing flat a spherical cap.

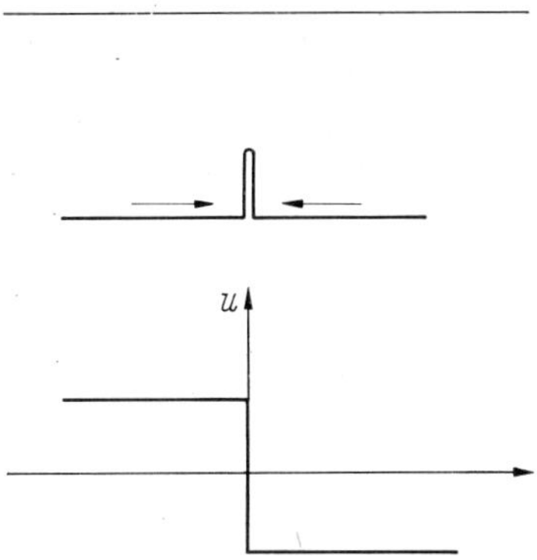

Fig. 3. Pleats and folds.

A simpler example (Fig. 4a) is provided by crushing a cone so that P falls on Q. Evidently there will again be overlap and folding even though the cone is isometric to the plane. A non-Euclidean metric is

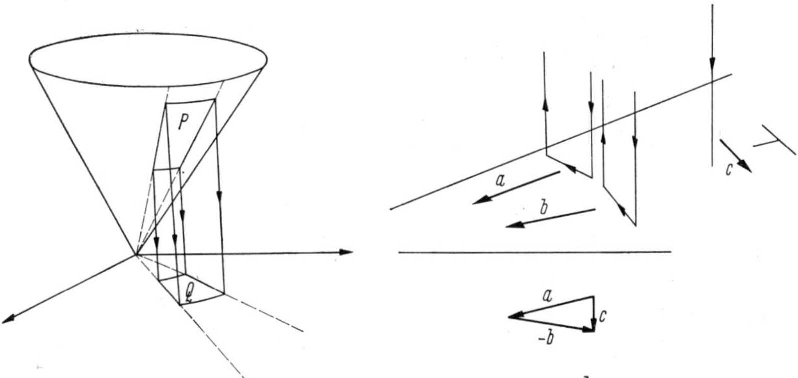

a b

Fig. 4. Crushing flat a cone.

thus not essential, as is still more obvious if an inclined plane is crushed into its projection by parallel straight folds. The uniform local pure strain occurring in the folding process here can be described by two orthogonal fold vectors equal to the eigen-vectors of the strain, and representing perpendicular folds[1]. For a more general surface we must

[1] The author is in debted to Dr. J. D. Eshelby for this remark.

always make such a specification *at one point* to establish an absolute relation between areas (cf. [35]); thereafter the usual apparatus of dislocation theory can give a complete account of the folding. In the notation of § 3, Λ_{cba} describes the incremental distortions dM_{ba} indicating the variation of folding required from point to point. If we introduce $\bar{\Gamma}^a_{cb}$ making the $\underset{A}{E^a}$ parallel, it is not metric with respect to a_{cb} and so we have for Λ [15, 17, 14, 18][1]

$$\Lambda_{cba} = S_{\{cab\}} - \tfrac{1}{2} Q_{\{cab\}} \tag{54}$$

Q determines the variation of strain in the surface [14].

In the cone (Fig. 4a) the incompatibility arises because we do not allow each neighbourhood to *roll* on to the plane but place P on Q. In polar coordinates (p) $\bar{L}_p^{\cdot q} \left(= \overset{-1}{F}{}^q_{\cdot p}\right)$ is evidently diagonal, and is

$$\bar{L}_p^{\cdot q} = \begin{pmatrix} \sqrt{1+t^2} & 0 \\ 0 & 1 \end{pmatrix} \tag{55}$$

where $\tan^{-1} t$ is the (constant) semi-angle of the cone; $\bar{\boldsymbol{L}}$ is not integrable since the integrability conditions are $\overset{a}{\nabla}_{[p} \bar{L}_{q]}^{\cdot r} = 0$, $\overset{a}{\nabla}$ referring to a^p_{qr} the Christoffel symbols of (p). For Γ^p_{rq} and $S_{rq}^{\cdot\cdot p}$ we have

$$\Gamma^p_{rq} = a^p_{rq} + L_s^{\cdot p} \overset{a}{\nabla}_r \bar{L}_q^{\cdot s}, \tag{56}$$

$$S_{rq}^{\cdot\cdot p} = L_s^{\cdot p} \nabla_{[r} \bar{L}_{q]}^{\cdot s} \tag{57}$$

and we find $S_{12}^{\cdot\cdot 1} = 0$,

$$S_{12}^{\cdot\cdot 2} = -\left(\sqrt{1+t^2} - 1\right)/r. \tag{58}$$

Obviously in this folding we insert rings of material so that as is evident from (55) there is no circumferential, but only radial, distortion. We may represent this by a series of small dislocations loops with radial Burgers vectors (Fig. 4b); these are the only fold vectors. Clearly $S_{rq}^{\cdot\cdot p}$ measures only the variation of the folding; it does not describe the dislocations *in* the surface. Fig. 4b shows that $\boldsymbol{c} = d\boldsymbol{f}/\partial\theta$. This is an example of a general differential relation, the third identity for the curvature tensor $R_{srq}^{\cdot\cdot\cdot p}$ of Γ^p_{rq} [12], which, since $R = 0$ relates S and Q

$$\overset{\Gamma}{\nabla}_{[d} Q_{c]ba} = -S_{dc}^{\cdot\cdot e} Q_{eba}. \tag{59}$$

The variation of the fold vector for the cone appears because it has a finite Euler-Schouten curvature, although its Riemannian curvature is zero.

If we fold by projection two-dimensional bodies in three dimensions, we can from the theory of Pfaffians always find an orthogonal curvi-

---

[1] We use the notation of [12], Ch. III.

linear coordinate system diagonalising $\bar{L}$ at each point; moreover, we can reconstruct from the fold matrix $\bar{L}$ the mapping $\zeta = \zeta(x)$ and so describe the unfolding. If we fold a three-dimensional body in six dimensions, the situation is more complicated; the eigen vectors of $\bar{L}$ may only define congruences, and only special unfolding maps can be reconstructed from the fold matrix.

The physical application of folding may perhaps give a formal description of the microscopic processes occurring when substances undergo deformations which are due to some unpacking or unfolding of the microstructure. Equally we can describe the contortions resulting from folding up the internal structure[1]. Examples are numerous organic and polymerised materials, and certain phenomena associated with radiation damage. Of course, since formally we are "pinching out" matter in the folding process the description of the resulting processes involves macroscopically an (incompatible!) density change. The situation here is analogous to the relation of the continuously dislocated crystal to macroscopic plasticity, where we seek to provide a more detailed microscopic picture of the plastic processes represented in the classical theory by shape changes.

We have tried to indicate that dislocation theory is a rich field for the interplay of different disciplines, often providing very concrete representations of abstract ideas, and emphasizing a unity of mathematical patterns of interest to us all. It is valuable both in understanding practical problems of materials and in illustrating mathematical ideas. The interaction of these different fields must surely be of profit to both of them, and it is to be hoped that there will be increasing application of these general theories in the future.

References

[1] Cartan, É.: Leçons sur la géométrie des espaces de Riemann. Paris: Gauthier-Villars.
[2] Cartan, É.: C. R. Acad. Sci. **174**, 593 (1922).
[3] Cartan, É.: C. R. Acad. Sci. **174**, 734 (1922).
[4] Cartan, É.: Ann. Éc. Norm. **40**, 325 (1923).
[5] Cartan, É.: Ann. Éc. Norm. **41**, 1 (1924).
[6] Cartan, É.: Ann. Éc. Norm. **42**, 17 (1925).
[7] Einstein, A.: S.-B. Preuss. Akad. Wiss. 1928, pp. 217–224.
[8] Cartan, É.: Math. Ann. **102**, 698 (1930).
[9] Weitzenböck, R.: Enzyklopädie **3**, Heft 6 (1922).

[1] The present author began to study the idea of folding in June, 1966. Dr. J. D. Eshelby kindly informed him in October 1966 of an unpublished example in a forthcoming book by Professor Nabarro, in which the production of a spherical surface from a plane by a tucking process is cited as an example of dislocations in incompatibility. Nabarro uses the graphic picture of forming tucks in chequered cloth.

[10] Weitzenböck, R.: Invariantentheorie. Groningen: Noordhoff **1923**, p. 320.
[11] Ricci, G.: Mem. Acc. Linc. (**5**) **2**, 276 (1895).
[12] Schouten, J. A.: Ricci-Calculus, 2nd Ed. Berlin/Göttingen/Heidelberg: Springer 1954.
[13] Bilby, B. A., and E. Smith: Proc. Roy. Soc. A **236**, 481 (1956).
[14] Bilby, B. A.: Prog. Solid Mech. **1**, 331 (1960).
[15] Bilby, B. A., R. Bullough and E. Smith: Proc. Roy. Soc., **A231**, 263—273, 1955.
[16] Gardner, L. R. T.: Ph. D. Thesis, Sheffield University 1958.
[17] Bilby, B. A., L. R. T. Gardner, A. Grinberg and M. Zorawski: Proc. Roy. Soc. A **292**, 105 (1966).
[18] Bilby, B. A.: Teoria Dyslokacji, Czesc III, Geometrical Aspects of Continuous Distributions of Dislocations. Warsaw: Polska Akademia Nauk 1966, pp. 1—150.
[19] Bilby, B. A.: 1954 Bristol Conference on Defects in Crystalline Solids. London: The Physical Society 1955, p. 123.
[20] Nye, J. F.: Acta Met. **1**, 153 (1953).
[21] Frank, F. C.: Plastic Deformation of Crystalline Solids. Pittsburgh, NAVEXOS-P-834 (1950) p. 150.
[22] West, C. D.: ONR No. N 7-onr 39102-Project NR 032313 (1955).
[23] Read, W. T.: Dislocations in Crystals. New York: McGraw Hill 1953.
[24] Bilby, B. A., L. R. T. Gardner and E. Smith: Acta Met. **6**, 29 (1958).
[25] Kröner, E.: Z. Phys. **141**, 386 (1955).
[26] Bullough, R., and B. A. Bilby: Proc. Phys. Soc. B **69**, 1276 (1956).
[27] Crocker, A. G., and B. A. Bilby: Acta Met., **9**, 678 (1961).
[28] Kondo, K.: Proc. 2nd Jap. Nat. Congr. Appl. Mech. 1952, Science Council of Japan 1953, p. 41.
[29] Eshelby, J. D.: New York: Academic Press, Solid State Physics 3/2, 1956, p. 79.
[30] Eisenhart, L. P.: Continuous Groups. New York: Dover 1961.
[31] Eisenhart, L. P.: Non-Riemannian Geometry. New York: American Math. Soc. Coll. 8, 1927.
[32] Bilby, B. A., and J. W. Christian: Monogr. Ser. Inst. Metals, No. 18, 1955, p. 121.
[33] Kröner, E., and G. Rieder: Z. Phys. **145**, 424 (1956).
[34] Bilby, B. A., L. R. T. Gardner and A. N. Stroh: IV. Int. Congr. Mecanique Appliquée, Actes, 7, 1957, p. 35.
[35] Bilby, B. A., R. Bullough, L. R. T. Gardner and E. Smith: Proc. Roy. Soc. A **244**, 538 (1958).
[36] Bilby, B. A., and L. R. T. Gardner: Proc. Roy. Soc. A **247**, 92 (1958).
[37] Whapham, A. D., and H. Witman: Proc. Roy. Soc. A **237**, 513 (1956).
[38] Read, W. T.: Acta Met. **5**, 83 (1957).
[39] Kröner, E.: Z. Phys. **142**, 463 (1955).
[40] Kröner, E.: Proc. Phys. Soc. A **68**, 53 (1955).
[41] Kröner, E.: Kontinuumstheorie der Versetzungen und Eigenspannungen. Berlin/Göttingen/Heidelberg: Springer 1958.
[42] Kröner, E., and A. Seeger: Arch. Rat. Mech. An. **3**, 97 (1959).
[43] Pfleiderer, H., A. Seeger and E. Kröner: Z. Naturf. **15a**, 758 (1960).
[44] Kröner, E.: Arch. Rat. Mech. An. **4**, 273 (1960).
[45] Stojanovitch, R.: Phys. Stat. Sol. **2**, 566 (1962).
[46] Kondo, K.: (Ed.) Memoirs of the Unifying Study of the Basic Problems in Eng. by Means of Geometry. Tokyo, Vol. I, 1955.

[47] Kondo, K.: ibid, Vol. II, 1958.
[48] Kondo, K.: ibid, Vol. III, 1962.
[49] Kröner, E.: J. Math. Phys. **42**, 27 (1963).
[50] Truesdell, C., and W. Noll: Handbuch der Physik III/3. Berlin/Heidelberg/New York: Springer 1965, pp. 1—591.
[51] Noll, W.: Materially Uniform Simple Bodies with Inhomogeneities. Carnegie Inst. of Techn., Math. Dept., 67-26, 1967.
[52] Gunther, W.: Abh. d. Braunschweigischen Wiss. Ges. **10**, 195 (1958).
[53] Green, A. E., and P. M. Naghdi: Arch. Rat. Mech. An. **18**, 251 (1965).
[54] Fox, N.: Quart. J. Mech. Appl. Math. **19**, **343** (1966).
[55] Holländer, E. F.: Czech. J. Phys. **10**, 409, 479, 551 (1960).
[56] Holländer, E. F.: Czech. J. Phys. **12**, 35 (1962).
[57] Ben Abraham: private communication.
[58] Simmons, J. A.: Thesis, University of California, Berkeley, 1961.
[59] Lemprière, B. M.: Tech. Rep. No. 2 ONR and Dept. Aeron. and Astron., Stanford University, 1962.
[60] Indenbom, W. L., and A. N. Orlov: Usp. fiz. Nauk **76**, 557 (1962).
[61] Indenbom and Orlov: this volume.
[62] Mura, T.: Phys. Stat. Sol. **10**, 447 (1965).
[63] Mura, T.: Phys. Stat. Sol. **11**, 683 (1965).
[64] Mura, T.: Int. J. Engng. Sc. **5**, 341 (1967).
[65] Mura, T.: this volume.
[66] Bross, H.: Phys. Stat. Sol. **5**, 329 (1964).
[67] Fox, N. J.: Inst. Math. Appl. **2**, 285 (1966); Q. T. M. A. M. **21**, 67 (1968).
[68] Sedov, L. I., and V. L. Berditchevski: this volume.
[69] Anthony, K., U. Essmann, A. Seeger and H. Träuble: this volume.
[70] Schaefer, H.: ZAMM **47**, 319 (1967), and this volume.
[71] Günther, H.: Schrift. Inst. Math. Deutsch. Acad. Wiss. A **4**, 1—75 (1967).
[72] Eshelby, J. D., and T. Laub: Can. J. Phys. **45**, 887 (1967).

On the Two Main Currents of the Geometrical Theory of Imperfect Continua

By

K. Kondo

Department of Mathematical Engineering and Instrumentation Physics
University of Tokyo

1. Introduction

The introductory part of this survey lecture could be a reminiscence. I shall first review what happened to me before my contact around 1955 with the European and American currents. My acquaintance with the field of plasticity theory started with my proposal of a Theory of Yielding in Riemannian terminology which I undertook twenty years ago a few years after the Second World War [1, 2]. I may be allowed to count this, with some partiality to myself, as one of the two main currents I want to emphasize. In my subsequent investigations to explore the geometrical features of the problem, I came in contact with another of the remarkable developments.

The first current has remained for many years an isolated endeavour by myself and some of my colleagues in Japan, and it is a pleasure for me to be able to outline, on this occasion, how its first vague conception fell upon me and what difficulties were met with during the subsequent development, as is inevitably the case with every heuristic approach.

For some time, I refrained intentionally from resorting to the well-established models used in the conventional classical theory of plasticity and the theory of dislocations.

Up to this point, the Riemannian terminology sufficed so that the curvature field was the main non-Euclidean object that was handled in the period of making. It was principally for the sake of formality afforded by pure mathematical consideration, that I needed to generalize the standpoint from the Riemannian to non-Riemannian geometry in terms of the torsion tensor which had been introduced by French Geometre Élie Cartan [3]. This second step was essentially a work of translation of the crystallographic dislocation theory into the terminology of differential geometry. This was started in 1952 and it was then pointed out that the torsion tensor is an invariant representation of the edge and screw-dislocations [4][1]. This was the beginning of the second main current I became next aware of. It was impressive to know how the French scholars E. and F. COSSERAT, on whose name the title of the present symposium is based, had anticipated by half a century the recent evolution in recognizing the possibility of a torsion-like anomaly. To them may be traced back the natural genealogical circumstance in Europe revived in the recent explicit introduction of the terminology of the space of teleparallelism by B. A. BILBY and his collaborators in Britain [6], followed closely by E. KRÖNER's similar work in Germany [7].

Considerable efforts followed to throw light on COSSERAT's improper continua from the recent viewpoints.[2]

In this connexion, the suggestion of a new technical term was impressive to me. In place of CARTAN's torsion tensor, sometimes "the structure curvature tensor" was favoured [9, 10]. It would serve to eliminate complication to restrict the latter expression to a specific kind of non-Riemannian property, attributed to some crystallographical model for illustration. COSSERAT's exposition is concerned largely with the incompatible rotation distribution which the lattice picture of the dislocation gives an impression to exclude. In the differential geometrical expression by the torsion tensor, the rotational and non-rotational objects are both feasible toyield discrepancyvectors of edge and screw types.

Another point which captured my interest from the point of view of terminology and definition was the remark which suggests an introduction of the conecpt of FINSLER's space. We shall return to it later (in Chap. 4). A proper orientation being assigned to each element to rotate independently of its neighbours, its generalization, "the concept of the directors" is appropriate to express independent deformations of material elements. Needless to say that this director terminology can be absorbed into that of anholonomic features connected with the deformation fields [11, 12]. This is an item I shall soon return to.

Rather than the task of translation, a heuristically constructive approach was needed in the theory of yielding where the distinction between the more macroscopic and the more microscopic degrees of freedom of deformations was instrumental. The multidimensional features which appear in the theory are disguises of the microscopic anomalous fluctuations under a certain statistical summary.

The yielding is argued to be a stability limit. The first heuristic approach to this recognition was based on the polycrystalline model of many grains of elastic character. The multidimensional rigidity, in generalization of the flexural rigidity of the two-dimensional plate,

[1] The interested reader may refer to reference [5], where the contributions up to the present are summarized.

[2] See i.a. [8].

was derived from the elastic constants under polygrain statistics. We have discussed the mathematical feasibility to reduce it to the relative curvature tensor of the immersion picture in multidimensional space which can account for more anomalies than can the terminology of the ordinary intrinsic curvature tensor alone. All these were brought under our analysis to afford the fundamental equations of yielding. More analytical and physical aspects of feasibility of this criterion for yielding will be treated later.

In the recent extension of the Theory of Yielding to that of *Dual Yielding,* I see a creative interaction between the two phases of the geometrical approach, which appear to be mutually contrasted. It was a *conscious* or *unconscious* extension of Cosserat's work as well as Boltzmann's that the possibility of *asymmetric stresses* and *couple stresses* was studied independently in different countries. This was combined with the formalism of the Riemannian stress-function space which had been proposed [13—15] to suggest a new physical possibility to exist in correspondence with a remarkable analytic feature that an asymmetric metric has, in the background, an anholonomic tearing and rotation of the element and a natural consequence thereof is an introduction of a further torsion tensor field. The duality between the stress and strain being apparent, the new torsion tensor stands dually to the torsion tensor of dislocations. This turns out what one calls the couple stress.

Hereupon, it should be pointed out that entirely the same formalism as the criterion of yielding is connected with the dual case. Such an analogue, as associated with the stress-function space, may be called the dual yielding.

The multidimensional picture has originated from a statistics of fluctuational microscopic deformations not excluding distributions of dislocation. Once such a construction has been made a remarkable kind of plastic feature is expressed as the deformation into the fourth and higher dimensions. Here is the instance that the initial approach to this idea was made by an analogy with the lower-dimensional theory of buckling of the flat plate to analyze the critical condition for the deformation into the third dimension from the flat two-dimensional state.

If we analyze the macroscopic behaviour of a two-dimensional microscopically non-uniform material, like a two-dimensional polycrystal, which cannot deform into three dimensions, it exhibits the same statistical mechanical character as the two-dimensional flat plate which is actually bent.

As is the concept of plate with flexural rigidity a certain mean feature of more detailed elastic characteristics of the material, so is the material manifold with plastic resistance to yielding also a mean feature of more detailed plastic aspects. The dislocation model is mainly concerned with such detailed structures. The proposed theory of yielding is, on the contrary, concerned very remarkably with how the ultimate plastic structures and reactions can more macroscopically be viewed.

A little consideration on the energy feature at yielding and dual yielding reveals that they are thermodynamically based. In other words, they are also afforded by a statistics of the distribution of the geodesics or paths in the material manifold under plastic changes. Depending on whether the geodesics or paths are drawn in the space of strain or in that of stress, the energy balance gives the criterion of yielding or of dual yielding. An important point is that, owing to the restricted capacity of the Euclidean observer, most of the non-Euclidean degrees of freedom are missed. These hidden degrees of freedom entails the inequality in the second law of thermodynamics. Hence it is obvious that the non-Euclidean geometry is an appropriate language for thermodynamics. The hidden freedom is partly revealed in the multidimensional expression of the phenomena of yielding etc.

If it is not too inappropriate I would point out on this occasion, that the formalism of general relativity theory can also similarly be based upon a statistical summary, which is the same as the statistics leading to the yield theory, provided the difference of the dimension numbers and of the intervention of a non-definite metric in the latter case are not concerned about. The recognition has an intrinsic relation to the anholonomic construction of anomalies in the terminology of continua. By the procedures of the general diakoptics, as was suggested in 1958 [16], a space can anholonomically been torn into another space. We have called such a process a tearing and its inverse retearing or refurbishments.

A further advance along such lines will inevitably bring about a more penetration into the physics of the microscopic world on one hand. On the other hand, it will entail an introduction of the analytical procedures of the geometries of higher order spaces of whatever kinds they are. Even within the restricted realm of the material physics in direct connexion with the mechanics of plastic bodies, it reminds us of various attempts for the formalism of mechanical spin degrees of freedom, in the so-called-multipolar theory of materials, etc. The deeper the penetration, the more complex will become the assumed multipolarity.

2. Anholonomic Aspects of Plastic Disturbances

The most important characteristic of plastic phenomena is afforded by their anholonomic features. Here a unifying means of analysis and interpretation can be found even between apparently conflicting viewpoints. Different representations can sometimes be mutually connected by the process of *tearing* from the point of view of the general diakoptics. Tearing can be performed at many different levels so that the same plastic or hyperplastic anomalies can be represented by more than one kind of non-Euclidean analytical terminology. More complicated physical objects are represented by the torsion-curvature tensors of higher orders which can be torn and retorn. Conversely the conception of the latter in mathematics suggests the summability of physics in the form of the former (cf. Chap. 4).

The anholonomic features are expressions of the fact that the plastic changes do not preserve the topological connexion relation between material elements. A regular lattice structure is not preserved in a plastic change. An arbitrary material vector is transformed from $dx^\varkappa$ $(\varkappa = 1, 2, 3)$ to

$$(dx)^{\varkappa'} = A^{\varkappa'}_{\varkappa}\, dx^\varkappa \quad (\varkappa' = 1, 2, 3) \tag{1}$$

where $x^\varkappa$ are arbitrary coordinates (fixed to the material point) and $A^{\varkappa'}_{\varkappa}$ are arbitrary analytic functions. Each material element undergoes a deformation-rotation from $dx^\varkappa$ to $(dx)^{\varkappa'}$, independently of its neighbours, viz. without preserving the topological invariance. Hence the Pfaffian (1) needs generally to be unintegrable so that the transformation from the frame $(\varkappa)$ to frame $(\varkappa')$ is anholonomic.

A Riemannian metric

$$ds^2 = g_{\lambda\varkappa}\, dx^\varkappa\, dx^\lambda \tag{2}$$

is naturally associated with this deformation, where

$$g_{\lambda\varkappa} = \delta_{\lambda'\varkappa'}\, A^{\varkappa'}_{\varkappa}\, A^{\lambda'}_{\lambda} \qquad (\varkappa', \lambda' = 1, 2, 3)$$

so that

$$ds^2 = \delta_{\lambda'\varkappa'}\, (dx)^{\varkappa'}\, (dx)^{\lambda'}$$

provided the element $(dx)^{\varkappa'}$ is measured in reference to a rectangular coordinate system fixed in space. However, it is not advantageous to define the strain solely in connexion with the metric (2). A more general and appropriate definition is afforded by (1), where one may put

$$A_{\lambda\varkappa} = \delta_{\lambda\varkappa} + \varepsilon_{\lambda\varkappa}, \qquad A_{\lambda\varkappa} = A^{\varkappa'}_{\varkappa}\, \delta_{\varkappa'\lambda}, \qquad \varepsilon_{\lambda\varkappa} = \varepsilon^{\varkappa'}_{\varkappa}\, \delta_{\varkappa'\lambda}$$

and let $\varepsilon_{(\lambda\varkappa)}$ and $\varepsilon_{[\lambda\varkappa]}$ represent respectively the deformation and rotation parts of the strain.

A Riemannian metric is said to be torn when an affine connexion other than the Levi-Cività parallelism defined by the metric tensor such as $g_{\lambda\varkappa}$ is assumed. If the corresponding intrinsic curvature tensor vanishes identically, i.e. if

$$R_{\nu\mu\lambda}^{\cdot\cdot\cdot\varkappa} \overset{\text{def}}{=} 2(\partial_{[\nu}\Gamma^{\varkappa}_{\mu]\lambda} + \Gamma^{\varkappa}_{[\nu|\varrho|}\Gamma^{\varrho}_{\mu]\lambda}) = 0, \tag{3}$$

where $\Gamma_{\mu\lambda}^{\varkappa}$'s are the parameters of connexion of the new parallelism, the tearing is said to be *perfect*. Evidently a perfectly torn manifold has teleparallelism. If the assumed connexion does not let the curvature tensor $R_{\nu\mu\lambda}^{\cdot\cdot\cdot\varkappa}$ vanish the tearing is *imperfect*.

I have often drawn attention to the fact that there are cases in which perfect tearing is not physically allowed.

The anomalies of an imperfect material manifold, such as imperfect crystal, can be classified into more than one phase. Part of which can be represented by the metric associated with an anholonomity, represented, for instance, by the perfect tearing of the Riemannian metric ds^2 of (2). There can be another represented by the perfect tearing of another Riemannian metric

$$dz^2 = b_{\lambda\varkappa}\, dx^{\varkappa}\, dx^{\lambda} \tag{4}$$

where

$$b_{\lambda\varkappa} = \delta_{\lambda''\varkappa''} B_{\varkappa}^{\varkappa''} B_{\lambda}^{\lambda''} \qquad (\varkappa'', \lambda'' = 1, 2, 3)$$

so that the anholonomic deformation-rotation

$$(dx)^{\varkappa''} = B_{\varkappa}^{\varkappa''}\, dx^{\varkappa}$$

is required to remove it.

The perfect tearing by the first anholonomy gives the torsion tensor

$$S_{\mu\lambda}^{\cdot\cdot\varkappa} \overset{(\varkappa)}{=} A_{\varkappa'}^{\varkappa}\, \partial_{[\mu} A_{\lambda]}^{\varkappa'}$$

implying a continuous distribution of dislocations. Two kinds of components of $S_{\mu\lambda}^{\cdot\cdot\varkappa}$ afford the well known types called the edge and screw dislocation. But the second phase implied by (4) is not perfectly torn by it unless another continuous distribution of torsion

$$Z_{\mu\lambda}^{\cdot\cdot\varkappa} \overset{(\varkappa)}{=} B_{\varkappa''}^{\varkappa}\, \partial_{[\mu} B_{\lambda]}^{\varkappa''}$$

is introduced.

If $dz^2 = ds^2$, i.e. the metric features of the two phases coincide, the transformation from $(dx)^{\varkappa'}$ to $(dx)^{\varkappa''}$ and vice versa are orthogonal:

$$(dx)^{\varkappa''} = O_{\varkappa'}^{\varkappa''}\, (dx)^{\varkappa'}, \qquad (dx)^{\varkappa'} = O_{\varkappa''}^{\varkappa'}\, (dx)^{\varkappa''}$$

where $[O_{\varkappa'}^{\varkappa''}]$ is an orthogonal matrix and $[O_{\varkappa''}^{\varkappa'}]$ its inverse. In such a case,

$$B_{\varkappa}^{\varkappa''} = O_{\varkappa'}^{\varkappa''}\, A_{\varkappa}^{\varkappa'}, \qquad B_{\varkappa''}^{\varkappa} = A_{\varkappa'}^{\varkappa}\, O_{\varkappa''}^{\varkappa'}.$$

Since $S_{\mu\lambda}^{\cdot\cdot\varkappa}$ and $Z_{\mu\lambda}^{\cdot\cdot\varkappa}$ are tensors,

$$Q_{\mu'\lambda'}^{\cdot\cdot\varkappa'} \overset{\text{def}}{=} S_{\mu'\lambda'}^{\cdot\cdot\varkappa'} - Z_{\mu'\lambda'}^{\cdot\cdot\varkappa'} \overset{(\varkappa')}{=} O_{\mu'}^{\mu''}\, O_{\lambda'}^{\lambda''}\, \partial_{[\mu''} O_{\lambda'']}^{\varkappa'}$$

is also a tensor. Obviously, a perfect tearing of a Riemannian metric anomaly is indefinite admitting an arbitrary orthogonal transformation (cf. [17]).

Even an anholonomically distributed field of orthogonal transformations alone is sufficient to produce closure failures, having also the character of the torsion tensor [18].

It could be illuminating to explore which specific kinds among the so-called ordinary, extended, imperfect and/or generalized dislocations, in the conventional crystallographical terminology, need to correspond to a purely deformation-type, or a purely rotation-type torsion field and how the different types coexist.

The tearing or retearing is a kind of reorganization of plastic anomalies from the picture in terms of dislocations to that of curvature picture. In either picture something is lost and something is more accurately represented than in the other. The curvature picture is more global. An illustration for this can be given by a statistical decomposition of the intrinsic curvature into a pair of dislocations [18]. It has been shown that a finite curvature is approximately replaced by a pair of edge dislocations arranged in the direction of climb, whereas no finite curvature corresponds to a pair arranged in the direction of glide. This is in agreement with the crystallographical inference that a particle or hole is transported by a climb of edge dislocation and not by its glide.

Many more anholonomic features can be pointed out to lead to important physical interpretations.

3. Yielding[1] and Dual Yielding[2]

A. Yielding. Several apparently different approaches to the mathematical formulation of the theory of yielding and dual yielding have been made.

i) The first approach was analogical, as has been said. We started with generalizing to three dimensions the well-established equation of equilibrium for the buckling of flat plate,

$$D\varDelta\varDelta w + P_{j\,i}\partial^i\partial^j w = 0, \tag{5}$$

where $i, j = 1, 2$, $P_{j\,i}$ are the components of the two-dimensional stress tensor and isotropy is assumed so that $\varDelta$ means the two-dimensional Laplacian operator, and D is the flexural rigidity. The function w is actually the normal deflexion of the plate. This equation needs to be solved under suitable boundary conditions. In the extension to yielding, i and j run over $1, 2, 3$, the Laplacian operator and the stress tensor P_{ji} are replaced by three-dimensional ones respectively, and a constant, which we denote by B is substituted for D.

This was initially a heuristic assumption. But it was soon found to stand on an energy criterion that every anomalous feature represented by a non-Euclidean geometrical object needs to absorb energy. The

[1] Cf. footnote 1, p. 201.

[2] See [20, 21].

intrinsic curvature or torsion is no exception. However, from a more macroscopic the standpoint torsion-like anomalies can be lumped into curvature-like one. It is a remarkable point that the relative curvature denoted by a three-index symbol $H_{\mu\lambda}^{\cdot\cdot\alpha}$ $(\lambda, \mu = 1, 2, 3; \alpha = 1, 2, 3, 4, \ldots)$ is more comprehensive than the four-index tensor $R_{\nu\mu\lambda}^{\cdot\cdot\cdot\varkappa}$ such as in (3). Since we have:

$$R_{\nu\mu\lambda}^{\cdot\cdot\cdot\varkappa} = \sum_{\alpha} H_{[\nu}^{\cdot\,\varkappa\alpha} H_{\mu]\lambda}^{\cdot\cdot\alpha} \qquad (6)^1$$

the vanishing of the $R_{\nu\mu\lambda}^{\cdot\cdot\cdot\varkappa}$ does not necessarily imply that of the $H_{\mu\lambda}^{\cdot\cdot\alpha}$. The plastic energy connected with the material element need not be referred to $R_{\nu\mu\lambda}^{\cdot\cdot\cdot\varkappa}$ if it is represented in terms of $H_{\mu\lambda}^{\cdot\cdot\alpha}$. For small deviations from the Euclidean, i.e. undisturbed state, it is allowed to put

$$H_{\mu\lambda}^{\cdot\cdot\alpha} = \partial_\mu \partial_\lambda w^\alpha.$$

If a linear resistance to this is assumed to exist an energy

$$u = \tfrac{1}{2} B_{\beta\alpha}^{\cdot\cdot\varkappa\lambda\mu\nu} H_{\mu\lambda}^{\cdot\cdot\beta} H_{\varkappa\nu}^{\cdot\cdot\alpha} = \tfrac{1}{2} B_{\beta\alpha}^{\cdot\cdot\varkappa\lambda\mu\nu} \partial_\varkappa \partial_\nu w^\alpha \partial_\mu \partial_\lambda w^\beta$$

per unit volume should be absorbed by this configuration. This should be balanced with the ordinary strain energy, which is, per unit volume,

$$v = \tfrac{1}{4} \sum_{\alpha} \sigma^{\varkappa\lambda} \partial_\lambda w^\alpha \partial_\varkappa w^\alpha$$

where $\sigma^{\varkappa\lambda}$ is the stress tensor. The total energy integrated over the entire material body need not change at the deformation so that we have the variational criterion

$$\delta \int (u + v)\, dX = 0 \qquad (7)$$

where dX denotes the volume element.

It is a routine analysis to derive from (7) the field equation and the boundary condition to be satisfied by the functions $w^\varkappa$. For an isotropic material, they are simplified to

$$\boxed{B \Delta\Delta w^\alpha + \sigma^{\varkappa\lambda} \partial_\lambda \partial_\varkappa w^\alpha = 0} \qquad (8)$$

which has the same form as (5) extended to three dimensions, and to the corresponding free-boundary conditions.

Since these equations are homogeneous, characteristic values are required for $\sigma^{\varkappa\lambda}/B$ in order that the system has a solution. Until that characteristic point is reached, no disturbance of such a plastic nature is allowed so that the material remains flat and holonomic. Once, a w^α has set in, it can no longer remain so. Analytically, it is equivalent to swelling out into a multidimensional condition. This is interpreted to be the *criterion for yielding*.

---

[1] See [12].

A number of examples of practical yield problems have been analyzed using these equations.

ii) The second step is based on a statistical consideration of the microscopically non-uniform structure of the material. It may more apparently clarify the connexion with Cosserat's problem with a physical interpretation of the function w^α.

It is indeed possible to construct by a certain statistical consideration a six-dimensional Euclidean space in which the disturbed material manifold with microscopic structure like a polycrystal is locally immersed. The disturbances are classified into a mean kind which is three-dimensional, and a fluctuational kind which is also three-dimensional. The material element vector $dx^\varkappa$ ($\varkappa = 1, 2, 3$) behaves like deforming into $dx^{\bar{\varkappa}} + dx^{\hat{\varkappa}}$ where $\bar{\varkappa}$ $(= 1, 2, 3)$ indicates the average and $\hat{\varkappa}$ $(= 4, 5, 6)$ the fluctuational part. Of the metric also we can know only the mean value

$$ds^2 = d\bar{s}^2 + d\hat{s}^2 \tag{9}$$

where

$$d\bar{s}^2 = \delta_{\bar{\lambda}\bar{\varkappa}}\, dx^{\bar{\varkappa}}\, dx^{\bar{\lambda}}, \qquad d\hat{s}^2 = \delta_{\hat{\lambda}\hat{\varkappa}}\, dx^{\hat{\varkappa}}\, dx^{\hat{\lambda}}.$$

The linear terms in $dx^{\hat{\varkappa}}$ etc. disappearing by averaging over all possibilities. The $\hat{\varkappa}$-degrees of freedom are orthogonal to the $\bar{\varkappa}$-degrees. The deviation w^α can evidently a displacement among the $\hat{\varkappa}$-degrees of freedom. The ordinary equations of equilibrium for the three-dimensional stress-deformation field are readily brought to the same form as (8).

An easily accessible model of the above statistical ensemble is a granular medium. It is often referred to as a model of the Cosserat continuum since each grain deforms with considerable independence from its neighbours. There was indeed an investigation based on this picture by N. OSHIMA [22], which was performed independently of the European and American schools and follows S. MORIGUTI's earlier essay [23]. However, the multidimensional formalism intrinsically connected with it used to be missed.

The intrinsic relation is immediately recognized by considering that a beam or a plate is often treated as a degenerate Cosserat continuum. The theory of yielding is an analogue of the theory of stability of a plate. The section of a beam or a plate deforms and rotates considerably independently of its neighbours but without being entirely freed (cf. the end of Chap. 1).

Unlike the endeavour elsewhere to classify the more microscopic features, the theory of yielding is more concerned with the cosmical aspects of the same objects. The couple stress problem can also been viewed from a cosmical standpoint, so that the theory of a dual yielding is entailed, as will be stated later.

iii) The third step is a kind of refinement. On one hand, we needed to study what relation can be pointed out between Einstein's principle of general relativity to the space-time extension of our statistical construction. This may clarify the statistical background of Einstein's formalism which has so far been lacking. On the other hand, it clarifies the physical constructional meaning of the material constants such as B and $\varkappa$.

If the formula (6) extended to four dimensions is inserted into the general relativistic variational criterion, then no material constants such as B or $\varkappa B$ come across the analysis. Evidently, the Einsteinian formalism is a special case of a more general one,

$$\delta \int (k^{\lambda\mu} K_{\mu\lambda} + g^{\lambda\mu} M_{\mu\lambda}) \sqrt{g}\, dX = 0, \tag{10}$$

where $K_{\mu\lambda}$ is the contracted curvature tensor, $k^{\lambda\mu}$ is an unknown tensor and $M_{\mu\lambda}$ comes from a generalization of the material-energy tensor. Einstein's standpoint is to assume the metric tensor $g^{\lambda\mu}$ itself as the $k^{\lambda\mu}$. For anisotropic materials, the general variational approach by (10) gives the field equation and a boundary conditions which are generalization of (8) and (9), where B and $\varkappa B$ are generalized to

$$\boxed{B^{\varkappa\lambda\mu\nu} = k^{\varkappa\mu} g^{\lambda\nu} - k^{\varkappa\nu} g^{\lambda\mu} - k^{\lambda\nu} g^{\varkappa\mu} + k^{\lambda\mu} g^{\varkappa\nu}}\,. \tag{11}$$

It has been shown that a microscopic non-uniformity such as the polycrystalline structure is indispensable in order that $B^{\varkappa\lambda\mu\nu}$ does not vanish in the case of an isotropic material [24].

iv) A certain physical statistical meaning of the constants $k^{\lambda\mu}$'s has been revealed in reference to the possibility of dual yielding. They are stress functions!

In place of the physical model of iii), a mathematical model can also be used. The geodesics or paths can be treated as the elements of statistics to produce the same results as the foregoing.

B. Dual Riemannian space. The dual yielding results where the rôles of u and v in (7), or those of $k^{\lambda\mu}$, $K_{\mu\lambda}$ and $g^{\lambda\mu}$, $M_{\mu\lambda}$ in (10) are reversed. Dually to the definition

$$u = k^{\lambda\mu} g^{\varkappa\nu} K_{\nu\mu\lambda\varkappa} \sqrt{g}, \qquad K_{\mu\lambda} = g^{\varkappa\nu} K_{\nu\mu\lambda\varkappa}$$

we have

$$v = M\sqrt{g} = g^{\lambda\mu} k^{\varkappa\nu} M_{\nu\mu\lambda\varkappa}, \qquad M_{\mu\lambda} = k^{\varkappa\nu} M_{\nu\mu\lambda\varkappa}.$$

Noting the dual correspondence between

$$g_{\lambda\varkappa},\ g^{\varkappa\lambda},\ K_{\nu\mu\lambda\varkappa}; \quad I_{\mu\lambda} = K_{\mu\lambda} - \tfrac{1}{2} K g_{\mu\lambda}, \quad K = g^{\lambda\mu} K_{\mu\lambda}$$

and

$$k_{\lambda\varkappa},\ k^{\varkappa\lambda},\ M_{\nu\mu\lambda\varkappa}; \quad J_{\mu\lambda} = M_{\mu\lambda} - \tfrac{1}{2} M k_{\mu\lambda}, \quad M = k^{\lambda\mu} M_{\mu\lambda}$$

we obtain a Riemannian space, with

$$dy^2 = k_{\lambda\varkappa}\, dx^\varkappa\, dx^\lambda$$

as the fundamental metric form, the Einstein tensor of which, namely $J_{\mu\lambda}$, coincides with the stress tensor. This is, of course, a special kind of stress field, and as has been shown by H. SCHAEFER [13], is derived, for small disturbances from Beltrami's stress functions which are nothing but

$$\chi_{\lambda\varkappa} = \tfrac{1}{2}(k_{\lambda\varkappa} - \delta_{\lambda\varkappa})$$

provided the reference frame is Cartesian and rectangular.

In the problem of yielding, the objects, connected with the dual Riemannian metric dy^2 were fixed while those connected with the Riemannian metric ds^2 were varied. In the dual problem, the objects connected with the latter need be fixed while those connected with the former are varied. We need not explain in detail how the fundamental equations perfectly dual to those of yielding are derived, where the incompatibility tensor $I^{\varkappa\lambda}$ takes over the place of the stress tensor $J^{\varkappa\lambda}$. S. AMARI has shown that there is a kind of gauge transformation for the Beltrami-Schaefer stress functions by which the stress field is not affected [25]. The gauge functions turn out to be the dual of the w and the equations of dual yielding are usually described in terms of them.

Which physical phenomena are described by the dual equations may be conjectured as follows.

For small disturbances one can mutually transform $U = \int u\, dX$ and $V = \int v\, dX$ by integration by parts

$$U = \tfrac{1}{2}\int \chi_{\lambda\varkappa}\, I^{\varkappa\lambda}\, dX = -\tfrac{1}{2}\int S_{\mu\lambda}^{\cdot\cdot\varkappa}\, S'^{\mu\lambda}_{\cdot\cdot\varkappa}\, dX = \tfrac{1}{2}\int \varepsilon_{\lambda\varkappa}\, J^{\varkappa\lambda}\, dX = V$$

where the strain components dually related to the stress functions $\chi_{\lambda\varkappa}$, $S_{\mu\lambda}^{\cdot\cdot\varkappa}$ is the torsion tensor representing ordinarily the density of dislocations whereas $S'^{\mu\lambda}_{\cdot\cdot\varkappa}$ the dual torsion tensor, is the representation of the *couple stress*.

To sum up,

a) *At the yielding, where the stress is fixed, incompatibilities or dislocations are created by tearing so that the material elements are observed mutually to slip.*

b) *At the dual yielding, where the incompatibility* (*or dislocation to be produced by tearing it*) *is fixed, dual dislocations or couple stresses are created so that the material elements are supposed mutually to be bent and twisted so as to entail a fracture* (*by fatigue for example*).

A practical application of the equations of dual yielding was attempted by S. MINAGAWA [20]. Assuming an isotropic incompatibility distribution he obtained a solution

$$I\sqrt{S} = \text{const.} \tag{12}$$

where the scalar dual curvature S represents the stress. The stress being a part of the material-energy tensor, the quantity S in (12) can be supplemented by $S - S_0$ where S_0 is the summary of other sorts of material energy. This is KONDO's interpretation of the endurance limit [21, 26].

C. Thermodynamical implication. Since the temperature entropy term appears in the thermodynamical equation in the same manner as the pressure which is a kind of stress, the critical stress problem of yielding can be extended to the critical temperature problem, such as the martensitic transformation. Whether this is a kind of yielding or of dual yielding does not seem to matter. Analysis has been made to determine the condition to produce circular and needle-form martensite embryos and the associated hysteresis [27]. The possibility of spherical martensite embryo was also studied. There seems to be far less chances for it in agreement with the empirical fact [28].

A more general thermodynamical implication of the foregoing formulation may be pointed out. In thermodynamical phenomena, there are always hidden degrees of freedom which are not recognized by the Euclidean observer. This is responsible for the non-Euclidean energy (cf. Chap. 1, see e.g. [29]).

4. More Microscopic Penetration

By the statement: "Each point is provided with a space direction", we are reminded of FINSLER's space each element of which is represented by an n-dimensional coordinates and direction parameters which can vary independently of one another. Although the theory of Cosserat continua of modern times, as of old, has mostly been described not in FINSLER's terminology but in that of linear affine connexion, the so-called non-Riemannian geometry, a Finslerian representation need not be excluded. The only difference in the latter case is that the dislocation or the torsion need not be fixed allowing the dislocational anomaly to vary in infinitely different ways. If the direction parameters are fixed as definite functions of the point coordinates, then the Finslerian non-linear parameters of connexion reduces to a linear one entailing a torsion of the ordinary kind. Two interpretations can be given to such an osculating picture.

i) Firstly, it can be treated as a more natural interpretation of the origin of the theory of dislocation distributions in terms of linear affine connexion.

ii) Secondly, it affords the possibility to describe a certain feature of the spin distribution in a continuum in terms of the ordinary torsion tensor of linear connexion degenerated from the Finslerian spin.

The two interpretations can be combined into a unified theory of the penetration into the non-local features [30].

On further penetrating more microscopic structures, more complicated space concepts of higher orders will be introduced. The mathematician has also presented more than one sort of systematical formulation of higher order objects. One may resort whichever one may like, as has been done. I shall refer only to our application of the system proposed by Professor Akitsugu Kawaguchi (see [31, 32]). This is applicable even to problems of elementary particles and is perhaps the most natural realistic of the analytical formulations of the human observations of the so-called microscopic as well as macroscopic worlds. By degeneration, those geometrical and physical objects of higher orders can be mapped onto simpler terminologies, loosing at the same time some deeper recognitions so penetrated. Among the maps of such a kind can be found the dislocations, the spins, the general relativistic space-time curvature or material energy, etc. Therefore, the language to be chosen may also reflect with different degrees of degeneration. They might not appear to be unique. They can be *cosmical* as well as *microscopic*.

5. Concluding Remark

I have tried to provide a relatively unrestrained exposition of the general feature of the principal problems rather than their analytical details. It is not for me to trace the historical development. Nor is it possible for me to evaluate and to look for a unified picture of those various contributions which have been instrumental in the creation of new thought. It seems to me as if something is only being started today and not to be concluded as yet, especially regarding the extent of the possibility of applications of the concept of Cosserat continua. What is appropriate to me is to state, within the reasonable limit, which parts of the general flow of evolving thought have impressed a mind working in a remote quarter of the world. It is hoped that such an essay supplements what might be missed otherwise.

References

[1] Kondo, K.: J. Jap. Soc. App. Mech. **2**, 123, 146 (1949).
[2] Kondo, K.: ibid. **3**, 184 (1950).
[3] Cartan, E.: Ann. sc., l'école norm. sup. **40**, 325 (1923).
[4] Kondo, K.: Proc. 2nd Jap. Nat. Congr. App. Mech. 1952, p. 41.
[5] Kondo, K.: (Ed.) RAAG Memoirs, Div. D, **1**, 1955, pp. 453–572; **2**, 1958, pp. 199–232; **3**, 1962, pp. 91–214.
[6] Bilby, B. A., R. Bullough and E. Smith: Proc. Roy. Soc. London A **231**, 263 (1955).
[7] Kröner, E.: Z. Naturforschung IIa. 12, 969 (1956).

[8] GÜNTHER, W.: Abh. d. Braunschweigischen Wiss. Ges. **10**, 195 (1958).
[9] KRÖNER, E., and A. SEEGER: Arch. Rat. Mech. Anal. **3**, 97 (1959).
[10] KRÖNER, E.: ibid. **4**, 273 (1960).
[11] VRANCEANU, G.: Les espaces non-holonomes et leurs applications méchaniques. Paris: Gauthier-Villars 1936.
[12] SCHOUTEN, J. A.: Ricci-Calculus, 2nd Ed. Berlin/Göttingen/Heidelberg: Springer 1954.
[13] SCHAEFER, H.: Z. angew. Math. Mech. **33**, 356 (1953).
[14] MINAGAWA, S.: RAAG Memoirs **3**, 1962, p. 69.
[15] STOJANOVITCH, R.: Int. J. Engng. Sci. **1**, 323 (1963).
[16] KONDO, K.: RAAG Memoirs **2**, 1958, p. 409.
[17] KONDO, K.: ibid. **3**, 1962, p. 109.
[18] KONDO, K., M. SHIMBO and S. AMARI: ibid. **4** (in Press) D-XXII.
[19] KONDO, K., et al.: ibid. **3**, 1962, p. 134.
[20] MINAGAWA, S.: ibid. **4** (in Press) D-XVIII.
[21] KONDO, K.: ibid. D-XIX.
[22] OSHIMA, N.: ibid. **1**, 563 (1955).
[23] MORIGUTI, S.: Fundamental Theory of Dislocation of Elastic Body (in Japanese), Oyosugaku-Rikigaku (App. Math. and App. Mech.) 1, 1947, p. 29; ed. RAAG R.N. 3 Ser. 65, 1963.
[24] KONDO, K.: J. Fac. Engng. Univ. of Tokyo (B) **27**, 183 (1964).
[25] AMARI, S.: RAAG Memoirs **4** (in Press) D-XVII.
[26] KONDO, K.: Proc. 1st Int. Conf. Fracture **1**, 1965, p. 35.
[27] KONDO, K.: RAAG Memoirs **3**, 1962, p. 173.
[28] SHIMBO, M., K. SATO and T. DATE: ibid. **4** (in Press) D-XXI.
[29] KONDO, K.: ibid. **4** (in Press) D-XX.
[30] KONDO, K., and S. AMARI: ibid. **4** (in Press) D-XXIII.
[31] KAWAGUCHI, M.: ibid. **3**, 1962, p. 718.
[32] KONDO, K.: (Ed.) ibid., Div. E, **3**, 1962, pp. 215—318; **4** (in Press) E-IX, E-X, E-XII.

A Dynamic Theory of Continual Dislocations

By

L. I. Sedov and **V. L. Berditchevski**

USSR National Committee of Theoretical and Applied Mechanics
USSR Academy of Sciences, Moscow

Introduction

In the continuum theory of dislocations a continuous medium with continuously distributed microstructure imperfections—dislocations, is investigated. The previous papers essentially dealt with the statics of dislocations and the relation between the theory of plasticity and the continuum theory of dislocations has not been clearly shown. In the following we shall construct a class of models of continuous media which includes many known models, as well as new ones, and within the framework of which viscous, elastic and plastic effects and the motion of the imperfections—dislocations, combine.

In particular models of plastic bodies are constructed by means of variational principles and by making use of the properties of the internal energy and of the dissipation function.

In order to describe the distribution of dislocations it is necessary to insert new additional characteristics in the number of determining parameters[1]. As these additional quantities are chosen in different ways, this gives rise to different theories. As a preliminary we give here a brief review of these theories.

In the papers of K. Kondo, B. Bilby, E. Kröner, L. I. Sedov, I. A. Kunin and others [1—8], the continuum is associated with an affinely connected manifold M, and the metric, curvature and torsion tensors of this manifold are taken as characteristics of the dislocations. The manifold M may be introduced on the basis of various conceptual processes. B. Bilby [3—5] constructs the manifold M starting from the theory of lattices and obtains as a result that the curvature tensor vanishes (nine new degrees of freedom). K. Kondo [1, 2] defines the

[1] By definition, the behavior of medium is known, if one knows the dependence of a system of quantities, which are called determining parameters, on spacial coordinates and time.

manifold M as a manifold of initial states which represents a metric affinely—connected manifold of the most general form. As independent parameters one may, choose the metric tensor $g_{\mu\nu}$ and the torsion tensor $S_{\mu\nu}{}^{\lambda}$ of the manifold (in all, 15 new parameters), while the curvature tensor is expressed through their first and second derivatives [9, 10]. For $g_{\mu\nu}$ and $S_{\mu\nu}{}^{\lambda}$ it is necessary to have additional equations. In statics, E. Kröner [6, 11] suggests obtaining part of these equations by giving curvature tensor in function of coordinates. The case with the vanished curvature tensor is called a "limited" theory. In this case there is an absolute parallelism, i.e., the basic assumptions of the theory of Bilby are fulfilled. Linearizing the equations of the "limited" theory one may obtain the equations of the so-called "elementary" theory [11].

If the torsion tensor vanishes, the manifold M becomes Riemannian and it can be imbedded in an Euclidean multidimensional space E. The time variation of the geometrical characteristics of the manifold M result in the displacements of M in E. As determining parameters one may take the components of the displacement vector of the manifold M in E. The corresponding equations for the reversible phenomena are obtained in [2] (of Vol. 3) by means of a variational principle.

In the papers [12, 13] the determining parameters were introduced without using notions of differential geometry. The dynamic equations for them were derived by averaging the equations of motion for an isolated dislocation. Although [12] the entire analysis did not fall outside the limits of the linear theory, complicated integro-differential equations were obtained, which renders it difficult to reconcile the relations with the models of the theory of plasticity.

Further it will be shown that in order to describe the distribution of the dislocations and include known plastic models in the framewack of the theory it is sufficient to introduce nine new (by comparison with the classical theory of elasticity) degrees of freedom (by three degrees of freedom more than in the ordinary theory of plasticity).

1. The Determining Parameters

We shall consider the motion of the medium with respect to a certain in general curvilinear observer's frame with spacial[1] coordinates x^{α}, time coordinate t and basis $\ni_{\alpha}$. Let us introduce also a comoving frame for the medium[2] with Lagrangian coordinates ξ^{μ} time coordinate t and basis $\hat{\ni}_{\mu}$.

[1] The greek superscripts correspond to spacial coordinates and run from 1 to 3.

[2] Quantities related to the comoving system of coordinates are denoted by the symbol $\wedge$.

The law of motion of the medium is determined by the relation between these two frames [14]

$$x^\alpha = x^\alpha(\xi^\mu, t).$$

The base vectors $\hat{Э}_\nu, \hat{Э}^\mu$ are obtained from $Э_\alpha, Э^\beta$ by means of the affine transformation (in every point)

$$\hat{Э}_\nu = x^\alpha{}_\nu Э_\alpha, \quad x^\alpha{}_\nu = \frac{\partial x^\alpha}{\partial \xi^\nu}, \quad \hat{Э}^\mu = \xi^\mu{}_\beta Э^\beta, \quad \xi^\mu{}_\beta = \frac{\partial \xi^\mu}{\partial x^\beta},$$

$$\hat{Э}_\nu = \hat{g}_{\nu\mu} \hat{Э}^\mu, \quad \hat{g}_{\nu\mu} = g_{\alpha\beta}\, x^\alpha{}_\nu\, x^\beta{}_\mu,$$

where $\hat{g}_{\nu\mu}$ and $g_{\alpha\beta}$ are the covariant components of the metric tensor in the comoving and observer's frames.

Let us consider two (isomorphic) groups of coordinate transformation

$$x^\alpha \to = y^\beta(x^\alpha), \quad \xi^\mu \to \eta^\nu(\xi^\mu), \tag{1.1}$$

where y^β are the new spacial coordinates in the observer's frame, η^ν—the new Lagrangian coordinates. In this connection

$$Э_\alpha \to Э_\beta \frac{\partial x^\beta}{\partial y^\alpha}, \quad \hat{Э}^\mu \to \hat{Э}^\nu \frac{\partial \eta^\mu}{\partial \xi^\nu}.$$

Let A be an invariant of the form

$$A = A^\alpha{}_\mu Э_\alpha \hat{Э}^\mu = \hat{C}^\nu{}_\mu \hat{Э}_\nu \hat{Э}^\mu.$$

It is obvious that the $A^\alpha{}_\mu$ behave as the components of a contravariant vector under the transformations of the observer's frame (for a fixed index μ) and as components of a covariant vector under the transformations of the comoving frame (for a fined index α). The components $\hat{C}^\nu{}_\mu$ form a second rank tensor under the transformations of the comoving frame. The quantities $A^\alpha{}_\mu$ and $\hat{C}^\nu{}_\mu$ are the representatives of the same tensor A, corresponding to two different ways of choosing the base vectors. In particular, for the metric tensor G we may consider the representatives

$$G = g_{\alpha\beta} Э^\alpha Э^\beta = x^\alpha{}_\mu Э_\alpha \hat{Э}^\mu = \hat{g}_{\mu\nu} \hat{Э}^\mu \hat{Э}^\nu.$$

In the following we shall consider the derivatives of the tensor A with respect to coordinates, determined, by condition, by the formulas

$$\frac{\partial A}{\partial x^\alpha} Э^\alpha = \frac{\partial A}{\partial \xi^\mu} \hat{Э}^\mu = \frac{\partial A^\beta{}_\mu}{\partial x^\alpha} Э_\beta \hat{Э}^\mu Э^\alpha + A^\beta{}_\mu \frac{\partial Э_\beta}{\partial x^\alpha} \hat{Э}^\mu Э^\alpha +$$

$$+ A^\beta{}_\mu Э_\beta \frac{\partial \hat{Э}^\mu}{\partial \xi^\nu} \xi^\nu{}_\alpha Э^\alpha = \left(\frac{\partial A^\beta{}_\mu}{\partial x^\alpha} + \Gamma_{\alpha\gamma}{}^\beta A^\gamma{}_\mu - \hat{\Gamma}_{\mu\nu}{}^\lambda A^\beta{}_\lambda \xi^\nu{}_\alpha \right) Э_\beta \hat{Э}^\mu Э^\alpha$$

$$= \nabla_\alpha A^\beta{}_\mu Э_\beta \hat{Э}^\mu Э^\alpha = \hat{\nabla}_\gamma A^\beta{}_\mu Э_\beta \hat{Э}^\mu \hat{Э}^\nu,$$

where $\Gamma_{\alpha\beta}{}^{\gamma}$ and $\hat{\Gamma}_{\mu\nu}{}^{\lambda}$ denote the components of the connections for the bases $\ni_\alpha$ and $\hat{\ni}^\mu$ respectively

$$\frac{\partial \ni_\alpha}{\partial x^\beta} = \Gamma_{\alpha\beta}{}^{\gamma} \ni_\gamma, \quad \frac{\hat{\partial} \ni_\mu}{\partial \xi^\nu} = \hat{\Gamma}_{\mu\nu}{}^{\lambda} \hat{\ni}_\lambda.$$

It is easily seen that the quantities $x^\alpha{}_\mu$ and $\xi^\mu{}_\alpha$ are covariantly constants

$$\nabla_\beta x^\alpha{}_\mu = 0, \quad \nabla_\beta \xi^\mu{}_\alpha = 0.$$

The derivatives of the tensor A with respect to time may be defined in various meanings [14]. In what follows we shall make use of the following individual derivative[1]:

$$DA = \frac{d}{dt}(A^\alpha{}_\mu \ni_\alpha \hat{\ni}^\mu)_{\xi^\lambda = \text{const},\ \hat{\ni}^\mu = \text{«const»}} = \left(\frac{dA^\alpha{}_\mu}{dt} + \Gamma_{\beta\gamma}{}^{\alpha} A^\beta{}_\mu v^\gamma\right) \ni_\alpha \hat{\ni}^\mu. \tag{1.2}$$

where $v^\gamma = dx^\gamma/dt$ are the velocity components of the point of the medium. The derivative D is calculated at constant Lagrangian coordinates and "constant" Lagrangian basis $\hat{\ni}^\mu$, taking into account the variation of the vectors of the basis $\ni_\alpha$ for a moving point of the medium.

It is obvious that the quantities $DA^\alpha{}_\mu$ change as $A^\alpha{}_\mu$ when passing to a new system of coordinates.

In order to construct the alternate version of the continuum theory of dislocations proposed in the following it is sufficient to restrict ourselves to the following set of invariant determining parameters

$$\left.\begin{aligned}
&\vec{v} = v^\alpha \ni_\alpha \\
&G = g_{\alpha\beta} \ni^\alpha \ni^\beta = x^\alpha{}_\mu \ni_\alpha \hat{\ni}^\mu = \hat{g}_{\mu\nu} \hat{\ni}^\mu \hat{\ni}^\nu, \\
&A = A^\alpha{}_\mu \ni_\alpha \hat{\ni}^\mu = \hat{C}^\nu{}_\mu \hat{\ni}_\nu \hat{\ni}^\mu, \\
&\frac{\partial A}{\partial \xi^\mu} \hat{\ni}^\mu, \quad DA, \\
&S, \\
L_{(p)} &= \hat{L}^{\nu_1 \ldots \nu_n}{}_{\mu_1 \ldots \mu_n} \hat{\ni}_{\nu_1} \hat{\ni}^{\mu_n} \quad (p = 1, 2, \ldots, N)
\end{aligned}\right\} \tag{1.3}$$

where S is the entropy, $L_{(p)}$ — a set of tensors, characterizing the physical and geometrical properties of the medium in the initial state (for instance, the anisotropy). Among the tensors $L_{(p)}$ there may be a tensor $\mathring{G} = \mathring{g}_{\mu\nu} \hat{\ni}^\mu \hat{\ni}^\nu$ determining the distance ds_0 between the particles in the initial state

$$ds_0^2 = \mathring{g}_{\mu\nu}\, d\xi^\mu\, d\xi^\nu.$$

---

[1] Here we have taken account [14]

$$d\ni_\alpha/dt = \Gamma_{\alpha\beta}{}^{\gamma} \ni_\gamma v^\beta.$$

By definition

$$d\hat{L}_{(p)}^{\nu_1\ldots\nu_k}{}_{\mu_1\ldots\mu_n}/dt = 0 .$$

In comparison with the classical theory of elasticity we have nine new degrees of freedom among the determining parameters: the components of the tensor A. Further, we may ascribe to them the following physical meaning.

Consider an infinitely small particle with Lagrangian coordinates ξ^μ cut it out from the body and release it from the external forces. In this connection the particle deforms and the vectors of the basis $\hat{\ni}_\nu$ convert to the vectors $\overset{*}{\ni}_\mu$. This deformation is, by condition, described by the tensor A:

$$\overset{*}{\ni}_\mu = \hat{C}^\nu{}_\mu \hat{\ni}_\nu = A^\alpha{}_\mu \ni_\alpha . \tag{1.4}$$

Since after the deformation the particle is in an unstrained state, the components $\hat{C}^\nu{}_\mu$ of the tensor A describe the elastic deformation[1]. Farther on as a basic condition, we assume that the components of the tensor A depend only on the coordinate ξ^μ and time t.

In connection with the tensor A we may introduce a sequence of physical parameters:

$$\hat{\varepsilon}_{\mu\nu}{}^{(e)} = \tfrac{1}{2}(\hat{g}_{\mu\nu} - \overset{*}{g}_{\mu\nu}), \qquad \overset{*}{g}_{\mu\nu} = g_{\alpha\beta} A^\alpha{}_\mu A^\beta{}_\nu \tag{1.5}$$

the plastic strain tensor

$$\hat{\varepsilon}_{\mu\nu}{}^{(p)} = \tfrac{1}{2}(\overset{*}{g}_{\mu\nu} - \overset{\circ}{g}_{\mu\nu}) \tag{1.6}$$

the plastic strain velocity tensor

$$\hat{e}_{\mu\nu}{}^{(p)} = \frac{d\hat{\varepsilon}_{\mu\nu}{}^{(p)}}{dt} = \frac{1}{2}\frac{d\overset{*}{g}_{\mu\nu}}{dt} = \frac{1}{2} g_{\alpha\beta}(A^\alpha{}_\mu DA^\beta{}_\nu + A^\beta{}_\nu DA^\alpha{}_\mu)$$

the elastic strain gradient tensor

$$\hat{\nabla}_\lambda \hat{\varepsilon}_{\mu\nu}{}^{(e)} = -\tfrac{1}{2}\hat{\nabla}_\lambda \overset{*}{g}_{\mu\nu} = -\tfrac{1}{2} g_{\alpha\beta}(A^\alpha{}_\mu \hat{\nabla}_\lambda A^\beta{}_\nu + A^\beta{}_\nu \hat{\nabla}_\lambda A^\alpha{}_\mu) .$$

Consider a closed circuit L in the body. It is obvious that

$$\oint d\vec{r} = \oint \hat{\ni}_\mu \, d\xi^\mu = 0 .$$

With the above defined elastic strain A each infinitely small element $d\vec{r} = \hat{\ni}_\mu \, d\xi^\mu$ becomes $d\vec{r}^{\,*} = \overset{*}{\ni}_\mu \, d\xi^\mu$. The integral

$$\oint d\vec{r}^{\,*} = \oint \overset{*}{\ni}_\mu \, d\xi^\mu$$

[1] The tensor (1.4) differs from the tensor $A = A^\alpha{}_a \ni_\alpha \bar{\ni}^a$ introduced by E. KRÖNER [6] by the law of transformation of the components $A^\alpha{}_a$ since the vectors $\bar{\ni}_a = A^\alpha{}_a \ni_\alpha$ form a different non-holonomic basis, not equal to $\overset{*}{\ni}_\alpha$.

in general, does not vanish and is equal to the vector $\vec{b}_{(L)}$ which connects the ends of the cut circuit L after the elastic deformation[1]

$$\vec{b}_{(L)} = \vec{b}_{(L)}{}^{\nu}\,\hat{\mathfrak{Z}}_{\nu} = \oint \overset{*}{\mathfrak{Z}}_{\mu}\, d\xi^{\mu} = \iint\limits_{S} \hat{V}_{[\mu}\, A^{\alpha}{}_{\nu]}\, \mathfrak{Z}_{\alpha}\, d\xi^{\mu}\, d\xi^{\nu}. \tag{1.7}$$

According to the formula (1.7) to each (finite) closed circuit L corresponds a vector $\vec{b}_{(L)}$ called the Burger's vector.

Let in a certain point ξ^{μ} an infinitely small circuit enclose an area element $d\sigma$ with a normal vector $\vec{n}$. It is associated with an infinitely small Burger's vector $\vec{b}_{(n)}$ which depends in the choice of the point ξ^{μ} the vector $\vec{n}$ and are a element $d\sigma$. In order to obtain the Burger's vector in the point ξ, let us consider the limit

$$\lim_{d\sigma\to 0} \frac{\vec{b}_{(n)}}{d\sigma} = \hat{S}^{\omega\lambda}\, \hat{n}_{\omega}\, \hat{\mathfrak{Z}}_{\lambda}, \qquad \hat{S}^{\omega\lambda} = \hat{\varepsilon}^{\omega\mu\nu}\, B^{\lambda}{}_{\alpha}\, \hat{V}_{\mu}\, A^{\alpha}{}_{\nu}. \tag{1.8}$$

In the formula (1.8) $\hat{\varepsilon}^{\omega\mu\nu}$ denote the components of the fully antisymmetric tensor and $\hat{\varepsilon}^{123} = 1/\sqrt{\hat{g}}$, $B^{\lambda}{}_{\alpha}$ — the component of the matrix reciprocal to $A = \|A^{\alpha}{}_{\mu}\|$. Thus, the components of the Burger's vector $\vec{b}_{(n)}$ in a point are determined by the normal $\vec{n}$ and tensor $\hat{S}^{\omega\lambda}(\xi^{\mu}, t)$ called the dislocation density tensor. Parallel to $\hat{S}^{\omega\lambda}$ we may consider the third rank tensor α antisymmetric with respect to the indexes $\mu\,\nu$

$$\alpha = \hat{S}_{\mu\nu}{}^{\lambda}\, \hat{\mathfrak{Z}}^{\mu}\, \hat{\mathfrak{Z}}^{\nu}\, \hat{\mathfrak{Z}}_{\lambda}, \qquad \hat{S}_{\mu\nu}{}^{\lambda} = B^{\lambda}{}_{\alpha}\, \hat{V}_{[\mu}\, A^{\alpha}{}_{\nu]} \tag{1.9}$$

and related to the tensor $\hat{S}^{\omega\lambda}$ with the reciprocal formulas

$$\hat{S}_{\mu\nu}{}^{\lambda} = \hat{\varepsilon}_{\mu\nu\omega}\, \hat{S}^{\omega\lambda}, \qquad \hat{S}^{\omega\lambda} = \hat{\varepsilon}^{\omega\mu\nu}\, \hat{S}_{\mu\nu}{}^{\lambda}.$$

The tensor $\hat{S}_{\mu\nu}{}^{\lambda}$ will be also called the dislocation density tensor.

If $\vec{b}_{(1)}, \vec{b}_{(2)}, \vec{b}_{(3)}$ denote the Burger's vectors for the area elements with normals $\mathfrak{Z}_1, \mathfrak{Z}_2, \mathfrak{Z}_3$ respectively, then for the Burger's vector, associated with an area element having a normal $\vec{n}$, we may write:

$$\vec{b}_{(n)} = \vec{b}_{(1)}\, n^1 + \vec{b}_{(2)}\, n^2 + \vec{b}_{(3)}\, n^3. \tag{1.10}$$

As it follows from (1.8) and (1.10) the Burger's vector is analogous to the vector of the surface force, acting on an area element $d\sigma$, while $\hat{S}^{\omega\lambda}$ is analogous to the stress tensor.

---

[1] As usual the square brackets denote the alternation operation, the others — the symmetrization operation.

The location of the indexes in $A^{\alpha}{}_{\mu}$ is essential. Let us agree to omit the indexes in $A^{\alpha}{}_{\mu}$ and $B^{\mu}{}_{\alpha}$ by means of the metric $g_{\alpha\beta}$ for the index α and by means of the metric $\overset{*}{g}_{\mu\nu}$ for the index μ. Since $A_{\alpha\mu} = g_{\alpha\beta}\, A^{\beta}{}_{\mu}$, $B_{\mu\alpha} = \overset{*}{g}_{\mu\nu}\, B^{\nu}{}_{\alpha}$ from (1.5) it follows that

$$A_{\alpha\mu} = B_{\mu\alpha}.$$

Introduce the tensor $\Pi = \hat{\pi}_{\mu\nu} \hat{\mathbf{Э}}^\mu \hat{\mathbf{Э}}^\nu$ by means of the equality

$$d\overset{*}{\mathbf{Э}}_\nu/dt = \hat{\pi}_{\mu\nu} \overset{*}{\mathbf{Э}}^\mu = DA^\alpha{}_\nu \mathbf{Э}_\alpha .$$

Hence

$$\hat{\pi}_{\mu\nu} = B_{\mu\alpha} DA^\alpha{}_\nu .$$

The angular velocity of rotation for the affine transformation (1.4), defined by the components $A^\alpha{}_\mu$ will be called the plastic vorticity. The corresponding antisymmetric tensor will be denoted by $\Omega = \hat{\Omega}_{\mu\nu} \hat{\mathbf{Э}}^\mu \hat{\mathbf{Э}}^\nu$. It is easily seen that

$$\hat{\Omega}_{\mu\nu} = \hat{\pi}_{[\mu\nu]} . \tag{1.11}$$

Let us note that the components of the plastic strain velocity tensor are expressed in terms of the components of the tensor Π by means of the formula

$$\hat{e}_{\mu\nu}{}^{(p)} = \hat{\pi}_{(\mu\nu)} . \tag{1.12}$$

In the foregoing the dislocation characteristics were introduced without making use of geometrical notions. In what follows, for the dynamic theory, the corresponding geometrical interpretation is not required. However, in order to establish contacts between the theory developed in this paper and the already known kinematic theories and to obtain additional informations, facilitating the comparison of the theory with experiment, we shall show how to construct, in connection with the tensor A, the metric affinely—connected manifold of the "initial" state.

Let us define in the comoving system of coordinates the geometrical object.

$$\overset{*}{\Gamma}_{\mu\nu}{}^\lambda = B^\lambda{}_\alpha \left(\frac{\partial A^\alpha{}_\nu}{\partial \xi^\mu} + \Gamma_{\beta\gamma}{}^\alpha A^\beta{}_\nu x^\gamma{}_\mu \right) . \tag{1.13}$$

It is easily verified, that $\overset{*}{\Gamma}_{\mu\nu}{}^\lambda$ change as a connection when passing from one comoving system of coordinates to the other, and that the tensor $\overset{*}{g}_{\mu\nu}$ (1.5) is covariantly constant with respect to the connection $\overset{*}{\Gamma}_{\mu\nu}{}^\lambda$, i.e.

$$\overset{*}{\nabla}_\mu \overset{*}{g}_{\nu\omega} = \partial \overset{*}{g}_{\nu\omega} / \partial \xi^\mu - \overset{*}{\Gamma}_{\mu\nu}{}^\lambda \overset{*}{g}_{\lambda\omega} - \overset{*}{\Gamma}_{\mu\omega}{}^\lambda \overset{*}{g}_{\nu\lambda} = 0 . \tag{1.14}$$

The dislocation density tensor $\hat{S}_{\mu\nu}{}^\lambda$ coincides with the antisymmetric (with respect to $\mu\nu$) part of the connection

$$\hat{S}_{\mu\nu}{}^\lambda = \Gamma_{[\mu\nu]}{}^\lambda . \tag{1.15}$$

Let us introduce the manifold of the initial state with residual plastic strains as a manifold with the connection[1] $\overset{*}{\Gamma}_{\mu\nu}{}^\lambda$, metric tensor $\overset{*}{g}_{\mu\nu}$ and

[1] In the "limited" theory of KRÖNER [11] the components of the connection vanish for the basis $\overline{\mathbf{Э}}_\alpha$ of the initial state manifold. In the theory developed below from (1.4) we have

$$\frac{\partial \overset{*}{\mathbf{Э}}_\mu}{\partial \xi^\nu} = \frac{\partial}{\partial \xi^\nu} (A^\alpha{}_\mu \mathbf{Э}_\alpha) = \left(\frac{\partial \xi^\alpha{}_\mu}{\partial A^\nu} + \Gamma^\alpha{}_{\beta\gamma} A^\beta{}_\mu x^\gamma{}_\nu \right) \mathbf{Э}_\alpha = \overset{*}{\Gamma}_{\mu\nu}{}^\lambda \overset{*}{\mathbf{Э}}_\lambda$$

torsion tensor $\hat{S}_{\mu\nu}{}^{\lambda}$. From (1.14) and (1.15) it follows that $\overset{*}{\Gamma}_{\mu\nu}{}^{\lambda}$ is expressed in terms of the metric and torsion tensors by means of the ordinary formulas [9, 10]

$$\overset{*}{\Gamma}_{\mu\nu}{}^{\lambda} = \{{}_{\mu\nu}{}^{\lambda}\} + \hat{S}^{\cdot\cdot\lambda}_{\mu\nu\cdot} - \hat{S}^{\cdot\lambda\cdot}_{\mu\cdot\nu} - \hat{S}^{\cdot\lambda\cdot}_{\nu\cdot\mu}. \tag{1.16}$$

Where

$$\{{}_{\mu\nu}{}^{\lambda}\} = \frac{1}{2}\overset{*}{g}{}^{\lambda\omega}\left(\frac{\partial \overset{*}{g}_{\nu\omega}}{\partial \xi^{\mu}} + \frac{\partial \overset{*}{g}_{\mu\omega}}{\partial \xi^{\nu}} - \frac{\partial \overset{*}{g}_{\mu\nu}}{\partial \xi^{\omega}}\right).$$

Calculation [taking into account (1.5) and (1.13)] shows that the curvature tensor of the manifold M vanishes

$$R_{\alpha\beta\gamma}{}^{\delta}\,(\overset{*}{\Gamma}_{\mu\nu}{}^{\lambda},\ \partial\overset{*}{\Gamma}_{\mu\nu}{}^{\lambda}/\partial\xi^{\omega}) = 0. \tag{1.17}$$

Substituting (1.16) in (1.17) we obtain a relation between the metric and the dislocation density tensors. The resulting equality

$$R_{\alpha\beta\gamma}{}^{\delta}\left(\overset{*}{g}_{\nu\omega},\ \frac{\partial \overset{*}{g}_{\nu\omega}}{\partial\xi^{\mu}},\ \frac{\partial^2 \overset{*}{g}_{\nu\omega}}{\partial\xi^{\mu}\,\partial\xi^{\lambda}},\ \hat{S}_{\mu\nu}{}^{\lambda},\ \frac{\partial \hat{S}_{\mu\nu}{}^{\lambda}}{\partial\xi^{\omega}}\right) = 0 \tag{1.18}$$

certain authors call the basic geometric law. Let us note that at each (fixed) point the components of the metric and dislocation density tensors are kinematically independent, since (1.18) contains derivatives of $\overset{*}{g}_{\mu\nu}$ and $\hat{S}_{\mu\nu}{}^{\lambda}$ with respect to coordinates.

After linearizing the Eqs. (1.18) we may isolate the parts of $R_{\alpha\beta\gamma}{}^{\delta}$ which individually depend only on the derivatives $\overset{*}{g}_{\mu\nu}$ and on $\hat{S}_{\mu\nu}{}^{\lambda}$ [8]

$$R_{\alpha\beta\gamma}{}^{\delta} = K_{\alpha\beta\gamma}{}^{\delta}\left(\frac{\partial^2 \overset{*}{g}_{\mu\nu}}{\partial\xi^{\omega}\,\partial\xi^{\lambda}}\right) + N_{\alpha\beta\gamma}{}^{\delta}\left(\frac{\partial \hat{S}_{\mu\nu}{}^{\lambda}}{\partial\xi^{\omega}}\right) = 0. \tag{1.19}$$

The tensor $N_{\alpha\beta\gamma}{}^{\delta}$ is called the incompatibility tensor. The Eqs. (1.19) are considered as basic in the static linear theory of dislocations [6, 8, 11]. The incompatibility tensor being given, it was supposed to find from the Eq. (1.19) the metric tensor $\overset{*}{g}_{\mu\nu}$. This formulation cannot be regarded as satisfactory, since naturally formulated problems the incompatibility tensor itself must be determined by the solution of the problem.

In the following the dynamic equations will be obtained for the quantities $A^{\alpha}{}_{\mu}$, the basic geometrical law (1.18) being always satisfied identically. The tensors $\overset{*}{g}_{\mu\nu}$ and $\hat{S}_{\mu\nu}{}^{\lambda}$ are determined in terms of known $A^{\alpha}{}_{\mu}$ by the formulas (1.5) and (1.9).

2. A Variational Principle

The construction of the models will be carried out on the basis of a variational principle[1] [7, 17, 15, 16]

$$\delta \int\limits_{V}\int\limits_{t_1}^{t_2} \Lambda\, d\tau\, dt + \delta W + \delta W^{*} = 0, \tag{2.1}$$

---

[1] The functional δW is determined for an arbitrary domain $V t$ when Λ and δW^{*} are known. In this connection we obtain the state and kinetic equations.

where Λ is the Langrangean, V — an arbitrary domain, associated with the particles of the medium, $d\tau$ — a volume element

$$d\tau = \sqrt{\hat{g}}\, d\xi^1\, d\xi^2\, d\xi^3 = \sqrt{g}\, dx^1\, dx^2\, dx^3, \quad \hat{g} = \det\|\hat{g}_{\mu\nu}\|, \quad g = \det\|g_{\alpha\beta}\|.$$

The Eq. (2.1) is considered as the basic relation when the variations of the determining parameters are arbitrary, in particular, when they do not vanish on the boundary of the integration domain. The functional δW is represented by an integral of the linear combination of the variations of the determining parameters, taken along the boundary of a fourdimensional domain in the space of variables ξ^μ, t and is subject to determination. δW^* is a prescribed functional. The Lagrangean Λ, by supposition, depends on the components of the tensors (1.3), has the dimension of the body energy density and is a scalar with respect to the transformation group (1.1). For simplicity we shall assume in what follows, that the quantities $\hat{\nabla}_\nu A^\alpha{}_\mu$ enter the Lagrangean only through the components of the dislocation density tensor α (1.9).

Let us define the variations of the system of arguments of Λ (1.3). By condition, in (2.1), the particle trajectories of the medium

$$\delta x^\alpha = x'^\alpha(\xi^\mu, t) - x^\alpha(\xi^\mu, t)$$

the independent parameters $A^\alpha{}_\mu$ and the entropy S are subject to variation. For the variations of the tensor A let us put, by definition[1]

$$\delta A = A'^\alpha{}_\mu \mathbf{\ni}_\alpha(x')\, \hat{\mathbf{\ni}}^\mu(\xi) - A^\alpha{}_\mu \mathbf{\ni}_\alpha(x)\, \hat{\mathbf{\ni}}^\mu(\xi) = \delta A^\alpha{}_\mu \mathbf{\ni}_\alpha \hat{\mathbf{\ni}}^\mu.$$

The quantities $\delta A^\alpha{}_\mu$ change as $A^\alpha{}_\mu$ when passing to another system of coordinates. For the variations of the derivatives of x^α and A we make use of the formulas:

$$\left.\begin{aligned} \delta\vec{v} &= v'^\alpha(x', t)\, \mathbf{\ni}_\alpha(x') - v^\alpha(x, t)\, \mathbf{\ni}_\alpha(x) \\ &= [d\,\delta x^\alpha/dt + \Gamma_{\beta\gamma}{}^\alpha v^\gamma\, \delta x^\beta]\, \mathbf{\ni}_\alpha = (D\,\delta x^\alpha)\mathbf{\ni}_\alpha, \\ \delta G &= x'^\alpha{}_\mu \mathbf{\ni}_\alpha(x')\, \hat{\mathbf{\ni}}^\mu(\xi) - x^\alpha{}_\mu \mathbf{\ni}_\alpha(x)\, \hat{\mathbf{\ni}}^\mu(\xi) = x^\beta{}_\mu \nabla_\beta\, \delta x^\alpha\, \mathbf{\ni}_\alpha \hat{\mathbf{\ni}}^\mu, \\ \delta DA &= \delta(DA^\alpha{}_\mu \mathbf{\ni}_\alpha \hat{\mathbf{\ni}}^\mu) = (D\delta A^\alpha{}_\mu)\, \mathbf{\ni}_\alpha \hat{\mathbf{\ni}}^\mu = D\delta A, \\ &\delta(\hat{S}_{\mu\nu}{}^\lambda\, \hat{\mathbf{\ni}}^\mu \hat{\mathbf{\ni}}^\nu \hat{\mathbf{\ni}}_\lambda) = (\delta\hat{S}_{\mu\nu}{}^\lambda)\, \hat{\mathbf{\ni}}^\mu \hat{\mathbf{\ni}}^\nu \hat{\mathbf{\ni}}_\lambda. \end{aligned}\right\} \quad (2.2)$$

[1] The arbitrariness of the variations allows us to determine them in different ways.

The assignment of δW on the boundary of the domain $V\,t$ leads also to the boundary conditions. The momentum energy tensor (in particular, the stress tensor) is determined by means of the functional δW and not only by means of the equations, as is usually done. The functional δW^* contain terms which take into account the variations of entropy and heat flux this allows to obtain the proper energy equation for irreversible processes. An equality of the form (1.2) was used by R. A. TOUPIN [23], R. D. MINDLIN [24] and others in obtaining models only with reversible processes, the functional δW being given only for the formulation of the boundary conditions.

From (1.9) we have

$$\delta \hat{S}_{\mu\nu}{}^{\lambda} = B^{\lambda}{}_{\alpha} \hat{V}_{[\mu} \delta A^{\alpha}{}_{\nu]} - \delta B^{\lambda}{}_{\alpha} \cdot \hat{V}_{[\mu} A^{\alpha}{}_{\nu]} = B^{\lambda}{}_{\alpha} \hat{V}_{[\mu} \delta A^{\alpha}{}_{\nu]} - \hat{S}_{\mu\nu}{}^{\omega} B^{\lambda}{}_{\alpha} \delta A^{\alpha}{}_{\omega}$$

since $\delta B^{\lambda}{}_{\beta} = - B^{\lambda}{}_{\alpha} B^{\mu}{}_{\beta} \delta A^{\alpha}{}_{\mu}$, which follows from the definition of $B^{\alpha}{}_{\mu}$: $B^{\lambda}{}_{\beta} A^{\beta}{}_{\mu} = \delta^{\lambda}{}_{\mu}$. The components of the tensors L are considered as given functions of ξ^{μ} and, therefore are not varied.

In the basic equality (2.1) besides varying the function (at constant ξ^{μ}) we shall vary the time t by shifting it by an infinitely small constant δt. For the limited aims achieved in the present paper such a variation of time is sufficient[1]. It is obvious, that for any quantity A we have

$$\delta_1 A = A'(\xi^{\mu}, t') - A(\xi^{\mu}, t) = \delta A + DA \cdot \delta t.$$

In what follows the symbol δ_1 will denote the total variation and the symbol δ the variation at constant t. In particular, for $\delta_1 x^{\alpha}$ we may write

$$\delta_1 x^{\alpha} = x'^{\alpha}(\xi^{\mu}, t') - x^{\alpha}(\xi^{\mu}, t) = \delta x^{\alpha} + v^{\alpha} \delta t.$$

Farther on, the variation of the integral of Λ in (2.1) will be understood as the variation δ_1. On the basis of (2.2) and equalities

$$\delta_1 \Lambda = \delta \Lambda + \frac{d\Lambda}{dt} \delta t, \qquad \delta_1 d\tau = (V_{\alpha} \delta_1 x^{\alpha})\, d\tau, \qquad \frac{d}{dt} \Lambda \sqrt{\hat{g}} = \varrho \sqrt{\hat{g}} \frac{d}{dt} \frac{\Lambda}{\varrho} \tag{2.3}$$

where, by definition, $\varrho = f(\xi^{\mu})/\sqrt{\hat{g}}$ is the density of the medium. Varying the first term in (2.1) we obtain in a familiar way

$$\begin{aligned} \delta_1 \int\limits_{V} \int\limits_{t_1}^{t_2} \Lambda\, d\tau\, dt = \int\limits_{V} \int\limits_{t_1}^{t_2} \Big\{ & X_{\alpha} \delta x^{\alpha} + \frac{\delta \Lambda}{\delta A^{\alpha}{}_{\mu}} \delta A^{\alpha}{}_{\mu} + \frac{\partial \Lambda}{\partial S} \delta S + \\ & + \left[\varrho \frac{d}{dt} \left(\frac{\Lambda}{\varrho} - v^{\alpha} \frac{\partial}{\partial v^{\alpha}} \frac{\Lambda}{\varrho} - DA^{\alpha}{}_{\mu} \frac{\partial}{\partial (DA^{\alpha}{}_{\mu})} \frac{\Lambda}{\varrho} \right) + \right. \\ & \left. + V_{\beta} (\sigma_{\alpha}{}^{\beta} v^{\alpha} + \hat{\sigma}^{\nu\mu\lambda} \hat{\pi}_{\nu\mu} x^{\beta}{}_{\lambda}) \right] \delta t \Big\} d\tau\, dt - \\ & - \int\limits_{\Sigma} \int\limits_{t_1}^{t_2} (\sigma_{\alpha}{}^{\beta} \delta_1 x^{\alpha} + \hat{\sigma}_{\nu}{}^{\mu\lambda} x^{\beta}{}_{\lambda} B^{\nu}{}_{\alpha} \delta_1 A^{\alpha}{}_{\mu})\, n_{\beta}\, d\sigma\, dt - \\ & - \int\limits_{V} \varrho \left[\delta_1 x^{\alpha} \frac{\partial}{\partial v^{\alpha}} \frac{\Lambda}{\varrho} + \delta_1 A^{\alpha}{}_{\mu} \frac{\partial}{\partial (DA^{\alpha}{}_{\mu})} \frac{\Lambda}{\varrho} \right]_{t_1}^{t_2} d\tau \end{aligned} \tag{2.4}$$

[1] In addition, it is assumed that t_1 and t_2 do not depend on ξ^{μ}.

where n_β are the components of the normal to the boundary of the spacial domain V, i.e. to the surface Σ. The following notations are used in (2.4):

$$\sigma_\alpha{}^\beta = -x^\beta{}_\mu \frac{\partial \Lambda}{\partial x^\alpha{}_\mu} - \Lambda\, \delta_\alpha{}^\beta, \qquad \hat{\sigma}_\nu{}^{\mu\lambda} = \frac{\partial \Lambda}{\partial \hat{S}_{\mu\lambda}{}^\nu}.$$

$$X^\alpha = -\varrho\, D \frac{\partial \Lambda/\varrho}{\partial v^\alpha} + \nabla_\beta\, \sigma_\alpha{}^\beta,$$

$$\frac{\delta \Lambda}{\delta A^\alpha{}_\mu} = -\varrho\, D \frac{\partial \Lambda/\varrho}{\partial (D A^\alpha{}_\mu)} - \hat{\nabla}_\nu \frac{\partial \Lambda}{\partial (\hat{\nabla}_\nu\, A^\alpha{}_\mu)} + \frac{\partial \Lambda}{\partial A^\alpha{}_\mu} =$$
$$= -\varrho\, D \frac{\partial \Lambda/\varrho}{\partial (D A^\alpha{}_\mu)} + \hat{\nabla}_\nu\, (\hat{\sigma}_\lambda{}^{\mu\nu}\, B^\lambda{}_\alpha) + \hat{\sigma}_\lambda{}^{\omega\nu}\, \hat{S}_{\nu\omega}{}^\mu\, B^\lambda{}_\alpha + \frac{\partial \Lambda}{\partial A^\alpha{}_\mu}. \tag{2.5}$$

3. The Basic Equations

Further, by definition we set

$$\delta W^* = \int_V \int_{t_1}^{t_2} \{\varrho\, \Theta\, \delta S + F_\alpha\, \delta_1 x^\alpha - \tau_\alpha{}^\beta\, \nabla_\beta\, \delta x^\alpha - \hat{Q}^{\mu\nu}\, B_{\mu\alpha}\, \delta A^\alpha{}_\nu -$$
$$- \hat{Q}^{\mu\nu\lambda}\, \hat{\nabla}_\lambda (B_{\mu\alpha}\, \delta A^\alpha{}_\nu) + N\, \delta t\}\, d\tau\, dt = \int_V \int_{t_1}^{t_2} \{\varrho\, \Theta\, \delta S +$$
$$+ (F_\alpha + \nabla_\beta\, \tau_\alpha{}^\beta)\, \delta x^\alpha + (-\hat{Q}^{\mu\nu} + \hat{\nabla}_\lambda\, \hat{Q}^{\mu\nu\lambda})\, B_{\mu\alpha}\, \delta A^\alpha{}_\nu +$$
$$+ [F_\alpha\, v^\alpha + N + \nabla_\beta\, (\tau_\alpha{}^\beta\, v^\alpha + \hat{Q}^{\mu\nu\lambda}\, \hat{\pi}_{\mu\nu}\, x^\alpha{}_\lambda)]\, \delta t\}\, d\tau\, dt -$$
$$- \int_\Sigma \int_{t_1}^{t_2} \{\tau_\alpha{}^\beta\, \delta_1 x^\alpha + \hat{Q}^{\mu\nu\lambda}\, x^\beta{}_\lambda\, B_{\mu\alpha}\, \delta_1 A^\alpha{}_\nu\}\, n_\beta\, d\sigma\, dt \tag{3.1}$$

where Θ is a certain scalar, which for the majority of the models has the meaning of the absolute temperature;

$$F_\alpha,\ \tau_\alpha{}^\beta,\ \hat{Q}^{\mu\nu},\ \hat{Q}^{\mu\nu\lambda},\ N -$$

are certain generalized forces and stresses, which for the given small particle determine the external actions and the internal irreversible effects. Putting first the total variations δ_1 of the variables on the boundary of the four-dimensional domain Vt equal to zero (in this connection, by definition, $\delta W = 0$), from the basic variational equality (2.1), taking into account (2.4), (2.5) and (3.1), we obtain a system of equations

$$-\varrho\, D J_\alpha + \nabla_\beta p_\alpha{}^\beta + F_\alpha = 0, \tag{3.2}$$

$$\varrho \frac{d}{dt}\left(\frac{\Lambda}{\varrho} - v^\alpha\, J_\alpha - A^{\beta\mu}\, D A^\alpha{}_\mu\, J_{\alpha\beta}\right) + \nabla_\beta\, (p_\alpha{}^\beta\, v^\alpha + \hat{q}^{\mu\nu\lambda}\, \hat{\pi}_{\mu\nu}\, x^\beta{}_\lambda) +$$
$$+ N + F_\alpha\, v^\alpha = 0, \qquad \frac{\partial \Lambda}{\partial \varrho} + \varrho\, \Theta = 0, \tag{3.3}$$

$$\frac{\delta \Lambda}{\delta A^{\alpha}{}_{\nu}} A^{\alpha\mu} = \hat{Q}^{\mu\nu} - \hat{V}_{\lambda} \hat{Q}^{\mu\nu\lambda}, \tag{3.4}$$

$$\left.\begin{aligned} p_{\alpha}{}^{\beta} &= \sigma_{\alpha}{}^{\beta} + \tau_{\alpha}{}^{\beta}, \\ \hat{q}^{\mu\nu\lambda} &= \hat{\sigma}^{\mu\nu\lambda} + \hat{Q}^{\mu\nu\lambda}, \\ J_{\alpha} &= \frac{\partial}{\partial v^{\alpha}} \frac{\Lambda}{\varrho}, \\ J_{\alpha\beta} &= A_{\beta\mu} \frac{\partial}{\partial (D A^{\alpha}{}_{\mu})} \frac{\Lambda}{\varrho}. \end{aligned}\right\} \tag{3.5}$$

Introducing arbitrary non-vanishing variations on the boundary we may derive the expression for δW

$$\delta W = \int_{\Sigma} \int_{t_1}^{t_2} (p_{\alpha}{}^{\beta} \delta_1 x^{\alpha} + \hat{q}^{\mu\nu\lambda} x^{\beta}{}_{\lambda} B_{\mu\alpha} \delta_1 A^{\alpha}{}_{\nu}) n_{\beta} \, d\sigma \, dt +$$

$$+ \int_{V} \varrho \, [J_{\alpha} \delta_1 x^{\alpha} + J_{\alpha\beta} A^{\beta\mu} \delta_1 A^{\alpha}{}_{\mu}]_{t_1}^{t_2} \, d\tau. \tag{3.6}$$

The Eqs. (3.2) are the equations of motion, the two Eqs. (3.3) determine energy and temperature. The system of Eqs. (3.2)—(3.4), together with the state Eqs. (2.6) and (3.5), form a closed system if Λ and δW^* are given (i.e. if Λ, F_{α}, $\tau_{\alpha}{}^{\beta}$, $\hat{Q}^{\mu\nu}$, $\hat{Q}^{\mu\nu\lambda}$, N are given[1]). When determining the solution of specific problems it is necessary to give the functional δW on the boundary of the medium, which reduces to the following boundary conditions:

$$\left.\begin{aligned} p_{\alpha}{}^{\beta} n_{\beta} &= T_{\alpha}, \\ \hat{q}^{\mu\nu\lambda} x^{\gamma}{}_{\lambda} n_{\gamma} &= \hat{q}^{\mu\nu} \end{aligned}\right\} \text{ on } \Sigma,$$

$$\begin{aligned} J_{\alpha} &= J_{\alpha}{}^{(1)}, \quad J_{\alpha\beta} = J_{\alpha\beta}{}^{(1)} \quad \text{at } t = t_2, \\ J_{\alpha} &= J_{\alpha}{}^{(2)}, \quad J_{\alpha\beta} = J_{\alpha\beta}{}^{(2)} \quad \text{at } t = t_2. \end{aligned} \tag{3.7}$$

One could derive from the variational principle (2.1) the continuity conditions [18]. If we assume that the Lagrangean is equal to the difference between the kinetic and internal energies

$$\Lambda = T - \varrho U \tag{3.8}$$

then the equations of motion (4.2) may be reduced to the ordinary form

$$\varrho D v_{\alpha} = V_{\beta} p_{\alpha}{}^{\beta} + F_{\alpha}. \tag{3.9}$$

---

[1] The quantities Θ and N in δW^* (3.1) are not given independently. For obtaining a closed system it would be sufficient to prescribe an energy inflow N or temperature Θ. If N is given, one may determine Θ from the second equation (3.3). If Θ is given, one may determine the suitable energy inflow N from the first equation (3.3).

From (3.6) and (3.9) it follows that $p_\alpha{}^\beta$ have the meaning of the components of the stress tensor, white F_α have the meaning of the components of the body forces vector. Using the equality

$$\frac{\partial \varrho}{\partial x^\alpha{}_\mu} = -\varrho\, \xi^\mu{}_\alpha$$

and taking into account (2.6) and (3.5), we obtain for $p_\alpha{}^\beta$

$$p_\alpha{}^\beta = x^\beta{}_\mu \frac{\partial U}{\partial x^\alpha{}_\mu} + \tau_\alpha{}^\beta. \tag{3.10}$$

Let us consider now the above established Euler equations (3.4) for the internal degrees of freedom, connected with the components $A^\alpha{}_\mu$. After obvious transformations the Eq. (3.4), taking into account (3.5), may be reduced to the form

$$\varrho\, D J_{\alpha\beta} = \nabla_\gamma\, K_{\alpha\beta}{}^\gamma + \varrho\, h_{\alpha\beta}, \qquad K_{\alpha\beta}{}^\gamma = \hat{q}^{\mu\nu\lambda} A_{\alpha\nu} A_{\beta\mu}\, x^\gamma{}_\lambda,$$

$$\varrho\, h_{\alpha\beta} = \frac{\partial \Lambda}{\partial A^\alpha{}_\mu} A_{\beta\mu} + \frac{\partial \Lambda}{\partial (D A^\alpha{}_\mu)} D A_{\beta\mu} - \hat{Q}^{\mu\nu} A_{\alpha\mu} A_{\beta\nu} -$$

$$- \hat{Q}^{\mu\nu\lambda}\, \hat{\nabla}_\lambda (A_{\alpha\mu} A_{\beta\nu}). \tag{3.11}$$

The Eqs. (3.11) are antisymmetrized with respect to indexes $\alpha\,\beta$ and represent the equation of the internal moment of momentum. The quantities $J_{[\alpha\beta]}$ may be considered as the components of the internal moment of momentum tensor, $K_{[\alpha\bar{\beta}]}{}^\gamma$ as the components of the torque stress[1].

4. The Entropy Balance Equation. The Phenomenological Theory of Irreversible Processes

Let us consider the energy Eq. (3.3). By means of (3.1) and (3.4) it may be transformed to the form

$$\varrho\, \Theta \frac{dS}{dt} = N + \hat{Q}^{\mu\nu}\, \hat{\pi}_{\mu\nu} + \hat{Q}^{\mu\nu\lambda}\, \hat{\nabla}_\lambda \hat{\pi}_{\mu\nu} + \tau^{\mu\nu}\, \hat{\nabla}_\nu \hat{v}_\mu. \tag{4.1}$$

The equality (4.1) is the entropy balance equation and is basic for assigning generalized forces and stresses in δW^*. We shall assume that the entropy increase, connected with N, is due only to the heat flux $\vec{q}$ and N satisfies the equality $N = -\operatorname{div} \vec{q}$. Let us denote by σ the

[1] This conclusion can be derived from the Noether theorem for the rotation group.

internal production of entropy $\varrho\, \Theta\, d_i S$ and isolate σ from the right side of (4.1) in the following way:

$$\sigma = \sigma_1 + \sigma_2 + \sigma_3,$$

$$\left.\begin{aligned} \sigma_1 &= -\Theta^{-1}\hat{q}^{\mu}\hat{\nabla}_{\mu}\Theta, \\ \sigma_2 &= \hat{Q}^{\mu\nu}\hat{\pi}_{\mu\nu} + \hat{Q}^{\mu\nu\lambda}\hat{\nabla}_{\lambda}\hat{\pi}_{\mu\nu}, \\ \sigma_3 &= \hat{\tau}^{\mu\nu}\hat{\nabla}_{\nu}\hat{v}_{\mu}. \end{aligned}\right\} \tag{4.2}$$

The quantity σ_1 characterizes the irreversible effects due to heat conduction, σ_2 — that due to plastic deformations and motion of dislocations, and σ_3 — due to viscous dissipation. According to the second law of thermodynamics the quantity σ satisfies the inequality $\sigma \geqq 0$, while σ_1, σ_2 and σ_3, generally speaking, may be both positive and negative.

As a basic hypothesis, let us assume that the quantity σ is a function of thermodynamic fluxes $\hat{\nabla}_{\mu}\Theta$, $\hat{\pi}_{\mu\nu}$, $\hat{\nabla}_{\lambda}\hat{\pi}_{\mu\nu}$, $\hat{\nabla}_{\nu}\hat{v}_{\mu}$, of quantities (1.3), and of certain additional (constant and possibly variable) parameters χ_s, which may appear when assigning δW^*. In what follows we imply that the parameters χ_s are given functionals of the determining parameters (1.3). According to the general theory of irreversible processes [19, 20], let us assign the dependence of the dissipation function σ (or $\sigma_1, \sigma_2, \sigma_3$) on its arguments and assume that for irreversible processes the generalized forces $\hat{q}^{\mu}$, $\hat{Q}^{\mu\nu}$, $\hat{Q}^{\mu\nu\lambda}$ and $\tau^{\mu\nu}$ are associated with the thermodynamic fluxes by relations of the form

$$\left.\begin{aligned} -\Theta^{-1}\hat{q}^{\mu} &= \mu_1 \frac{\partial\sigma}{\partial(\hat{\nabla}_{\mu}\Theta)}, \\ \hat{Q}^{\mu\nu} &= \mu_2 \frac{\partial\sigma}{\partial\hat{\pi}_{\mu\nu}}, \\ \hat{Q}^{\mu\nu\lambda} &= \mu_2 \frac{\partial\sigma}{\partial(\hat{\nabla}_{\lambda}\hat{\pi}_{\mu\nu})}, \\ \hat{\tau}^{\mu\nu} &= \mu_3 \frac{\partial\sigma}{\partial(\hat{\nabla}_{\nu}\hat{v}_{\mu})}. \end{aligned}\right\} \tag{4.3}$$

where the partial derivatives are determined at $\chi_s = \text{const}$, while μ_1, μ_2, μ_3 are certain factors[1]. Instead of postulating the equality (4.3) we could make other equivalent assumptions, for example, assumptions similar to Drucker's postulate, Ziegler's and Prigogine's hypotheses

[1] In the special cases of known models of the theory of plasticity the following quantities may be taken as χ_s

$$\chi_1 = \int p^{\alpha\beta}\, d\varepsilon_{\alpha\beta}^{(p)}, \quad \chi_2 = \int \sqrt{d\varepsilon_{\alpha\beta}^{(p)}\, d\varepsilon^{(p)\alpha\beta}}.$$

The previous theory was developed in the assumption that the Lagrangean Λ does not depend on χ_s.

about the maximum rate of entropy generation, and so on. The addition of the parameters χ_s and the use of various factors μ_1, μ_2, μ_3 for various thermodynamic fluxes in (4.3) relax the usually assumed hypotheses.

Besides σ_1, σ_2 and σ_3, it is necessary to assign μ_1, μ_2, μ_3 in such a manner as to satisfy the equation

$$\left.\begin{aligned} \sigma &= \mu_1 \gamma_1 + \mu_2 \gamma_2 + \mu_3 \gamma_3, \\ \gamma_1 &= \hat{V}_\mu \Theta \frac{\partial \sigma}{\partial(\hat{V}_\mu \Theta)}, \\ \gamma_2 &= \hat{\pi}_{\mu\nu} \frac{\partial \sigma}{\partial \hat{\pi}_{\mu\nu}} + \hat{V}_\lambda \hat{\pi}_{\mu\nu} \frac{\partial \sigma}{\partial(\hat{V}_\lambda \hat{\pi}_{\mu\nu})}, \\ \gamma_3 &= \hat{V}_\nu \hat{v}_\mu \frac{\partial \sigma}{\partial(\hat{V}_\nu \hat{v}_\mu)}. \end{aligned}\right\} \tag{4.4}$$

In particular, if the thermodynamic fluxes $\hat{V}_\mu \Theta$ are contained only in σ_1, $\hat{\pi}_{\mu\nu}$ and $\hat{V}_\lambda \hat{\pi}_{\mu\nu}$ only in σ_2, $\hat{V}_\nu \hat{v}_\mu$ only in σ_3, then the following relations hold for μ_1, μ_2, μ_3

$$\mu_1 \gamma_1 = \sigma_1, \qquad \mu_2 \gamma_2 = \sigma_2, \qquad \mu_3 \gamma_3 = \sigma_3. \tag{4.5}$$

From (4.2) it follows, that the absence of dissipation is associated with the equality $\sigma = 0$. Physically it is obvious that when

$$\hat{V}_\mu \Theta = 0, \qquad \hat{\pi}_{\mu\nu} = 0, \qquad \hat{V}_\lambda \hat{\pi}_{\mu\nu} = 0, \qquad \hat{V}_\nu \hat{v}_\mu = 0 \tag{4.6}$$

then each of the quantities $\sigma_1, \sigma_2, \sigma_3$ also vanishes. The inverse statement is dependent of the properties of the prescribed functions $\sigma_1, \sigma_2, \sigma_3$ when constructing the models. In the general case the equalities (4.6) may not follow from $\sigma_1 = \sigma_2 = \sigma_3 = 0$.

The relations (4.3) are essentially connected with the existence of irreversible dissipative processes when $\sigma_\alpha \neq 0$. In reversible processes the second and third equalities in (4.3) must be replaced by the corresponding relations for $\hat{\pi}_{\mu\nu}$ and $\hat{V}_\lambda \hat{\pi}_{\mu\nu}$ in reversible processes.

In the following we shall show, that by the above developed way we may construct and generalize many of the practically used models, in particular, the models of plastic media. And every time homogeneous functions of a certain degree k_α of corresponding thermodynamic fluxes are chosen as functions $\sigma_1, \sigma_2, \sigma_3$. The factors μ_α are then determined from (4.15) and turn out to be constants: $\mu_\alpha = k^{-1}{}_\alpha$. In particular, if σ is a quadratic form with respect to its arguments, $\mu_1 = \mu_2 = \mu_3 = \frac{1}{2}$, and the relation between the thermodynamic forces and fluxes will be linear. In this case, from (4.3), the Onsager's relations are obtained automatically.

In obtaining models of plastic media one cannot use linear relations between thermodynamic forces and fluxes. In this case the dissipation function σ_2 may be chosen as an homogeneous function of first degree ($\mu_2 = 1$) with respect to $\hat{\pi}_{\mu\nu}$, $\hat{V}_\lambda \hat{\pi}_{\mu\nu}$. It is obvious, that in this connection the components of the tensors $\hat{Q}^{\mu\nu}$ and $\hat{Q}^{\mu\nu\lambda}$ lie on certain surfaces in the space of the variables $\{\hat{Q}^{\mu\nu}, \hat{Q}^{\mu\nu\lambda}\}$

$$f_k(\hat{Q}^{\mu\nu}, \hat{Q}^{\mu\nu\lambda}, \chi_s) = 0 \quad k = 1, \dots m < 36 \tag{4.7}$$

which may be called yield surfaces[1]. Indeed, in the case considered, the quantities $\hat{Q}^{\mu\nu}, \hat{Q}^{\mu\nu\lambda}$ determined by (4.3) will be homogeneous functions of zero degree and, therefore, will depend not from all the quantities $\hat{\pi}_{\mu\nu}$ and $\hat{V}_\lambda \pi_{\mu\nu}$, but only on independent arguments of the form $\hat{\pi}_{22}/\hat{\pi}_{11}, \dots$, the number of which is less than the number of the components of the tensors $\hat{Q}^{\mu\nu}, \hat{Q}^{\mu\nu\lambda}$. Therefore at least one relation of the form (4.7)[2] should exist between the components of the generalized forces $\hat{Q}^{\mu\nu}$ and $\hat{Q}^{\mu\nu\lambda}$.

The foregoing theory is developed within the framework of the finite deformation theory; the assumptions of small deformations may only simplify and linearize the relations obtained above.

5. Classical Models

1) A model of an ideal fluid (gas). $\Lambda = \frac{1}{2}\varrho v^2 - \varrho U$ where the internal energy is considered as a function of density and entropy only, and the functional δW^* is given in the form:

$$\delta W^* = \int_V \int_t \{\varrho \Theta \delta S + F_\alpha \delta_1 x^\alpha - \operatorname{div}\vec{q}\, \delta t\} \, d\tau \, dt. \tag{5.1}$$

[1] In the cases of classical models of the theory of plasticity when $k = 1$, the relation between the dissipation function and the loading function was considered by D. D. Ivlev [21].

[2] If $\sigma(x^i)$ is an homogeneous function of first degree, then there exists a relation between the generalized stresses determined by the formulas:

$$X_i = \frac{\partial \sigma}{\partial x^i} = \varphi_i(v^j), \quad v^j = \frac{x^j}{x^1} \quad i = 1, 2, \dots n; \quad j = 2, 3, \dots n.$$

Indeed, assuming that these relations, when $i = 2, 3, \dots, n$ may be solved with respect to v^j, we may write

$$v^j = \psi^j(X_2, \dots, X_n).$$

Substituting these functions in the equality when $i = 1$

$$X_1 - \varphi_1(v^j) = X_1 - \varphi_1(\psi^j(X_2 \dots X_n)) = f(X_i) = 0 \tag{A}$$

we obtain the desired relation, connecting the stresses X_i. This relation may be considered as the equation of the yield surface in the space $\{X_i\}$. If among n functions $\varphi_i(v_2^2, \dots, v^n)$ only $s < n - 1$ are independent, then the above stated assumption (corresponding to $s = n - 1$) about the solvability do not hold. In this case in place of one equation A we obtain $n - s$ equations of the form

$$f_k(X_i) = 0 \quad k = 1, 2, \dots, n - s.$$

The stress tensor is determined by (3.10). It turns out to be spherical

$$p_\alpha{}^\beta = -\varrho^2 \frac{\partial U}{\partial \varrho} \delta_\alpha{}^\beta = -p\,\delta_\alpha{}^\beta .$$

The Euler equations (3.2) are converted to the equations of motion of an ideal fluid. Further we have

$$\Theta = \left(\frac{\partial U}{\partial S}\right)_{\varrho = \mathrm{coust}} .$$

From (5.1) we obtain an equation for determining S if $\vec{q}$ is given

$$\varrho\,\Theta \frac{dS}{dt} = -\operatorname{div} \vec{q} .$$

In this case in solving problems we may consider and determine residual strains which by their nature are reversible.

2) A model of a viscous heat conducting fluid (gas). The difference from the ideal fluid consists only in a more complicated form of the functional δW. For a model of a viscous fluid it is necessary to put

$$\delta W^* = \int\limits_V \int\limits_t [\varrho\,\Theta\,\delta S + F_\alpha\,\delta_1 x^\alpha - \tau_\alpha{}^\beta\,\nabla_\beta\,\delta x^\alpha - \operatorname{div} \vec{q}\,\delta t]\,d\tau\,dt .$$

The entropy balance equation in this case is of the form

$$\varrho\,\Theta \frac{dS}{dt} = -\Theta \operatorname{div} \frac{\vec{q}}{\Theta} - \frac{\vec{q}}{\Theta} \operatorname{grad} \Theta + \hat{\tau}^{\mu\nu}\,\hat{\nabla}_\nu\,\hat{v}_\mu .$$

If, by definition, we assume that a dissipation function exists

$$\sigma(\hat{\nabla}_\mu \Theta, \hat{\nabla}_\nu \hat{v}_\mu, L_{(p)}) = -\frac{\hat{q}^\mu}{\Theta} \hat{\nabla}_\mu \Theta + \hat{\tau}^{\mu\nu}\,\hat{\nabla}_\nu\,\hat{v}_\mu$$

where $L_{(p)}$ are certain quantities determined in § 2, then in the presence of the laws (4.3) the following relations will hold:

$$-\frac{1}{\Theta}\hat{q}^\mu = \mu_1 \frac{\partial\sigma}{\partial(\hat{\nabla}_\mu \Theta)}, \qquad \hat{\tau}^{\mu\nu} = \mu_3 \frac{\partial\sigma}{\partial(\hat{\nabla}_\nu \hat{v}_\mu)} . \tag{5.2}$$

If the dissipation function depends on $\hat{\nabla}_\nu \hat{v}_\mu$ only through the components of the strain rate tensor, then the tensor $\hat{\tau}^{\mu\nu}$ is symmetric.

The classical model of the Navier-Stokes viscous fluid may be obtained, if it is assumed that the dissipation function σ is a positive definite quadratic form with respect to $\hat{\nabla}_\mu \Theta$ and $\hat{\nabla}_\nu \hat{v}_\mu$. In this case the relations (5.2) are linear and satisfy the Onsager principle.

The assumption of isotropy simplifies essentially the aspect of the quadratic form σ, the thermal and viscous cross effects being absent in this case.

3) A model of an elastic body. In this case we have

$$\Lambda = \tfrac{1}{2}\varrho\,v^2 - \varrho\,U(\hat{\varepsilon}_{\mu\nu}{}^{(e)}, S, L_{(p)}); \qquad \hat{\varepsilon}_{\mu\nu}{}^{(e)} = \tfrac{1}{2}(\hat{g}_{\mu\nu} - \overset{*}{g}_{\mu\nu})$$

$$\delta W^* = \int\limits_V \int\limits_t [\varrho\,\Theta\,\delta S + F_\alpha \delta_1 x^\alpha - \operatorname{div} \vec{q}\,\delta t]\,d\tau\,dt .$$

In order to obtain all the relations as special cases of the system of equations in §§ 3 and 4 one should set

$$\overset{*}{g}_{\mu\nu} = \overset{\circ}{g}_{\mu\nu}(\xi^\lambda), \qquad A^\alpha{}_\mu(\xi^\lambda) = x^\alpha{}_\mu(\xi^\lambda, t_0).$$

Since $A^\alpha{}_\mu$ determines the transition from the observer's system of coordinate x^α to the fixed basis of the "initial state", the quantities $A^\alpha{}_\mu$ are excluded from the set of unknown variable quantities and become parameters of type $L_{(p)}$.

The system of momentum equations reduces to the Eq. (3.9), in which the components of the stress tensor $p^{\gamma\beta} = g^{\gamma\alpha} p_\alpha{}^\beta$, according to (3.10), may be written in the form

$$p^{\gamma\beta} = \varrho \frac{\partial U}{\partial x^\alpha{}_\mu} x^\beta{}_\mu \, g^{\gamma\alpha} = \varrho \frac{\partial V}{\partial \hat{\varepsilon}_{\mu\nu}{}^{(e)}} x^\gamma{}_\mu \, x^\beta{}_\nu = \hat{p}^{\mu\nu} x^\gamma{}_\mu \, x^\beta{}_\nu. \tag{5.3}$$

The formula for Θ and the energy equation, which is equivalent to the entropy balance equation, are of the form

$$\Theta = \left(\frac{\partial U}{\partial S}\right)_{\hat{\varepsilon}_{\mu\nu}{}^{(e)} = \text{const}} \quad \text{and} \quad \varrho\, \Theta \frac{dS}{dt} = -\operatorname{div} \vec{q}.$$

4) Models of plastic bodies. Under certain additional partial assumptions many of the known models of the theory of plasticity and plasticity with strain hardening may be derived from the general theory, developed in §§ 3 and 4. For this, it is sufficient to assume

$$\Lambda = \tfrac{1}{2} \varrho\, v^2 - \varrho\, U(\overset{*}{g}_{\mu\nu} \hat{g}_{\mu\nu}, S, L_{(p)}),$$

$$\delta W^* = \int\limits_{Vt}\!\int [\varrho\, \Theta\, \delta S + F_\alpha \, \delta_1 x^\alpha - \hat{Q}^{\mu\nu} \, \delta\, \hat{\varepsilon}_{\mu\nu}{}^{(p)} - \operatorname{div} \vec{q}\, \delta t] \, d\tau \, dt. \tag{5.4}$$

As opposed to the general case, the component $A^\alpha{}_\mu$ enter Λ and δW^* only through the components of the metric tensor in the initial state

$$\overset{*}{g}_{\mu\nu} = g_{\alpha\beta} A^\alpha{}_\mu A^\beta{}_\nu \; [\hat{\varepsilon}_{\mu\nu}{}^{(p)} = \tfrac{1}{2}(\overset{*}{g}_{\mu\nu} - \overset{\circ}{g}_{\mu\nu})].$$

It is obvious that $\hat{Q}^{\mu\nu}$ may be considered as symmetric; in applying the general theory, it is necessary to consider the identity

$$\hat{Q}^{\mu\nu} \, \delta \hat{\varepsilon}^{\mu\nu(p)} = \hat{Q}^{\mu\nu} B_{\nu\alpha} \, \delta A^\alpha{}_\mu.$$

The arbitrariness of the variations $\delta \hat{\varepsilon}_{\mu\nu}{}^{(p)} = \frac{1}{2} \delta \overset{*}{g}_{\mu\nu}$, instead of eight Eqs. (3.4), leads now to six equations only

$$-\varrho \frac{\partial U}{\partial \overset{*}{g}_{\mu\nu}} = \frac{1}{2} Q^{\mu\nu}. \tag{5.5}$$

In just the same way as we derived the formula (5.3) for the stress tensor, from the formula (3.10), we obtain

$$\hat{p}^{\mu\nu} = \varrho \frac{\partial U}{\partial \hat{\varepsilon}_{\mu\nu}{}^{(e)}}. \tag{5.6}$$

In the important case, when $\overset{*}{g}_{\mu\nu}$ and $\hat{g}_{\mu\nu}$ are contained in (5.4) through the difference $\hat{g}_{\mu\nu} - \overset{*}{g}_{\mu\nu} = 2\hat{\varepsilon}_{\mu\nu}{}^{(e)}$ (a medium "without memory"), we obtain the equation

$$\hat{p}^{\mu\nu} = \hat{Q}^{\mu\nu}. \tag{5.7}$$

It should be emphasized that the equality (5.7) does not hold for a medium with "memory", when the internal energy depends on the components of elastic strain and on the components of plastic strains $\hat{\varepsilon}_{\mu\nu}{}^{(p)}$. According to (4.2) for σ_2 we have

$$\sigma_2 = \hat{Q}^{\mu\nu}\,\hat{e}_{\mu\nu}{}^{(p)}. \tag{5.8}$$

If the equality (5.7) holds, then the formula (5.8) gives

$$\sigma_2 = \hat{p}^{\mu\nu}\,\hat{e}_{\mu\nu}{}^{(p)}. \tag{5.9}$$

However, in the general case the formula (5.9) is incorrect. For a medium with memory the Eqs. (5.5) may be written in the form:

$$-2\varrho\frac{\partial U}{\partial \overset{*}{g}_{\mu\nu}} = \varrho\frac{\partial U}{\partial \hat{\varepsilon}_{\mu\nu}{}^{(e)}} - \varrho\frac{\partial U}{\partial \hat{\varepsilon}_{\mu\nu}{}^{(p)}} = \hat{Q}^{\mu\nu}.$$

Therefore in this case, instead of the equality (5.7), on the basis of (5.6), we obtain

$$\hat{Q}^{\mu\nu} = \hat{p}^{\mu\nu} - \varrho\frac{\partial U}{\partial \hat{\varepsilon}_{\mu\nu}{}^{(p)}}. \tag{5.10}$$

For a medium with memory the formula (5.9) is replaced by the formula[1]

$$\sigma_2 = \left(\hat{p}^{\mu\nu} - \varrho\frac{\partial U}{\partial \hat{\varepsilon}_{\mu\nu}{}^{(p)}}\right)\hat{e}_{\mu\nu}{}^{(p)}. \tag{5.11}$$

The above given discussion of the quantity σ is essential, since in scientific literature wide use is made of the equality (5.9).

The complete entropy balance Eq. (4.4) for the model considered has the form

$$\varrho\,\Theta\frac{dS}{dt} = -\Theta\,\hat{\nabla}_\mu\frac{\hat{q}^\mu}{\Theta} + \frac{\hat{q}^\mu}{\Theta}\hat{\nabla}_\mu\,\Theta + \hat{Q}^{\mu\nu}\,\hat{e}_{\mu\nu}{}^{(p)}. \tag{5.12}$$

For the magnitude of the dissipation σ we have

$$\sigma = -\frac{\hat{q}^\mu}{\Theta}\hat{\nabla}_\mu\,\Theta + \sigma_2, \quad \sigma_2 = \hat{Q}^{\mu\nu}\,\hat{e}_{\mu\nu}{}^{(p)}. \tag{5.13}$$

The formula (5.13) shows that, the requirement $\sigma_2 \geqq 0$ is fully justified only for isothermal states.

From the condition of independency of the entropy increment on the deformation time, we obtain that σ_2 is an homogeneous function

[1] The explanation of the difference and relation of the magnitude of the dissipation with the magnitude of the work done by the stress tensor over the plastic strains is given in the book [14] (see p. 251).

of first degree of the quantities $\hat{e}_{\mu\nu}{}^{(p)}$, which perhaps depends also on the corresponding strain hardening parameters χ_s.

From the theory developed in § 4 it follows, that for irreversible processes

$$\hat{Q}^{\mu\nu} = \frac{\partial \sigma_2}{\partial \hat{e}_{\mu\nu}{}^{(p)}}. \tag{5.14}$$

The relations (5.14) together with (5.5) may be considered as equations for determining the components of the plastic strain rate tensor $\hat{e}_{\mu\nu}{}^{(p)}$.

These relations replace the associated law. After determining from (5.14) the loading function we may derive the associated law, which in the general case, contains the components $\hat{Q}^{\mu\nu}$, and not the components of the stress tensor $\hat{p}^{\mu\nu}$, as it is assumed in the paper [21].

Let us note that in the case considered, we may write for $\hat{p}^{\mu\nu}$, as well as for $Q^{\mu\nu}$, relations of the type (5.14)

$$\hat{p}^{\mu\nu} = \frac{\partial \varphi}{\partial \hat{e}_{\mu\nu}{}^{(p)}}.$$

However, the function $\varphi \gtrless 0$ differs from the dissipation function σ_2 and, as it follows from (5.10), (5.14) and (5.4), is related to it by the equality

$$\varphi = \sigma_2 + \varrho \frac{\partial U}{\partial \hat{\varepsilon}_{\mu\nu}{}^{(p)}} e_{\mu\nu}{}^{(p)} + \omega$$

where ω is an arbitrary function which does not depend on $\hat{e}_{\mu\nu}{}^{(p)}$.

For a medium with "memory" as argument of the loading function we may introduce, according to (5.10), the quantities $\hat{p}^{\mu\nu}$ instead of the quantities $\hat{Q}^{\mu\nu}$. For irreversible processes we may assume that $\sigma_2 = 0$. In most typical cases from the equality $\sigma_2 = 0$ it follows that $\hat{e}_{\mu\nu}{}^{(p)} = 0$. However, we may consider also such functions σ_2 when residual and plastic strains may arise in reversible processes.

By properly assigning the internal energy and the dissipation function, from (3.2), (4.1) and (5.14) we may obtain specific models of the theory of plasticity, in which the strain hardening parameters χ_s are given functions or functionals of the determining parameters.

If the dissipation function σ_2 has the form[1]

$$\sigma_2 = k \sqrt{e_{\mu\nu}{}^{(p)} e^{(p)\mu\nu}}$$

where k does not depend on $\hat{e}_{\mu\nu}{}^{(p)}$, the quantities $\hat{Q}^{\mu\nu}$, determined by (5.14), as it is easily seen, lie on the surface

$$Q_{\mu\nu} Q^{\mu\nu} = k^2. \tag{5.15}$$

[1] Here instead of $\hat{e}^{(p)}_{\mu\nu}$ one can take the deviator of $\hat{e}^{(p)}_{\mu\nu}$. Suitable changes in the following are obvious.

In the model of ideal plasticity of MISES

$$U = U(\hat{\varepsilon}_{\mu\nu}{}^{(e)}, S, L_{(p)}), \quad k = \text{const.}$$

In the model of plasticity with strain hardening of SCHMIDT-OSGOOD

$$U = U(\hat{\varepsilon}_{\mu\nu}{}^{(e)}, S, L_{(p)}), \qquad k = k(\chi)$$

where χ is determined by the equality $d\chi = \sqrt{d\varepsilon_{\mu\nu}{}^{(p)}\, d\varepsilon^{(p)\mu\nu}}$ and $k(\chi)$ are empirically established functions. Since in these cases the internal energy do not depend on the plastic strain, the equality (5.7) holds and the equation of the yield surface may be written in the form

$$p^{\mu\nu}\, p_{\mu\nu} = k^2.$$

In the model with translational strain hardening we may put

$$U = U_0(\hat{\varepsilon}_{\mu\nu}{}^{(e)}, S, L_{(p)}) + \tfrac{1}{2} c\, \hat{\varepsilon}^{(p)\mu\nu}\, \hat{\varepsilon}_{\mu\nu}{}^{(p)}, \qquad k = \text{const.}$$

Now, instead of (5.7) we obtain

$$\hat{Q}^{\mu\nu} = \hat{p}^{\mu\nu} - c\, \hat{\varepsilon}^{(p)\mu\nu}.$$

Therefore the equation of the yield surface (5.15) has the form:

$$(\hat{p}^{\mu\nu} - c\, \hat{\varepsilon}^{(p)\mu\nu})(\hat{p}_{\mu\nu} - c\, \hat{\varepsilon}_{\mu\nu}{}^{(p)}) = k^2.$$

6. An Example of a Model of Continuum Theory of Dislocations

The variational principle allows the construction of models of media with continuously distributed dislocations. In the following we shall consider a specific example within the framework of the theory of small deformations[1] in which the kinetic energy is related only to the inertial properties of the actual state $T = \frac{1}{2}\varrho v^2$ while the internal energy is a quadratic form of the components of elastic strain tensor $\varepsilon_{\mu\nu}{}^{(e)}$, of the entropy difference $S - S_0$ of the deformed and initial states and, as opposed to the classical theory of elasticity, also of nine internal parameters, i.e., of the components of the dislocation density tensor $S^{\alpha\beta}$.

$$U = \tfrac{1}{2} A^{\alpha\beta\gamma\delta}\, \varepsilon_{\alpha\beta}{}^{(e)}\, \varepsilon_{\gamma\delta}{}^{(e)} + B^{\alpha\beta}\, \varepsilon_{\alpha\beta}{}^{(e)}(S - S_0) + C^{\alpha\beta}{}_{\gamma\delta}\, \varepsilon_{\alpha\beta}{}^{(e)}\, S^{\gamma\delta} +$$
$$+ D_{\alpha\beta}\, S^{\alpha\beta}(S - S_0) + \tfrac{1}{2} E_{\alpha\beta\gamma\delta}\, S^{\alpha\beta}\, S^{\gamma\delta} \qquad (6.1)$$

where $A^{\alpha\beta\gamma\delta}$, $B^{\alpha\beta}$, $C^{\alpha\beta}{}_{\gamma\delta}$, $D_{\alpha\beta}$, $E_{\alpha\beta\gamma\delta}$ are given non-varying parameters (in the general theory they correspond to the parameters $L_{(p)}$).

[1] As is known, in the theory of small deformations we may assume that the components of the small tensors in the observer's and the comoving frames are the same. In this connection the symbol ^ is dropped farther on, and no distinction is made between the operators D, d/dt and $\partial/\partial \xi^\mu$, $\partial/\partial x^\mu$ when they are applied to the components of the small tensors. Besides, it is further assumed that $\mathring{g}_{\mu\nu}(\xi^\lambda) = g_{\mu\nu}(\xi^\lambda)$.

Let us put

$$\Lambda = T - \varrho\, U,$$

$$\delta W^* = \int_V\int [\varrho\, \Theta\, \delta S + F_\alpha\, \delta_1 x^\alpha - Q^{\mu\nu}\, \delta \varepsilon_{\mu\nu}{}^{(p)} - Q^{\mu\nu\lambda}\, \nabla_\lambda (B_{\mu\alpha}\, \delta A^\alpha{}_\nu) -$$
$$- \operatorname{div} \vec{q}\, \delta t]\, d\tau\, dt. \qquad (6.2)$$

According to the general theory of § 5 the components $Q^{\mu\nu}$ and q^α are determined by the formulas (4.3). Let us give the dissipation functions by the following expressions:

$$\sigma_1 = \Theta^{-1}\, F^{\alpha\beta} (\nabla_\alpha \Theta)\, (\nabla_\beta \Theta), \qquad (6.3)$$

$$\sigma_2 = \sqrt{K^{\alpha\beta\gamma\delta}\, e_{\alpha\beta}{}^{(p)}\, e_{\gamma\delta}{}^{(p)} + 2\, L^{\alpha\beta\gamma\delta}\, e_{\alpha\beta}{}^{(p)}\, \frac{d S_{\gamma\delta}}{dt} + M^{\alpha\beta\gamma\delta}\, \frac{d S_{\alpha\beta}}{dt}\, \frac{d S_{\gamma\delta}}{dt}} \qquad (6.4)$$

where $F^{\alpha\beta}$, $K^{\alpha\beta\gamma\delta}$, $L^{\alpha\beta\gamma\delta}$, $M^{\alpha\beta\gamma\delta}$ are the components of the tensors—the physical characteristics of the medium, which by virtue of the positive definiteness of σ_1 and σ_2 must satisfy the well-known inequalities. When using the formula (4.3) we should borne in mind that for small deformations are valid the following equalities

$$\frac{d S_{\alpha\beta}}{dt} = \varepsilon^{\gamma\sigma}{}_\alpha\, \nabla_\gamma\, \pi_{\beta\sigma}. \qquad (6.5)$$

From (4.3), (6.3)—(6.5) it follows that

$$q^\alpha = - F^{\alpha\beta}\, \nabla_\beta\, \Theta, \qquad (6.6)$$

$$Q^{\alpha\beta} = Q^{\beta\alpha} = \frac{\partial \sigma_2}{\partial e_{\alpha\beta}{}^{(p)}} = \frac{1}{\sigma_2} \left(K^{\alpha\beta\gamma\delta}\, e_{\gamma\delta}{}^{(p)} + L^{\alpha\beta\gamma\delta}\, \frac{dS}{dt^{\gamma\delta}} \right), \qquad (6.7)$$

$$Q^{\beta\sigma\gamma} = \frac{\partial \sigma_2}{\partial (\nabla_\gamma\, \pi_{\beta\sigma})} = \frac{\partial \sigma_2}{\partial \left(\frac{d S_{\alpha\beta}}{dt} \right)}\, \varepsilon^{\gamma\sigma}{}_\alpha = R^{\alpha\beta}\, \varepsilon^{\gamma\sigma}{}_\alpha. \qquad (6.8)$$

From the equalities (6.8) follow eighteen relations

$$Q^{\beta\sigma\gamma} + Q^{\beta\gamma\sigma} = 0 \qquad (6.9)$$

which represent a part of the Eqs. (4.7), determining the yield surface in the space of the variables $\{Q^{\alpha\beta}, Q^{\alpha\beta\gamma}\}$. The remaining nine components $Q^{\beta[\sigma\gamma]}$ are expressed in terms of nine components of the tensor

$$R^{\alpha\beta} = \frac{\partial \sigma_2}{\partial \left(\frac{d S_{\alpha\beta}}{dt} \right)} = \frac{1}{\sigma_2} \left(L^{\gamma\delta\alpha\beta}\, e_{\alpha\beta}{}^{(p)} + M^{\alpha\beta\gamma\delta}\, \frac{d S_{\gamma\delta}}{dt} \right). \qquad (6.10)$$

In the space of variables $\{Q^{\alpha\beta}, R^{\alpha\beta}\}$ the equation of the yield surface has the form

$$f(Q^{\alpha\beta}, R^{\alpha\beta}, K^{\alpha\beta\gamma\delta}, L^{\alpha\beta\gamma\delta}, M^{\alpha\beta\gamma\delta}) = 0. \qquad (6.11)$$

In the case considered, from the general theory of §§ 4 and 5 we obtain the following system of equations, consisting of the equations of motion (3.9), the entropy balance equation

$$\varrho\, \Theta \frac{dS}{dt} = -\Theta \nabla_\alpha \frac{q^\alpha}{\Theta} + \sigma_1 + \sigma_2$$

the equations for the internal parameters (3.4)

$$\varepsilon^{\gamma\beta}{}_\sigma \nabla_\gamma (R^{\sigma\alpha} + \Sigma^{\sigma\alpha}) + p^{\alpha\beta} = Q^{\alpha\beta} \tag{6.12}$$

state equations

$$\varrho^{-1} p^{\alpha\beta} = A^{\alpha\beta\gamma\delta} \varepsilon_{\gamma\delta}{}^{(e)} + B^{\alpha\beta}(S - S_0) + C^{\alpha\beta}{}_{\gamma\delta} S^{\gamma\delta}, \tag{6.13}$$

$$\varrho^{-1} \Sigma_{\gamma\delta} = C^{\alpha\beta}{}_{\gamma\delta}\, \varepsilon_{\alpha\beta}{}^{(e)} + D_{\gamma\delta}(S - S_0) + E_{\alpha\beta\gamma\delta} S^{\gamma\delta} \tag{6.14}$$

and also of the equalities (6.6), (6.7), (6.10). The Eqs. (6.12) are symmetrized with respect to the indexes $\alpha\,\beta$ and differ from (5.7) in the theory of plasticity by the term $\varepsilon^{\gamma\beta}{}_\sigma \nabla_\gamma (R^{\sigma\alpha} + \Sigma^{\sigma\alpha})$. The alternated with respect to $\alpha\,\beta$ Eqs. (6.12) have the form

$$-\tfrac{2}{3} \nabla_\alpha (R_\beta{}^\beta + \Sigma_\beta{}^\beta) + \nabla_\beta (R'_\alpha{}^\beta + \Sigma'_\alpha{}^\beta) = 0 \tag{6.15}$$

where the prime denotes the operation of taking the deviator[1].

In problems, where there is homogeneity with respect to the particles, i.e., when $\hat{\nabla}_\gamma A^\alpha{}_\mu = 0$ and $\partial S / \partial \xi^\nu = 0$ we obtain $S^{\mu\nu} = 0$, therefore in this case the solution within the framework of the proposed model coincides with that of the same problem within the framework of the model of the plastic medium without "memory" in § 6.

The relations (6.12) may be used to substitute in (6.11) $Q^{\alpha\beta}$ by $p^{\alpha\beta}$, $R^{\sigma\alpha}$ and $\Sigma^{\sigma\alpha}$ and by $\nabla_\gamma R^{\sigma\alpha}$ and $\nabla_\gamma \Sigma^{\sigma\alpha}$. Hence, it is clear that in formulating the loading conditions it is more convenient to use the quantities $Q^{\alpha\beta}$ than the quantities $p^{\alpha\beta}$.

Such is the general linear theory allowing for anisotropy. All the previous formulas are essentially simplified if the anisotropy properties appear only in the quantities $e_{\alpha\beta}{}^{(p)}$ and $S^{\alpha\beta}$. In this case instead of (6.6), (6.7) and (6.10) we obtain

$$q^\alpha = -\varkappa \nabla^\alpha \Theta,$$

$$Q_{\alpha\beta} = \frac{1}{\sigma_2}\left[l_1 e_{\alpha\beta}{}^{(p)} + \left(l_2 e_{\gamma\delta}{}^{(p)} + l_3 \frac{dS_{\gamma\delta}}{dt}\right) g^{\gamma\delta} g_{\alpha\beta} + l_4 \frac{dS_{(\alpha\beta)}}{dt}\right], \tag{6.16}$$

$$R_{\alpha\beta} = \frac{1}{\sigma_2}\left[l_4 e_{\alpha\beta}{}^{(p)} + \left(l_3 e_{\gamma\delta}{}^{(p)} + l_5 \frac{dS_{\gamma\delta}}{dt}\right) g^{\gamma\delta} g_{\alpha\beta} + l_6 \frac{dS_{\alpha\beta}}{dt} + l_7 \frac{dS_{\beta\alpha}}{dt}\right]$$

where $\varkappa, l_1, \ldots, l_7$, are scalars which, in the general case, may be considered as functions of scalar invariants of the determining parameters and in the linear theory as constants. The loading function (6.11)

[1] Let us note that the consideration of the inertial properties of the initial state in the kinetic energy would complicate the equality (6.15) by a term of the form dK_α/dt, where K_α is the generalized momentum associated with the motion of dislocations.

corresponding to the dissipation function (6.4), is found by substituting in the right hand side of the equality

$$\sigma_2 = Q^{\alpha\beta} e_{\alpha\beta}{}^{(p)} + Q^{\alpha\beta\gamma} \nabla_\gamma \pi_{\alpha\beta} = Q^{\alpha\beta} e_{\alpha\beta}{}^{(p)} + R^{\alpha\beta} \frac{dS_{\alpha\beta}}{dt}$$

the quantities $e_{\alpha\beta}{}^{(p)}$ and $dS_{\alpha\beta}/dt$ expressed from (6.16) in terms $\sigma_2 Q^{\alpha\beta}$ and $\sigma_2 R^{\alpha\beta}$, and have the form

$$\begin{aligned} f(Q^{\alpha\beta}, R_{\alpha\beta}, l_1, \ldots, l_7) &= (Q'_{\alpha\beta} - c_1 R_{\alpha\beta})(Q'^{\alpha\beta} - c_1 R^{\alpha\beta}) - \\ &- c_2 R'_{(\alpha\beta)} R'^{(\alpha\beta)} - c_3 R_{[\alpha\beta]} R^{[\alpha\beta]} - c_4 (g_{\alpha\beta} R^{\alpha\beta})^2 - c_5 (g_{\alpha\beta} Q^{\alpha\beta})^2 - \\ &- c_6 (g_{\alpha\beta} R^{\alpha\beta})(g_{\alpha\beta} Q^{\alpha\beta}) - c_0 . \end{aligned} \tag{6.17}$$

where $c_0, c_1, \ldots, c_6$, are determined in terms of $l_1, \ldots, l_7$ in an obvious manner. The loading function is a second order surface in the space of variables $\{Q^{\alpha\beta}, R^{\alpha\beta}\}$. This surface is stationary if $c_i = \text{const}$. The plastic strains are incompressible if $c_5 = c_6 = 0$. The case $l_1 = \text{const}$, $l_2 = l_3 = \cdots = l_7 = 0$ ($c_1 = \cdots = c_6 = 0$, $c_0 = \text{const}$) corresponds to the perfectly plastic body considered in § 5 and we have $R^{\alpha\beta} = 0$.

In the general case $R^{\alpha\beta}$ are different from zero and describe the strain hardening phenomenon. In the stress space $\{Q'^{\alpha\beta}\}$ the yield surface is a sphere with a time vary center $- c_1 R^{\alpha\beta}$ and a radius

$$r^2 = c_0 + c_2 R'_{(\alpha\beta)} R'^{(\alpha\beta)} + c_3 R_{[\alpha\beta]} R^{[\alpha\beta]} + c_4 (g_{\alpha\beta} R^{\alpha\beta})^2$$

which varies also in the process of deformation. Specific models are stated by the choice of the coefficients l_i or c_i.

References

[1] Kondo, K.: On the geometrical and physical foundations of the theory of yielding. Proc. 2. Japan Mat. Congr. Appl. Mech., 1952, pp. 41—47.

[2] Kondo, K.: Memoirs of the unifying study of the basic problems in engineering by means of geometry. Tokyo 1955—1962, Vol. 1—3.

[3] Bilby, B. A., R. Bullough and E. Smith: Continuous distributions of dislocations: a new application of the methods of non-Riemannian geometry. Proc. Roy. Soc. A 231, 1955, pp. 263—273.

[4] Bilby, B. A., and E. Smith: Continuous distribution of dislocations. Proc. Roy. Soc. A 236, 1956, pp. 481—505.

[5] Bilby, B. A.: Continuous distribution of dislocations. Progr. Sol. Mech. 1960, Vol. 1.

[6] Kröner, E.: Arch. Rat. Mech. Anal. 4, 273 (1960); Russ. Transl. Moscow: MIR 1965.

[7] Sedov, L. I.: UMN 20, 125 (1965).

[8] Kunin, I. A.: Theory of Dislocations, Addition to the Russ. Transl. of J. A. Schouten's book Tensor Analysis for Physicists. Moscow: Nauka 1965, p. 373.

[9] Cartan, E.: Spaces of Affine, Projective and Conform Connection, Russ. Transl. Kasan: Univ. Press 1962.

[10] Schouten, J. A., and D. J. Struik: Einführung in die neueren Methoden der Differentialgeometrie, Vol. 1. Groningen: Noordhoff 1935. Russ. Transl. Moscow: MIR 1939.

[11] Kröner, E.: Kontinuumstheorie der Versetzungen und Eigenspannungen. Berlin/Göttingen/Heidelberg: Springer 1958.
[12] Kosevitch, A. M.: Dynamical Theory of Dislocations. UFN **84**, 579 (1964).
[13] Landau, L. D., and E. M. Lifshitz: Theory of Elasticity. Moscow: FIZMATGIZ 1965.
[14] Sedov, L. I.: Introduction in the Mechanics of Continuous Media. Moscow: FIZMATGIZ 1962. Engl. Transl. New York: Addison and Wesley 1965, Pergamon Press 1966.
[15] Berdichevski, W. L.: Prikl. Mat. Mech. **30**, 510 (1966).
[16] Berdichevski, W. L.: Prikl. Mat. Mech. **30**, 1081 (1966).
[17] Sedov, L. I.: Dokl. Acad. Nauk SSSR **164**, 519 (1965).
[18] Lur'e, M. W.: Prikl. Mat. Mech. **30**, 747 (1966).
[19] Ziegler, H.: Über ein Prinzip der größten spezifischen Entropieproduktion und seine Bedeutung für die Rheologie. Rheol. Acta **2**, 3 (1962).
[20] Besseling, J. F.: A Thermodynamic Approach to Rheology. Rep. IUTAM-Symposium in Vienna 1966.
[21] Ivlev, D. D.: Dokl. Acad. Nauk SSSR **176**, 1037 (1967).
[22] Lochin, W. W., and L. I. Sedov: Prikl. Mat. Mech. **27**, 393 (1963).
[23] Toupin, R. A.: Theories of elasticity with couple-stress. Arch. Rat. Mech. and Anal. **17**, 2, 85 (1964).
[24] Mindlin, R. D.: Micro-structure in linear elasticity. Arch. Rat. Mech. and Anal. **17**, 1, 51 (1964).

Inhomogeneities in Materially Uniform Simple Bodies

By

Walter Noll

Department of Mathematics
Carnegie-Mellon University, Pittsburgh

1. Introduction

Many physical characteristics of continuously distributed matter are *local* in the sense that they pertain to individual material points and their immediate neighborhoods, rather than to a body as a whole. Examples of such local characteristics are: elastic response, viscosity, heat capacity, electrical conductivity, magnetic response, optical refraction, chemical composition, color.

The *theory of simple bodies* deals only with such local characteristics. A given set of physical phenomena is then governed by appropriate physical responses at the various material points of the body. The term "simple" expresses the assumption that deformations whose gradient is the identity at a given material point do not alter the physical response at that point with respect to the set of phenomena under consideration. Roughly speaking, a body is simple if only first spatial gradients occur in the constitutive description of its physical properties. A simple body is called *materially uniform* if the physical response is the same at all points of the body.

I shall show that the theory of the structure of materially uniform simple bodies leads, by the force of logic alone, to the possibility of inhomogeneities. Under sufficient smoothness assumptions, such inhomogeneities can be described locally in terms of a certain third order tensor field, which I call the *inhomogeneity* of the body. This inhomogeneity corresponds to what is called "dislocation density" in the theory of continuous distributions of dislocations as described in other contributions to this Symposium.

A particular type of inhomogeneity is what I call *contorted aeolotropy*, which includes the more familiar curvilinear aeolotropy as a special case. Certain types of laminated and fibrous bodies give intuitive physical examples of bodies with contorted aeolotropy.

If one deals with mechanical material properties one must take into account the principle of balance of forces, which gives rise to Cauchy's familiar equations of balance. The classical form of these equations, however, is unsuited for dealing with inhomogeneous uniform bodies. I shall give a new form for these equations, a form tailored to bodies with a given distribution of inhomogeneities.

The presentation I shall give here will be somewhat informal and I shall omit almost all proofs. A more complete and detailed description of the theory will be found in my paper "Materially Uniform Simple Bodies with Inhomogeneities", Archive for Rational Mechanics and Analysis, Vol. 27, pp. 1—32. That paper contains also the relevant references.

2. The Concept of a Smooth Body. Local Configurations. References

In continuum physics, a *body* $\mathscr{B}$ is described mathematically as a set of *material points* $X, Y, \ldots$ The *configurations* of $\mathscr{B}$ in space are mappings

$$\varkappa : \mathscr{B} \to \mathscr{E},$$

where $\mathscr{E}$ is a fixed three-dimensional Euclidean space. The translation space of $\mathscr{E}$, i.e., the space of *spatial vectors*, is denoted by $\mathscr{V}$.

We say that $\mathscr{B}$ is a *smooth* body if all of its configurations can be obtained from one of them by smooth deformations, i.e. by one-to-one mappings having suitable differentiability properties.

Let $f : \mathscr{B} \to \mathscr{R}$ ($\mathscr{R}$ = real line) be a scalar field on $\mathscr{B}$, i.e. a function that assigns a real number to each material point. Given a configuration $\varkappa$ of $\mathscr{B}$, we can consider the composition $f \circ \overset{-1}{\varkappa} : \varkappa(\mathscr{B}) \to \mathscr{R}$, which is a function that assigns a real number to each point in the region $\varkappa(\mathscr{B})$ occupied by $\mathscr{B}$ in the configuration $\varkappa$. We assume that $\varkappa(\mathscr{B})$ is an open set. It may happen that $f \circ \overset{-1}{\varkappa}$ is differentiable. Its gradient $\nabla\left(f \circ \overset{-1}{\varkappa}\right) : \varkappa(\mathscr{B}) \to \mathscr{V}$ is then a function which assigns to each point in the region $\varkappa(\mathscr{B})$ a spatial vector. The composition of this function with $\varkappa$ is denoted by $\nabla_\varkappa f : \mathscr{B} \to \mathscr{V}$, so that

$$\nabla_\varkappa f|_X = \nabla\left(f \circ \overset{-1}{\varkappa}\right)\Big|_{\varkappa(X)}, \quad X \in \mathscr{B}. \tag{2.1}$$

($|_X$ means that the function in question is to be evaluated at X.) The vector field $\nabla_\varkappa f$ on $\mathscr{B}$ defined by (2.1) is called the *gradient of f relative to the configuration $\varkappa$*.

Now let $\boldsymbol{\lambda} : \varkappa(\mathscr{B}) \to \mathscr{E}$ be any smooth deformation. The gradient $\nabla \boldsymbol{\lambda}|_{\varkappa(X)}$ at the point $\varkappa(X)$ in the region $\varkappa(\mathscr{B})$ is an invertible linear transformation of $\mathscr{V}$. The composition $\gamma = \boldsymbol{\lambda} \circ \varkappa : \mathscr{B} \to \mathscr{E}$ is again a configuration. The chain rule for the differentiation of compositions yields

$$\nabla_\varkappa f|_X = (\nabla \boldsymbol{\lambda}|_{\varkappa(X)})^T (\nabla_\gamma f)|_X, \tag{2.2}$$

where the upper index T denotes transposition.

It is clear from (2.2) that

$$\nabla_\varkappa f|_X = \nabla_\gamma f|_X \quad \text{if} \quad (\nabla \boldsymbol{\lambda})|_{\varkappa(X)} = \mathbf{1}$$

($\mathbf{1}$ = identity transformation of $\mathscr{V}$). Thus, the gradient $\nabla_\varkappa f|_X$ remains unaltered if $\varkappa$ is replaced by a configuration which differs from $\varkappa$ only by a deformation whose gradient at $\varkappa(X)$ is the identity.

Let $X \in \mathscr{B}$ be a specific material point. Two configurations $\varkappa$ and γ are said to be *equivalent at* X if the gradient at $\varkappa(X)$ of the deformation $\boldsymbol{\lambda} = \gamma \circ \overset{-1}{\varkappa}$ relating $\varkappa$ and γ is the identity, i.e. if $(\nabla \boldsymbol{\lambda})|_{\varkappa(X)} = \mathbf{1}$. The equivalence classes for the equivalence relation thus defined are called *local configurations at* X. A function K which associates with each material point $X \in \mathscr{B}$ a local configuration at X is called a *reference* on $\mathscr{B}$. If $\varkappa$ is a configuration, we denote by $\nabla\varkappa$ the reference which associates with each $X \in \mathscr{B}$ the equivalence class (i.e. the local configuration at X) to which $\varkappa$ belongs. References of the form $\nabla\varkappa$ are said to be *homogeneous*.

To visualize a reference K, one has to cut the body into infinitesimal pieces and to view the piece corresponding to the material point X in a configuration that belongs to the equivalence class $K(X)$. The pieces will *not* fit together to form a coherent region in space if K is an inhomogeneous reference. If $K = \nabla\varkappa$ is homogeneous, however, then the pieces can be fit together to form the region $\varkappa(\mathscr{B})$.

The fact that $\nabla_\varkappa f|_X = \nabla_\gamma f|_X$ holds if $\boldsymbol{\lambda} = \gamma \circ \overset{-1}{\varkappa}$ satisfies $(\nabla \boldsymbol{\lambda})|_{\varkappa(X)} = \mathbf{1}$ can be expressed by saying that the value at X of the gradient relative to $\varkappa$ of the scalar field f depends on $\varkappa$ only through the local configuration at X defined by $\varkappa$. Hence, if K is any reference, it is meaningful to define the *gradient* $\nabla_K f: \mathscr{B} \to \mathscr{V}$ *of* f *relative to the reference* K by

$$\nabla_K f|_X = \nabla_\varkappa f|_X \quad \text{for all } \varkappa \in K(X). \tag{2.3}$$

Gradients relative to a reference K can be defined not only for scalar fields, but in an analogous manner also for vector-fields, i.e. functions $\boldsymbol{h}: \mathscr{B} \to \mathscr{V}$, and tensor fields, i.e. functions $\boldsymbol{T}: \mathscr{B} \to \mathscr{L}_1$, where $\mathscr{L}_1$ is the space of all linear transformations of $\mathscr{V}$ into itself. In the same way one can also define the relative gradient $\nabla_K \gamma$ of a configuration γ. Its value $\nabla_K \gamma|_X$ is an invertible member of $\mathscr{L}_1$ which depends only on $K(X)$ and the local configuration $\nabla\gamma(X)$ at X defined by γ. We express this fact by using the notation

$$\nabla_K \gamma|_X = (\nabla\gamma(X))\,(K(X))^{-1}, \qquad \nabla_K \gamma = (\nabla\gamma)\,K^{-1}. \tag{2.4}$$

Conversely, given any invertible $\boldsymbol{L} \in \mathscr{L}_1$, there exists configurations γ such that $\boldsymbol{L} = \nabla_K \gamma|_X$. The class of all these configurations constitute a local configuration at X, which we denote by $\boldsymbol{L}K(X)$. Thus, every local configuration at X can be written in the form $\boldsymbol{L}K(X)$ with a suitable choice of an invertible $\boldsymbol{L} \in \mathscr{L}_1$.

3. The Concept of a Materially Uniform Simple Body

The physical properties of a body $\mathscr{B}$ in a given configuration $\varkappa$ can often be described by functions that assign to every material point $X \in \mathscr{B}$ some quantity with a particular physical meaning. For example, the inertial and gravitational properties can be described by giving the mass density at each $X \in \mathscr{B}$; the forces that hold the body together can be described by giving a stress tensor at each $X \in \mathscr{B}$; the thermal properties can be described by giving the heat capacity and heat conductivity at each $X \in \mathscr{B}$.

In general, given a configuration $\varkappa$ of $\mathscr{B}$, the physical response at $X \in \mathscr{B}$ is described by specifying some member of a set $\mathfrak{R}$. We denote this member of $\mathfrak{R}$ by $\mathfrak{G}_X(\varkappa)$ and call it the *response descriptor* of the material at X. The nature of the set $\mathfrak{R}$ depends on the physical phenomena under consideration. For example, if $\mathscr{B}$ is an elastic body, $\mathfrak{R}$ is the space of all symmetric tensors and $\mathfrak{G}_X(\varkappa) \in \mathfrak{R}$ is the stress tensor at X if $\mathscr{B}$ is in the configuration $\varkappa$. If we consider only the inertial and gravitational behavior of $\mathscr{B}$, then $\mathfrak{R}$ is the set of all positive real numbers and $\mathfrak{G}_X(\varkappa)$ is the mass density at X if $\mathscr{B}$ is in the configuration $\varkappa$.

It may happen that $\mathfrak{G}_X(\varkappa)$ depends on $\varkappa$ only through the local configuration $\nabla\varkappa(X)$ at X defined by $\varkappa$. In this case, we say that $\mathscr{B}$ is a *simple body* and we write $\mathfrak{G}_X(\nabla\varkappa(X))$ instead of $\mathfrak{G}_X(\varkappa)$. We can then regard $\mathfrak{G}_X$ as a mapping from the set of all local configurations at X into the set $\mathfrak{R}$ of all response descriptors.

Now let K be a reference. We have seen in the previous section that every local configuration at X is of the form $\boldsymbol{L}K(X)$, where $\boldsymbol{L}$ is an invertible member of $\mathscr{L}_1$. The response descriptor of the body at X in any configuration belonging to $\boldsymbol{L}K(X)$ is given by $\mathfrak{G}_X(\boldsymbol{L}K(X))$. The physical response to deformations of the body at $X \in \mathscr{B}$, relative to the reference K, is described by the nature of the relation between $\boldsymbol{L}$ and the descriptor $\mathfrak{G}_X(\boldsymbol{L}K(X))$. Therefore, it appears reasonable to say that the response of the material at X is the same as the response of the material at Y if

$$\mathfrak{G}_X(\boldsymbol{L}\,K(X)) = \mathfrak{G}_Y(\boldsymbol{L}\,K(Y)) \tag{3.1}$$

holds for all invertible $\boldsymbol{L} \in \mathscr{L}_1$. The response would be the same everywhere if (3.1) holds identically in $\boldsymbol{L}$ for all material points X and Y.

However, whether or not (3.1) holds for all invertible $\boldsymbol{L}$ and all $X, Y \in \mathscr{B}$ depends on the choice of the reference K. The references, if any, for which (3.1) does hold will be called *uniform references.*

We say that a body is *materially uniform* if it admits uniform references. We say that the body is *homogeneous* if it admits homogeneous uniform references, i.e. uniform references of the form $\nabla \varkappa$, where $\varkappa$ is a (global) configuration. Intuitively, if a body is homogeneous, we can view it in a configuration such that all parts of the body respond in the same way. If the body is inhomogeneous but materially uniform, we must first cut it into infinitesimal pieces that do not fit together before we can make all parts respond in the same way.

A materially uniform body $\mathscr{B}$ always admits infinitely many uniform references. However, from now on we shall fix the attention to one particular uniform reference K. Also, we shall assume that K is smooth in the sense that the invertible tensor field $\nabla_K \gamma$ is smooth for every configuration γ.

4. The Inhomogeneity Field

Let K be a uniform reference and γ an arbitrary configuration. We write

$$\boldsymbol{F} = \nabla_K \gamma = (\nabla \gamma)\, K^{-1} \tag{4.1}$$

for the gradient of γ relative to K. This gradient $\boldsymbol{F}$ is a smooth tensor field on $\mathscr{B}$ with invertible values in $\mathscr{L}_1$. The gradient $\nabla_K \boldsymbol{F}$ relative to K is a field on $\mathscr{B}$ with values in the space $\mathscr{L}_2$ of all linear transformations of $\mathscr{V}$ into $\mathscr{L}_1$. Thus, if $\boldsymbol{u}, \boldsymbol{v} \in \mathscr{V}$ are spatial vectors, then $(\nabla_K \boldsymbol{F})\, \boldsymbol{u} \in \mathscr{L}_1$ and $\big((\nabla_K \boldsymbol{F})\, \boldsymbol{u}\big)\, \boldsymbol{v} \in \mathscr{V}$.

Consider now the case when $K = \nabla \varkappa$ is a homogeneous uniform reference. In this case, we have $\boldsymbol{F} = \nabla_\varkappa \gamma$ and $\nabla_K \boldsymbol{F} = \nabla_\varkappa \boldsymbol{F} = \nabla_\varkappa \nabla_\varkappa \gamma$. In view of the well-known symmetry property of second gradients, it follows that

$$\big((\nabla_K \boldsymbol{F})\boldsymbol{u}\big)\, \boldsymbol{v} - \big((\nabla_K \boldsymbol{F})\, \boldsymbol{v}\big)\, \boldsymbol{u} = \boldsymbol{O} \tag{4.2}$$

holds for all $\boldsymbol{u}, \boldsymbol{v} \in \mathscr{V}$ when K is a homogeneous reference. By classical theorems of analysis, the converse is true locally: if (4.2) holds for all $\boldsymbol{u}, \boldsymbol{v} \in \mathscr{V}$, then every material point has a (finite) neighborhood such that K is a homogeneous reference of that neighborhood.

The consideration just shown proves that the field $\boldsymbol{S} : \mathscr{B} \to \mathscr{L}_2$ defined by

$$(\boldsymbol{S}\, \boldsymbol{u})\, \boldsymbol{v} = \boldsymbol{F}^{-1} \big[\big((\nabla_K \boldsymbol{F})\, \boldsymbol{u}\big)\, \boldsymbol{v} - \big((\nabla_K \boldsymbol{F})\, \boldsymbol{v}\big)\, \boldsymbol{u}\big] \tag{4.3}$$

vanishes if and only if the uniform reference K is locally homogeneous. Now, it turns out that $\boldsymbol{S}$ depends only on the choice of K and not on the choice of the configuration γ, even though $\boldsymbol{F}$ and $\nabla_K \boldsymbol{F}$ depend on both K and γ. Therefore, the field $\boldsymbol{S}$ defined by (4.3) gives an intrinsic measure of the deviation of the uniform reference K from being homogeneous, at least locally. We call the field $\boldsymbol{S}$ the *inhomogeneity* of the uniform reference K.

5. Contorted Aeolotropy

A uniform reference K may happen to have the following property: There exists a (global) configuration γ such that the gradient

$$\boldsymbol{Q} = \nabla_K \gamma = (\nabla \gamma)\, K^{-1} \tag{5.1}$$

of γ relative to K has only *orthogonal* values. If this is the case, we say that K is a reference of *contorted aeolotropy*. Intuitively, if we view the body in the configuration γ and then cut it into infinitesimal pieces, mere rotations of these pieces will bring them into configurations such that they all respond in the same way.

A real body which can be expected to possess contorted aeolotropy can be manufactured in the following manner: Take very many thin sheets of a homogeneous material and bend them into cylindrical shape in such a way that they can be stacked snugly. Then glue them together with a homogeneous glue (see Figure). One thus obtains a body in a certain configuration γ. If we consider small pieces of the body at material points X and Y, we see that these pieces can be brought into alignment by rotating one of them (see Figure).

For the body we have just described, we can introduce a cylindrical coordinate system and choose the orthogonal tensor field $\boldsymbol{Q}$ in such a way that $\boldsymbol{Q}(X)\, \boldsymbol{e}_i(X)$ is independent of X when the $\boldsymbol{e}_i(X)$, $i = 1, 2, 3$, are unit vectors pointing in the direction of the coordinate lines. When an orthogonal curvilinear coordinate system with the property just mentioned exists, we say that the resulting uniform reference is a reference of *curvilinear aeolotropy*. Not all references of contorted aeolotropy are also of curvilinear aeolotropy. An example is given by taking many very thin fibers of homogeneous material, twisting them together as in a rope, and then glueing them together.

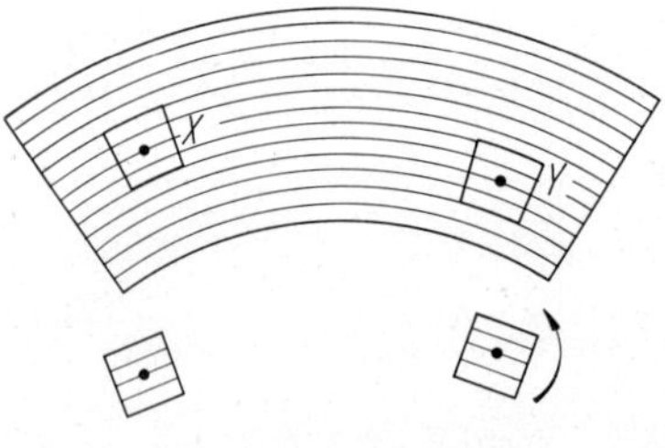

Contorted aeolotropy

The rate at which the orthogonal tensor field $\boldsymbol{Q}$ on $\mathscr{B}$ changes as one proceeds in the direction $\boldsymbol{u}$ is described by

$$\boldsymbol{D}\,\boldsymbol{u} = -\,\boldsymbol{Q}^{\mathsf{T}}\big((\nabla_K \boldsymbol{Q})\, \boldsymbol{u}\big). \tag{5.2}$$

The field $\boldsymbol{D}: \mathscr{B} \to \mathscr{L}_2$ defined by (5.2) is called the *contortion* of the reference of contorted aeolotropy given by (5.1). It is easily seen that this reference is homogeneous if and only if its contortion is zero.

One can prove that the inhomogeneity $\boldsymbol{S}$ and the contortion $\boldsymbol{D}$ determine one another by the following formulas:

$$(\boldsymbol{S}\,\boldsymbol{u})\,\boldsymbol{v} = (\boldsymbol{D}\boldsymbol{u})\,\boldsymbol{v} - (\boldsymbol{D}\boldsymbol{v})\,\boldsymbol{u}, \tag{5.3}$$

$$(\boldsymbol{D}\boldsymbol{u})\,\boldsymbol{v} = \tfrac{1}{2}\{[(\boldsymbol{S}\,\boldsymbol{u}) - (\boldsymbol{S}\,\boldsymbol{u})^T]\,\boldsymbol{v} - (\boldsymbol{S}\,\boldsymbol{v})^T\,\boldsymbol{u}\}. \tag{5.4}$$

These formulas are meaningful even if K is not a reference of contorted aeolotropy. Hence, if K is *any* uniform reference, we may define its contortion $\boldsymbol{D}$ by (5.4). If K is a reference of contorted aeolotropy, the field $\overset{*}{\boldsymbol{R}}$ on $\mathscr{B}$ defined by

$$\begin{aligned}\overset{*}{\boldsymbol{R}}(\boldsymbol{u},\boldsymbol{v}) = {} & \big((\nabla_K \boldsymbol{D})\,\boldsymbol{v}\big)\,\boldsymbol{u} + (\boldsymbol{D}\boldsymbol{u})\,(\boldsymbol{D}\boldsymbol{v}) + \boldsymbol{D}\big((\boldsymbol{D}\boldsymbol{v})\,\boldsymbol{u}\big) - \\ & - \big[\big((\nabla_K \boldsymbol{D})\,\boldsymbol{u}\big)\,\boldsymbol{v} + (\boldsymbol{D}\boldsymbol{v})\,(\boldsymbol{D}\boldsymbol{u}) + \boldsymbol{D}\big((\boldsymbol{D}\boldsymbol{u})\,\boldsymbol{v}\big)\big]\end{aligned}$$

can be shown to vanish. Conversely, if $\overset{*}{\boldsymbol{R}}$ vanishes, then K is a reference of contorted aeolotropy, at least locally. Thus, $\overset{*}{\boldsymbol{R}}$ is an intrinsic measure of the deviation of K from being a reference of contorted aeolotropy, at least locally.

6. Cauchy's Equation of Balance

Let us assume that the body $\mathscr{B}$ is held together by internal contact forces and subject to external body forces. If sufficient smoothness conditions are satisfied and if a configuration $\varkappa$ of $\mathscr{B}$ is given, the internal contact forces are determined by a stress tensor field $\boldsymbol{T}_\varkappa$ and the external body forces by a body force density vector field $\boldsymbol{b}_\varkappa$. The balance of forces is expressed by Cauchy's equation

$$\operatorname{div}_\varkappa \boldsymbol{T}_\varkappa + \boldsymbol{b}_\varkappa = \boldsymbol{O} \tag{6.1}$$

where $\operatorname{div}_\varkappa$ is defined in the obvious way in terms of gradients relative to the configuration $\varkappa$.

Under a change of configuration from $\varkappa$ to $\gamma = \boldsymbol{\lambda} \circ \varkappa$, the stress and the body force density transform according to the formulas

$$\boldsymbol{T}_\gamma = \frac{1}{J}\,\boldsymbol{T}_\varkappa \boldsymbol{F}^T, \qquad \boldsymbol{b}_\gamma = \frac{1}{J}\,\boldsymbol{b}_\varkappa, \tag{6.2}$$

where

$$\boldsymbol{F} = \nabla \boldsymbol{\lambda}, \qquad J = |\det \boldsymbol{F}|. \tag{6.3}$$

It is clear from (6.2) and (6.3) that $\boldsymbol{T}_\gamma(X) = \boldsymbol{T}_\varkappa(X)$ and $\boldsymbol{b}_\gamma(X) = \boldsymbol{b}_\varkappa(X)$ hold when $\nabla \boldsymbol{\lambda}|_{\varkappa(X)} = \boldsymbol{1}$, i.e., when γ and $\varkappa$ belong to the same local configuration at X. Thus, it is meaningful to define a stress tensor field $\boldsymbol{T}_K$ and a body force vector field $\boldsymbol{b}_K$ relative to a reference K

of $\mathscr{B}$ by the condition that

$$\boldsymbol{T}_K(X) = \boldsymbol{T}_\varkappa(X), \qquad \boldsymbol{b}_K(X) = \boldsymbol{b}_\varkappa(X) \tag{6.4}$$

hold whenever $\varkappa$ belongs to $K(X)$.

The balance equation (6.1) does not remain valid if we replace the configuration $\dot{\varkappa}$ by a smooth uniform reference K. Rather, the balance of forces is expressed, in terms of K, by

$$\operatorname{div}_K \boldsymbol{T}_K + \boldsymbol{T}_K\, \mathbf{s} + \boldsymbol{b}_K = \boldsymbol{O}, \tag{6.5}$$

where $\mathbf{s}$ is a vector field that is obtained from the inhomogeneity field $\boldsymbol{S}$ of K by the condition that

$$\mathbf{s} \cdot \boldsymbol{u} = \operatorname{trace}(\boldsymbol{S}\,\boldsymbol{u}) \tag{6.6}$$

hold for all $\boldsymbol{u} \in \mathscr{V}$. When $\mathscr{B}$ is a materially uniform but inhomogeneous simple body, one must use the modified balance equation (6.5) rather than (6.1) in order to solve specific initial value and boundary value problems.

On the Geometric Structure of Simple Bodies, a Mathematical Foundation for the Theory of Continuous Distributions of Dislocations[1]

By

C.-C. Wang

Department of Mechanics
The Johns Hopkins University, Baltimore, Ma.

Abstract

This paper consists of three major parts. In part 1, I introduce the concept of a *materially uniform simple body*. My basic hypotheses concerning such bodies are three

i) A materially uniform simple body is a differentiable manifold, called the *body manifold*, whose elements are called *particles*.

ii) The particles are *simple particles* in the sense of NOLL; moreover, they are *all of the same kind*.

iii) The mechanical response of the particles varies *smoothly* over the body manifold.

For brevity, the terms *simple body* or *body* will refer only to a materially uniform simple body in the above sense. The mathematical meanings of hypotheses i) and ii) are self-evident. The hypothesis iii) represents a physical concept generalizing the notion of *continuous distributions of dislocations* for crystalline bodies. More specifically, this hypothesis means that in a neighborhood of every particle in a body, there exists a *smooth* field of local reference configurations relative to which the response functional is independent of the particle. In particular, for a solid crystal body this condition requires that the crystal axes form smooth local fields on the body manifold, and for a fluid body it requires that the mass density be a smooth field. This condition

[1] This abstract is based on a full length paper published in the Archive for Rational Mechanics and Analysis **27**, 33—94 (1967).

also has precise mathematical meanings in the cases the body is a transversely isotropic solid, or an isotropic solid, or some kind of fluid crystal.

An important feature in my smoothness condition is that the smooth field of local reference configurations is assumed to exist only in a *neighborhood* of every particle. Thus I generalize the definition of a materially uniform simple body given by NOLL[1]. In particular, I do *not* exclude such bodies as the so-called *Moebius crystals.*

My theory differs also from the conventional theory of continuous distributions of dislocations in lattices in many aspects especially in the following three:

i) *There is no appropriate lattice model for an arbitrary simple body, since the symmetry of a simple particle, in general, cannot be described by a discrete (point) group.* In particular, the geometry characterizing the material structure on the body does not always admit a *smooth* distant parallelism.

ii) *The physical properties of a simple body are laid down once and for all in their entirety by the constitutive equations specifying the mechanical response of its particles*—namely, by the contact force that arises at a given particle in consequence of the deformation it has suffered. In particular, the material geometric structure of a simple body is *implied* mathematically by these physical properties.

iii) *I derive the exact field equations*[2] *for elastic bodies based on the constitutive equations and the geometrical structures characterizing the distribution of the mechanical response of the particles.* No mathematical approximation of any kind is made in this derivation. These exact field equations seem not to have been found by any of the physical approaches to dislocation theory.

To determine the material structure from the constitutive equations of the particles is a definite mathematical problem. That is the problem set and solved in this paper, for simple bodies. Therefore, the material structure, some cases of which were motivated by the lattice models in the conventional theories, is the *outcome* in here. Since the three basic hypotheses defining a materially uniform simple body are very general, all continuous distributions of dislocations so far considered by others, insofar as they fit upon the framework of Noll's theory of simple particles, seem to be covered here.

---

[1] Cf. TRUESDELL, C. A., and W. NOLL: The Non-linear Field Theories of Mechanics. Encyclopedia of Physics, Vol. III/3, Sect. 34. Berlin/Heidelberg/New York: Springer 1965.

[2] Equations of this kind were first found by NOLL (1, Sect. 44). My equations are more general than NOLL's, since the simple bodies considered here are more general than those treated by him.

In part 2, I consider the geometric structure of a simple body. I apply some concepts and results from the modern theory of differentiable manifolds. The major idea they represent in the context of simple bodies are sketched as follows:

In differential geometry, the overall coherence of the tangent spaces of a manifold M is characterized by two fibre bundles: the tangent bundle $T(M)$ and the bundle of linear frames $E(M)$. Thus for a simple body B there exist bundles $T(B)$ and $E(B)$ associated with the body manifold. The general structure of these two bundles, however, cannot characterize the exact material structure of the body, since they do not depend on the distribution of the response functionals on the body manifold. In order to describe the material structure of a body, I introduce the notion of *material tangent bundles*. Roughly speaking, the material tangent bundles are tangent bundles whose structure groups are the isotropy groups of the body. If we call the usual tangent bundle of the body manifold the *geometric tangent bundle* of the body, then in the terminology of fibre bundles the geometric tangent bundle is *reducible* to the material tangent bundles. As usual, we can construct the *associated principal bundles* of the material tangent bundles. These last are subbundles of the bundle of linear frames $E(B)$ of the body manifold and are called *bundles of reference frames*.

Next I apply the theory of G-connections for the bundles of frames of the body. Of special interest are those connections on $E(B)$ which are *reducible* to connections on the bundles of reference frames. Such connections are called *material connections*, since the parallel transports relative to them are always material isomorphisms. For a material connection, it is necessary and sufficient that the values of the connection form, restricted to the bundles of reference frames, belong to the fundamental fields induced by the Lie algebras of the isotropy groups of the body.

For a crystalline solid body, whose isotropy groups are discrete, the material connection is unique and coincides with the classical one motivated by the lattice models. Such material connections, evidently, are always completely integrable. Hence their curvature tensors vanish. A simple body in general, however, need *not* possess any curvature-free material connection at all.

In part 3, I apply the geometric theory for simple bodies to achieve the following results:

i) I give a precise description of the local inhomogeneity of a simple body in terms of the curvatures and the torsions of the material connections. I generalize the notions of Burger's vector and the dislocation density.

ii) I show that a simple body is a solid if and only if it can be equipped with an intrinsic Riemannian metric, which is invariant under the parallel transports relative to all material connections. In the case the body is an isotropic solid, the Riemannian connection of the intrinsic metrics is unique and is a material connection. Thus the material structure of an isotropic solid body can be described by a Riemannian geometry. I show also that a solid body is locally homogeneous if and only if it has a flat material connection. Such a condition for local inhomogeneity is false if the body is a fluid crystal.

iii) I derive exact field equations of motion for elastic bodies. As remark earlier, the derivation is based on the response functions of the particles and the geometry characterized by a material connection. No linearization of any nature is used in this analysis. Consequently, any exact solution of these field equations represents precisely a mechanical response of an elastic body which bears a specific continuous distribution of dislocations. Such solutions are reserved for a future paper.

Differential Geometry of a Nonlinear Continuum Theory of Dislocations[1]

By

R. deWit

Metallurgy Division, Institute for Materials Research
National Bureau of Standards, Washington, D. C.

Abstract. The differential geometric aspects of the limited nonlinear continuum theory of crystal dislocations are developed in terms of a non-Riemannian geometry with vanishing Riemann-Christoffel curvature. The emphasis is on a general notation and a covariant formulation of the theory. Comparisons are made between the work of Kröner and classical continuum mechanics. The drawback of distant parallelism in passing to the general theory with non-vanishing curvature is discussed.

Key Words: Dislocation, Continuum, Non-Riemannian geometry, Torsion, Non-linear, Deformation, Defect, Affine connection.

Introduction

This paper is to be regarded as a progress report of work by Dr. R. Bullough (Harwell), Dr. J. A. Simmons (NBS), and myself on a review of the *Nonlinear Continuum Theory of Crystal Dislocations.* Our aim is threefold: 1) Develop a general all-inclusive notation and formulate the theory in covariant form. 2) Review, possibly extend, and correlate the work of Kröner [1], Bilby [2], and Kondo [3]. 3) Present an introduction to the field for the beginner.

This paper deals only with the geometric aspects of the theory. Our present notation and development derives from the work of Kröner [1] and that of Eringen [4]. The correspondence is shown in Table 1.

Theory

The material points of the medium are described by three sets of curvilinear coordinate systems (Fig. 1), one for the *initial* state (K), one for the *intermediate* state ($\varkappa$), and one for the *final* state (k) of the

---

[1] Contribution of the National Bureau of Standards, not subject to copyright.

body. These coordinate systems can always be characterized through their relations to any Cartesian (rectangular) frame of reference. For generality, the coordinates may be anholonomic (i.e. non-integrable) and the initial state may contain dislocations. When the initial state does not contain dislocations, then the three states are frequently also referred to as *ideal* for (K), *natural* for $(\varkappa)$ and *deformed* for (k).

The relative position of two material points r and r', which are separated by an infinitesimal distance, is described by the local coordinates dx^K, $dy^\varkappa$, and dz^k or by the *local position vectors* $\boldsymbol{dr}_0$, $\boldsymbol{dr}_n$, and $\boldsymbol{dr}_f$, respectively, in the three states (see Table 1). The dx^K are

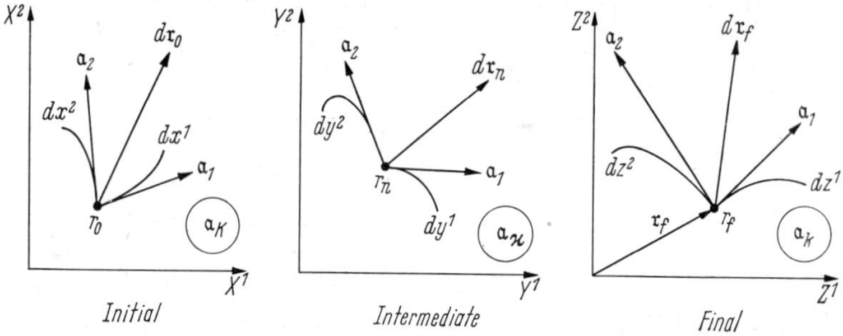

Fig. 1. The three states of the medium with their respective coordinates, bases, and position vectors. (Only two dimensions are shown for simplicity.)

usually called the *material* or *Lagrangean* coordinates and the dz^k the *spatial* or *Eulerian* coordinates. The three reference frames will in general be nonidentical, so that there is associated with each state a distinct set of base vectors $\boldsymbol{a}_K$, $\boldsymbol{a}_\varkappa$, and $\boldsymbol{a}_k$. There are two independent conditions that may be imposed on the base vectors. Expressed in holonomic coordinates, the first is $\partial_{[m}\boldsymbol{a}_{l]} = 0$,[1] which may be called the holonomity condition for $\boldsymbol{a}_k$, since it implies that there exists a whole (= holo or unbroken) net for the coordinate system. The second condition $\partial_{[n}\partial_{m]}\boldsymbol{a}_l = 0$ is called the integrability condition, since it implies that the $\boldsymbol{da}_k$ are integrable along the coordinate net. For generality the three bases may be anholonomic in the above sense. The coordinates, position vectors, and bases in the three states refer to the *same* material point, but to three different points of the geometric manifold.

The above formulation and notation is very general, but can be reduced to many different useful special cases for particular applications, e.g.:

1) Intrinsic or convected coordinates are obtained by setting $dx^K = dy^\varkappa = dz^k$ for $K = \varkappa = k$.

[1] $\partial_{[m}\boldsymbol{a}_{l]} = \frac{1}{2}(\partial_m\boldsymbol{a}_l - \partial_l\boldsymbol{a}_m)$.

2) The other limiting case might be called extrinsic coordinates, i.e. when $\boldsymbol{a}_K = \boldsymbol{a}_\varkappa = \boldsymbol{a}_k$ for $K = \varkappa = k$. A special case of this is a Cartesian reference frame, which is distinguished by using capital kernel letters.

3) Our notation in the table reduces to KRÖNER's restricted theory if $dy^\varkappa$ is anholonomic, dx^K and dz^k holonomic, and the bases rectilinear (and hence also holonomic).

4) Our notation reduces to ERINGEN's if we ignore the natural state and the coordinates and bases are holonomic.

The local position vectors $d\boldsymbol{r}_0$ or $d\boldsymbol{r}_n$ are non-integrable if the initial or final states contain dislocations, respectively, but $d\boldsymbol{r}_f$ is always integrable. For non-integrable $d\boldsymbol{r}$ the corresponding state is also called a *natural state*. The vector $d\boldsymbol{r}_f$ represents the Euclidean distance between the points r_f and r_f' in the final state, but $d\boldsymbol{r}_0$ or $d\boldsymbol{r}_n$ represent a non-Riemannian distance in their respective natural states. This point can be further clarified by viewing the deformation in two steps as illustrated in Fig. 2. The (K) and (k) states could be obtained from an undeformed, dislocation-free, ideal reference crystal by deformations A and B, respectively. Then the plastic deformation, denoted by P, takes us from

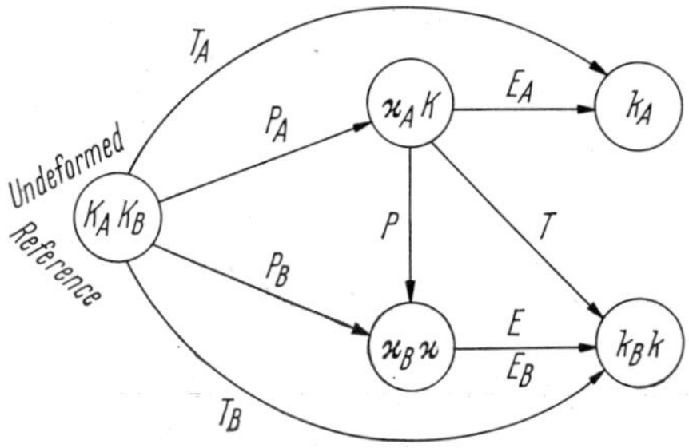

Fig. 2. Two-step deformation process. Deformations A and B lead from an undeformed, dislocation-free, reference state, to the initial and final states, respectively, of a medium undergoing a general deformation $T = EP$ from an initial state with dislocations to a final state with dislocations.

the $(\varkappa)$ state of A to the $(\varkappa)$ state of B. The elastic deformation, denoted by E, takes us from the $(\varkappa)$ state of B to the (k) state of B, and the total deformation denoted by T, takes us from the $(\varkappa)$ state of A to the (k) state of B. In other words the $(\varkappa)$ state of A is our (K) state.

The *metric tensors* are the dot-products of the base vectors, and the distance square between the two points r and r' is the square of the position vector (Table 1).

We assume now that the bases in the three states are related to each other by *shifters*, which are two-state tensors (Table 1). Since the shifters relate vectors at different points, it is necessary to haev a geo-

Table 1. *Notation of* Eringen [4], Kröner [1], *and our work*

Description	Eringen	Kröner	Ours
Ideal or initial state	B	$(\mathfrak{K})$	(K)
Natural or intermediate state	—	$(\varkappa)$	$(\varkappa)$
Deformed or final state	b	(k)	(k)
Material point in (K), $(\varkappa)$, (k)	P, —, p	—	r_0, r_n, r_f
Coordinate systems in (K)	dX^K, —, dx^k	$dx^{\mathfrak{K}}$, $dx^{\varkappa}$, dx^k	dx^K, $dy^{\varkappa}$, dz^k
Base vectors in (K), $(\varkappa)$, (k)	$\boldsymbol{G}_K$, —, $\boldsymbol{g}_k$	—	$\boldsymbol{a}_K$, $\boldsymbol{a}_{\varkappa}$, $\boldsymbol{a}_k$
Cartesian coordinates in (K), $(\varkappa)$, (k)	dZ^K, —, dz^k	—	dX^K, $dY^{\varkappa}$, dZ^k
Cartesian base in (K), $(\varkappa)$, (k)	$\boldsymbol{I}_K$, —, $\boldsymbol{i}_k$	—	$\boldsymbol{I}_K$, $\boldsymbol{I}_{\varkappa}$, $\boldsymbol{I}_k$
Position vector in (K)	$d\boldsymbol{P} = \boldsymbol{G}_K\, dX^K$	—	$d\boldsymbol{r}_0 = \boldsymbol{a}_K\, dx^K$
Position vector in $(\varkappa)$	—	—	$d\boldsymbol{r}_n = \boldsymbol{a}_{\varkappa}\, dy^{\varkappa}$
Position vector in (k)	$d\boldsymbol{p} = \boldsymbol{g}_k\, dx^k$	—	$d\boldsymbol{r}_f = \boldsymbol{a}_k\, dz^k$
Metric in (K)	$G_{KL} = \boldsymbol{G}_K \cdot \boldsymbol{G}_L$	$b_{\mathfrak{K}\mathfrak{L}}$	$a_{KL} = \boldsymbol{a}_K \cdot \boldsymbol{a}_L$
Metric in $(\varkappa)$	—	$g_{\varkappa\lambda}$	$a_{\varkappa\lambda} = \boldsymbol{a}_{\varkappa} \cdot \boldsymbol{a}_{\lambda}$
Metric in (k)	$g_{kl} = \boldsymbol{g}_k \cdot \boldsymbol{g}_l$	a_{kl}	$a_{kl} = \boldsymbol{a}_k \cdot \boldsymbol{a}_l$
Distance square in (K)	$dS^2 = G_{KL}\, dX^K\, dX^L$	$ds^2_{(\mathfrak{K})} = b_{\mathfrak{K}\mathfrak{L}}\, dx^{\mathfrak{K}}\, dx^{\mathfrak{L}}$	$ds_0^2 = a_{KL}\, dx^K\, dx^L$
Distance in square $(\varkappa)$	—	$ds^2_{(\varkappa)} = g_{\varkappa\lambda}\, dx^{\varkappa}\, dx^{\lambda}$	$ds_n^2 = a_{\varkappa\lambda}\, dy^{\varkappa}\, dy^{\lambda}$
Distance square in (k)	$ds^2 = g_{kl}\, dx^k\, dx^l$	$ds^2_{(k)} = a_{kl}\, dx^k\, dx^l$	$ds_f^2 = a_{kl}\, dz^k\, dz^l$
Shifter for T	$\boldsymbol{g}_k = g^K{}_k\, \boldsymbol{G}_K$	—	$\boldsymbol{a}_k = \boldsymbol{a}_K\, S^K_k$
Shifter for P	—	—	$\boldsymbol{a}_{\varkappa} = \boldsymbol{a}_K\, S^K_{\varkappa}$
Shifter for E	—	—	$\boldsymbol{a}_k = \boldsymbol{a}_{\varkappa}\, S^{\varkappa}_k$
Distortion for T	$dx^k = x^k{}_{,K}\, dX^K$	$dx^k = A^k{}_{\mathfrak{K}}\, dx^{\mathfrak{K}}$	$dz^k = A^k_K\, dx^K$
Distortion for P	—	$dx^{\varkappa} = A^{\varkappa}{}_{\mathfrak{K}}\, dx^{\mathfrak{K}}$	$dy^{\varkappa} = A^{\varkappa}_K\, dx^K$

Distortion for E	—	$d x^k = A^k_\varkappa \, d x^\varkappa$	$d z^k = A^k_\varkappa \, d y^\varkappa$
Lagrangian strain for T	$2 E_{KL} = C_{KL} - G_{KL}$	—	$2 \overset{T}{\eta}_{KL} = c_{KL} - a_{KL}$
Lagrangian strain for P	—	—	$2 \overset{P}{\eta}_{KL} = h_{KL} - a_{KL}$
Lagrangian strain for E	—	—	$2 \eta_{KL} = c_{KL} - h_{KL}$
Eulerian strain for T	$2 e_{kl} = g_{kl} - c_{kl}$	$2 \overset{G}{\varepsilon}_{kl} = a_{kl} - b_{kl}$	$2 \overset{T}{\varepsilon}_{kl} = a_{\kappa l} - b_{kl}$
Eulerian strain for P	—	$2 \overset{P}{\varepsilon}_{kl} = g_{kl} - b_{kl}$	$2 \overset{P}{\varepsilon}_{kl} = g_{kl} - b_{kl}$
Eulerian strain for E	—	$2 \varepsilon_{kl} = a_{kl} - g_{kl}$	$2 \varepsilon_{kl} = a_{kl} - g_{kl}$
Connection for (K) in (K)	—	—	$A^K_{ML} = \boldsymbol{a}^K \cdot \partial_M \boldsymbol{a}_L$
Connection for $(\varkappa)$ in $(\varkappa)$	—	$\Gamma^\varkappa_{\mu\lambda}$	$A^\varkappa_{\mu\lambda} = \boldsymbol{a}^\varkappa \cdot \partial_\mu \boldsymbol{a}_\lambda$
Connection for (k) in (k)	—	A^k_{ml}, a'^k_{ml}	$A^k_{ml} = \boldsymbol{a}^k \cdot \partial_m \boldsymbol{a}_l$
Connection for $(\varkappa)$ in (k)	—	Γ^k_{ml}	$\Gamma^k_{ml} = \boldsymbol{g}^k \cdot \partial_m \boldsymbol{g}_l$
Covariant derivative for (k)	—	$\nabla_m v_l = \partial_m v_l - v_k A^k_{ml}$	$\overset{f}{\nabla}_m v_l$
Covariant derivative for $(\varkappa)$	—	$\overset{*}{\nabla}_m v_l = \partial_m v_l - v_k \Gamma^k_{ml}$	$\overset{n}{\nabla}_m v_l$
Initial dislocation density in (k)	—	—	$\overset{0}{\alpha}{}^k_{ml} = B^k_{[ml]} + \Omega^k_{ml}$
Final dislocation density in (K)	—	$\alpha^{\mathfrak{k}}_{\mathfrak{m}\mathfrak{l}} = A^{\mathfrak{k}}_\varkappa \, \partial_{[\mathfrak{m}} A^\varkappa_{\mathfrak{l}]}$	$\alpha^K_{ML} = H^K_{[ML]} + \Omega^K_{ML}$
Final dislocation density in $(\varkappa)$	—	—	$\alpha^\varkappa_{\mu\lambda} = A^\varkappa_{[\mu\lambda]} + \Omega^\varkappa_{\mu\lambda}$
Final dislocation density in (k)	—	α^k_{ml}	$\alpha^k_{ml} = \Gamma^k_{[ml]} + \Omega^k_{ml}$
Integrability condition	—	$\partial_{[\mathfrak{m}} A^k_{\mathfrak{l}]} = 0$	$\tilde{T}^K_{K'} \overset{0}{\nabla}_{[M} T^{K'}_{L]} = - \overset{0}{\alpha}{}^K_{ML}$
Dislocation transformation	—	$\alpha^k_{ml} = A^k_{\mathfrak{k}} \, \alpha^{\mathfrak{k}}_{\mathfrak{m}\mathfrak{l}} \, A^{\mathfrak{m}}_m \, A^{\mathfrak{l}}_l$	$\alpha^k_{ml} = A^k_K \, \alpha^K_{ML} \, \tilde{A}^M_m \, \tilde{A}^L_l$
Basic geometric equation	—	$A^k_\varkappa \, \partial_{[m} A^\varkappa_{l]} = \alpha^k_{ml}$	$E^k_{k'} \overset{f}{\nabla}_{[m} E^{k'}_{l]} = \alpha^k_{ml}$

metry with distant parallelism. Hence the bases are integrable and the *Riemann-Christoffel curvature* tensor vanishes in this formulation. The deformation of the material is described by relating the local coordinates of the three states by the *distortions*, which are also two-state tensors. The total distortion A^k_K is also called the *deformation gradient* in Case 4 above. The distortions must be finite to insure the indestructability and impenetrability of matter.

We have defined the (one-state) tensors[1]

$$\left.\begin{array}{lll} T^{K'}_K = S^{K'}_k A^k_K, & P^{K''}_K = S^{K''}_\varkappa A^\varkappa_K, & E^{K'}_{K''} = S^{K'}_k A^k_\varkappa \tilde{S}^\varkappa_{K''}, \\ T^{\varkappa''}_{\varkappa'} = S^{\varkappa''}_k A^k_K S^K_{\varkappa'}, & P^\varkappa_{\varkappa'} = A^\varkappa_K S^K_{\varkappa'}, & E^{\varkappa''}_\varkappa = S^{\varkappa''}_k A^k_\varkappa, \\ T^k_{k''} = A^k_K S^K_{k''}, & P^{k'}_{k''} = \tilde{S}^{k'}_\varkappa A^\varkappa_k S^K_{k''}, & E^k_{k'} = A^k_\varkappa S^\varkappa_{k'}. \end{array}\right\} \quad (1)$$

We have not yet found these described in the literature and so we have provisionally named them deformations. They give the relation between the components of a vector that has undergone T, P, or E in (K), $(\varkappa)$, or (k), as illustrated below. The rationale underlying our notation is that shifted vectors retain the same letter for kernel, while deformed vectors change their kernel letter:

$$\left.\begin{array}{ll} \boldsymbol{u} = \boldsymbol{a}_K u^K = \boldsymbol{a}_\varkappa u^\varkappa = \boldsymbol{a}_k u^k, & u^K = S^K_\varkappa u^\varkappa = S^K_k u^k, \\ \boldsymbol{v} = \boldsymbol{a}_K v^K = \boldsymbol{a}_\varkappa v^\varkappa = \boldsymbol{a}_k v^k, & v^K = S^K_\varkappa v^\varkappa = S^K_k v^k, \\ \boldsymbol{w} = \boldsymbol{a}_K w^K = \boldsymbol{a}_\varkappa w^\varkappa = \boldsymbol{a}_k w^k, & w^K = S^K_\varkappa w^\varkappa = S^K_k w^k, \end{array}\right\} \quad (2)$$

If $\boldsymbol{u} -(P)\rightarrow \boldsymbol{v} -(E)\rightarrow \boldsymbol{w}$, $\boldsymbol{u} -(T)\rightarrow \boldsymbol{w}$, then

$$w^k = A^k_\varkappa v^\varkappa = A^k_K u^K, \qquad v^\varkappa = A^\varkappa_K u^K, \quad (3)$$

and it follows from Eq. (1) that

$$\left.\begin{array}{ll} w^{K'} = E^{K'}_{K''} v^{K''} = T^{K'}_K u^K, & v^{K''} = P^{K''}_K u^K, \\ w^{\varkappa''} = E^{\varkappa''}_\varkappa v^\varkappa = T^\varkappa_{\varkappa'} u^{\varkappa'}, & v^\varkappa = P^\varkappa_{\varkappa'} u^{\varkappa'}, \\ w^k = E^k_{k'} v^{k'} = T^k_{k''} u^{k''}, & v^{k'} = P^{k'}_{k''} u^{k''}. \end{array}\right\} \quad (4)$$

In order to compare the position vectors of the three states, we have also written them in terms of the three coordinates:

$$\left.\begin{array}{l} d\boldsymbol{r}_0 = \boldsymbol{a}_K dx^K = \boldsymbol{f}_\varkappa dy^\varkappa = \boldsymbol{b}_k dz^k, \\ d\boldsymbol{r}_n = \boldsymbol{h}_K dx^K = \boldsymbol{a}_\varkappa dy^\varkappa = \boldsymbol{g}_k dz^k, \\ d\boldsymbol{r}_f = \boldsymbol{c}_K dx^K = \boldsymbol{e}_\varkappa dy^\varkappa = \boldsymbol{a}_k dz^k. \end{array}\right\} \quad (5)$$

By this procedure we have introduced the *deformed bases*:

$$\begin{array}{lll} \boldsymbol{h}_K = \boldsymbol{a}_\varkappa A^\varkappa_K, & \boldsymbol{f}_\varkappa = \boldsymbol{a}_K \tilde{A}^K_\varkappa, & \boldsymbol{b}_k = \boldsymbol{a}_K \tilde{A}^K_k, \\ \boldsymbol{c}_K = \boldsymbol{a}_k A^k_K, & \boldsymbol{e}_\varkappa = \boldsymbol{a}_k A^k_\varkappa, & \boldsymbol{g}_k = \boldsymbol{a}_\varkappa \tilde{A}^\varkappa_k. \end{array} \quad (6)$$

---

[1] We use the notation $\tilde{S}^k_\varkappa$ for the inverse of $S^\varkappa_k$.

Fig. 3 illustrates all the relations between the 9 bases[1]. If we choose $\boldsymbol{a}_\varkappa$ along the crystallographic directions then $\boldsymbol{e}_\varkappa = \boldsymbol{a}_{\varkappa''} E^{\varkappa''}_\varkappa$ gives BILBY's lattice correspondence relation. Metrics can also be defined from the deformed bases. The metric b_{kl} is *Cauchy's deformation tensor* for Case 4

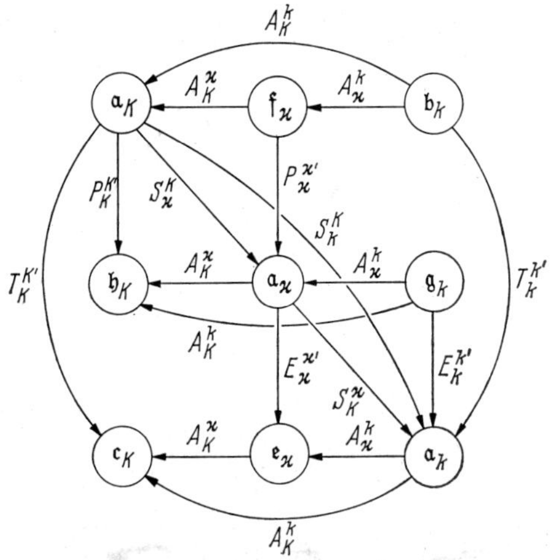

Fig. 3. Relations between the bases.

above and c_{KL} *Green's deformation tensor*. The metrics allow us to write the strains in a simple form (Table 1).

The *linear connection coefficients* are defined from the base vectors, using the corresponding capital letter as kernel. (The only exception is Γ^k_{ml}, instead of G^k_{ml}, to make this particular connection agree with KRÖNER.) So there are 9 connections, and each has a covariant derivative associated with it (Table 1):

$$\left.\begin{array}{llllll} A^K_{ML}, & F^\varkappa_{\mu\lambda}, & B^k_{ml}, & \overset{0}{\nabla}_M, & \overset{0}{\nabla}_\mu, & \overset{0}{\nabla}_m, \\ H^K_{ML}, & A^\varkappa_{\mu\lambda}, & \Gamma^k_{ml}, & \overset{n}{\nabla}_M, & \overset{n}{\nabla}_\mu, & \overset{n}{\nabla}_m, \\ C^K_{ML}, & E^\varkappa_{\mu\lambda}, & A^k_{ml}, & \overset{f}{\nabla}_M, & \overset{f}{\nabla}_\mu, & \overset{f}{\nabla}_m. \end{array}\right\} \tag{7}$$

Only three connections are essentially independent, one for each state; the others occur by reflection in the other states (i.e. expressing them in the coordinates of the other three states). For holonomic final coordinates A^k_{ml} represents the ordinary Christoffel symbol. For holonomic coordinates without initial or final dislocations, A^K_{ML} or $A^\varkappa_{\mu\lambda}$ also represent Christoffel symbols, respectively. In these cases the bases will

[1] We are currently considering a change in notation to give this array a more logical appearance.

Table 2. *Relations between connections*

$H^K_{ML} = \tilde{A}^K_\varkappa A^\varkappa_{\mu\lambda} A^\mu_M A^\lambda_L + \tilde{A}^K_\varkappa \partial_M A^\varkappa_L$	$F^\varkappa_{\mu\lambda} = A^\varkappa_K A^K_{ML} \tilde{A}^M_\mu \tilde{A}^L_\lambda + A^\varkappa_K \partial_\mu \tilde{A}^K_\lambda$	$B^k_{ml} = A^k_K A^K_{ML} \tilde{A}^M_m \tilde{A}^L_l + A^k_K \partial_m \tilde{A}^K_l$
$= \tilde{P}^K_{K'} A^{K'}_{ML'} P^{L'}_L + \tilde{P}^K_{K'} \partial_M P^{K'}_L$	$= P^\varkappa_{\varkappa'} A^{\varkappa'}_{\varkappa\lambda'} \tilde{P}^{\lambda'}_\lambda + P^\varkappa_{\varkappa'} \partial_\varkappa \tilde{P}^{\varkappa'}_\lambda$	$= T^k_{k'} A^{k'}_{ml'} \tilde{T}^{l'}_l + T^k_{k'} \partial_m \tilde{T}^{k'}_l$
$= A^K_{ML} + \tilde{A}^K_\varkappa \overset{0}{\nabla}_M A^\varkappa_L = A^K_{ML} + \tilde{P}^K_{K'} \overset{0}{\nabla}_M P^{K'}_L$	$= A^\varkappa_{\mu\lambda} + A^\varkappa_K \overset{n}{\nabla}_\mu \tilde{A}^K_\lambda = A^\varkappa_{\mu\lambda} + P^\varkappa_{\varkappa'} \overset{n}{\nabla}_\mu \tilde{P}^{\varkappa'}_\lambda$	$= A^k_{ml} + A^k_K \overset{f}{\nabla}_m \tilde{A}^K_l = A^k_{ml} + T^k_{k'} \overset{f}{\nabla}_m \tilde{T}^{k'}_l$
	$= \tilde{A}^\varkappa_k B^k_{ml} A^m_\mu A^l_\lambda + \tilde{A}^\varkappa_k \partial_\mu A^k_\lambda$	
$C^K_{ML} = \tilde{A}^K_k A^k_{ml} A^m_M A^l_L + \tilde{A}^K_k \partial_M A^k_L$	$E^\varkappa_{\mu\lambda} = \tilde{A}^\varkappa_k A^k_{ml} A^m_\mu A^l_\lambda + \tilde{A}^\varkappa_k \partial_\mu A^k_\lambda$	$\Gamma^k_{ml} = A^k_\varkappa A^\varkappa_{\mu\lambda} \tilde{A}^\mu_m \tilde{A}^\lambda_l + A^k_\varkappa \partial_m \tilde{A}^\varkappa_l$
$= \tilde{T}^K_{K'} A^{K'}_{ML'} T^{L'}_L + \tilde{T}^K_{K'} \partial_M T^{K'}_L$	$= \tilde{E}^\varkappa_{\varkappa'} A^{\varkappa'}_{\mu\lambda'} E^{\lambda'}_\lambda + \tilde{E}^\varkappa_{\varkappa'} \partial_\mu E^{\varkappa'}_\lambda$	$= E^k_{k'} A^{k'}_{ml'} \tilde{E}^{l'}_l + E^k_{k'} \partial_m \tilde{E}^{k'}_l$
$= A^K_{ML} + \tilde{A}^K_k \overset{0}{\nabla}_M A^k_L = A^K_{ML} + \tilde{T}^K_{K'} \overset{0}{\nabla}_M T^{K'}_L$	$= A^\varkappa_{\mu\lambda} + A^\varkappa_k \overset{n}{\nabla}_\mu A^k_\lambda = A^\varkappa_{\mu\lambda} + E^\varkappa_{\varkappa'} \overset{n}{\nabla}_\mu E^{\varkappa'}_\lambda$	$= A^k_{ml} + A^k_\varkappa \overset{f}{\nabla}_m \tilde{A}^\varkappa_l = A^k_{ml} + E^k_{k'} \overset{f}{\nabla}_m \tilde{E}^{k'}_l$
	$= A^\varkappa_K C^K_{ML} \tilde{A}^M_\mu \tilde{A}^L_\lambda + A^\varkappa_K \partial_\mu \tilde{A}^K_\lambda$	$= A^k_K H^K_{ML} \tilde{A}^M_m \tilde{A}^L_l + A^k_K \partial_m \tilde{A}^K_l$
$A^\varkappa_{\mu\lambda} = \tilde{S}^\varkappa_K A^K_{ML} \tilde{A}^M_\mu S^L_\lambda + \tilde{S}^\varkappa_K \partial_\mu S^K_\lambda$	$\Gamma^k_{ml} - B^k_{ml} = A^k_K (H^K_{ML} - A^K_{ML}) \tilde{A}^M_m \tilde{A}^L_l = A^k_\varkappa (A^\varkappa_{\mu\lambda} - F^\varkappa_{\mu\lambda}) \tilde{A}^\mu_m \tilde{A}^\lambda_l$	
$A^k_{ml} = \tilde{S}^k_\varkappa A^\varkappa_{\mu\lambda} \tilde{A}^\mu_m S^\lambda_l + \tilde{S}^k_\varkappa \partial_m S^\varkappa_l$	$A^k_{ml} - \Gamma^k_{ml} = A^k_K (C^K_{ML} - H^K_{ML}) \tilde{A}^M_m \tilde{A}^L_l = A^k_\varkappa (E^\varkappa_{\mu\lambda} - A^\varkappa_{\mu\lambda}) \tilde{A}^\mu_m \tilde{A}^\lambda_l$	
$= \tilde{S}^k_K A^K_{ML} \tilde{A}^M_m S^L_l + \tilde{S}^k_K \partial_m S^K_l$	$A^k_{ml} - B^k_{ml} = A^k_K (C^K_{ML} - A^K_{ML}) \tilde{A}^M_m \tilde{A}^L_l = A^k_\varkappa (E^\varkappa_{\mu\lambda} - F^\varkappa_{\mu\lambda}) \tilde{A}^\mu_m \tilde{A}^\lambda_e$	

also be holonomic. The covariant derivatives $\overset{0}{\nabla}_M$, $\overset{n}{\nabla}_\mu$, and $\overset{f}{\nabla}_m$ of the shifters vanish. To each of the 18 relations in Fig. 3 corresponds a relation between connections, as shown in Table 2. Some of these have been rewritten in terms of the covariant derivatives. Table 2 also shows 3 relations for differences of connections, which can also be deduced intuitively from a certain symmetry in Fig. 3.

The initial *dislocation density* is given by the torsions of the connections A^K_{ML}, $F^\varkappa_{\mu\lambda}$, and B^k_{ml} and the final dislocation density by the torsions of H^K_{ML}, $A^\varkappa_{\mu\lambda}$, and Γ^k_{ml} in the three states, respectively, while the torsions of C^K_{ML}, $E^\varkappa_{\mu\lambda}$, and A^k_{ml} vanish. For anholonomic coordinates the anholonomic objects Ω must be included (Table 1), which are defined by their transformation equations:

$$\left.\begin{aligned} \Omega^\varkappa_{\mu\lambda} &= A^\varkappa_K\,\Omega^K_{ML}\,\tilde{A}^M_\mu\,\tilde{A}^L_\lambda - A^\varkappa_K\,\partial_{[\mu}\tilde{A}^K_{\lambda]}\,,\\ \Omega^k_{ml} &= A^k_\varkappa\,\Omega^\varkappa_{\mu\lambda}\,\tilde{A}^\mu_m\,\tilde{A}^\lambda_l - A^k_\varkappa\,\partial_{[m}\tilde{A}^\varkappa_{l]}\,,\\ &= A^k_K\,\Omega^K_{ML}\,\tilde{A}^M_m\,\tilde{A}^L_l - A^k_K\,\partial_{[m}\tilde{A}^K_{l]}\,. \end{aligned}\right\} \tag{8}$$

From these equations and Table 2 follow the transformation (deformation) equations between the dislocation densities (Table 1), and the following set of *basic geometric equations*:

$$\left.\begin{aligned} \alpha^K_{ML} - \overset{0}{\alpha}{}^K_{ML} &= \tilde{A}^K_\varkappa\,\overset{0}{\nabla}_{[M}A^\varkappa_{L]} &&= \tilde{P}^K_{K'}\,\overset{0}{\nabla}_{[M}P^{K'}_{L]}\,,\\ -\overset{0}{\alpha}{}^K_{ML} &= \tilde{A}^K_k\,\overset{0}{\nabla}_{[M}A^k_{L]} &&= \tilde{T}^K_{K'}\,\overset{0}{\nabla}_{[M}T^{K'}_{L]}\,,\\ -\alpha^\varkappa_{\mu\lambda} &= \tilde{A}^\varkappa_k\,\overset{n}{\nabla}_{[\mu}\tilde{A}^k_{\lambda]} &&= \tilde{E}^\varkappa_{\varkappa'}\,\overset{n}{\nabla}_{[\mu}E^{\varkappa'}_{\lambda]}\,,\\ \overset{0}{\alpha}{}^\varkappa_{\mu\lambda} - \alpha^\varkappa_{\mu\lambda} &= A^\varkappa_K\,\overset{n}{\nabla}_{[\mu}\tilde{A}^K_{\lambda]} &&= P^\varkappa_{\varkappa'}\,\overset{n}{\nabla}_{[\mu}\tilde{P}^{\varkappa'}_{\lambda]}\,,\\ \overset{0}{\alpha}{}^k_{ml} &= A^k_K\,\overset{f}{\nabla}_{[m}\tilde{A}^K_{l]} &&= T^k_{k'}\,\overset{f}{\nabla}_{[m}\tilde{T}^{k'}_{l]}\,,\\ \alpha^k_{ml} &= A^k_\varkappa\,\overset{f}{\nabla}_{[m}\tilde{A}^\varkappa_{l]} &&= E^k_{k'}\,\overset{f}{\nabla}_{[m}\tilde{E}^{k'}_{l]}\,. \end{aligned}\right\} \tag{9}$$

The second and third lines are actually integrability conditions, since the curvature tensor vanishes. They are in fact equivalent to the integrability of $d\boldsymbol{r}_f$:

$$\left.\begin{aligned} T^K_L &= \overset{0}{\nabla}_L\,r^K_f\,, & r^K_f &= \boldsymbol{a}^K\cdot\boldsymbol{r}_f\,,\\ A^k_L &= \overset{0}{\nabla}_L\,r^k_f\,, & r^\varkappa_f &= \boldsymbol{a}^\varkappa\cdot\boldsymbol{r}_f\,,\\ E^\varkappa_\lambda &= \overset{n}{\nabla}_\lambda\,r^\varkappa_f\,, & r^k_f &= \boldsymbol{a}^k\cdot\boldsymbol{r}_f\,,\\ A^k_\lambda &= \overset{n}{\nabla}_\lambda\,r^k_f\,. \end{aligned}\right\} \tag{10}$$

By performing a *Cartan cycle* or *Burgers circuit* it is seen that the upper index of the dislocation density tensor refers to the Burgers vector, while the lower two indices refer to the surface area. The final density α^k_{ml} may also be called the *local* dislocation density, since the final coordinates dz^k may be regarded as local coordinates (cf. BILBY). We may call

$$\alpha^{\varkappa}_{ml} = \tilde{A}^{\varkappa}_k \, \alpha^k_{ml} \tag{11}$$

the *true* dislocation density since it refers to the Burgers vector in its natural or invariant state. We have the following simple form for the continuity condition on dislocations

$$\overset{f}{\nabla}_{[n}\alpha^{\varkappa}_{ml]} = 0\,, \tag{12}$$

which is easily verified from the last of Eq. (9). Alternatively, it can be regarded as a consequence of the vanishing of the Riemann-Christoffel curvature tensor, using the identity

$$\begin{aligned} \Gamma^k_{[nml]} &= 2\overset{n}{\nabla}_{[n}\alpha^k_{ml]} - 4\alpha^r_{[nm}\alpha^k_{l]r} \\ &= 2\overset{f}{\nabla}_{[n}\alpha^k_{ml]} + 2\alpha^r_{[ml}(\Gamma^k_{n]r} - A^k_{n]r}). \end{aligned}$$

Comments

Our approach introduces quite a few different quantities, some of which are of course redundant; but we feel that they help round out the geometrical picture. We could work exclusively in the final state coordinates (cf. KRÖNER), and so eliminate all quantities involving K and $\varkappa$.

It is interesting to consider the creation of torsion from this point of view. The pair $(dz^k, \boldsymbol{a}_k)$ determines the position vector $d\boldsymbol{r}_f$ of the final state, which represents a real distance in Euclidean space. So this pair describes a Euclidean geometry with vanishing torsion and curvature. By deforming the base to $\boldsymbol{g}_k = \boldsymbol{a}_{k'} \tilde{E}^{k'}_k$ we obtain another pair $(dz^k, \boldsymbol{g}_k)$, which determines the position vector $d\boldsymbol{r}_n$ of the natural state. This change in the pair implies no change in the curvature, but a change $\Gamma^k_{[ml]} - A^k_{[ml]}$ in the torsion, which will not vanish if $\overset{f}{\nabla}_{[m}\tilde{E}^{k'}_{l]} \neq 0$ (Table 2). Thus the torsion $\alpha^k_{ml} = \Gamma^k_{[ml]} - A^k_{[ml]}$ has been created and we have a non-Riemannian (flat) geometry in Euclidean space.

Discussion

In our development, the existence of shifters precludes a non-vanishing curvature tensor. This is a drawback of the present description, and we do not yet know how to remedy this.

Fig. 4 shows an example of a defect that is not properly described by our theory. This type of defect could be caused by a temperature inhomogeneity or by a concentration of substitutional foreign atoms whit a different lattice spacing. There are three differential geometric ways in which this defect could be described:

1) By a distribution of torsion, e.g. a distribution of dislocation loops, that is a distribution of stress free (plastic) strain. However, since no

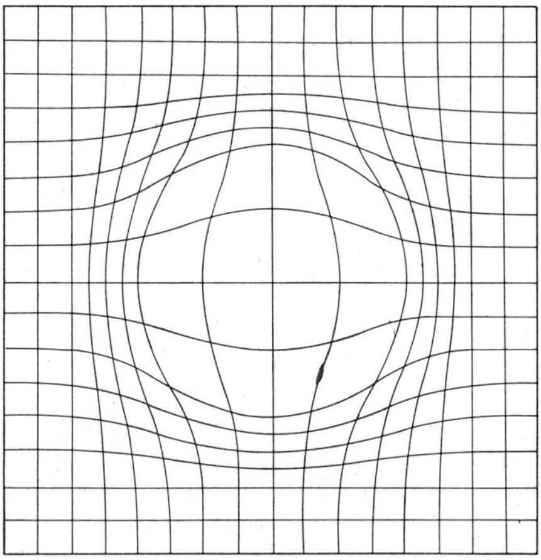

Fig. 4. Type of defect that can be described in three ways, by 1) torsion; 2) curvature, or 3) non-metric connection.

dislocations *as such* are visible in the figure, this would not seem the best approach.

2) By a non-vanishing curvature. It is possible to create this in a manner analogous to the one for torsion above. Deform the differential base in a non-integrable way $d\boldsymbol{g}_k = d\boldsymbol{a}_{k'} \tilde{M}_k^{k'}$. Then, if the pair $(dz^k, d\boldsymbol{a}_k)$ determines a vanishing curvature $A^k_{nml} = 0$, the pair $(dz^k, d\boldsymbol{g}_k)$ determines the non-vanishing curvature $\Gamma^k_{nml} = 2(\Gamma^k_{mr} M^r_s \partial_n \tilde{M}^s_l)_{[nm]}$. However, for this approach all our results involving shifters would have to be abandoned. This idea is currently being developed by Fong [6].

3) By a non-metric connection, Bilby [5]. We have not yet fully investigated this possibility, but it would seem the most promising approach, for the defect of Fig. 4 appears to have a non-metric character.

References

[1] Kröner, E.: Arcn. Rat. Mech. Anal. **4**, 273 (1960).
[2] Bilby, B. A.: Progr. Solid Mechanics **1**, 329 (1960).
[3] Kondo, K.: RAAG (Japan) Memoirs, Vols. 1—4, 1955—1967.
[4] Eringen, A. C.: Nonlinear Theory of Continuous Media. New York: McGraw-Hill 1962, Chap. 1.
[5] Bilby, B. A.: Proc. Roy. Soc. **A 292**, 105 (1966).
[6] Fong, J. T.: Unpublished note dated 7-31-67, National Bureau of Standards.

Statistical Theory of Dislocations

By

H. Zorski

Institute of the Basic Technical Problems
Polish Academy of Sciences, Warszawa

Abstract[1]

The paper is devoted to a statistical derivation of the equations governing a continuous distribution of dislocations in a linear elastic medium. We begin with a system of a finite number of infinitesimal Somigliana dislocations (infinitesimal dislocation loops) moving in accordance with the laws of dynamics of discrete dislocations. By introducing the classical phase space with its Liouville and transport equations and defining the expectation values of various physical quantities connected with the motion of the system, we introduce the Kirkwood formalism making an extensive use of the Dirac delta functions. This formalism excellently serves our purpose, since the delta functions appear initially in our problem in the right-hand sides of the Lamé equations, expressing the influence of dislocations on the generation of the elastic displacement field. By ordinary means, as in the statistical hydrodynamics, on the basis of the transport equation we now derive the equations for the density $\nu(\boldsymbol{x}, t)$ and the velocity $\boldsymbol{v}(\boldsymbol{x}, t)$ of the dislocation fluid—the continuity and linear momentum equations; the influence of the displacement (or rather the stresses and the velocity) of the elastic body is expressed by certain terms in the above equations vanishing in the case of a static homogeneous stress field. Further, we introduce the expectation value of the elastic displacement, the latter being a random function since it depends on the state of motion of the dislocations by which it is produced. This makes it possible to perform the averaging procedure on the Lamé equations and to derive an equation for the expectation value of the displacement. The procedure here is similar to that used in deriving the equations for a macroscopic Maxwell field. Incidentally, the averaging of the

[1] The full text of the paper will be published in International Journal of Solids and Structures, 1968.

Lamé equations results in a very natural definition of quantities describing the continuous distribution of dislocations such as the dislocation density tensor, introduced earlier by other authors.

The analogy to the statistical derivation of the hydrodynamics equations is evident, as a result however we obtain a compound continuous medium D_R constituting a mixture (in the sense that each geometric point in the space is occupied by two particles; in our case one of the particles is material, the other may not be material) of the material elastic body and the dislocation fluid. The final system contains seven quasi-linear partial differential equations

$$\mu \nabla^2 u_i + (\lambda + \mu) \nabla_i \nabla_p u^p - \varrho \frac{\partial^2 u_i}{\partial t^2} + \\ + \mu \varkappa_{pq} \left[\sigma^{pqr}{}_{;i} \nabla_r \nu + c_2^{-2} \delta_i^p \frac{\partial}{\partial t} (\nu v^q) \right] = -X_i ,$$

$$\frac{\partial \nu}{\partial t} + \nabla_p (\nu v^p) = 0 ,$$

$$\frac{D v_i}{D t} + \frac{c^2}{\nu} \nabla_i \nu + 2 \mu^{-1} m_{i}^{\cdot n} \varkappa_{pq} \left(\sigma^{pqrs} \nabla_n \nabla_r u_s - c_2^{-2} \delta_n^q \frac{\partial^2 u^p}{\partial t^2} \right) = 0$$

where $\boldsymbol{u}(\boldsymbol{x}, t)$ is the total displacement of the body, $\nu(\boldsymbol{x}, t)$ is the density of the dislocation fluid and $\boldsymbol{v}(\boldsymbol{x}, t)$ its velocity, $\varkappa_{ij}$ is a constant tensor characterizing the structure of the dislocations assumed to be the same throughout the whole body, σ^{ijpq} is a numerical tensor, $\overset{-1}{m}{}^{ij}$ the inverse of the tensorial mass of a single dislocation, and c is the sound velocity in the dislocation fluid. In deriving the above system it was assumed that the fluid is barotropic and its motion adiabatic. We do not carry out a general analysis of the derived system, confining ourselves to some general remarks and an example of a one-dimensional motion of tangential dislocations, for only in the case of one spatial variable a fairly complete theory of such equations exists. We believe that this example illustrates the basic features of the D_R medium, such as the creation and propagation of slip planes and shock waves, various types of wave motion, etc., and indicates the relation of the D_R medium to the classical elastic-plastic body. Obviously, only a further investigation, first of all thermodynamic considerations can decide whether it is possible to construct by statistical methods on the basis of the dynamics of discrete dislocations or other defects, a rational theory of plasticity. There are strong indications that the answer is positive.

The structure of the D_R medium examined in this paper is restricted in the following sense. In the general case the dislocations are characterized by the vectorial intensity $\underset{\alpha}{\boldsymbol{U}}$ (the discontinuity of the elastic displacement) and the director $\underset{\alpha}{\boldsymbol{n}}$, different for each dislocation; the general theory constructed on this basis is rather complicated, the fundamental system of differential equations containing 42 equations with the following unknowns: the displacement vector $\boldsymbol{u}$, the velocity of the dislocation fluid $\boldsymbol{v}$, the dislocation density tensor $\boldsymbol{\varkappa}$ and the dislocation velocity tensor $\boldsymbol{\varepsilon}$; moreover, instead of one constitutive equation at least three are required. Needless to say, the complexity of the system of equations makes any conclusions or practical applications almost impossible; this is emphasized by the fact that presently very little is known about the properties of the dislocation fluid and no data are known for establishing the required constitutive equations. We assume therefore that $\underset{\alpha}{\boldsymbol{U}} = \boldsymbol{U}$ and $\underset{\alpha}{\boldsymbol{n}} = \boldsymbol{n}$, i.e. these two vectors are the same for all dislocations. Then $\boldsymbol{n} = \boldsymbol{U}\,\boldsymbol{n}\,\boldsymbol{\nu}$ and $\boldsymbol{\varepsilon} = \boldsymbol{U}\,\boldsymbol{n}\,\boldsymbol{v}\,\boldsymbol{\nu}$. In other words we endow the medium with a homogeneous structure $\{\boldsymbol{U}, \boldsymbol{n}\}$ and seek only the density and velocity of the dislocation fluid. A generalization to a mixture of several such fluids is straightforward.

We do not attempt at this stage to compare our theory with the existing theories of continuous distribution of dislocations. First of all the basic model of the defect is different; secondly our variables are different; in fact, we introduce from the very beginning a displacement vector, the existence of which in most of the theories is denied. It seems that only a comparison of the final results will be possible.

There are two Appendices. Appendix A contains the equations for a fluid composed of vacancies; Appendix B is devoted to a generalization of the theory to the case when the defects have mass and there occurs its transfer.

Some Remarks about High Velocity Dislocations

By

H. Günther

Institut für Reine Mathematik
Deutsche Akademie der Wissenschaften, Berlin

It is known from many works that the speed of a dislocation is limited at least by the sonic velocity of the medium, see e.g. J. I. Frenkel and T. A. Kontorova [1], F. C. Frank [2], J. D. Eshelby [3], G. Leibfried and H. D. Dietze [4], J. Weertman [5] and others (cf. the references of Weertman).

What is the reason? Let us consider in an infinite elastic medium (for simplicity the linear approximation and isotropic case; for the anisotropic medium, however, see R. Bullough and B. A. Bilby [6], and A. W. Saenz [7]) a straight and therefore infinite dislocation that moves with constant velocity v. Except for a strange behaviour of the Rayleigh velocity v (see the discussions of Eshelby [3], Weertman [5]), the result is a strain field which goes to infinity if the dislocation velocity approaches the sonic velocity. Moreover, the elastic energy per unit length of a screw dislocation goes to infinity with the characteristic relativistic factor $1/\sqrt{1-(v/c_T)^2}$, where c_T is the transverse sonic velocity. The supersonic case has been considered by Eshelby [8] in the Peierls-Nabarro-model.

For edge dislocations the situation is somewhat complicated because both sonic velocities, the longitudinal c_L and the transverse c_T, enter. In the expressions for the strain both relativistic factors $1/\sqrt{1-(v/c_T)^2}$ and $1/\sqrt{1-(v/c_L)^2}$ appear, cf. Eshelby [3], Weertman [5]. But the formulas for the displacement field given there are only valid if the condition $v < c_T < c_L$ holds. In the linear elastic theory, if the dislocation velocity is greater, the strain (or stress) field has not only singularities at $v = c_T$ or $v = c_L$, as Weertman [5] states, but the field has a singular part (resp. is a singular one) in the whole region $v > c_T$ (resp. $v > c_L$). The reason for this will soon become clear.

The problem of moving dislocations can be handled very easily by the time dependent generalization of E. Kröner's [9], B. A. Bilby,

R. BULLOUGH and E. SMITH [10], and K. KONDO's [11] continuum theory of dislocations as proposed by many authors, see e.g. S. AMARI [12], H. BROSS [13], A. M. KOSEWICH [14], T. MURA [15], H. GÜNTHER [16].

Without going into details, it can be stated that in the linear approximation one obtains the following inhomogeneous wave equations for the elastic deformations ε_{kl}

$$\Box_L \varepsilon = -2\left[\frac{1}{c_L^2} T^0_{rr,0} + 2\frac{c_T^2}{c_L^2} T_{rss,r}\right], \tag{1a}$$

$$\Box_T \varepsilon_{kl} = -\left[\frac{1}{c_T^2}(T^0_{kl} + T^0_{lk})_{,0} + (T_{rkl} + T_{rlk})_{,r} + T_{krr,l} + T_{lrr,k}\right] + \left(1 - \frac{c_L^2}{c_T^2}\right)\varepsilon_{,kl} \tag{1b}$$

with $\varepsilon = \varepsilon_{rr}$ = dilatation, T_{rkl} = dislocation density, T^0_{kl} = dislocation fluxes,

$$\Box_L = \Delta - \frac{1}{c_L^2}\frac{\partial^2}{\partial t^2}, \qquad \Box_T = \Delta - \frac{1}{c_T^2}\frac{\partial^2}{\partial t^2}.$$

The integration can now easily be performed with the help of retarded GREEN's functions.

For an edge dislocation lying in the x-direction and moving in z-direction the essential feature of the solution is illustrated schematically by the three Fig. 1a) to c). For the exact analytical expression see GÜNTHER [17]. Responsible for the distinction of the three cases is the fact that the Eqs. (1a) and (1b) alter their character from elliptic to hyperbolic equations in going from Fig. 1a) to c) [first both elliptic, than (1a) elliptic and (1b) hyperbolic, finally both hyperbolic]. The field consists of two parts according to the two sonic velocities. For $c_T < c_L < v$ both these are "LORENTZ" contracted; this corresponds to ESHELBY's [3] solution. In the case $c_T < v < c_L$ for the transverse part of the field one Mach cone is produced, but in the whole inner region of the Mach cone the field vanishes. It is different from zero only on the cone itself and singular there; see Fig. 1b). For $c_T < c_L < v$ two Mach cones arise; the field vanishes everywhere except on the two Mach cones themselves and is singular there.

The situation is essentially changed if we consider not infinite but finite dislocations, e.g. dislocation loops as considered by F. KROUPA [18]. The solution of this problem gives a complicated expression for ε_{kl} and will not be discussed here; for details see GÜNTHER [17]. We again illustrate the field schematically by three Fig. 2a) to c). In the region $v < c_T < c_L$ the field is double "LORENTZ" contracted as before (for $v = 0$ the strain field is that of KROUPA [18]). But in the supersonic region $c_T < v < c_L$ of Fig. 2b, [$c_T < c_L < v$ of Fig. 2c)] there arise(s) one (two) radiation field(s) that is (are) different from zero in the one (two) Mach cone(s). First one part is "LORENTZ" contracted whereas the other is a pure radiation field [$c_T < v < c_L$, Fig. 2b)] and lastly for $c_T < c_L < v$ two radiation fields arise. The radiation fields are finite and different from zero in the whole Mach cone. The radiated

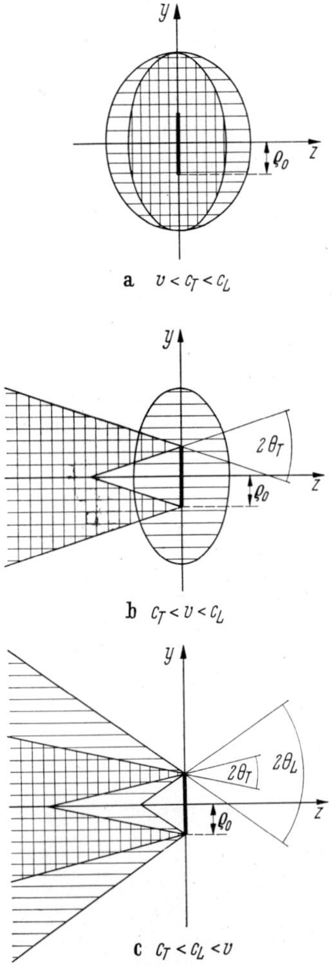

Fig. 1. The schematical field of a straight moving edge dislocation (dislocation line in x-direction, motion in z-direction). The Mach angles are determined by $\sin \Theta_T = v/c_T$, $\sin \Theta_L = v/c_L$.

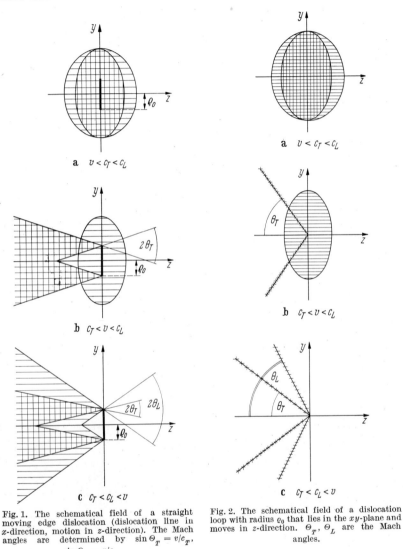

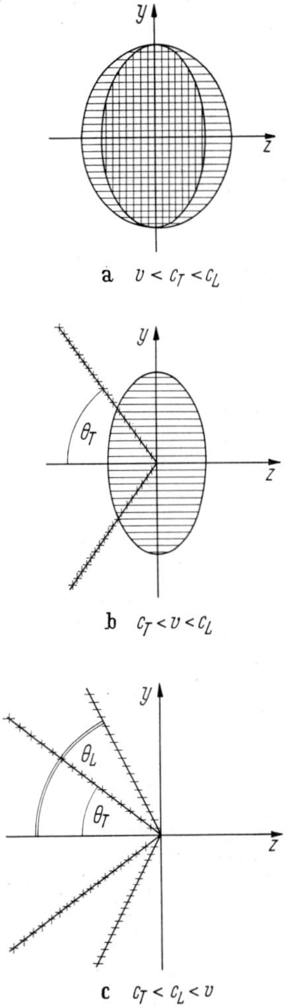

Fig. 2. The schematical field of a dislocation loop with radius ϱ_0 that lies in the xy-plane and moves in z-direction. Θ_T, Θ_L are the Mach angles.

energy per unit time will be finite too. Therefore here is a total elastic analogy to the Tscherenkov effect of electrodynamics, as stated first by ESHELBY [8].

The different behaviour of the fields for straight and closed dislocations has the following explanation: The Tscherenkov effect can be explained with the help of Huygens principle. Now the validity or violation of Huygens principle essentially depends upon whether the dimensionality of the wave equation is even or odd. But for straight dislocations the dimensionality of the problem is lowered by one. Therefore here the Mach cone is not filled up.

The above considerations are purely kinematical ones. The possibility of accelerating finite dislocations into the supersonic region is not discussed here. When this is not possible, the supersonic singularities of linear dimensions could perhaps be regarded as dynamical singularities of a new kind which appear only in high energy dynamical problems.

References

[1] FRENKEL, J., and T. KONTOROVA: Phsy. Z. Sowjet. **13**, 1 (1938).
[2] FRANK, F. C.: Proc. Phys. Soc. (London) **A 62**, 131 (1949).
[3] ESHELBY, J. D.: Proc. Phys. Soc. (London) **A 62**, 307 (1949).
[4] LEIBFRIED, G., and H. D. DIETZE: Z. Physik **126**, 790 (1949).
[5] WEERTMAN, J.: in: Response of Metals to High Velocity Deformation, ed. by P. G. SHEWMON and V. F. ZACKAY. New York/London: Interscience Publ., p. 205.
[6] BULLOUGH, R., and B. A. BILBY: Proc. Roy. Soc. (London) **B 67**, 615 (1954).
[7] SAENZ, A. W.: J. Rat. Mech. Anal. **2**, 83 (1953).
[8] ESHELBY, J. D.: Proc. Phys. Soc. (London), **B 69**, 1013 (1956).
[9] KRÖNER, E.: Arch. Rat. Mech. Anal. **4**, 273 (1959/60).
[10] BILBY, B. A., R. BULLOUGH and E. SMITH: Proc. Roy. Soc. (London) **A 231**, 263 (1955).
[11] KONDO, K.: RAAG Memoirs, Vol. I and II, Tokyo: Gakujutsu Bunken Fukyu-Kai 1955 and 1958.
[12] AMARI, S.: RAAG Research Notes 3. Series 52 (1962).
[13] BROSS, H.: Phys. Stat. Sol. **5**, 329 (1964).
[14] KOSEVICH, A. M.: Zh. eksper. teor. Fiz. **42**, 152 (1962); **43**, 637 (1962).
[15] MURA, T.: Phil. Mag. **8**, 843 (1963).
[16] GÜNTHER, H.: Zur nichtlinearen Kontinuumstheorie bewegter Versetzungen. Berlin: Akademie-Verlag 1967.
[17] GÜNTHER, H.: Annalen der Physik (in preparation).
[18] KROUPA, F.: Czech. J. Phys. **B 10**, 284 (1960).

Continuum Theory of Dislocations and Plasticity[1]

By

T. Mura

Department of Civil Engineering
Northwestern University, Evanston, Ill.

1. Introduction

Dislocations as line imperfections in crystals have already found extensive use in the description and interpretation of the mechanical properties of materials. Very little work however has been done towards their application to problems in structural analysis and continuum mechanics. This paper summarizes the author's previous work [1—6] and extends these ideas to the field of continuum plasticity.

Stress fields in elasto-plastic materials are considered as the sum of the applied elastic stress and the dislocation stress. The Mises yield criterion is interpreted as a critical value of the Peach-Koehler force resolved into the direction of motion of dislocations. The Peach-Koehler force on continuous dislocations is defined by introducing the dislocation flux (velocity) tensor. This tensor is related to the time derivative of the dislocation density tensor, and the relation describes the multiplication of dislocations and Frank-Read sources. The Prandtl-Reuss relation is derived from the relation between the plastic strain rate and the dislocation flux tensor.

A new concept of impotent distributions of dislocations is also proposed.

2. Dislocation Density and Flux Tensors

In this section dislocation density and flux tensors are introduced and the dislocation multiplication is discussed by the use of these tensors.

---

[1] The research upon which this paper was based was supported by the Advanced Research Projects Agency of the Department of Defense, through the Northwestern University Materials Research Center.

Consider an arbitrary point P in a deformed material at time t. In the neighborhood of point P there are several dislocation lines with various Burgers vectors $\overset{(1)}{\boldsymbol{b}}, \overset{(2)}{\boldsymbol{b}}, \ldots$ with directions $\overset{(1)}{\boldsymbol{\nu}}, \overset{(2)}{\boldsymbol{\nu}}, \ldots$, and these dislocations are moving with velocities $\overset{(1)}{\boldsymbol{V}}, \overset{(2)}{\boldsymbol{V}}, \ldots$ If a small surface $\Delta \boldsymbol{S}$ is considered in the neighborhood of point P, the number of dislocations threading the surface depends upon the direction and area of $\Delta \boldsymbol{S}$. The total Burgers vector of n dislocations threading the surface $\Delta \boldsymbol{S}$ will be

$$\alpha_{h i} \eta_h \Delta S = \sum_{a=1}^{n} \overset{(a)}{\nu_h} \overset{(a)}{b_i} \eta_h . \tag{2.1}$$

where $\boldsymbol{\eta}$ is the normal vector of $\Delta \boldsymbol{S}$. The $\alpha_{h i}$ is called the dislocation density tensor. The total Burgers vector is changing with time due to the motion of dislocations. The flux of the dislocation transport will be

$$V_{l h i} \eta_h \Delta S = \sum_{a=1}^{n} \overset{(a)}{V_l} \overset{(a)}{\nu_h} \overset{(a)}{b_i} \eta_h . \tag{2.2}$$

The tensor $V_{l h i}$ is called the dislocation flux (or velocity) tensor. The growth rate of the total Burgers vector of dislocations threading an arbitrary surface $\boldsymbol{S}$, is $\int_S \dot{\alpha}_{h i} \eta_h \, dS$, where $(\dot{\ })$ denotes the differentiation with respect to time. On the other hand, the growth rate can be considered as the result of the motion of dislocations moving across

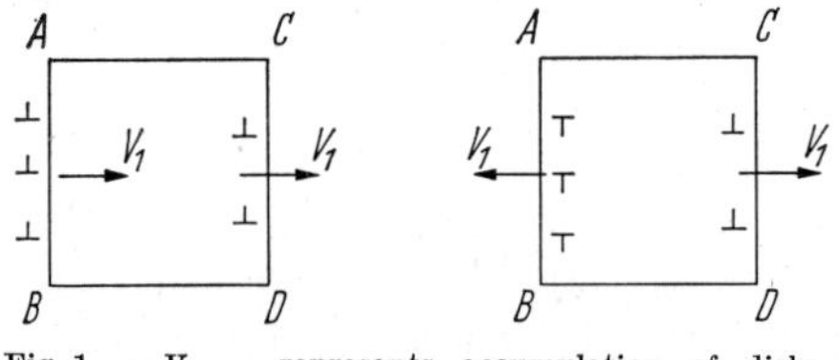

Fig. 1. $-V_{131,1}$ represents accumulation of dislocations caused by the difference of dislocation flux at AB and CD. The negative direction motion of negative dislocations has the same contribution to V_{131} as the positive direction motion of positive dislocations.

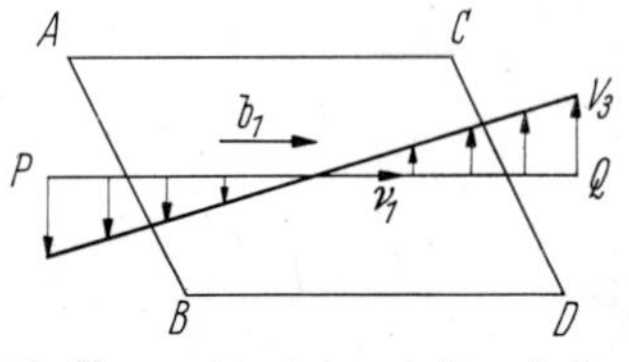

Fig. 2. $V_{311,1}$ represents rotation of dislocation PQ caused by the x_1 direction change of velocity V_3.

the boundary of the surface $\boldsymbol{S}$. Since only the dislocation velocity component which is normal to the boundary line of the surface and to the direction of the dislocation lines can contribute to the growth, we have

$$\int_S \dot{\alpha}_{h i} \eta_h \, dS = \oint \varepsilon_{l h k} V_{l h i} \, dl_k , \tag{2.3}$$

where $\boldsymbol{dl}$ is the line element of the boundary and $\varepsilon_{l h k}$ is the unit permutation tensor. Applying Stokes' theorem to the right side of (2.3) and comparing the integrands on both sides, we have

$$\dot{\alpha}_{h i} = \varepsilon_{h l k} \varepsilon_{m n k} V_{m n i, l} = V_{h l i, l} - V_{l h i, l} , \tag{2.4}$$

where $(_{,l})$ denotes the differentiation with respect to the coordinate x_l. The term $-V_{lhi,l}$ is the net flux of α_{hi} dislocations accumulated in the unit area normal to the x_h axis and the term $V_{hli,l}$ represents a creation of dislocations α_{hi} caused by the rotational motion of the dislocation lines. Figs. 1 and 2 illustrate how the terms $-V_{131,1}$ and $V_{311,1}$, respectively, contribute to a change in α_{31}.

3. Strains and Rotations

Strains and rotations during plastic deformations are closely related to the dislocation density and flux tensors. Replacing a finite number of slip surfaces by an infinite number of slip surfaces with infinitesimal gliding, we have a continuous displacement field U_i. The gradient of the displacement components is called a distortion. The distortion of a deformed body $U_{i,j}$ is not necessarily equal to the plastic distortion β^*_{ji} which is the infinitesimal plastic gliding of the x_j plane in the direction of x_i. The difference between them is called the elastic distortion by KRÖNER [7] or the lattice distortion by BILBY [8] and it is denoted by β_{ji} here, that is,

$$U_{i,j} = \beta^*_{ji} + \beta_{ji}. \tag{3.1}$$

The strain and rotation are defined as the symmetric and antisymmetric parts of the distortions respectively. The plastic and elastic strains are

$$e^*_{ij} = \tfrac{1}{2}(\beta^*_{ij} + \beta^*_{ji}), \quad e_{ij} = \tfrac{1}{2}(\beta_{ij} + \beta_{ji}). \tag{3.2}$$

The plastic and lattice rotations are

$$\omega^*_{ij} = \tfrac{1}{2}(\beta^*_{ij} - \beta^*_{ji}), \quad \omega_{ij} = \tfrac{1}{2}(\beta_{ij} - \beta_{ji}). \tag{3.3}$$

And

$$\beta^*_{ij} = e^*_{ij} + \omega^*_{ij}, \quad \beta_{ij} = e_{ij} + \omega_{ij}. \tag{3.4}$$

The time rate of the plastic distortion $\dot\beta^*_{21}$, for instance, is caused by the motion of α_{31} in the x_1 direction or by the motion of α_{11} in the $-x_3$ direction, that is,

$$\dot\beta^*_{21} = V_{131} - V_{311} \tag{3.5}$$

or more generally,

$$\dot\beta^*_{ij} = -\varepsilon_{imn} V_{mnj}. \tag{3.6}$$

Substituting (3.6) into (2.4) leads to

$$\dot\alpha_{hi} = -\varepsilon_{hlk}\dot\beta^*_{ki,l} \tag{3.7}$$

or, after integrating,

$$\alpha_{hi} = -\varepsilon_{hlk}\beta^*_{ki,l}, \tag{3.8}$$

which has been considered by KONDO [9], KRÖNER [7] and BILBY [8]. The following relations which are obtained from (3.8), (3.4), and (3.1) are also well known:

$$\alpha_{hi} = -\varepsilon_{hlk} e^*_{ki,l} - \varepsilon_{hlk} \omega^*_{ki,l} = \varepsilon_{hlk} e_{ki,l} + \varepsilon_{hlk} \omega_{ki,l}, \tag{3.9}$$

and

$$\tfrac{1}{2}\varepsilon_{ijk} \alpha_{hk,j} + \tfrac{1}{2}\varepsilon_{hmn} \alpha_{in,m} = \varepsilon_{hmn} \varepsilon_{ijk} e_{nk,jm}. \tag{3.10}$$

4. Field Equations

Consider a problem to seek displacement and stress fields for a given $\beta^*_{ij}(x, t)$ in an infinite body. The solution is expressed by the use of Green's tensor functions G_{ij} (MURA [2], KOSEVICH and NATSIK [10]),

$$U_m(x', t') = -\int\int_{-\infty}^{\infty} C_{klij}\, G_{km,l}(x - x', t - t')\, \beta^*_{ji}(x, t)\, dD(x)\, dt \tag{4.1}$$

and

$$\begin{aligned} \sigma_{pq}(x', t') &= C_{pqmn}\, \beta_{nm}(x', t'), \\ \beta_{nm}(x', t') &= \int\int_{-\infty}^{\infty} C_{klij}\, G_{km,ln}(x - x', t - t')\, \beta^*_{ji}(x, t)\, dD(x)\, dt - \\ &\qquad - \beta^*_{nm}(x', t'), \end{aligned} \tag{4.2}$$

where C_{ijkl} are the elastic constants, and $dD(x)$ is the volume element. If the plastic distortion is given as

$$\beta^*_{nm}(x, t) = \bar{\beta}^*_{nm} \exp\{\mathrm{i}(\omega t + \boldsymbol{k} \cdot \boldsymbol{r})\}, \tag{4.3}$$

where

$$\boldsymbol{k} \cdot \boldsymbol{r} = k_1 x_1 + k_2 x_2 + k_3 x_3, \quad \mathrm{i} = \sqrt{-1}$$

and k_i and ω are arbitrary constants and $\bar{\beta}^*_{nm}$ are arbitrary functions of k_i and ω, the solution is (MURA [3])

$$U_m(x, t) = -\mathrm{i}\, k_l\, C_{klij}\, L_{mk}\, \bar{\beta}^*_{ji} \exp\{\mathrm{i}(\omega t + \boldsymbol{k} \cdot \boldsymbol{r})\}, \tag{4.4}$$

and

$$\begin{aligned} \sigma_{pq}(x, t) &= C_{pqmn}\, \beta_{nm}(x, t), \\ \beta_{nm}(x, t) &= (k_n\, k_l\, C_{klij}\, L_{mk}\, \bar{\beta}^*_{ji} - \bar{\beta}^*_{nm}) \exp\{\mathrm{i}(\omega t + \boldsymbol{k} \cdot \boldsymbol{r})\}, \end{aligned} \tag{4.5}$$

where

$$L_{mk} = \frac{\delta_{mk}\{(\lambda + 2\mu) k^2 - \varrho\,\omega^2\} - k_m k_k (\lambda + \mu)}{(\mu k^2 - \varrho\,\omega^2)\{(\lambda + 2\mu) k^2 - \varrho\,\omega^2\}} \tag{4.6}$$

for an isotropic material with Lamé constants λ, μ and density ϱ, $k^2 = k_1^2 + k_2^2 + k_3^2$, and δ_{ij} is the Kronecker delta.

It can be seen from (4.1), (4.2) or (4.4), (4.5) that the displacement and stress fields are caused only by e^*_{ij} in (3.4) and are not influenced by ω^*_{ij} since $C_{klij}\, \omega^*_{ji} = 0$. However, the ω^*_{ij} does give a contribution to the dislocation density according to (3.9). A special case in which $\beta^*_{ij} = \omega^*_{ij}$ yields $U_m = \sigma_{pq} = 0$ and $\beta_{nm} = -\beta^*_{nm}$. Therefore, the total rotation also vanishes. The associated distribution of dislocations will be called an *impotent* distribution of dislocations. Consider, for instance, an edge dislocation with Burgers vector b_1 at $x_1 = x_2 = 0$ which is caused by the plastic distortion

$$\beta^*_{21} = b_1\, \delta(x_2)\, H(-x_1), \qquad \beta^*_{12} = 0, \tag{4.7}$$

where δ is the Dirac function and H the heaviside step function. The well-known displacement and stress fields are easily obtained from (4.4) and (4.5) by the use of Fourier integrals. The corresponding dislocation density is obtained from (3.8)

$$\alpha_{31} = b_1\, \delta(x_2)\, \delta(x_1). \tag{4.8}$$

On the other hand, another plastic distortion

$$\beta^*_{21} = \beta^*_{12} = \tfrac{1}{2} b_1\, \delta(x_2)\, H(-x_1), \tag{4.9}$$

which has no plastic rotation but the same plastic strain as that obtianed from (4.7) yields the same displacement and stress fields as (4.7). But the corresponding dislocation distribution becomes from (3.8)

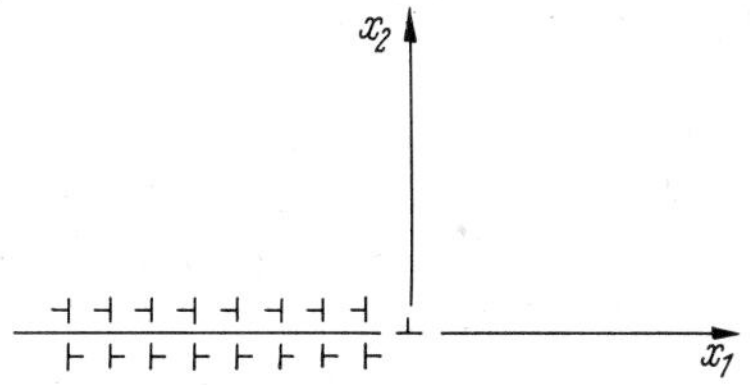

Fig. 3. The dislocation distribution caused by plastic distortion (4.9) which yields the same displacement and stress fields as (4.7).

$$\begin{aligned} \alpha_{31} &= \tfrac{1}{2} b_1\, \delta(x_2)\, \delta(x_1), \\ \alpha_{32} &= \tfrac{1}{2} b_1\, \delta'(x_2)\, H(-x_1), \end{aligned} \tag{4.10}$$

as shown in Fig. 3. The difference of (4.8) and (4.10) yields the impotent distribution of dislocations which causes neither stress nor displacement.

5. Plastic Yielding and Flow

The Mises yield criterion and the Prandtl-Reuss flow law in the continuum theory of plasticity will now be interpreted in terms of dislocation behavior.

By taking a unit area normal to η in (2.1) and (2.2), we write

$$\alpha_{hi} = \nu_h\, b_i\, B + \nu'_h\, b'_i\, B' + \cdots, \tag{5.1}$$

$$V_{lhi} = v_l\, \nu_h\, b_i\, VB + v'_l\, \nu'_h\, b'_i\, V'\, B' + \cdots, \tag{5.2}$$

where ν_h is the direction cosine of dislocation lines which have the strength $b_i\, B$ and velocity $v_i\, V$, and ν'_h is referred to dislocation lines having the strength $b'_i\, B'$ and velocity $v'_i V'$ and so on. The directions of the velocity and Burgers vectors are indicated by the unit vectors v_i and b_i, respectively, while the positive scalars V and B represent their magnitudes, that is,

$$v_i\, v_i = 1, \qquad b_i\, b_i = 1, \qquad \nu_i\, \nu_i = 1. \tag{5.3}$$

Expression (5.2) includes the motion of dislocation dipoles or loops, since two dislocations of opposite signs moving in opposite directions give the same contribution to the expression; namely $(-v_l)\,\nu_h(-b_i) = v_l\,\nu_h\,b_i$.

Substituting (5.2) into (3.6) leads to

$$\dot{\beta}^*_{ij} = n_i\,b_j\,V B + n'_i\,b'_j\,V'\,B' + \cdots, \tag{5.4}$$

and, therefore,

$$\dot{e}^*_{ij} = n_{ij}\,V B + n'_{ij}\,V'\,B' + \cdots, \tag{5.5}$$

where

$$n_i = \varepsilon_{inm}\,\nu_n\,v_m, \qquad n_{ij} = \tfrac{1}{2}(n_i\,b_j + n_j\,b_i), \tag{5.6}$$

with similar relations (5.6) for n'_{ij} and so on. The n_i is the normal vector of the slip plane on which the dislocation line and the dislocation velocity vector are lying. Since b_i should be on the slip plane

$$n_{ii} = n_i\,b_i = 0, \qquad n_{kl}\,n_{kl} = \tfrac{1}{2}. \tag{5.7}$$

The irreversibility of plasticity requires that

$$\sigma_{ij}\,\dot{e}^*_{ij} > 0, \tag{5.8}$$

where σ_{ij} is the sum of an applied stress and the dislocation stress. The applied stress is defined as that stress field which is in elastic equilibrium with the applied boundary forces. Substituting (5.5) into (5.8) leads to

$$\sigma_{ij}(n_{ij}\,V B + n'_{ij}\,V'\,B' + \cdots) > 0, \tag{5.9}$$

or

$$\sigma_{ij}\,\varepsilon_{inm}(\nu_n\,v_m\,b_j\,V B + \nu'_n\,v'_m\,b'_j\,V'\,B' + \cdots) > 0. \tag{5.10}$$

We see that $\sigma_{ij}\,\varepsilon_{inm}\,\nu_n\,v_m\,b_j$ is the component of Peach-Koehler force [11] in the direction of dislocation motion and also the resolved shear stress on the slip plane in the slip direction, that is, $\sigma_{ij}\,n_{ij}$.

An assumption[1] is introduced here that for polycrystals all dislocations have a constant resistance, $\bar{k}$, for motion, namely

$$\sigma_{ij} n_{ij} = \sigma_{ij}\,n'_{ij} = \cdots = \bar{k}, \tag{5.11}$$

or

$$s_{ij}\,n_{ij} = s_{ij}\,n'_{ij} = \cdots = \bar{k}, \tag{5.12}$$

by using the reduced stress $s_{ij} = \sigma_{ij} - (\frac{1}{3})\,\delta_{ij}\,\sigma_{kk}$ and (5.7). Since V, $B, V', B', \ldots, \bar{k} > 0$, condition (5.8) is identically satisfied.

Consider s_{ij} and n_{ij} as vectors in the 6-dimensional Euclidean space[2]. According to (5.12), vectors $n_{ij}\,n'_{ij}, \ldots,$ make an acute angle θ with s_{ij} and constitute a circular cone when an infinite number of these

---

[1] Essentially the same assumption was made by Taylor [12].

[2] $(s_{11}, \sqrt{2}s_{12}, \ldots, s_{33})$, $(n_{11}, \sqrt{2}n_{12}, \ldots, n_{33})$.

vectors is taken by considering a polycrystal. Denoting $\bar{n}_{ij}$ to be the component of n_{ij} parallel to s_{ij}, we have

$$\bar{n}_{ij} = \bar{n}'_{ij} = \cdots = \mu^* s_{ij}, \qquad (\mu^* > 0) \tag{5.13}$$

and

$$s_{ij}\bar{n}_{ij} = (s_{ij}\, s_{ij})^{\frac{1}{2}} (n_{kl}\, n_{kl})^{\frac{1}{2}} \cos\theta = \bar{k}, \tag{5.14}$$

or, from (5.7),

$$(\tfrac{1}{2}) s_{ij}\, s_{ij} = k^2, \tag{5.15}$$

where

$$k = \bar{k}/\cos\theta. \tag{5.16}$$

Eq. (5.15) is the Mises yield criterion and the value of $\cos\theta$ is $\frac{3}{8}$ which is calculated from a statistical average of all possible orientations of crystal grains (MURA [6]). From (5.13), (5.14), (5.15) and (5.16),

$$\mu^* = \cos\theta/(2k). \tag{5.17}$$

For isotropic polycrystals, the components of $n_{ij}, n'_{ij}, \ldots,$ normal to s_{ij} are assumed to cancel each other in Eq. (5.5), then

$$\dot{e}^*_{ij} = s_{ij}(V B + V' B' + \cdots) \cos\theta/(2k). \tag{5.18}$$

It should be remembered that the magnitude of dislocation velocity is defined for a single dislocation. According to GILMAN [13] it is a function of a shear stress which may be taken as k. Since all dislocations are subjected to the same k, we have,

$$V = V' = \cdots = V_0. \tag{5.19}$$

Thus, we have finally

$$\dot{e}^*_{ij} = s_{ij}\, V_0\, N \cos\theta/(2k), \tag{5.20}$$

where

$$N = B + B' + \cdots. \tag{5.21}$$

Eq. (5.20) is nothing more than the Prandtl-Reuss relation and the coefficient $V_0 N$ has a clearer physical meaning. The N is the total sum of dislocation strengths $B, B' \ldots,$ disregarding the signs of the dislocations. It means that N takes into account dislocation dipoles and loops. Eq. (5.20) may be used not only for static loadings but also for dynamic loadings and in creep considerations. For work-hardening materials, the k may be a function of deformation.

6. An Example

The stress fields in elasto-plastic materials can be considered as the sum of the applied elastic stress and the dislocation stress. After a suitable choice of dislocation distribution, if the sum of these stresses satisfies the Mises yield criterion in plastic domains, the stress field is a statically admissible field. If, in addition, plastic strain satisfying (3.9) for such a dislocation distribution also satisfies the Prandtl-Reuss relation, the above stress field may be the true one.

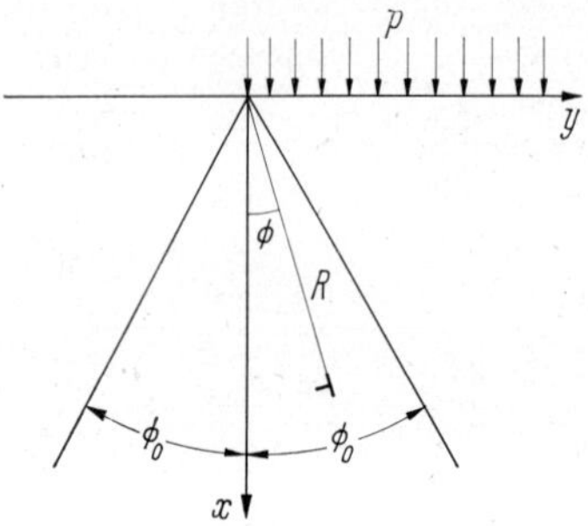

Fig. 4. Dislocations α_{ZR} distributed in the plastic domain $\phi_0 \leqq \phi \leqq \phi_0$ with density b_0/R.

Consider an example (MURA and KÖNIG [14]) of a half-infinite domain loaded by a semi-infinite band of uniform pressure p on the surface as shown in Fig. 4. The problem is assumed as a plane strain problem and the material is assumed to be incompressible.

The applied elastic stress is

$$\left.\begin{aligned} \sigma_R^A &= -\frac{p}{\pi}\left(\frac{\pi}{2} + \phi - \tfrac{1}{2}\sin 2\phi\right), \\ \sigma_\phi^A &= -\frac{p}{\pi}\left(\frac{\pi}{2} + \phi + \tfrac{1}{2}\sin 2\phi\right), \\ \sigma_{R\phi}^A &= \frac{p}{2\pi}(1 + \cos 2\phi), \end{aligned}\right\} \tag{6.1}$$

where the polar coordinates (R, ϕ) are used.

Choose the following dislocation distribution in domain $-\phi_0 \leqq \phi \leqq \phi_0$ which is assumed to be a plastic domain,

$$\alpha_{ZR} = b_0/R, \quad \alpha_{Z\phi} = 0. \tag{6.2}$$

The corresponding dislocation stress is obtained from the result of DUNDURS and MURA [15], as follows

$$\sigma_R^D = \frac{G\,b_0}{2\pi}(2f + f''), \quad \sigma_\phi^D = \frac{G\,b_0}{2\pi}\,2f, \quad \sigma_{R\phi}^D = -\frac{G\,b_0}{2\pi}\,f', \tag{6.3}$$

where G is the shear modulus and

$$\begin{aligned} f(\phi) = {} & (2\phi_0 + \sin 2\phi_0)\sin 2\phi - \pi \sin 2\phi_0 \cos 2\phi + \\ & + 2(2\phi_0 + \sin 2\phi_0)\,\phi + 2\pi\,\phi_0 \quad \text{for} \quad -\pi/2 \leqq \phi \leqq -\phi_0; \end{aligned} \tag{6.4}$$

$$\begin{aligned} f(\phi) = {} & (2\phi_0 + \sin 2\phi_0 + \pi \cos 2\phi_0)\sin 2\phi + \\ & + 2(2\phi_0 + \sin 2\phi_0 - \pi)\,\phi \quad \text{for} \quad -\phi_0 \leqq \phi \leqq \phi_0; \end{aligned} \tag{6.5}$$

and

$$\begin{aligned} f(\phi) = {} & (2\phi_0 + \sin 2\phi_0)\sin 2\phi + \pi \sin 2\phi_0 \cos 2\phi + \\ & + 2(2\phi_0 + \sin 2\phi_0)\,\phi - 2\pi\,\phi_0 \quad \text{for} \quad \phi_0 \leqq \phi \leqq \pi/2. \end{aligned} \tag{6.6}$$

It can be shown that if we choose

$$b_0 = k/\{G(1 + \cos 2\phi_0)\} \tag{6.7}$$

and

$$p = 2k(2\phi_0 + \sin 2\phi_0 + \pi \cos 2\phi_0)/(1 + \cos 2\phi_0), \tag{6.8}$$

the stress field

$$\sigma_R = \sigma_R^A + \sigma_R^D, \quad \sigma_\phi = \sigma_\phi^A + \sigma_\phi^D, \quad \sigma_{R\phi} = \sigma_{R\phi}^A + \sigma_{R\phi}^D, \tag{6.9}$$

is a statically admissible stress field. Namely, (6.9) satisfies the Mises yield criterion[1]

$$\tfrac{1}{4}(\sigma_R - \sigma_\phi)^2 + \sigma_{R\phi}^2 \leqq k^2, \tag{6.10}$$

where the equality holds in the plastic domain and the inequality holds in the elastic domain. The stress field (6.9) and pressure (6.8) agree with SOKOLOVSKY's result [16]. However, he did not show any field of plastic strain. The following is a method of finding a kinematically admissible plastic strain.

The polar coordinate expression of (3.9) becomes

$$\begin{aligned} b_0/R &= -\frac{\partial e^*_{\phi R}}{\partial R} - \frac{2}{R} e^*_{\phi R} + \frac{\partial e^*_R}{R\,\partial\phi} + \frac{\partial \omega^*_z}{\partial R}, \\ 0 &= -\frac{\partial e^*_\phi}{\partial R} + \frac{e^*_R - e^*_\phi}{R} + \frac{\partial e^*_{R\phi}}{R\,\partial\phi} + \frac{\partial \omega^*_z}{R\,\partial\phi}, \end{aligned} \tag{6.11}$$

by the use of (6.2), where ω^*_z is a plastic rotation about the z axis. The condition of incompressibility is

$$e^*_R + e^*_\phi = 0. \tag{6.12}$$

The plastic strain tensors cannot be determined uniquely from (6.11) and (6.12) without any further assumption. If we assume[2], however, $\omega^*_z = 0$, the solution is obtained easily as

$$\begin{aligned} e^*_{\phi R} &= -\frac{b_0}{2} + A \sin 2\phi + B \cos 2\phi, \\ e^*_R &= -e^*_\phi = B \sin 2\phi - A \cos 2\phi \end{aligned} \tag{6.13}$$

where A and B are constants of integration.

If the boundary condition

$$e^*_{\phi R} = 0 \quad \text{at} \quad \phi = \pm\phi_0, \tag{6.14}$$

is imposed in order to determine A and B, we have

$$e^*_{\phi R} = \frac{b_0}{2}\left(\frac{\cos 2\phi}{\cos 2\phi_0} - 1\right), \qquad e^*_R = -e^*_\phi = \frac{b_0}{2}\,\frac{\sin 2\phi}{\cos 2\phi_0}, \tag{6.15}$$

which leads to a kinematically admissible displacement field.

[1] The stress field also satisfies the boundary condition and the equilibrium condition.

[2] Since two distributions of dislocations can be considered equivalent when their difference is only an impotent distribution of dislocations, this assumption is always justified for a non-oriented continuum.

References

[1] MURA, T.: Phil. Mag. **8**, 843 (1963).
[2] MURA, T.: Int. J. Engng. Sci. **1**, 371 (1963).
[3] MURA, T.: Proc. Roy. Soc. **A 280**, 528 (1964).
[4] MURA, T.: Phys. Stat. Sol. **10**, 447 (1965); **11**, 683 (1965).
[5] MURA, T.: Int. J. Engng. Sci. **5**, 341 (1967).
[6] MURA, T.: The Work Hardening Symposium, to be published by Gordon and Breach as a volume of the AIME book series.
[7] KRÖNER, E.: Kontinuumstheorie der Versetzungen und Eigenspannungen. Berlin/Göttingen/Heidelberg: Springer 1958.
[8] BILBY, B. A.: Progress in Solid Mechanics, Vol. 1, ed. by I. N. SNEDDON and R. HILL. Amsterdam: North-Holland 1960.
[9] KONDO, K.: Memoirs of the unifying study of the basic problems in engineering sciences by means of geometry, Vol. 1. Tokyo: Gakujutsu Bunken Fukyu-Kai 1955.
[10] KOSEVICH, A. M., and V. D. NATSIK: Soviet Physics-Solid State **6**, 181 (1964).
[11] PEACH, M. O., and J. S. KOEHLER: Phys. Rev. **80**, 436 (1950).
[12] TAYLOR, G. I.: J. Inst. Metals **62**, 307 (1938).
[13] GILMAN, J. J.: Aust. J. Phys. **13**, 327 (1960).
[14] MURA, T., and J. KÖNIG: unpublished work.
[15] DUNDURS, J., and T. MURA: J. Mech. Phys. Solids **12**, 177 (1964).
[16] SOKOLOVSKY, V. V.: PMM **14**, 391 (1950).

Continuous Distributions of Dislocations in Hyperelastic Materials of Grade 2

By

C. Teodosiu

Institute of Mathematics
Academy of S. R. Rumania, Bucharest

The present work aims at the elaboration of a theory of grade 2 materials with initial stresses and hyperstresses induced by dislocations or quasi-dislocations. Some of the results presented here were first published in [1] and a full account will be given in [2].

The material continuum M is considered in three states:

the reference state (K), in which the material continuum is free of any external loads, but may contain sources of initial stresses and hyperstresses, whose engendering had an inelastic character;

the natural state $(\varkappa)$, which may be obtained from the reference state by tearing the material continuum into very small particles and by releasing them individually;

the deformed state (k), which results from the reference state by the action of the external loads upon the interior and the boundary of the material continuum.

In the reference and deformed states we choose general curvilinear co-ordinates X^K and x^k. In the natural state we refer each particle to a rectangular Cartesian co-ordinate system. In this state, a single co-ordinate system, holonomic with respect to those of (K)- and (k)-states, generally fails to exist. We summarize here the geometric equations relating the three states, as described by Kondo [3], Bilby, Bullough and Smith [4) and Kröner [5].

If $(d\xi)^\varkappa$, dX^K and dx^k denote the components of a material vector uniting two points of the same particle in the three states, we have

$$dX^K = A^K_\varkappa (d\xi)^\varkappa, \qquad dx^k = A^k_K\, dX^K, \tag{1}$$

where $A^k_K \equiv x^k_{,K}$ are the gradients of the deformation $x^k = x^k(X^K)$ induced by the action of the external loads. We suppose the distortions $A^K_\varkappa$ and A^k_K are three times continuously differentiable as functions of X^K and admit the inverses $A^\varkappa_K$, A^K_k. Let now $d\sigma^2$, dS^2, ds^2 be the square lengths of the same material vector in the three states and $\delta_{\varkappa\lambda}$, G_{KL}, g_{kl} the components of the metric tensors. Then

$$d\sigma^2 = \delta_{\varkappa\lambda}(d\xi)^\varkappa (d\xi)^\lambda, \quad dS^2 = G_{KL}\, dX^K dX^L, \quad ds^2 = g_{kl}\, dx^k dx^l, \tag{2}$$

$$d\sigma^2 = C_{KL}\, dX^K dX^L, \qquad C_{KL} \equiv \delta_{\varkappa\lambda} A^\varkappa_K A^\lambda_L. \tag{3}$$

Two material vectors V^K and $V^K + dV^K$ with origins at (X^M) and $(X^M + dX^M)$ become parallel and have equal lengths after releasing if

$$d V^K = -\Gamma^K_{ML} V^L dX^M, \quad \Gamma^K_{ML} \equiv A^K_\varkappa \, \partial_M A^\varkappa_L. \tag{4}$$

It is easily proved that the connexion Γ^K_{ML} is metric with respect to C_{KL} and that the curvature tensor corresponding to this connexion vanishes, that is

$$R_{\dot N \dot M \dot L}{}^K \equiv 2(\partial_N \Gamma^K_{ML} + \Gamma^K_{NR} \Gamma^R_{ML})_{[NM]} = 0. \tag{5}$$

The torsion tensor corresponding to the connexion Γ^K_{ML},

$$S_{\dot M \dot L}{}^K \equiv \Gamma^K_{[ML]} = A^K_\varkappa \, \partial_{[M} A^\varkappa_{L]}, \tag{6}$$

represents the dislocation density, which is supposed to be known from physical measurements.

In developing the present theory we assume as in the classical theory of elasticity, that the energy stored in a volume v of M, in the deformed state, is

$$\mathcal{W} = \int_v j \, W \, dv,$$

where W is the energy per unit volume of the natural state, and $j = \varrho/\varrho_0$ is the ratio of the mass densities in the deformed and the natural states.

Now, we suppose that W depends upon the distortions A^k_λ and their total covariant derivatives, that is

$$W = W(A^k_\lambda, A^k_{\lambda;\mu}). \tag{7}$$

In other words, we assume the material uniform and of grade 2.

To obtain the equilibrium equations and the boundary conditions for a part of the material continuum, filling in the deformed state a volume v, bounded by a smooth surface s we shall adopt the principle of virtual work postulated by Toupin [6]. This principle asserts that for any variations $\delta x^k(X^K)$, for which $\delta(\varrho \, dv) = 0$, and $\delta(A^\varkappa_K) = 0$, the equality

$$\delta \int_v j \, W \, dv = \int_v f^k \, \delta x_k \, dv + \int_s (t^k \, \delta x_k + h^k \, D \, \delta x_k) \, da \tag{8}$$

holds, where f^k is the body force per unit deformed volume, t^k and h^k are the tractions and hypertractions per unit deformed area and

$$D \, \delta x_k = (\delta x_k)_{,l} \, n^l.$$

By introducing (7) into (8) and calculating the variations δA^k_λ, $\delta A^k_{\lambda;\mu}$ we obtain after some transformations the equilibrium equations

$$\sigma^{kl}{}_{,l} - h^{lkm}{}_{,lm} + f^k = 0, \quad \boldsymbol{x} \in v, \tag{9}$$

and the boundary conditions[1]

$$\sigma^{kl} n_l - D h^{lkm} n_l n_m - 2 D_l h^{(l|k|m)} n_m + h^{lkm} (b_{lm} - b^p_p n_l n_m) = t^k,$$
$$h^{lkm} n_l n_m = h^k, \quad \boldsymbol{x} \in s \tag{10}$$

where the *stress tensor* σ^{kl} and the *hyperstress tensor* h^{lkm} are defined by

$$\sigma^{kl} = j\, g^{ks} \left(\frac{\partial W}{\partial A^s_\lambda} A^l_\lambda + \frac{\partial W}{\partial A^s_{\lambda;\mu}} A^l_{\lambda;\mu} \right), \qquad h^{lkm} = j\, g^{ks} \frac{\partial W}{\partial A^s_{\lambda;\mu}} A^l_\lambda A^m_\mu . \tag{11}$$

Now, by making use of the principle of material indifference we deduce that W is expressible as a function of the *strain tensor*

$$\varepsilon_{\varkappa\lambda} \equiv \tfrac{1}{2}(\gamma_{\varkappa\lambda} - \delta_{\varkappa\lambda}), \qquad \gamma_{\varkappa\lambda} \equiv g_{kl} A^k_\varkappa A^l_\lambda, \tag{12}$$

and of the *hyperstrain tensor*

$$\varrho_{\varkappa\lambda\mu} \equiv g_{kl} A^k_\varkappa A^l_{\lambda;\mu} . \tag{13}$$

Consequently, we derive from (11) the reduced constitutive equations

$$\begin{aligned} \sigma^{kl} &= j \left(\frac{\partial W}{\partial \varepsilon_{\varkappa\lambda}} A^k_\varkappa A^l_\lambda + 2 \frac{\partial W}{\partial \varrho_{\varkappa\lambda\mu}} A^{(k}_\varkappa A^{l)}_{\lambda;\mu} \right) \equiv \sigma^{(kl)}, \\ h^{lkm} &= j \frac{\partial W}{\partial \varrho_{\varkappa\lambda\mu}} A^k_\varkappa A^l_\lambda A^m_\mu . \end{aligned} \tag{14}$$

Alternatively, we may characterize the strain and the hyperstrain by the dual tensors

$$e_{kl} \equiv \tfrac{1}{2}(g_{kl} - c_{kl}), \qquad r_{klm} \equiv -\delta_{\varkappa\lambda} A^\varkappa_k A^\lambda_{l;\,m} \tag{15}$$

related by the equations

$$r_{klm} = e_{\{kl,m\}} + S_{\{mkl\}} . \tag{16}$$

Finally, we get from (5) the generalized equations of strain compatibility, as derived by KRÖNER and SEEGER [7]

$$-\varepsilon^{ikl} \varepsilon^{jmn} e_{ln,km} = \eta^{ij} + q^{ij} \tag{17}$$

where η^{ij} represents the symmetric known tensor of the incompatibility, and q^{ij} stands for the non-linear term of the equation.

The Eqs. (9), (14), (16), (17), form a system of 66 equations with 6 boundary conditions (10) for the determination of 66 unknown functions: σ^{kl}, h^{lkm}, e_{kl}, r_{klm}.

In order to linearize the constitutive equations, we assume for the distortions $A^k_\varkappa$ the form

$$A^k_\varkappa = g^k_\varkappa + \beta^k_\varkappa, \tag{18}$$

where $g^k_\varkappa$ are shifters, and $\beta^k_\varkappa$ are "small", and we suppose that the strain energy density is a homogeneous, quadratic function of its arguments.

---

[1] b_{lm} is the three-dimensional extension of the second fundamental tensor of the surface s, and $D_l h^{lkm} = h^{lkm}{}_{,l} - D h^{lkm} n_l$.

After some manipulation, we obtain

$$\sigma^{kl} = F^{klst}\, e_{st} + H^{klstn}\, r_{stn}, \qquad h^{lkm} = H^{klmst}\, e_{st} + G^{klmstn}\, r_{stn}, \quad (19)$$

and we put $q^{ij} = 0$ into (17).

To get the solution of the linearized boundary-value problem, we write the solution of the Eq. (17) under the form

$$e_{kl} = \bar{e}_{kl} + \tfrac{1}{2}(u_{k,l} + u_{l,k}), \quad (20)$$

where $\bar{e}_{kl}$ is a particular solution and u_k is the displacement vector of the associated boundary-value problem. By introducing (20) and (16) into (19) and then into (9) we obtain, for instance in the case of isotropic materials,

$$(\lambda + 2\mu)\,(1 - l_1^2 \Delta)\, u^l{}_{,kl} - \mu(1 - l_2^2 \Delta)\,(u^l{}_{,kl} - \Delta u_k) + \tilde{f}_k = 0, \quad (21)$$

where l_1 and l_2 are material constants dependent on the supplementary constants which intervene in the present theory. To solve the associated boundary-value problem (21), (10), one can use the representations of the solutions of the Eq. (21) given by MINDLIN [8].

Finally, we mention other results obtained in the framework of the present theory: the demonstration of the uniqueness theorem for the linearized boundary-value problem, the determination in closed form of the initial stresses and hyperstresses engendered by a uniform distribution of edge dislocations in an infinite plate and by a screw dislocation in an infinite cylindrical bar and the generalization of the algorithm given by KRÖNER and SEEGER [7] for the solving by successive approximations of the non-linear boundary-value problems.

References

[1] TEODOSIU, C.: Bull. Acad. Polon. Sci., sér. sci. techn. **15**, 103 and 95 (1967).
[2] TEODOSIU, C.: Rev. Roum. Sci. Techn. **12**, 961, 1061 and 1291 (1967).
[3] KONDO, K.: RAAG Memoirs, 1, 1955, C-I, p. 361.
[4] BILBY, B. A., R. BULLOUGH and E. SMITH: Proc. Roy. Soc., London, Ser. A, **231**, 263 (1955).
[5] KRÖNER, E.: Arch. Rat. Mech. Anal. **4**, 273 (1959).
[6] TOUPIN, R. A.: Arch. Rat. Mech. Anal. **17**, 85 (1964).
[7] KRÖNER, F., and A. SEEGER: Arch. Rat. Mech. Anal. **3**, 97 (1959).
[8] MINDLIN, R. D.: Arch. Rat. Mech. Anal. **16**, 51 (1964).

On the Dual Yielding and Related Problems

By

S. Minagawa and **S.-I. Amari**

Institute for Strength and Fracture of Materials, Tohoku University, Sendai

University of Tokyo

We report on several topics explored by us in Japan. First, we explore, as the main subject, the duality between strain space and stress-function space as well as the multidimensional expressions connected thereto. The recognition of the dual yielding which is responsible for the fatigue fracture is its important part. The other topics include the theory of moving dislocations, Finslerian approaches to ferromagnetic substances, etc. These investigations have been carried out under Professor KAZUO KONDO's guidance as part of his general scheme for a geometrical theory of plasticity [1, 2].

1. Dual Spaces and Dual Yielding

The principal thought of our investigation concerns the dual correspondence between the Riemannian space of strain-incompatibility (by KONDO and others [2]) and that of stress-function-stress (by SCHAEFER [3], KONDO [4], MINAGAWA [5], AMARI [6]). Noting that in the former the Euclidean features (strain) and in the latter more non-Euclidean features (stresses) have been classically recognized, we connect the other phases of this duality, i.e., the non-Euclidean facet of the former (incompatibility) and the Euclidean facet of the latter (dual displacement).

The Riemannian metric tensor of the dual space constructed from Beltrami's stress-functions remains indeterminate within a gauge transformation in the sense that a contribution from arbitrary three-dimensional Euclidean dual displacements can be added to it. This is part of the multidimensional displacements in an enveloping Euclidean space in which the Beltrami-Schaefer Riemannian space is immersed. The gauge functions which usually escape the observation play the principal roles in the theory of dual yielding as a more extended disturbance in the generalized theory corresponding to the general yielding criterion formulated in earlier papers by KONDO.

The condition of the dual yielding is assumed by MINAGAWA [7] to have the same form as the condition of the yielding which has been given by KONDO. On this basis, the fundamental equations of dual yielding have been formulated as

$$\partial_l \, \partial_j (c^{ijkl} \, \partial_k \, \partial_i \, v) - \partial_i (J^{ij} \, \partial_j v) = 0 \quad \text{(in the interior)},$$

$$c^{ijkn} \, \partial_k \, \partial_i v = 0,$$

$$\partial_k (c^{injk} \, \partial_j \, \partial_i v) - J^{ni} \, \partial_i v = 0 \quad \text{(on the boundary)},$$

where c^{ijkl}'s are the matter constants, J^{ij} the incompatibility tensor and v is the multidimensional deflection of the material point. The analytical and physical meaningfulness of this formulation was proved by KONDO [8]. It indicates the possibility of existence of a critical phenomenon such that:

when the growing incompatibility reaches a value which is obtained as an eigenvalue of the field equation, the multidimensional deflection perpendicular to the physical 3-space needs to appear.

This dual of yielding was assumed by one of us to be comparable with the process of the fracture of materials by fatigue. The comparison is completed by identifying the incompatibility with an *atmosphere* which is created through the process of fatigue, and the multidimensional deflection to an unobserved object to be set in correspondence with the *genesis* of fracture, so that,

when the growing atmosphere reaches a certain amount, the genesis of fracture begins to develop into visible size.

A heuristic analysis of the characteristic value for the incompatibility was carried out by MINAGAWA [7] assuming an isotropy of the plastic properties of the material. An $S-N$ curve for fatigue was obtained.

2. Moving Dislocations, Finslerian Approach and Membrane Stresses

i) *Moving dislocations.* We can construct a four-dimensional non-Riemannian material-time manifold, in such a manner that the moving dislocation field is represented by its torsion tensor (AMARI [9]). By the motion of dislocations, the intrinsic change of natural metric and the incompatibility flow are entailed. The former is represented by the torsion and the latter by the curvature structure of the manifold.

ii) *A Finslerian approach to the plastic feature of ferromagnetic substances.* In the ferromagnetic substance, the plastic state depends not only on the crystallographic imperfections but also on the spin directions of the crystal points. Therefore, in order to analyze its plastic features, one needs the geometry responsible for the rotation of points (AMARI [10]). Such geometry has been proposed by P. FINSLER and called the Finsler geometry.

It should be noted that, in the general Cosserat continuum, the rotation does not affect the metric of the space but only contributes to the torsion. In this case, the conventional non-Riemannian approach is sufficient. In the Finsler model, the rotations of points affect the metric as well as the torsion. Therefore, the Finsler approach refers to a more general model than the Cosserat continuum.

iii) *Membrane stresses.* The theory of Riemannian stress-function space can be extended to cover problems of two-dimensional curved manifolds which have so far been excluded from the consideration because the Einstein tensor identically vanishes (MINAGAWA [11]). In such a case, the equilibrium equation of stresses is given as the equation of Codazzi for the surface assumed in three-dimensional Euclidean space. The expression for stresses is given in terms of geometrical objects of the assumed surface.

References

[1] KONDO, K.: J. Jap. Soc. Appl. Mech. **3**, 188 (1950); **4**, 5 and 35 (1951).
[2] KONDO, K.: RAAG Memoirs of the Unifying Study of Basic Problems in Engineering and Physical Sciences by Means of Geometry, 1, 2, 3, 4. Tokyo: Gakujutsu Bunken Fukyu-Kai 1955, 1958, 1962 (in press).
[3] SCHAEFER, H.: ZAMM **33**, 356 (1953).
[4] KONDO, K.: RAAG Memoirs 3, 1962, C-X, p. 82.
[5] MINAGAWA, S.: ibid. C-IX, p. 69.
[6] AMARI, S.: RAAG Memoirs 4 (in press), D-XVII.
[7] MINAGAWA, S.: ibid. D-XIX.
[8] KONDO, K.: ibid. D-XIX.
[9] AMARI, S.: ibid. D-XVI.
[10] AMARI, S.: ibid. D-XV.
[11] MINAGAWA, S.: ibid. C-XI.

The Plane Boundary in Anisotropic Elasticity

By

M. O. Tucker and **A. G. Crocker**

Department of Physics
University of Surrey, London

The problem of a three-dimensional state of stress in an anisotropic medium in which the stress is invariant in one direction has been fully considered when the medium is infinite and homogeneous. The problem is essentially an extension of the two-dimensional problem in anisotropic elasticity theory, the solution to which was given independently by LEKHNITSKY [1] and by GREEN [2]. The generalisation to the three-dimensional case was developed by ESHELBY [3] and ESHELBY et al. [4] for the purpose of investigating the elastic properties of dislocations in anisotropic materials. Later STROH [5] treated this problem more fully and also considered dislocation interactions and cracks. In the present note, which is based on STROH's analysis, we first report solutions to the standard boundary value problems for the elastic half-space. This is followed by solutions for a two phase material consisting of two different elastic half-spaces welded together at a plane boundary. Finally some applications of these solutions are discussed.

Following STROH [5], we let the displacements u_k be independent of x_3 and seek a solution, in the form of a function of the complex variable $z = x_1 + p\,x_2$, of the equations of elastic equilibrium. Then if $f(z)$ is an analytic function of z, $u_k = A_k\, f(z)$ provided p satisfies the sextic equation

$$|c_{i1k1} + p(c_{i1k2} + c_{i2k1}) + p^2\, c_{i2k2}| = 0 \qquad (1)$$

where c_{ijkl} are the elastic stiffness constants. Eq. (1) has roots occurring in complex conjugate pairs. The three roots with positive imaginary parts are denoted by p_α and the corresponding vector components A_k are represented by $A_{k\alpha}$, which may be considered to be a 3×3 matrix.

Acknowledgment. The support of this work by the Ministry of Technology is acknowledged.

Thus, writing $z_\alpha = x_1 + p_\alpha x_2$, we have u_k given by the real part of

$$u_k = \sum_\alpha A_{k\alpha} f_\alpha(z_\alpha). \tag{2}$$

The stresses σ_{i1} and σ_{i2} are now given by the real parts of

$$\sigma_{i1} = -\sum_\alpha L_{i\alpha} p_\alpha f'_\alpha(z_\alpha), \tag{3}$$

$$\sigma_{i2} = \sum_\alpha L_{i\alpha} f'_\alpha(z_\alpha) \tag{4}$$

where the primes denote first derivatives and

$$L_{i\alpha} = (c_{i2k1} + p_\alpha c_{i2k2}) A_{k\alpha}.$$

STROH [5] considers in detail the case of singularities, described by a function $f_{0\alpha}(z_\alpha)$, in an infinite homogeneous medium. We have obtained the corresponding solutions for singularities in the elastic half-space $x_2 > 0$. We let the displacements and stresses in this region be described by a function $f_\alpha(z_\alpha)$, given by

$$f_\alpha(z_\alpha) = f_{0\alpha}(z_\alpha) + f_{1\alpha}(z_\alpha) \tag{5}$$

and determine the forms of $f_{1\alpha}(z_\alpha)$ satisfying the two classical boundary value problems. For the first of these the tractions $\tau_i(x_1)$ on $x_2 = 0$ are known but the displacements are arbitrary, and we obtain

$$f_\alpha(z_\alpha) = f_{0\alpha}(z_\alpha) - \sum_\beta M_{\alpha i} \bar{L}_{i\beta} \bar{f}_{0\beta}(z_\alpha) + \\ + (\pi i)^{-1} M_{\alpha i} \int dz_\alpha \int_{-\infty}^{\infty} \tau_i(t)\,(t - z_\alpha)^{-1}\, dt \tag{6}$$

where $M_{\alpha i}$ is the matrix reciprocal to $L_{i\alpha}$ and bars indicate complex conjugates. The definite integral in Eq. (6) may be evaluated by means of residue theory and hence, using Eqs. (2)—(4), expressions obtained for the displacements and stresses. A case of particular physical interest arises when the applied stresses $\tau_i(x_1)$ are zero, so that the last term in Eq. (6) vanishes. We then obtain explicit expressions for the displacements and stresses due to any known singularities in the half-space $x_2 > 0$ bounded by a free surface. Similarly for the second boundary value problem, in which we have known surface displacements $r_k(x_1)$ on $x_2 = 0$, we obtain

$$f_\alpha(z_\alpha) = f_{0\alpha}(z_\alpha) - \sum_\beta N_{\alpha k} \bar{A}_{k\beta} \bar{f}_{0\beta}(z_\alpha) + \\ + (\pi i)^{-1} N_{\alpha k} \int_{-\infty}^{\infty} r_k(t)\,(t - z_\alpha)^{-1}\, dt \tag{7}$$

where $N_{\alpha k}$ is reciprocal to $A_{k\alpha}$. Again the integral in Eq. (7) may be evaluated using residue theory.

We consider now two anisotropic elastic half-spaces, labelled I and II, having in general different elastic constants, which are welded together along the surface $x_2 = 0$ such that I occupies the region $x_2 > 0$. Since the interface can transmit the displacements u_k and the tractions σ_{i2}, these quantities must be continuous across the interface. Imposing these boundary conditions and assuming that all singularities of stress are in medium I we obtain the following expressions for the functions $f_\alpha^{(I)}(z_\alpha)$ and $f_\alpha^{(II)}(z_\alpha)$, defining the stress fields in medium I and medium II respectively.

$$f_\alpha^{(I)}(z_\alpha) = f_{0\alpha}^{(I)}(z_\alpha) - \sum_\beta M_{\alpha i}^{(I)} G_{ik}^{(I,II)} \bar{A}_{kj}^{(I,II)} \bar{L}_{j\beta}^{(I)} \bar{f}_{0\beta}^{(I)}(z_\alpha), \tag{8}$$

$$f_\alpha^{(II)}(z_\alpha) = \sum_\beta M_{\alpha i}^{(II)} L_{i\beta}^{(I)} f_{0\beta}^{(I)}(z_\alpha) - \\ - \sum_\beta M_{\alpha i}^{(II)} \bar{G}_{ik}^{(I,II)} A_{kj}^{(I,II)} L_{j\beta}^{(I)} f_{0\beta}^{(I)}(z_\alpha). \tag{9}$$

Superscripts (I) and (II) on the right hand sides of Eqs. (8) and (9) indicate previously defined quantities for medium I and medium II respectively, assuming them to be infinite and homogeneous,

$$A_{kj}^{(I,II)} = i \sum_\alpha \{A_{k\alpha}^{(I)} M_{\alpha j}^{(I)} - A_{k\alpha}^{(II)} M_{\alpha j}^{(II)}\} \tag{10}$$

and $G_{ik}^{(I,II)}$ is the matrix reciprocal to $B_{ki}^{(I,II)}$ where

$$B_{ki}^{(I,II)} = i \sum_\alpha \{A_{k\alpha}^{(I)} M_{\alpha i}^{(I)} - \bar{A}_{k\alpha}^{(II)} \bar{M}_{\alpha i}^{(II)}\}. \tag{11}$$

These solutions of the equations of elastic equilibrium may now be used to investigate the interactions of line defects and plane boundaries in anisotropic materials. For example, for the case of an infinitely long straight dislocation at $(0, X_2)$ and parallel to the axis Ox_3, STROH [5] has shown that

$$f_{0\alpha}(z_\alpha) = (2\pi)^{-1} M_{\alpha j} d_j \log(z_\alpha - p_\alpha X_2), \tag{12}$$

d_j being defined by $b_i = B_{ij} d_j$, where b_i are the Burgers vector components and

$$B_{ij} = \tfrac{1}{2} i \sum_\alpha (A_{i\alpha} M_{\alpha j} - \bar{A}_{i\alpha} \bar{M}_{\alpha j}).$$

Substituting from (12) into (6), with $\tau_i(x_1)$ zero, and (8) we obtain the required forms of $f_\alpha(z_\alpha)$ for an infinitely long straight dislocation lying parallel to, and at a distance X_2 from a free surface and an intercrystalline boundary respectively. Hence, using a similar procedure to that adopted by COTTRELL [6] for dislocation interactions in an isotropic medium, the force F_2 normal to the boundary experienced by such a dislocation is found to be

$$F_2 = -b_i d_i (4\pi X_2)^{-1}$$

for the free surface, and the real part of

$$F_2 = -b_i \{G_{ik}^{(\mathrm{I},\,\mathrm{II})} \bar{A}_{kj}^{(\mathrm{I},\,\mathrm{II})}\} d_j (4\pi X_2)^{-1}$$

for the intercrystalline boundary. These forces may be interpreted as resulting from image dislocations at $(0, -X_2)$ with Burgers vectors $-b_i$ and $-\mathrm{Re}\{G_{ik}^{(\mathrm{I},\,\mathrm{II})} \bar{A}_{kj}^{(\mathrm{I},\,\mathrm{II})}\} b_j$ respectively. However, care must be excercised in using this concept as in general these image dislocations imply forces parallel to the boundary which in practice are not present.

We are applying these results to interactions between dislocations and intercrystalline boundaries, and in particular twin boundaries, in metals with the close packed hexagonal crystal structure. Explicit solutions have been obtained for certain symmetrical boundaries for which the sextic equation (1) can be solved analytically for both the dislocation and its image. For more general boundaries Eq. (1) has to be solved numerically and the analysis is at present being programmed or an electronic digital computer to facilitate this part of the work.

References

[1] Lekhnitsky, S. G.: Theory of Elasticity of an Anisotropic Body. Moscow: 1950, in Russian; English translation by P. Fern. San Francisco: Holden-Day 1963.

[2] Green, A. E.: Proc. Camb. Phil. Soc. **41**, 224 (1945).

[3] Eshelby, J. D.: Phil. Mag. **40**, 903 (1949).

[4] Eshelby, J. D., W. T. Read and W. Shockley: Acta Met. **1**, 251 (1953).

[5] Stroh, A. N.: Phil. Mag. **3**, 625 (1958).

[6] Cottrell, A. H.: Dislocations and Plastic Flow in Crystals. London: Oxford University Press 1953.

Line Sources of Internal Stresses with zero Burgers Vector

By

F. Kroupa

Institute of Physics, Czechosl. Ac. Sci., Prague

A simplified treatment of the elastic fields of elongated defects with zero Burgers vector in a classical isotropic continuum is proposed using special line singularities.

1. "Inflation" Cylinder and "Inflation" Straight Line

The main idea can be seen from the following example (Fig. 1): an infinitely long cylinder of original radius R_0 increases its radius plastically to R_1 and is then placed into a cylindrical hole of radius R_0 in an infinite medium. For $\gamma = (R_1 - R_0)/R_1 \ll 1$ the classical theory

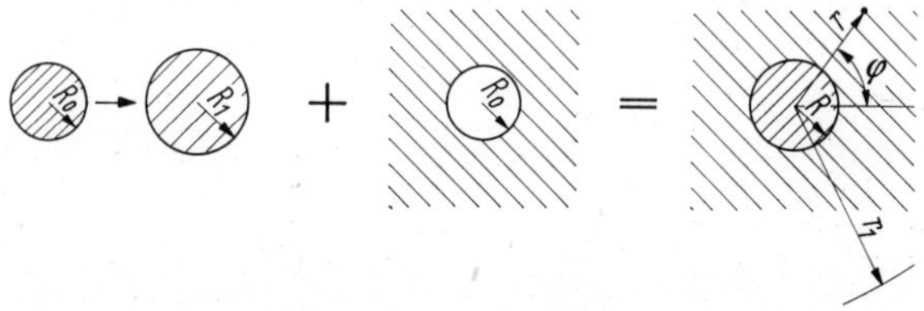

Fig. 1. Formation of "inflation" cylinder.

of elasticity can be used. The solution of this special plane strain problem is well known. When the two bodies have the same Poisson's ratio ν, and shear modulus μ, the equilibrium radius R is given by:

$$R = 2R_0\, R_1(1 - \nu)/[R_0 + (1 - 2\nu)\, R_1].$$

Inside the cylinder there is a plastic displacement $u_r^P = \gamma\, r$ and elastic displacement $u_r = -r(R_1 - R)/R_0$; outside the cylinder the displacement is only elastic and is of magnitude $u_r = R_0(R - R_0)/r$.

The total volume dilatiaton ΔV^∞ per unit length of the inflation cylinder in an infinite body is a sum of plastic and elastic dilatations inside the cylinder, $\Delta V^{\infty} = 2\pi R_0 (R - R_0) \approx \gamma \pi R_0^2/(1-\nu)$. However, real bodies are finite and the true total volume dilatation ΔV per unit length should be calculated assuming a body with a free surface, e.g. a cylinder with the outer radius $r_1 \gg R_0$; for $r_1 \to \infty$ it follows $\Delta V = 2\pi R_0 (R_1 - R_0) \approx 2\gamma \pi R_0^2$. An analogous relation was found for dilatation centres by ESHELBY [1].

The cylinder can also be treated as a straight line for the limiting case $R_0 \to 0$, $\gamma \to \infty$ so that $\Delta V = 2\gamma \pi R_0^2$ remains constant. The displacement field is then $u_r = \Delta V/[4\pi(1-\nu) r]$ and the corresponding non-zero stress components are $\sigma_{rr} = -\sigma_{\varphi\varphi} = -\mu \Delta V/[2\pi(1-\nu) r^2]$.

2. General Line Singularities of Second Order

A more general class of line sources of internal stresses with zero Burgers vector for lines of arbitrary shape will now be introduced.

It was shown [2, 3] that the total displacement du_i at x_i from a distribution of infinitesimal dislocation loops (characterized by a tensor of dislocation loop density $\gamma_{ij} = -\beta_{ij}^P$, where β_{ij}^P is the Kröner's tensor of plastic distortion [4]) in a volume element dV' at x_i' (Fig. 2) is given by

$$du_i = -\frac{1}{8\pi(1-\nu)}\,\frac{1}{\varrho^2}\left[(1-2\nu)\,\frac{2\gamma_{ik}^{(S)}\varrho_k - \varrho_i\gamma_{kk}^{(S)}}{\varrho} + \frac{3\varrho_i\varrho_k\gamma_{km}^{(S)}\varrho_m}{\varrho^3}\right] dV' \quad (1)$$

where $\gamma_{ik}^S = \frac{1}{2}(\gamma_{ik} + \gamma_{ki})$, $\varrho_i = x_i' - x_i$, $\varrho = \sqrt{\varrho_i \varrho_i}$; the Einstein's summation convention is used. Eq. (1) gives the elastic field of a class of point defects whose stress fields decrease with distance ϱ as $1/\varrho^3$.

For a line distribution of such point defects along a line L we may write

$$dV' = \delta A\, dl \quad (2)$$

where δA is the constant cross-section of a thin elongated region (e.g. for a circular cross-section, $\delta A = \pi R_0^2$) and dl is the line element. The displacement field of the line defect is then given by the line integral

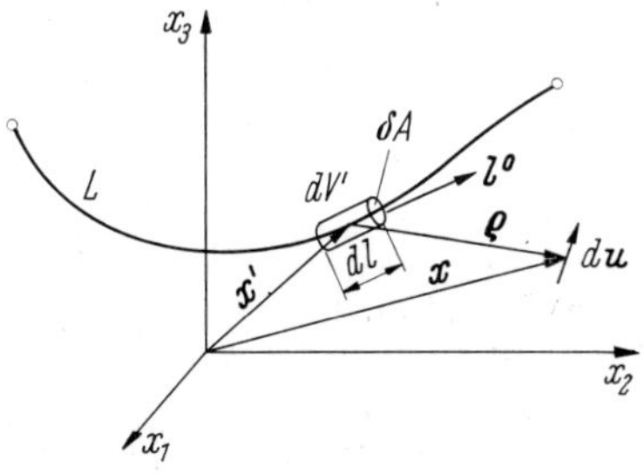

Fig. 2. Linear distribution of point defects.

$$u_i = \int_L du_i \quad (3)$$

where du_i is given by Eqs. (1) and (2); γ_{ij} is, in general, a function of position x_i' and $\gamma_{ij}\,\delta A$ may be regarded as the line density of point defects (or of dislocation loops).

Such a line L will be called a line defect of second order as the stress field in the vicinity of the line decreases with distance ϱ as

$1/\varrho^2$. This is in contrast with a dislocation whose stress field decreases with $1/\varrho$ and which we shall call a line defect of first order. In general, line defects of higher order n may also be introduced.

Three of the special types of line defect of second order will now be discussed in more detail.

a) Dilatation lines. For an isotropic volume dilatation, we have

$$\gamma_{ij}^{D} = -\gamma\, \delta_{ij} \tag{4}$$

where γ is the constant linear dilatation and δ_{ij} the Kronecker symbol; for this case, the displacement field of a dilatation centre follows from (1) and, after integration over L,

$$u_i^{D} = \frac{1+\nu}{4\pi(1-\nu)}\,\gamma\,\delta A \int\limits_L \frac{\varrho_i}{\varrho^3}\, dl\,. \tag{5}$$

Although the hydrostatic stress $\sigma^D = \frac{1}{3}\sigma_{ii}^D$ is zero, there is a total volume dilatation per unit length, $\Delta V^D = 3\gamma\,\delta A$.

b) Elongation lines. For a linear elongation γ only in the direction of the unit vector $\boldsymbol{l}^0$ (which is locally tangent to the line L) the tensor γ_{ij} has the value

$$\gamma_{ij}^{E} = -\gamma\, l_i^0\, l_j^0\,. \tag{6}$$

After integration we have the displacement for an "elongation line",

$$u_i^{E} = \frac{1}{8\pi(1-\nu)}\,\gamma\,\delta A \int\limits_L \left[(1-2\nu)\frac{2 l_i^0\, l_k^0\, \varrho_k - \varrho_i}{\varrho^3} + \frac{3\,\varrho_i\, l_k^0\, \varrho_k\, l_m^0\, \varrho_m}{\varrho^5}\right] dl \tag{7}$$

whence the hydrostatic stress is

$$\sigma^{E} = \frac{1}{3}\,\sigma_{ii}^{E} = \mu\,\frac{1+\nu}{6\pi(1-\nu)}\,\gamma\,\delta A \int\limits_L \left[\frac{1}{\varrho^3} - \frac{3\,(l_k^0\, \varrho_k)^2}{\varrho^5}\right] dl\,. \tag{8}$$

The total volume dilatation per unit length is $\Delta V^E = \gamma\,\delta A$.

c) Inflation lines. The tensor γ_{ij}, for an isotropic elongation γ perpendicular to the line, is given as $\gamma_{ij}^I = \gamma_{ij}^D - \gamma_{ij}^E$ i.e.

$$\gamma_{ij}^{\mathrm{I}} = -\gamma\,(\delta_{ij} - l_i^0\, l_j^0) \tag{9}$$

and, for an inflation line, it follows that the displacements are

$$u_i^{\mathrm{I}} = \frac{1}{8\pi(1-\nu)}\,\gamma\,\delta A \int\limits_L \left[3\,\frac{\varrho_i}{\varrho^3} - (1-2\nu)\,\frac{2 l_i^0\, l_k^0\, \varrho_k}{\varrho^3} - \frac{3\,\varrho_i\, l_k^0\, \varrho_k\, l_m^0\, \varrho_m}{\varrho^5}\right] dl \tag{10}$$

and the hydrostatic stress $\sigma^I = -\sigma^E$ where σ^E is given by Eq. (8). The total dilatation per unit length is $\Delta V^I = 2\gamma\,\delta A$. For the special case when the line L is an infinite straight line (taken as z axis), a simple integration in Eq. (10) leads to the results obtained directly in § 1.

3. Discussion

The line defects we have discussed above may be used to describe, for example, the elastic field of an elongated precipitate. All integrals can be easily evaluated for a straight segment. Taking the case of isotropic dilatation in such a long precipitate (Fig. 3) it follows from Eq. (5) that the displacements are (in cylindrical coordinates):

$$u_r = \frac{1+\nu}{4\pi(1-\nu)} \gamma \, \delta A \, \frac{1}{r} \left[\frac{z+a}{\varrho_2} - \frac{z-a}{\varrho_1} \right],$$

$$u_z = \frac{1+\nu}{4\pi(1-\nu)} \gamma \, \delta A \left[\frac{1}{\varrho_1} - \frac{1}{\varrho_2} \right],$$

where $\varrho_1 = \sqrt{r^2 + (z-a)^2}$, $\varrho_2 = \sqrt{r^2 + (z+a^2)}$.

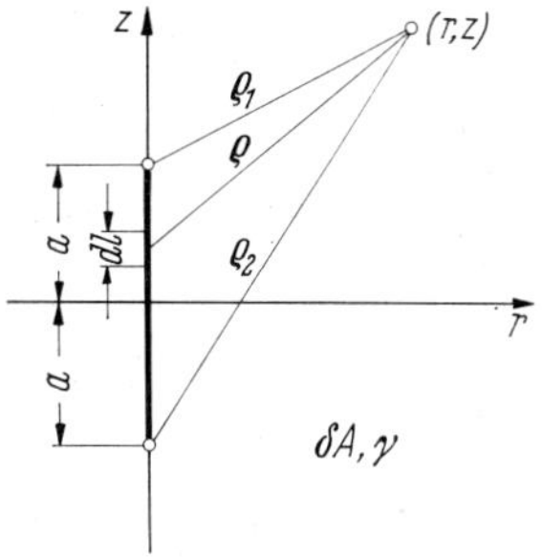

Fig. 3. Long thin precipitate as segment of dilatation line.

Whence, of course, $\sigma_{rr} + \sigma_{\varphi\varphi} + \sigma_{zz} = 0$; the total dilatation due to the whole segment is $\Delta V_0^D = 6 a \gamma \, \delta A$.

A dislocation dipole [5] or a jog line [6] may also be treated as line defects of second order.

References

[1] Eshelby, J. D.: Solid State Physics **3**, 79 (1956).
[2] Kroupa, F.: Czech. J. Phys. **B 12**, 191 (1962).
[3] Kroupa, F.: in: Theory of Crystal Defects. Prague: Academia 1966, p. 275.
[4] Kröner, E.: Kontinuumstheorie der Versetzungen und Eigenspannungen. Berlin/Göttingen/Heidelberg: Springer 1958.
[5] Kroupa, F.: Czech. J. Phys. **B 17**, 220 (1967).
[6] Seeger, A., and H. Bross: J. Phys. Chem. Solids **16**, 253 (1960).

On the Screw Dislocation in Finite Elasticity

By

Z. Wesołowski[1] and **A. Seeger**

Institut für Physik am Max-Planck-Institut für Metallforschung, Stuttgart, and Institut für theoretische und angewandte Physik der Universität Stuttgart

There exists only one deformation that is possible in all homogeneous compressible elastic materials, namely the homogeneous deformation [1]. In the case of incompressible isotropic bodies exact solutions for five families of deformations have so far been found. One of these deformations is the deformation corresponding to a screw dislocation. The existence of such a solution suggested the possibility of finding a solution for a moderately compressible elastic material with the aid of a perturbation method.

In the present paper we present the equation for determining the additional displacement for arbitrary isotropic materials with moderate compressibility. Further we specialize the equation to the case of the material proposed by MURNAGHAN. We propose to use the theory for a material which for small deformations has the response of MURNAGHAN's material, but for large deformations is better suited for applications to crystals.

1. General Theory

We divide the total deformation into two parts: The finite initial deformation and a small additional deformation. We denote the finitely deformed body by B and introduce in B the cylindrical coordinate system $\theta^i = (r, \vartheta, z)$. Let us assume that the initial deformation is that of a screw dislocation of strength b in an incompressible material, i.e. (cf. [1])

$$\mathring{r} = r, \quad \mathring{\vartheta} = \vartheta, \quad \mathring{z} = z + \frac{b}{2\pi}\vartheta, \tag{1.1}$$

[1] On leave of absence from Institute of Basic Technical Problems of the Polish Academy of Sciences, Warszawa, Poland.

where $\overset{\circ}{r}$, $\overset{\circ}{\vartheta}$ and $\overset{\circ}{z}$ are initial coordinates of the point θ^i. The strain tensor γ_{ij} corresponding to (1.1) is (for the necessary formulae cf. [2]) with

$$\gamma_{ij} = \tfrac{1}{2} \begin{vmatrix} 0 & 0 & 0 \\ 0 & -c^2 & -c \\ 0 & -c & 0 \end{vmatrix} \tag{1.2}$$

$c = b/2\pi$. The corresponding strain invariants are

$$I_1 = I_2 = 3 + c^2/r^2, \qquad I_3 = 1. \tag{1.3}$$

The deformation (1.1) produces in the body B the stress τ^{ij} determined by the relations

$$\left.\begin{aligned} \tau^{11} &= \Phi + (2 + c^2/r^2)\,\Psi + p, \\ r^2\,\tau^{22} &= \Phi + 2\Psi + p, \\ \tau^{33} &= (1 + c^2/r^2)\,\Phi + (2 + c^2/r^2)\,\Psi + p, \\ r\,\tau^{23} &= -c/r\,(\Phi + \Psi), \\ \tau^{31} &= \tau^{12} = 0. \end{aligned}\right\} \tag{1.4}$$

Φ, Ψ and p denote first derivatives of the strain energy W with respect to the invariants I_1, I_2, I_3:

$$\Phi = 2\,\frac{\partial W}{\partial I_1}, \qquad \Psi = 2\,\frac{\partial W}{\partial I_2}, \qquad p = 2\,\frac{\partial W}{\partial I_3}. \tag{1.5}$$

Although the deformation is isochoric ($I_3 = 1$), the material is compressible and the stress tensor τ^{ij} is uniquely determined. Therefore the stresses given by (1.5) do in general not satisfy the equilibrium equations. If the compressibility is sufficiently small, the correct displacement may be expected to differ from that described by (1.1) by a small radial displacement u. Such an additional displacement produces increments τ'^{ij} in the stress tensor τ^{ij}. On the basis of the equations given by A. E. Green, R. S. Rivlin and R. T. Shield [3], these increments are calculated to be ($u_r \equiv du/dr$)

$$\left.\begin{aligned} \tau'^{11} &= \Phi' + (2 + c^2/r^2)\,\Psi' + p' + 2\Psi\,u/r - 2p\,u_r, \\ r^2\,\tau'^{22} &= \Phi' + 2\Psi' + p' + 2\Psi\,u_r - 2p\,u/r, \\ \tau'^{33} &= (1 + c^2/r^2)\,\Phi' + (2 + c^2/r^2)\,\Psi' + p' + \\ &\quad + 2\Psi(u_r + u/r + (c^2/r^2)\,u_r), \\ r\,\tau'^{23} &= -(c/r)\,(\Phi' + \Psi') - 2(c/r)\,\Psi\,u_r, \\ \tau'^{31} &= \tau'^{12} = 0, \end{aligned}\right\} \tag{1.6}$$

where

$$\left.\begin{aligned}\Phi' &= (2A + 2E - \Phi)(u_r + u/r) + 2F[(2 + c^2/r^2)(u_r + u/r) - (c^2/r^2)(u/r)],\\ \Psi' &= (2F + 2D - \Psi)(u_r + u/r) + 2B[(2 + c^2/r^2)(u_r + u/r) - (c^2/r^2)(u/r)],\\ p' &= (2E + 2C + p)(u_r + u/r) + 2D[(2 + c^2/r^2)(u_r + u/r) - (c^2/r^2)(u/r)],\end{aligned}\right\} \quad (1.7)$$

$$\begin{aligned} A &= 2\frac{\partial^2 W}{\partial I_1^2}, & B &= 2\frac{\partial^2 W}{\partial I_2^2}, & C &= 2\frac{\partial^2 W}{\partial I_3^2},\\ D &= 2\frac{\partial^2 W}{\partial^2 I_2\, \partial I_3}, & E &= 2\frac{\partial^2 W}{\partial I_3\, \partial I_1}, & F &= 2\frac{\partial^2 W}{\partial I_1\, \partial I_2}. \end{aligned} \quad (1.8)$$

The last functions are to be calculated for I_1, I_2 given by (1.3).

Substituting the stresses τ^{ij} (1.4) and the stress increments τ'^{ij} (1.6) into the equilibrium equations gives for zero body forces

$$\begin{aligned} &\frac{\partial}{\partial r}\tau^{11} + \frac{\partial}{\partial r}\tau'^{11} - 2\frac{c^4}{r^5}B\frac{u}{r} + \frac{c^2}{r^3}\Psi' + \\ &+ 2(c^2/r^3)[F + (2 + c^2/r^2)B + D](u_r + u/r) + 2\Phi(u_{rr} + u_r/r - u/r^2) + \\ &+ 2\Psi[(2 + c^2/r^2)u_{rr} + u_r/r - (1 + c^2/r^2)u/r^2] + 2p\, u_{rr} = 0. \end{aligned} \quad (1.9)$$

This equation is valid for arbitrary forms of the function W. Its solution is the additional displacement u, provided u is small when compared with c. In the next paragraph we discuss special forms of Eq. (1.9) for particular forms of the function W.

2. Special Materials

Assume that the elastic potential W is that proposed by MURNAGHAN [4]

$$\begin{aligned} W_M = &\frac{l + 2m}{24}(I_1 - 3)^3 + \frac{\lambda + 2\mu + 4m}{8}(I_1 - 3)^2 + \frac{8\mu + n}{8}(I_1 - 3) - \\ &- \frac{m}{4}(I_1 - 3)(I_2 - 3) - \frac{4\mu + n}{8}(I_2 - 3) + \frac{n}{8}(I_3 - 1), \end{aligned} \quad (2.1)$$

where λ, μ, l, m, n are elastic constants. The expression (2.1) allows us to calculate the stresses τ^{ij} (1.4) and stress increments τ'^{ij} (1.6) as well as the functions $\Phi, \Psi, A, \ldots, F$.

Substituting these into the equilibrium equation (1.9) gives

$$\begin{aligned} &(\lambda + 2\mu)\left(u_{rr} + \frac{1}{r}u_r - \frac{u}{r^2}\right) + \\ &+ \frac{c^2}{r^2}\left(\frac{2\lambda + 2m + 4l - n}{4}u_{rr} + \frac{6\lambda + 6m - 4l + 8\mu - 3n}{4}\frac{u_r}{r} + \right.\\ &\left. + \frac{2\lambda - 12l - 6m - n}{4}\frac{u}{r^2}\right) + \frac{c^4}{r^4}\left(\frac{l}{4}u_{rr} + \frac{4m + 5l}{4}\frac{u_r}{r} + \frac{3}{4}l\frac{u}{r^2}\right) \\ &= \frac{c^2}{r^3}\left(\lambda + \mu + m - \frac{n}{4}\right) + \frac{c^4}{r^5}\left(l + \frac{m}{2}\right). \end{aligned} \quad (2.2)$$

The general solution of the above equation cannot be expressed in terms of known functions. The equation resulting from (2.2) by neglecting c^2/r^2 when compared with unity was obtained and solved by A. SEEGER and E. MANN [5]. They found the solution

$$u = -\frac{\lambda + \mu + m - n/4}{2(\lambda + 2\mu)} \frac{c^2}{r} \ln \frac{r}{c} + C_1 r + C_2/r, \qquad (2.3)$$

where C_1 and C_2 are integration constants.

We have emphasized that the treatment in Sect. 1 holds for arbitrary functions W. One may choose a functional form of W which coincides for small deformations with (2.1) and which takes into account some features of crystals. Due to the periodicity of the arrangement of atoms in crystals, W is not a monotonously rising function for increasing shear but returns to zero for homogeneous deformations that restore the atomic arrangement of the undeformed crystal. An appropriate choice of W should enable us to give a continuum description of this feature of crystals and to permit an approximate treatment of screw dislocations in such a material.

References

[1] TRUESDELL, C., and W. NOLL: The Non-linear Field Theories of Mechanics. Encyclopedia of Physics, Vol. III/3, Ed. S. FLÜGGE. Berlin/Heidelberg/New York: Springer 1965.

[2] GREEN, A., E., and W. ZERNA: Theoretical Elasticity. Oxford 1954.

[3] GREEN, A. E., R. S. RIVLIN and R. T. SHIELD: General theory of small elastic deformations superposed on finite elastic deformations. Proc. Roy. Soc. **A 211**, 128 (1952).

[4] MURNAGHAN, F. D.: Finite Deformations of an Elastic Solid. New York 1951.

[5] SEEGER, A., and E. MANN: Anwendung der nichtlinearen Elastizitätstheorie auf Fehlstellen in Kristallen. Z. Naturf. **14a**, 154 (1959).

Some Considerations of the Relation between Solid State Physics and Generalized Continuum Mechanics

By

J. A. Krumhansl

Laboratory of Atomic and Solid State Physics
Cornell University, Ithaca

1. Introduction

In general, the concern of this paper is the relation between what physicists know about condensed matter and what continuum theories say about its behavior. Both physicists and those studying continuum mechanics have some familiarity with the limiting macroscopic behavior described by conventional elasticity. By contrast, the mathematical developments of generalized continuum [1—3] theory are hardly known to the average physicist, while the recent extensive advance in measurement of microscopic and macroscopic physical properties is frequently not familiar to the theoretical mechanician.

It is interesting to look at history briefly. The Cosserat generalization [1] of continuum elasticity was suggested in 1909. What was the state of physics at that time? Sir J. J. Thomson [4] and Rutherford [5] had not yet settled the question whether the atom was homogeneously corpuscular or nuclear; the real beginning of the quantum theory of matter was not to come until the Bohr concept of the atom in 1913; and the fundamental description of the elastic motion of solids was to come still later with the Born-von Karman theory.

We all realize the considerable advance since that time. But it must be recognized that while the mathematical methodology and formalism have been substantially generalized on the one hand, on the other the physical facts now known delineate much more specifically the admissable constitutive behavior of real systems. In this sense, we do not

---

Acknowledgement. I would like to express my appreciation to Professor Kröner, to the organizing committee of this IUTAM conference, to Professors Mindlin and Toupin for stimulating my interest in these matters, and to the German government whose generous grant made my attendance at this conference possible.

have the freedom to postulate the variety of material behavior which the Cosserats could. In the final analysis we must see at least in principle how any postulated continuum variables relate to real physical observables; otherwise the arguments, though "global", are reminiscent of the issue of the round versus the flat earth—significant experimental evidence plays the vital role.

It seemed to me that my principal purpose should be to report on certain recent experimental methods and descriptive techniques to describe the vibrational excitations of solids; this will constitute Sect. 2. It will be seen that there is now very detailed information about homogeneous simple solids so in Sect. 3 there is a review of the principal theoretical features of lattice vibrations. Next, the question of how these microscopic facts are related to continuum approximations is examined; a formally exact method developed by the writer for a special purpose (1958) was generalized to answer this question [7]; it has been used independently by Kunin [8], and is related to Mindlin's recent work [9]. In Sect. 4 some special systems will be discussed as examples of particular physical conditions which might be described approximately by a Cosserat field, once their determining physical features are known. Finally in Sect. 5 I would like to make some general remarks on the relationship of continuum mechanics to real physical systems.

The discussion will have no direct relevance to plastic behavior, which is much more difficult to treat dynamically.

2. Experimental and Mathematical Characterization

If we knew the *positions* of each of the atoms in a material as a function of time we would certainly know what primarily determines the mechanics of continua, i.e. we are interested in the field of "matter" motion. Of course each of the atomic units may carry other attributes i.e. "physical observables" such as electron and nuclear spin angular momentum, electronic dipole moment; these are of considerable interest in magnetism and other subjects but will not be discussed further here.

In fact, the techniques of modern experimental physics can supply essentially all of the information needed on a microscopic and macroscopic scale, so we begin with the experimental side of the story.

A philosophy has developed—that of the "probe"—in which a test excitation is applied to a system in the form of a neutron, laser photon, x-ray photon or bombarding electron and the scattering of that probe by the system gives direct information about those degrees of freedom of the system which couple with the probe. This technique is not very different than that familiar in mechanical testing of materials;

the applied loading clamps can be regarded as the "probe" and as the sample deforms their movement (or that of indices on the sample) may be regarded as the response.

To generalize the information which can be obtained in this way one may develop a collection of probes each of which can apply excitations to the system on a different wavelength scale or different time scale. Thus, it is common to characterize such a "probe" by the quantities $K = (2\pi/\lambda)$ and $\omega = 2\pi f$. A static testing machine has $K \to 0$, $\omega \to 0$; a neutron scattering covers the range $K \cong 10^{-7}\ \text{cm}^{-1}$, $\omega \cong 10^{12}$ c.p.s. Using other methods the spectral range in between may be explored.

Of all of these methods neutron scattering experiments have yielded the most information about the atom dynamics of materials, since the wavelength range and frequency (energy) of thermal (reactor) neutrons just match the scale and frequency of displacement motions in condensed material; in particular the neutron scattering measures the matter density fluctuations directly. Details may be found in many recent reviews [10—12].

We state here what is measured, using the language developed by PLACZEK and VAN HOVE—which has been extremely fruitful because it also makes explicit what "theoretical" quantity should be calculated. This formulation may be applied to matter in any form, crystalline, liquid, amorphous, and so on. Both theory and experiment can be reduced to a determination of:

1) The time dependent pair density correlation function

$$G(r, r', t) = \sum_{i \neq j}\sum \delta(\boldsymbol{r} - \boldsymbol{r}_i(t))\, \delta(\boldsymbol{r}' - \boldsymbol{r}_j(0)). \tag{1}$$

2) The time dependent "self" correlation function

$$G_S(\boldsymbol{r}, \boldsymbol{r}', t) = \sum_i \delta(\boldsymbol{r} - \boldsymbol{r}_i(t))\, \delta(\boldsymbol{r}' - \boldsymbol{r}_i(0)) \tag{2}$$

where the δ are the usual Dirac functions; $\boldsymbol{r}_i(t)$ is the *dynamic* position coordinate of the ith particle at time t. (These correlations are those for a system for which the $t = 0$ configuration is a representative one.)

In the language of continuum mechanics the $\boldsymbol{r}, \boldsymbol{r}'$ are *spatial* coordinates which specify the motion of the medium. Obviously the first of these correlation functions measures *relative* position of different particles in time and therefore *deformation* in the material, while the second measures diffusion of a given particle with time. The notation here has been simplified to refer to monatomic materials; the generalization to multicomponent systems presents no problem of principle. For a discussion of the magnitude limitations on the measurements the references should be consulted.

The coherent scattering experiment gives directly the Fourier transform $S(k, \omega)$ of the correlation function

$$S(k, \omega) = \frac{1}{(2\pi)^4} \iiiint\limits_{-\infty}^{\infty} dt\, d(\boldsymbol{r} - \boldsymbol{r}')\, e^{-i[\omega t - \boldsymbol{k}\cdot(\boldsymbol{r}-\boldsymbol{r}')]}\, G(\boldsymbol{r}, \boldsymbol{r}', t). \tag{3}$$

The angle through which the neutrons are scattered determines K and their energy loss determines ω. The self correlation $S_s(\boldsymbol{K}, \omega)$ is similarly defined and is given experimentally by the incoherent scattering.

If the material motion is well described by collective modes of wave number $\boldsymbol{q}$ and frequency $\omega_{\boldsymbol{q}}$ then $S(K, \omega)$ is sharply peaked at

$$\begin{aligned} &\boldsymbol{K} = \boldsymbol{q} + \text{(Bragg condition in periodic systems)},\\ &\omega = \omega_q \end{aligned} \tag{4}$$

and the *dispersion relation* is determined directly. The lattice vibrations of harmonic crystals are such modes of excitation and have now

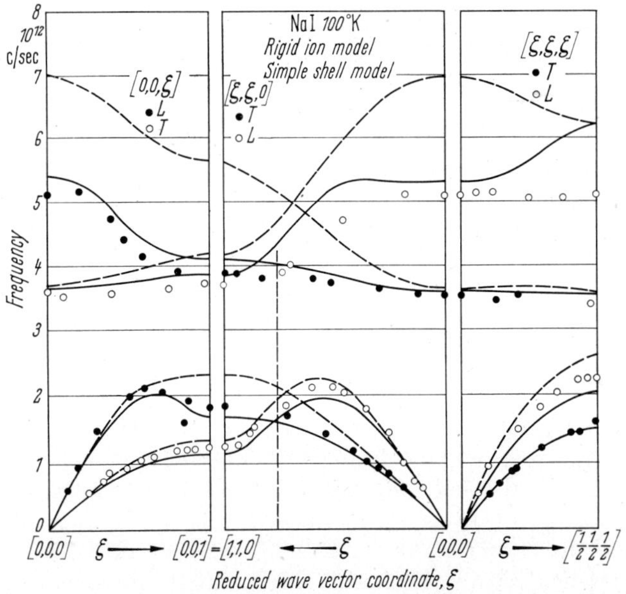

Fig. 1. Dispersion curves for NaI at 100 °K. (Woods, A. D. B., W. Cochran, and B. N. Brockhouse: Phys. Rev. **119**, 980 (1960).)

been studied extensively for a variety of systems [10—12]. As representative examples we show dispersion curves for NaI, Fig. 1, and Si, Fig. 2, obtained this way.

Microscopic Born-von Karman theoretical models agree well with measurements, if interatomic potentials are properly chosen. Much clarification of the nature of atomic interactions in simple homogeneous crystals has thus resulted. Taken together with ultrasonic and light scattering measurements it is fair to say that the dynamic elastic behavior at both the macroscopic and atomic scale are now fully characterized in these systems. So we may proceed with some confidence in lattice theory to see how continuum theory results as a limiting description. (We refer, of course, only to the elastic regime.)

Before so doing I believe a very important point must be made from these dispersion curves. At best, ultrasonic measurements of

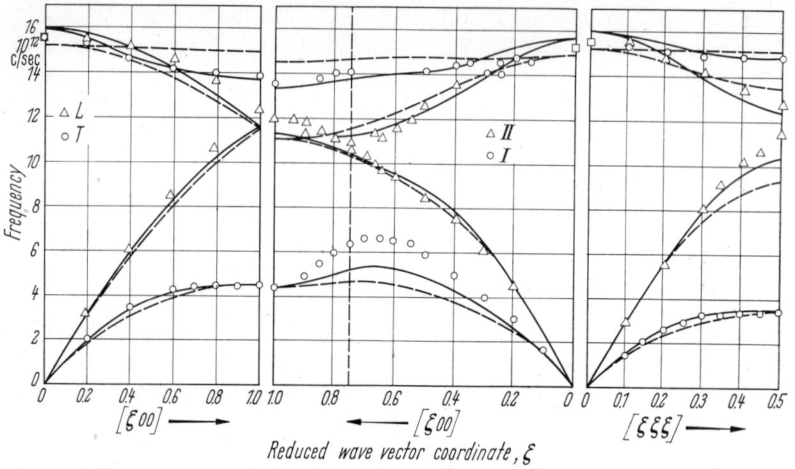

Fig. 2. Experimental phonon dispersion relation for the directions [ζ00], [0ζζ], and [ζζζ] in a single crystal of silicon at 296 °K (open circles — transverse (*T*) modes, open triangles — longitudinal (*L*) modes, solid points — undetermined polarization). Dashed and solid curves represent calculated dispersion curves for the shell model with nearest neighbors and for the next to nearest neighbors forces respectively. (Dolling, G.: Lattice Vibrations in Crystals with the Diamond Structure, Chalk River Rep., AECL-1573 (1962).)

waves cover only a few percent of the lowest q, ω_q spectrum. It is immediately apparent from a lattice model that there is no way of extrapolating this limited data to obtain the high frequency behavior. Mindlin [9] also discusses the point from a detailed model. No significant information about atomic scale dynamics can be found from low order generalized continuum theories. (Except for the rather artificial case where simply parameterized *model* potentials are assumed and the parameters are matched to continuum limits.)

3. Lattice Dynamics and Continuum Limits

In the first part of this section we state the elements of Born-von Karman theory; in the second part we use mathematical methods previously developed [7] to place lattice theory into an exact correspondence with a continuum representation.

In a crystal [11, 12] the atoms are enumerated by referring them to a lattice whose sites are labeled by a "space-like" coordinate $\boldsymbol{x}\binom{l}{k}$, denoting the l-th unit cell and the k-th species of atom for crystals containing more than one atom per unit cell. The displacements $\hat{\boldsymbol{u}}\binom{l}{k}$ denote the oscillatory displacements of the respective atoms from their reference sites; the karat notation is used to explicitly indicate that $\hat{\boldsymbol{u}}$ are dynamic physical variables, whereas $\boldsymbol{x}$ is only a counting variable. Physical laws (i.e. quantum or classical mechanics) apply to $\hat{\boldsymbol{u}}$, but the $\boldsymbol{x}$ are only *topological* in nature.

The "ground state" of the interacting many particle system is such that the given lattice configuration minimizes the free energy. To study the motion about this ground state we take the classical approach, remarking in passing that for the harmonic oscillator the Heisenberg position and momentum operators satisfy the classical equations of motion exactly. Then for small displacements about the equilibrium positions we may expand the potential energy function Φ. (For the rest of the discussion I restrict to a monatomic system with one atom per unit cell; the extension to more complicated systems is straightforward but algebraically complicated.) No special assumption is made on Φ.

$$\begin{aligned}\Phi &= \Phi_0 + 0 + \sum_{\alpha,\beta}\sum_{l,l'}\frac{1}{2}\left(\frac{\partial^2\Phi}{\partial u_{\alpha,l}\,\partial u_{\beta,l'}}\right)\hat{u}_{\alpha,l}\,\hat{u}_{\beta,l'} + \cdots \\ &= \Phi_0 + 0 + \sum_{\alpha,\beta}\sum_{l,l'}\tfrac{1}{2}\,V_{\alpha,l;\,\beta,l'}\,\hat{u}_{\alpha,l}\,\hat{u}_{\beta,l'}\end{aligned} \tag{5}$$

using Greek subscripts for Cartesian coordinates. The equations of motion follow directly

$$m_l\,\ddot{\hat{u}}_{\alpha,l} = -\sum_{\beta,l'} V_{\alpha,l;\,\beta,l'}\,\hat{u}_{\beta,l'}. \tag{6}$$

It follows from translational symmetry that solutions are of the form (assuming periodic boundary conditions for N unit cells)

$$(m_l)^{\frac{1}{2}}\,\hat{u}_{\alpha,l} = \frac{\hat{e}_{\alpha,q}}{N^{\frac{1}{2}}}\exp[i(\boldsymbol{q}\cdot\boldsymbol{x}(l) - \omega_{\boldsymbol{q}}\,t)]. \tag{7}$$

This amounts to a unitary transformation between the $\hat{u}_{\alpha,l}$ and the $\hat{e}_{\alpha,q}$ so the equation of motion (6) is transformed to

$$\omega_{\boldsymbol{q}}^2\,\hat{e}_{\alpha,\boldsymbol{q}} = \sum_{\beta}\Omega_{\alpha,\beta}(\boldsymbol{q})\,\hat{e}_{\beta,\boldsymbol{q}} \tag{8}$$

an eigenvalue problem for the ω_q^2, where

$$\Omega_{\alpha,\beta}(q) = \sum_{l,l'}\frac{e^{-i\boldsymbol{q}\cdot\boldsymbol{x}l}}{(N)^{\frac{1}{2}}}\,\frac{V_{\alpha,l;\,\beta,l'}}{(m_l\,m_{l'})^{\frac{1}{2}}}\,\frac{e^{i\boldsymbol{q}\cdot\boldsymbol{x}l'}}{(N)^{\frac{1}{2}}}. \tag{9}$$

Mathematically this is straightforward and familiar; but a few comments on the physical implications may be made.

The "frequency matrix" $\Omega(q)$ defines the dispersion relation, but it is also directly the Fourier transform (on $\boldsymbol{q}$) of the potential or the mass normalized[1] potential $[V_{\alpha l;\beta,l'}/(m_l m_{l'})^{\frac{1}{2}}]$. Because of translational periodicity in the monatomic crystal

$$\Omega_{\alpha,\beta}(q) = \sum_{(l'-0)} \frac{V_{\alpha,0;\beta,l'}}{m} e^{i\boldsymbol{q}\cdot(\boldsymbol{x}_{l'}-\boldsymbol{x}_0)}. \tag{10}$$

This is *exactly* a Fourier series in $\boldsymbol{q}$ space, periodic in the so-called "Brillouin zone" with harmonic components $\exp[i\,\boldsymbol{q}\cdot(x_{l'}-\boldsymbol{x}_0)]$ whose respective amplitudes depend on the potential $V_{\alpha,0;\beta,l'}$. Whence, *the longer the range of the potential the more Fourier components must be included to properly represent the dispersion curve.*

What are typical real potentials? In ionic crystals they are coulomb and polarization which are long-range and must be treated by special (EWALD, KELLERMAN [11, 12]) summing methods to obtain coulomb

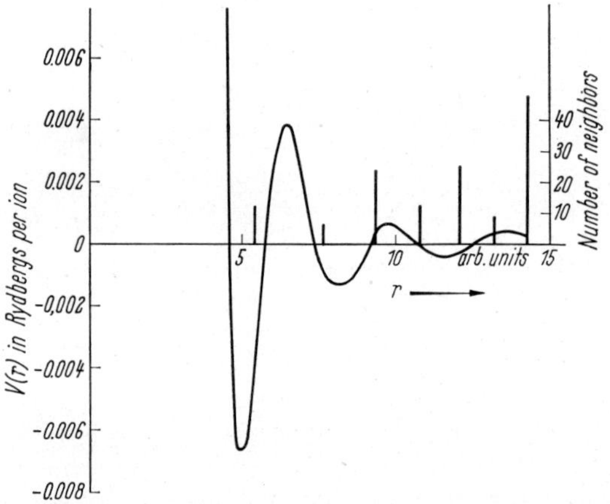

Fig. 3. The effective interaction between ions in aluminum. Also shown is the distribution of neighbors as a function of distance in the face-centered cubic structure. (HARRISON, W. A.: Phys. Rev. **136**, A 1107 (1966).)

plus short range (BORN-MAYER) terms. In metals we now know conclusively that the potential energy for ion displacement is also long-range, due to the polarization of the electron gas. Experimentally, in lead for example, BROCKHOUSE et al. [11], found that the *effective* range extended at least out to eleventh neighbors; theoretically similar results have been obtained and in Fig. 3 is shown the theoretical effective interaction potential between two ions in Al which is seen to extend out to many interionic separations.

[1] The mass normalization has been formally written in general form; actually, having simplified for discussion to a monatomic crystal all $m_l = m_{l'} = m$.

The purpose of this emphasis is to draw the following conclusion concerning real solids: since the forces are long range the dispersion curves can only be properly represented in q space by many Fourier components, and it can be stated unequivocally that low order continuum theories cannot possibly represent (by algebraic expressions in q^2, q^4) the dispersion relations of a crystal.

In fact lattice theory, which is an *exact* description of the dynamics of crystalline system, can only be properly represented by an infinite order continuum theory, which we now show. The method is discussed in greater detail in Ref. [7], so only the essentials are repeated. (The approach is equivalent to KUNIN's, developed independently.)

The vibration problem, in common with many other problems of crystals, is referred to a discrete (i.e. lattice) basis. The dynamic variables $\boldsymbol{u}_l$ are thus indexed by a *discrete* "space-like" coordinate $\boldsymbol{x}_l$. Is it possible to relate the $\boldsymbol{x}_l$ to a continuous space $\boldsymbol{x}$, and at the same time define a field $\boldsymbol{u}(\boldsymbol{x})$ which at every $\boldsymbol{x} = \boldsymbol{x}_l$ takes on the value $\boldsymbol{u} = \boldsymbol{u}_l$? The answer is yes; by using sampling representations, which have been employed in "information theory".

Define the "sampling function"

$$S(\boldsymbol{x} - \boldsymbol{x}_l) = \frac{\Omega_0}{(2\pi)^3} \int\limits_{\text{B.Z.}} d\boldsymbol{q}\, e^{i\boldsymbol{q}\cdot(\boldsymbol{x}-\boldsymbol{x}_l)} \tag{11}$$

where Ω_0 is the volume of the unit (lattice) cell and the integral is over the Brillouin zone (B.Z.) of reciprocal space; $\boldsymbol{x}$ is a continuous space line variable, and $\boldsymbol{x}_l$ is a lattice site. Then[1]

$$\hat{u}_\alpha(\boldsymbol{x}) = \sum_l \hat{u}_{\alpha,l}\, S(\boldsymbol{x} - \boldsymbol{x}_l) \tag{12}$$

together with the inverse relation

$$\hat{u}_{\alpha,l} = \frac{1}{\Omega_0} \int\limits_{\infty} d\boldsymbol{x}\, S(\boldsymbol{x} - \boldsymbol{x}_l)\, \hat{u}_\alpha(\boldsymbol{x}) \tag{13}$$

define a dynamic variable on a continuum $\hat{\boldsymbol{u}}(\boldsymbol{x})$ which is precisely equal to $\hat{\boldsymbol{u}}_l$ at $\boldsymbol{x} = \boldsymbol{x}(\boldsymbol{l})$; however, $\boldsymbol{u}(\boldsymbol{x})$ has special analytic properties determined by the fact that its Fourier representation in $\boldsymbol{q}$ space extends only over the B.Z. (In fact, it is just this restriction which allows precise determination of a continuous function from its values on only a discrete lattice of points.)

---

[1] The normalized quantity a_α appearing in Ref. [7] is redundant and is omitted here.

Several properties of this representation follow:

$$\text{Normalization:} \quad S(x_l - x_{l'}) = \delta_{l,l'}, \tag{14}$$

$$\text{Orthogonality:} \quad \frac{1}{\Omega_0}\int dx\, S(x - x_l)\, S(x - x_{l'}) = \delta_{l,l'}, \tag{15}$$

$$\text{Quasi-orthogonality:} \quad \sum_l S(\boldsymbol{x} - \boldsymbol{x}_l)\, S(\boldsymbol{x}' - x_l) = S(\boldsymbol{x} - \boldsymbol{x}'), \tag{16}$$

$$\text{Shift of Field} \quad \hat{\boldsymbol{u}}_{l'} = \frac{1}{\Omega_0}\int dx\, S(\boldsymbol{x} - x_l)\, e^{(\boldsymbol{x}_{l'} - \boldsymbol{x}_l)\cdot\nabla_x}\, \hat{u}(x). \tag{17}$$

All of these may be regarded as the transformation properties from the basis $\hat{\boldsymbol{u}}_l$ to $\hat{\boldsymbol{u}}(\boldsymbol{x})$.

Using these transformation properties it follows that the dynamical problem is exactly equivalent to that of a continuous medium having non-local kinetic and potential energy densities:

$$T = \frac{1}{2}\iint \frac{d\boldsymbol{x}\, d\boldsymbol{x}'}{(\Omega_0)^2}\, S(\boldsymbol{x} - \boldsymbol{x}')\, \hat{\boldsymbol{u}}(\boldsymbol{x}) \cdot \hat{\boldsymbol{u}}(\boldsymbol{x}'), \tag{18}$$

$$V = \frac{1}{2}\int \frac{d\boldsymbol{x}\, d\boldsymbol{x}'}{(\Omega_0)^2}\, S(x - x')\, \hat{\boldsymbol{u}}(\boldsymbol{x})\, \mathscr{V}(\nabla_{x'})\, \hat{\boldsymbol{u}}(x') \tag{19}$$

where

$$\mathscr{V}_{\alpha,\beta}(\nabla_x) = \sum_{l'} \frac{V_{\alpha,0;\beta,l'}}{m}\, e^{(\boldsymbol{x}_{l'} - 0)\cdot\Delta_{x'}} \tag{20}$$

an infinite order differential operator. And the "equation of motion" becomes the infinite order differential equation

$$\ddot{\hat{u}}_\alpha(\boldsymbol{x}) + \sum_\beta \mathscr{V}_{\alpha,\beta}(\nabla_x)\, \hat{u}_\beta(x) = 0. \tag{21}$$

This is an exact representation of the lattice motion; its solution can be carried out to all orders in some cases, as the author and others have done; the result is identical to lattice theory. Alternatively, an integral equation form can be constructed. But it is instructive to indicate the form given by the series expansion of the exponential operator, for a lattice with a center of symmetry [7]:

$$-\omega^2 \boldsymbol{u}(\boldsymbol{x}) + \left\{L^{(2)}\left(\alpha, \beta; \frac{\partial^2}{\partial x_\mu\, \partial x_\nu}\right) + \right.$$
$$\left. + L^{(4)}\left(\alpha, \beta; \frac{\partial^4}{\partial x_\mu\, \partial x_\nu\, \partial x_\pi\, \partial x_\varrho}\right) + \cdots\right\} \boldsymbol{u}(x) = 0. \tag{22}$$

Retaining only the second order terms gives just the usual elastic equations; indeed the elastic constants may be defined from the interatomic potentials by

$$L^{(2)} = \sum_{\mu,\nu}\left[\sum_{l'} \frac{1}{2} V_{\alpha 0;\beta l'}\big(x_\mu(l') - 0\big)\big(x_\nu(l') - 0\big)\right] \frac{\partial}{\partial x_\mu}\, \frac{\partial}{\partial x_\nu}.$$

The terms in bracket are equal to linear combinations of elastic constants.

The next terms from (22) are fourth order partial derivatives and *appear* to be what the strain gradient continuum theories would introduce. However, these are obviously only mathematical consequences of the replacement of difference operators by differential operators, and do not derive from any physically real potential energy demanding strain gradients. *The resemblance of the exact continuum representation of the motion of a crystal lattice to general continuum theories is entirely mathematical in nature, and is a consequence of dispersion in atomically discrete systems.*

Having thus reached the conclusion that generalized continuum theories would be artificial for single crystal systems which can be handled exactly from microscopic to macroscopic, I wish to conclude this section with three remarks. They are trite, in the sense that they are immediately apparent; but they are also intended to be profound with regard to the purposes of this conference, in view of the difficulties which many of us have had in relating generalized continuum "mechanics" to real systems.

The first remark is that the examples just considered envisage the kind of material for which there is a "faithful mapping" of the smallest atomic unit, i.e. the unit cell into the macroscopic. There is no intermediate scale at which the atomic level structure has a natural termination as in polycrystalline metal, polymeric crystals, and molecular crystals.

The second remark is that the example considered assumed an infinite block of crystal, and ignored the consequences of boundary conditions. For almost every experiment in solid state physics this is a completely acceptable procedure.

The third remark is directed to a much more subtle point, one which often leads not only to physical and mathematical inconsistencies, but which also colors our whole semantic approach to this subject. Let us very carefully realize that the dynamic variable which we have always used intuitively to talk about material mechanics is *"position-like"*, and thus is only one of many allowable dynamic variables (electronic structure, electron and nuclear spin, etc.) which could also characterize the elements of matter, and many of these *are in no sense "position-like"* in character. Although these latter may be denoted mathematically as vectors and called "Cosserat directors" they must a priori be placed in a different category physically. The mathematics must never carry one into the other except for individual cases determined on physical grounds.

But having made these remarks, we will now pass on to the next section in which examples are given where generalized continuum descriptions may in fact be useful because of particular physical circumstances.

4. Selected Examples Requiring Additional Considerations Beyond Crystal Theory

1. Consider the complex crystal $(NH_4)^+Cl^-$ rather than Na^+Cl^-. Clearly, it is possible in principle to consider N, the four H, and Cl interacting all with each other and carry out Born-von Karman theory directly.

Physical experience tells us immediately that this is not all at the best approach; rather, we first combine NH_4 as a molecule and then consider that as a unit. This is not only the valid but the relevant approach to the dynamics as soon as the wavelength exceeds the molecular diameter. In short, there is not an uninterrupted carry over from the atomic to macroscopic scale.

Note that when this is done, the NH_4 unit is characterized both by its "position-like" dynamic variable "center of mass" and by angular orientation variables, the molecular Euler angles. The latter may be represented by Cosserat directors; indeed when the l-th ammonium radical carries a dipole moment $\boldsymbol{m}_l$ we may also expect an energy term proportional to $A_{ll'}\,\boldsymbol{m}_l \cdot \boldsymbol{m}_{l'}$.

If there were no "dipole-ion" interaction i.e. no $(NH_4)^+$ *orientation* to (Cl^-) *displacement* coupling the $(NH_4)^+$ moments could be used as elements of a pure (but non-mechanical) Cosserat field, whereas with interaction the "position-like" ion displacement mechanical field becomes coupled with the Cosserat director field. At what wavelengths one makes the transition form one to another in the dispersion relation is determined entirely from physical magnitudes.

2. Next consider some aspects of covalent bonded systems. In molecular chemistry it is entirely commonplace to consider two kinds of interatomic forces: the first depends primarily on interatomic distances; the second depends on "bond-angles". The former are two center forces and the latter involve at last three centers, and arise because the electronic distribution is not structureless. There is no doubt whatsoever that this decomposition of a many body problem (i.e. molecule) is often an excellent approximation and in good agreement with experiments.

Since a bond is a *directed line* joining two atoms it has the character of a Cosserat director; moreover there is real *physical reason* to assign an energy term which depends on the angles between successive bonds and therefore the gradient of this director (i.e. bond orientations) as one moves through covalent matter.

In this case when bonds bend the atoms move so the motion of the "directors" is topologically coupled to position dynamic variables as well.

The author [7] and K. KOMATSU [13] used two center central forces as well as bond-bending forces in graphite (a highly anisotropic and covalent material) and dispersion relations similar to those of a couple stress model for layered structures were obtained. At all but the longest wavelengths $\omega \propto q^4$ was obtained. Similarly, HERMAN [14] introduced bond-bending forces in addition to pair interactions with considerable success in computing the vibration spectrum of germanium and silicon. Why are these effects not seen more commonly? The bond-bending energies are usually an order of magnitude smaller than the two center forces, so that except in open structures like graphite they do not dominate the behavior.

3. Finally we call attention to another factor which does not appear in lattice dynamic models, that of finite size and boundary condition effects. Consider first an infinite crystal of NaCl; propagate a long wave along (1, 0, 0) with transverse polarization. The frequency obeys $\omega^2 = c_{T,100}{}^2\, q^2$.

Now cut a small square cross section NaCl rod with (1, 0, 0) direction along the rod. Again propagate a transverse wave along (1, 0, 0); the frequency now obeys the Euler beam solution $\omega^2 = A\, q^4$ where A depends on cross section and moduli.

The long wavelength limit of the crystal dispersion relations does not give the actual dispersion!! From this example one sees that *apparent* "couple stress" behavior can arise when boundary effects dominate over bulk effects; but it is not correct to consider this behavior to arise from a Cosserat property in the *material*.

In this example when the wavelength of the transverse Euler waves decreases to the cross sectional dimension a transition to the bulk dispersion relation will occur.

On the other hand if a composite material is made up of rods, plates, etc. it seems quite certain that on a scale *larger* than the dominant boundary condition of these intermediate size element we can expect couple stress descriptions to apply; while on the smaller scale ordinary crystal dynamics will apply within the units. It follows that mathematical problems in the couple stress formulation, which assume mathematical point loadings, ect. will show singularities irrelevant to real composite medium.

The macroscopic resemblance to generalized continuum behavior would clearly be a consequence of the composite inhomogeneity, and would not imply any fundamentally unusual material behavior at the microscopic level.

5. Some Closing Remarks and Conclusions

To end at the beginning we must remind ourselves first that there is no such thing as a continuum. The "elements" are not "points" but real, discrete, physical objects.

It may be a coincidence and a cruel semantic joke that some of the true physical dynamic variables are of necessity space-like (i.e. particle position) and can take on a continuum of values in Euclidean three space; while at the same time the other non-dynamic counting variables which simply identify the particles have *for convenience* also been mapped onto Euclidean three space and have then loosely been allowed to cover a continuum. In fact, it is apparent that the counting variable which is the "material coordinate" in continuum mechanics is hardly a coordinate at all, and certainly not continuous as it enters the governing physical laws.

From a study of crystal dynamics and its continuum limit, we concluded that where simple "position-like" variables (i.e. atom positions) are observed as "mechanical" variables, and the crystal reproduces the atomic symmetry continuously up to the macroscopic level, then there is no physical significance of higher order continuum theories used to represent the lattice dispersion curves.

But for heterogeneous systems and those with some natural intermediate level of milli-structure (neither micro nor macro) I believe non-classical continuum theories such as described by RIVLIN [15] or ERINGEN [16] will have a great utility and relevance in real material problems—as long as inferences are not drawn at the atomic scale, nor singularities on a scale smaller than the milli-structure scale are incorporated into model problems.

Put formally, suppose for physical reasons it is apparent that the Hamiltonian can be effectively decomposed into parts each depending on a well defined set of coordinates,

$$H = H_1 + H_2 + \cdots + H_{\text{int}}$$

together with some interaction terms H_{int} which may couple these parts. Then, particularly when H_{int} is small, and some of the H_i depend on "position like" variables (e.g. center of mass coordinates) while others of the H_i refer to "internal variables" (e.g. orientation), the total behavior is well approximated by generalized continuum theories on a sufficiently macroscopic basis. The complete treatment of this problem from first (physical) principles would require use of thermodynamic many body methods.

Perhaps, in closing, a mild plea for gradual evolution of a physically accurate terminology should be made: mechanics of *macroscopic* continua, Cosserat *media* (instead of continuum), and *elementary physical unit* instead of "point" or "particle".

References

[1] COSSERAT, E., and F. COSSERAT: Théorie des Corps Déformable. Paris: Hermann 1909.

[2] TOUPIN, R. A.: Arch. Rat. Mech. Anal. **11**, 385 (1962).

[3] MINDLIN, R. D., and H. F. TIERSTEN: Arch. Rat. Mech. Anal. **11**, 413 (1962).
[4] THOMSON, J. J.: Phil. Mag. **21**, 648 (1911).
[5] LORD RUTHERFORD: Phil. Mag. **21**, 669 (1911).
[6] BOHR, N.: Phil. Mag. **26**, 1 (1913).
[7] KRUMHANSL, J. A.: in: Lattice Dynamics, Copenhagen 1963. Ed. R. F. WALLIS. Oxford: Pergamon Press 1965.
[8] KUNIN, I. A.: this conference.
[9] MINDLIN, R. D.: this conference.
[10] Symposium on Inelastic Scattering of Neutrons by Condensed Systems. Brookhaven 1965, BNL 940 (C-45).
[11] Phonons. Ed. R. W. H. STEVENSON. Edinburgh: Oliver and Boyd 1966.
[12] Phonons and Phonon Interactions. Ed. T. BAK. New York: Benjamin 1964.
[13] KOMATSU, K.: J. Chem. Phys. **6**, 380 (1956).
[14] HERMAN, F.: J. Phys. Chem. Solids **8**, 405 (1959).
[15] RIVLIN, R. S.: this conference.
[16] ERINGEN, A. C.: this conference.

Theories of Elastic Continua and Crystal Lattice Theories[1]

By

R. D. Mindlin

Department of Civil Engineering
Columbia University, New York

Introduction

This paper is concerned with the relations between discrete and continuum theories of the elastic behavior of perfect crystals. The point of view adopted is that the validity of any extension of the classical theory of elasticity, intended to accommodate effects of the atomic structure of crystalline solids, can be tested by comparison with an appropriate lattice theory. The particular test to be applied is how well and to how short wave lengths the dispersion relation for plane waves, deduced from the continuum theory, reproduces that for the lattice.

In the case of a simple Bravais lattice of mass particles, the difference equations and dispersion relation reduce, in the long wave limit, to the differential equations and dispersionless velocities of classical elasticity. Extensions of the continuum theory to accommodate short wave lengths and, consequently, effects of the atomic structure of solids, began with Cauchy in 1851 [1]. Although Cauchy did not carry his work very far, it is now known that his theory, if completed, would correspond to an augmentation of classical linear elasticity through the inclusion of all the gradients of strain, in addition to the strain, in the potential energy density. Interest in this type of "gradient theory" was stimulated, in 1960, by Aero and Kuvshinskii [2], Grioli [3], Rajagopal [4] and Truesdell and Toupin [5] who took into account the first gradient of the rotation, i.e. eight of the eighteen components of the first gradient of the strain. This is what Toupin [6] has termed the "Cosserat theory with constrained rotations". Further extension to the complete first gradient was accomplished by Toupin [7], to the second gradient by the writer [8] and to gradients of all orders by Green and Rivlin [9].

---

[1] Paper prepared for IUTAM "Symposium on the Generalized Cosserat Continuum ...", August, 1967.

In 1963, KRUMHANSL [10] showed, by expansion of the differences in the expression for the potential energy of a general Bravais lattice, that the second order terms, in the resulting series of derivatives, contain the rotation gradient and, hence, the essence of the Cosserat theory with constrained rotations. In the first part of the present paper, it is shown in what way successively higher order gradient theories correspond to successively shorter-wave approximations to the equations of a simple Bravais lattice and how the additional material constants, which appear in the gradient theories, are related to the force constants of the lattice model. It is also shown why the requirement of positive definiteness of the potential energy density should not be applied to the strain-gradient terms when the theory is viewed as an approximation to a lattice model. The example chosen for illustration is a GAZIS-HERMAN-WALLIS [11] monatomic simple cubic lattice of mass particles.

In the case of a lattice with a basis, the dispersion relation exhibits optical branches in addition to the acoustic branches. It has been conjectured [12] that the differential equations of a Cosserat continuum [13] with a deformable Cosserat *trièdre* (micro-structure) may be a long wave to moderately long wave approximation to the difference equations of a lattice with a basis—inasmuch as such a Cosserat-type theory does exhibit both transverse and longitudinal optical branches in the dispersion relation. However, it is shown, in the second part of this paper, that the long wave limit of the difference equations of a lattice with a basis corresponds to an elastic continuum of a different type. It is also indicated how to construct higher order continuum approximations, to such lattice theories, accommodating shorter wave lengths. The lattice chosen for illustration is a GAZIS-WALLIS [14] NaCl-type lattice.

In the final part of the paper, it is shown how the difference equations of motion of lattices may be converted to differential equations which yield the same dispersion relations, as do the difference equations, for all wave lengths.

To simplify the exposition and to cope with space limitation, the lattices chosen to illustrate the methods presented are simple ones which do not correspond to real crystals. However, the same methods have been applied to face centered and body centered cubic lattices and to the NaCl-type lattice with polarizable atoms for which there are dispersion data from neutron diffraction measurements. These results will be reported elsewhere.

Simple Bravais Lattice

For one particle of mass M at each point $(l\,h, m\,h, n\,h)$ of a cubic lattice having central force interactions between nearest and next nearest neighbor particles with force constants α and β, respectively, and angular interactions, with force constant γ, between three consecutive non-collinear nearest neighbors, Gazis, Herman and Wallis [11] give three difference equations of motion of the type

$$\begin{aligned} M\ddot{u}_1^{l,m,n} &= \alpha\Big(\sum_{\lambda} u_1^{l+\lambda,m,n} - 2u_1^{l,m,n}\Big) + \\ &+ \beta\sum_{\lambda\mu}(u_1^{l+\lambda,m+\mu,n} + u_1^{l+\lambda,m,n+\mu} - 8u_1^{l,m,n}) + \\ &+ (\beta+\gamma)\sum_{\lambda\mu}\lambda\,\mu\,(u_2^{l+\lambda,m+\mu,n} + u_3^{l+\lambda,m,n+\mu}) + \\ &+ 4\gamma\sum_{\lambda}(u_1^{l,m+\lambda,n} + u_1^{l,m,n+\lambda} - 4u_1^{l,m,n}), \end{aligned} \tag{1}$$

where $\lambda, \mu = \pm 1$ and the $u_i^{l,m,n}$ are the three rectangular components of displacement of the particle at $(l\,h, m\,h, n\,h)$. Displacements

$$u_j^{l,m,n} = A_j \exp[i(l\,\theta_1 + m\,\theta_2 + n\,\theta_3 - \omega\,t)], \tag{2}$$

when substituted in (1), yield the dispersion relation

$$|d_{ij}| = 0, \quad (i, j = 1, 2, 3) \tag{3}$$

where

$$\begin{aligned} d_{ij} &= M\omega^2 - 2\alpha(1-\cos\theta_i) - 4\beta\Big(2 - \cos\theta_i\sum_{k\neq i}\cos\theta_k\Big) - \\ &\qquad - 8\gamma\Big(2 - \sum_{k\neq i}\cos\theta_k\Big) \quad i = j, \\ d_{ij} &= -(\beta+\gamma)\cos\theta_i\cos\theta_j \quad (i \neq j). \end{aligned}$$

For the purpose, it is convenient to rewrite (1) in terms of difference operators defined as follows:

$$\left.\begin{aligned} \Delta_1^+ u_i^{l,m,n} &\equiv h^{-1}(u_i^{l+1,m,n} - u_i^{l,m,n}), \\ \Delta_1^- u_i^{l,m,n} &\equiv h^{-1}(u_i^{l,m,n} - u_i^{l-1,m,n}), \\ \Delta_1^2 u_i^{l,m,n} &\equiv \Delta_1^+\Delta_1^- u_i^{l,m,n} = h^{-2}\sum_{\lambda}(u_i^{l+\lambda,m,n} - 2u_i^{l,m,n}), \\ \Delta_1\Delta_2 u_i^{l,m,n} &\equiv \tfrac{1}{4}(\Delta_1^+\Delta_2^+ + \Delta_1^-\Delta_2^- + \Delta_1^+\Delta_2^- + \Delta_1^-\Delta_2^+)\,u_i^{l,m,n}, \end{aligned}\right\} \tag{4}$$

and analogous definitions, e.g. $\Delta_2^+ u_i^{l,m,n} \equiv h^{-1}(u_i^{l,m+1,n} - u_i^{l,m,n})$, for forward, backward, second central and cross differences in the remaining coordinate directions.

Then, noting that

$$\left.\begin{aligned} \sum_{\lambda\mu} u_i^{l+\lambda,m+\mu,n} - 4u_i^{l,m,n} &= h^2(h^2\Delta_1^2\Delta_2^2 + 2\Delta_1^2 + 2\Delta_2^2)\,u_i^{l,m,n}, \\ \sum_{\lambda\mu}\lambda\,\mu\,u_i^{l+\lambda,m+\mu,n} &= 4h^2\Delta_1\Delta_2 u_i^{l,m,n}, \end{aligned}\right. \tag{5}$$

we can write (1) in the form

$$\begin{aligned} \varrho\,\ddot{u}_1^{l,m,n} &= [c_{11}\Delta_1^2 + c_{44}(\Delta_2^2 + \Delta_3^2) + \tfrac{1}{2}h^2 c_{12}\Delta_1^2(\Delta_2^2 + \Delta_3^2)]\,u_1^{l,m,n} + \\ &+ (c_{12} + c_{44})(\Delta_1\Delta_2 u_2^{l,m,n} + \Delta_1\Delta_3 u_3^{l,m,n}), \end{aligned} \tag{6}$$

where

$$\varrho = M/h^3, \quad h\,c_{11} = \alpha + 4\beta, \quad h\,c_{12} = 2\beta, \quad h\,c_{44} = 2(\beta + 2\gamma). \quad (7)$$

Now, expand the difference operations, on the $u_i^{l,m,n}$, in Taylor series of partial derivatives, ∂_i, of continuous displacement functions $u_i(x_j)$:

$$\left.\begin{aligned} \Delta_i^2 u_j^{l,m,n} &= \left(1 + \frac{1}{12} h^2 \partial_i^2 + \cdots\right) \partial_i^2 u_j, \\ \Delta_i \Delta_j u_k^{l,m,n} &= (1 + \tfrac{1}{6} h^2 \partial_i^2 + \tfrac{1}{6} h^2 \partial_j^2 + \cdots)\, \partial_i \partial_j u_k, \\ \Delta_i^2 \Delta_j^2 u_k^{l,m,n} &= \left(1 + \frac{1}{12} h^2 \partial_i^2 + \frac{1}{12} h^2 \partial_j^2 + \cdots\right) \partial_i^2 \partial_j^2 u_k. \end{aligned}\right\} \quad (8)$$

If only second derivatives are retained, (6) reduces to

$$\varrho\, \ddot{u}_1 = [c_{11} \partial_1^2 + c_{44}(\partial_2^2 + \partial_3^2)]\, u_1 + (c_{12} + c_{44})\, (\partial_1 \partial_2 u_2 + \partial_1 \partial_3 u_3), \quad (9)$$

which is the equation of motion of classical elasticity for materials with the constants c_{11}, c_{12}, c_{44} of cubic symmetry. Retaining up to fourth derivatives we find

$$\begin{aligned} \varrho\, \ddot{u}_1 = {} & [c_{11} \partial_1^2 + c_{44}(\partial_2^2 + \partial_3^2)]\, u_1 + (c_{12} + c_{44})\, (\partial_1 \partial_2 u_2 + \partial_1 \partial_3 u_3) + \\ & + \frac{1}{12} h^2 [c_{11} \partial_1^4 + c_{44}(\partial_2^4 + \partial_3^4) + 6 c_{12} \partial_1^2 (\partial_2^2 + \partial_3^2)]\, u_1 + \\ & + \tfrac{1}{6} h^2 (c_{12} + c_{44})\, [(\partial_1^3 \partial_2 + \partial_1 \partial_2^3)\, u_2 + (\partial_1^3 \partial_3 + \partial_1 \partial_3^3)\, u_3], \end{aligned} \quad (10)$$

which has the fourth derivatives characteristic of the equations of motion of the first-strain-gradient theory. In general, the potential energy density of the first-gradient theory for the crystal class $m3m$ [15], to which the simple cubic lattice belongs, has the form

$$W = \tfrac{1}{2} c_{ijkl}\, \varepsilon_{ij}\, \varepsilon_{kl} + \tfrac{1}{2} c_{ijkpqr}\, \varepsilon_{ijk}\, \varepsilon_{pqr}, \quad (11)$$

where

$$\varepsilon_{ij} = \tfrac{1}{2}(\partial_i u_j + \partial_j u_i), \qquad \varepsilon_{ijk} = \partial_i \partial_j u_k, \quad (12)$$

$$c_{ijkl} = (c_{11} - c_{12} - 2c_{44})\, \delta_{ijkl} + c_{12}\, \delta_{ij}\, \delta_{kl} + c_{44}(\delta_{ik}\, \delta_{jl} + \delta_{il}\, \delta_{jk}), \quad (13)$$

$$\begin{aligned} c_{ijkpqr} = {} & a_1\, \delta_{jk}\, \delta_{ip}\, \delta_{qr} + a_2\, \delta_{ij}\, \delta_{kp}\, \delta_{qr} + a_3\, \delta_{kr}\, \delta_{ij}\, \delta_{pq} + \\ & + a_4\, \delta_{ip}\, \delta_{jq}\, \delta_{kr} + a_5\, \delta_{kp}\, \delta_{ir}\, \delta_{jq} + a_1'\, \delta_{jk}\, \delta_{ipqr} + \\ & + a_2'\, \delta_{ij}\, \delta_{kpqr} + a_3'\, \delta_{kr}\, \delta_{ijpq} + a_4'\, \delta_{ip}\, \delta_{jqkr} + \\ & + a_5'\, \delta_{kp}\, \delta_{irjq} + a''\, \delta_{ijkpqr}, \end{aligned} \quad (14)$$

in which the $\delta_{i\ldots}$ are unity if all indices are alike and zero otherwise. The $a_1, a_2 \ldots, a_1', a_2' \ldots$ and a'' are eleven additional material constants. If

$$\left.\begin{aligned} a'' &= -\frac{1}{12} h^2 (c_{11} - 10 c_{12} - 5 c_{44}) = -\frac{1}{12} h (\alpha - 26\beta - 20\gamma), \\ a_1' &= a_5' = -\tfrac{1}{6} h^2 (c_{12} + c_{44}) = -\tfrac{2}{3} h (\beta + \gamma), \\ a_3' &= -\frac{1}{12} h^2 c_{44} = -\frac{1}{6} h (\beta + 2\gamma), \quad a_4' = -\frac{1}{2} h^2 c_{12} = -h\beta, \\ a_2' &= a_1 = a_2 = a_3 = a_4 = a_5 = 0, \end{aligned}\right\} \quad (15)$$

then (10) is the Euler equation of TOUPIN's ([7], § 7) variational principle for the first-gradient theory. It should be observed that (15) are not unique as there are combinations of terms in (14) which do not contribute to the equations of motion.

Whereas the equations of motion (9), of classical elasticity, give ω^2, in the dispersion relation (7), to the order θ_i^2, the first-gradient Eq. (10) gives ω^2 to the order θ_i^4, i.e. to shorter wave lengths. However, the latter is true only if the gradient terms in the energy (11) are not required to conform to positive definiteness of W. Consider, for example, a longitudinal wave in the [100] direction. The lattice equation, (1) or (6), and the first-gradient approximation (10) give

$$\omega^2 = \frac{2c_{11}}{\varrho h^2}(1-\cos\theta_1) = \frac{2c_{11}}{\varrho h^2}\left(\frac{\theta_1^2}{2!} - \frac{\theta_1^4}{4!} + \frac{\theta_1^6}{6!} - \cdots\right), \tag{16}$$

$$\omega^2 = \frac{2c_{11}}{\varrho h^2}\left(\frac{\theta_1^2}{2!} - \frac{\theta_1^4}{4!}\right), \tag{17}$$

respectively, for the dispersion relation. Complete positive definiteness of W would require the sign of the θ_i^4 term in (17) to be positive so as to prevent instability, i.e. imaginary ω. However, with the negative sign, instability occurs only at wave numbers far beyond the range in which (17) approximates (16). Thus, if the strain-gradient equations are regarded as approximations to the lattice equations, the requirement of positive definiteness need not be applied to the gradient terms in the potential energy as long as solutions are restricted to the range of wave lengths in which the approximation is valid. The requirement of positive definiteness must, of course, be applied to the strain terms in the energy—with the usual result

$$c_{11} > 0, \quad c_{11} - c_{12} > 0, \quad c_{11} + 2c_{12} > 0, \quad c_{44} > 0.$$

The same method as that employed for the first-gradient theory can be applied to obtain shorter-wave approximations by retaining sixth, eighth, etc. derivatives in the expansions of the difference operators, resulting in second-, third-, etc. gradient approximations.

Lattice with a Basis

We now consider a lattice of NaCl-type with particles at points $(l\,h, m\,h, n\,h)$—of mass $\overset{1}{M}$ for $l+m+n$ even and mass $\overset{2}{M}$ for $l+m+n$ odd. The interactions are taken to be the same as for the simple cubic lattice considered in the preceding section. Next nearest neighbor force constants, $\overset{1}{\beta}$ and $\overset{2}{\beta}$ are assumed to be different for particles with mass $\overset{1}{M}$ and $\overset{2}{M}$. The angular force constant, γ, is the same whether a particle of mass $\overset{1}{M}$ or $\overset{2}{M}$ is at the apex. Then GAZIS and WALLIS [14] find, for the particles at $l+m+n$ even, three equations of motion of the type (1) with M and β replaced by $\overset{1}{M}$ and $\overset{2}{\beta}$; and, for particles at $l+m+n$ odd, three equations of motion of type (1) with M and β replaced by $\overset{2}{M}$ and $\overset{2}{\beta}$. When the displacements

$$\overset{\nu}{u}{}_j^{l,m,n} = \overset{\nu}{A}_j \exp[i(l\,\theta_1 + m\,\theta_2 + n\,\theta_3 - \omega\, t)],$$

with $\nu = 1, 2$ for $l+m+n$ even and odd, respectively, are substituted in the six difference equations of motion, there results the dispersion relation

$$\begin{vmatrix} \overset{1}{d}_1 & \overset{1}{d}_{12} & \overset{1}{d}_{13} & d_1 & 0 & 0 \\ \overset{1}{d}_{21} & \overset{1}{d}_2 & \overset{1}{d}_{23} & 0 & d_2 & 0 \\ \overset{1}{d}_{31} & \overset{1}{d}_{32} & \overset{1}{d}_3 & 0 & 0 & d_3 \\ d_1 & 0 & 0 & \overset{2}{d}_1 & \overset{2}{d}_{12} & \overset{2}{d}_{13} \\ 0 & d_2 & 0 & \overset{2}{d}_{21} & \overset{2}{d}_2 & \overset{2}{d}_{23} \\ 0 & 0 & d_3 & \overset{2}{d}_{31} & \overset{2}{d}_{32} & \overset{2}{d}_3 \end{vmatrix} = 0, \tag{18}$$

where

$$\overset{\nu}{d}_i = \overset{\nu}{M}\omega^2 - 2(\alpha + 8\gamma) - 4\overset{\nu}{\beta}(2 - \cos\theta_i \sum_{k \neq i} \cos\theta_k),$$

$$\overset{\nu}{d}_{ij} = -4(\overset{\nu}{\beta} + \gamma)\sin\theta_i \sin\theta_j, \qquad d_i = 2\alpha\cos\theta_i + 8\gamma \sum_{k \neq i} \cos\theta_k.$$

If we designate $\overset{1}{u}_i$ and $\overset{2}{u}_i$ as the displacements of particles of mass $\overset{1}{M}$ and $\overset{2}{M}$, respectively, and write

$$\overset{\nu}{u}{}_j^{l,m,n} = \overset{\nu}{A}_j \exp[i(l\,\theta_1 + m\,\theta_2 + n\,\theta_3 - \omega\, t)] \tag{19}$$

for all l, m, n, the same dispersion relation as (18) results from three equations of the type

$$\begin{aligned} \overset{1}{M}\,\overset{1}{u}{}_{1,tt}^{l,m,n} &= 2(\alpha + 8\gamma)\left(\overset{2}{u}{}_1^{l,m,n} - \overset{1}{u}{}_1^{l,m,n}\right) + 2h^2\,\overset{1}{\beta}(2\Delta_1^2 + \Delta_2^2 + \Delta_3^2)\,\overset{1}{u}{}_1^{l,m,n} + \\ &+ 4(\overset{1}{\beta} + \gamma)\,h^2\left(\Delta_1\Delta_2\,\overset{1}{u}{}_2^{l,m,n} + \Delta_1\Delta_3\,\overset{1}{u}{}_3^{l,m,n}\right) + \\ &+ h^4\,\overset{1}{\beta}(\Delta_1^2\Delta_2^2 + \Delta_1^2\Delta_3^2)\,\overset{1}{u}{}_1^{l,m,n} + \\ &+ h^2\left[\alpha\,\Delta_1^2\,\overset{2}{u}{}_1^{l,m,n} + 4\gamma(\Delta_2^2 + \Delta_3^2)\,\overset{2}{u}{}_1^{l,m,n}\right] \end{aligned} \tag{20}$$

and three more obtained by interchange of superscripts 1 and 2. This form is more convenient for passing to continuum approximations since the necessity for distinguishing between $l+m+n$ even and odd is dispensed with.

At long wave lengths and low frequency, $\overset{1}{u}_i = \overset{2}{u}_i$ $(= u_i$, say). Then, employing the expansions (8), retaining only second derivatives and adding corresponding members of the two sets of three equations of motion, we recover the equations of classical elasticity (9) with $\varrho = (\overset{1}{M} + \overset{2}{M})/2h^3$ and stiffness constants given by (7) with $\beta = \frac{1}{2}(\overset{1}{\beta} + \overset{2}{\beta})$.

At long wave lengths but not necessarily low frequency, (20) becomes, (with $\overset{1}{\varrho} = \overset{1}{M}/2h^3$)

$$\begin{aligned} h\overset{1}{\varrho}\overset{1}{u}_{1,tt} &= h^{-2}(\alpha + 8\gamma)(\overset{2}{u}_1 - \overset{1}{u}_1) + \overset{1}{\beta}(2\partial_1^2 + \partial_2^2 + \partial_3^2)\overset{1}{u}_1 + \\ &\quad + 2(\beta + \gamma)(\partial_1\partial_2\overset{1}{u}_2 + \partial_1\partial_3\overset{1}{u}_3) + \\ &\quad + \tfrac{1}{2}\alpha\,\partial_1^2\overset{2}{u}_1 + 2\gamma(\partial_1^2 + \partial_2^2)\overset{2}{u}_1. \end{aligned} \tag{21}$$

Thus, for the long wave approximation, there are three equations of the type (21) and three more obtained by interchange of superscripts 1 and 2. The six equations yield the long wave region of the dispersion relation (18) for both the acoustic and optical branches. The equations do not have the same form as those for a Cosserat continuum with a deformable *trièdre* [12]. However, the appropriate continuum theory may be constructed without difficulty. We consider two interpenetrating continua representing the two face centered cubic sub-lattices of the NaCl structure. The potential energy density is taken as a quadratic function of the strains of the two continua and their relative displacement and rotation. For crystal class $m3m$, this is (with $\varkappa = 1, 2$; $\lambda = 1, 2$)

$$W = c_0^*(\overset{2}{\varepsilon}_{ii} - \overset{1}{\varepsilon}_{ii}) + \tfrac{1}{2}a^* u_i^* u_i^* + c^*\omega_{ij}^*\omega_{ij}^* + \tfrac{1}{2}\sum_{\varkappa\lambda} c_{ijkl}^{\varkappa\lambda}\overset{\varkappa}{\varepsilon}_{ij}\overset{\lambda}{\varepsilon}_{kl}, \tag{22}$$

where

$$u_i^* \equiv (\overset{2}{u}_i - \overset{1}{u}_i), \qquad \overset{\varkappa}{\varepsilon}_{ij} \equiv \tfrac{1}{2}(\partial_i\overset{\varkappa}{u}_j + \partial_j\overset{\varkappa}{u}_i), \qquad \omega_{ij}^* \equiv \tfrac{1}{2}(\partial_i u_j^* - \partial_j u_i^*),$$

$$\begin{aligned} c_{ijkl}^{\varkappa\lambda} = c_{ijkl}^{\lambda\varkappa} &= (c_{11}^{\varkappa\lambda} - c_{12}^{\varkappa\lambda} - 2c_{44}^{\varkappa\lambda})\,\delta_{ijkl} + c_{12}^{\varkappa\lambda}\delta_{ij}\delta_{kl} + \\ &\quad + c_{44}^{\varkappa\lambda}(\delta_{ik}\delta_{jl} + \delta_{il}\delta_{jk}). \end{aligned}$$

With kinetic energy density $\frac{1}{2}\sum_{\varkappa}\overset{\varkappa}{\varrho}\overset{\varkappa}{u}_{i,t}\overset{\varkappa}{u}_{i,t}$ and the potential energy density (22), Hamilton's variational principle, for independent variations $\delta \overset{1}{u}_i$ and $\delta \overset{2}{u}_\iota$, yields the six Euler equations

$$\overset{\varkappa}{\varrho} \overset{\varkappa}{u}_{j,tt} = -(-1)^\varkappa \left[a^*(\overset{2}{u}_j - \overset{1}{u}_j) + c^* \sum_\lambda \partial_i(\partial_i \overset{\lambda}{u}_j - \partial_j \overset{\lambda}{u}_i)\right] + \\ + \sum_\lambda \left[(c_{11}^{\varkappa\lambda} - c_{12}^{\varkappa\lambda} - 2c_{44}^{\varkappa\lambda})\, \delta_{ijkl}\, \partial_i \partial_k \overset{\lambda}{u}_l + \right. \\ \left. + c_{12}^{\varkappa\lambda} \partial_i \partial_j \overset{\lambda}{u}_i + c_{44}^{\varkappa\lambda} \partial_i(\partial_i \overset{\lambda}{u}_j + \partial_j \overset{\lambda}{u}_i)\right]. \tag{23}$$

These become the six equations of the type (21) if

$$\begin{aligned} &h\, c_{11}^{\varkappa\varkappa} = 2h\, c_{44}^{\varkappa\varkappa} = 2\overset{\varkappa}{\beta}, \quad && h\, c_{12}^{\varkappa\varkappa} = \overset{\varkappa}{\beta} + 2\gamma, \quad && c^* = 0, \\ &h\, c_{44}^{12} = -h\, c_{12}^{12} = 2\gamma, \quad && h\, c_{11}^{12} = \tfrac{1}{2}\,\alpha, \quad && h^3 a^* = \alpha + 8\gamma. \end{aligned} \tag{24}$$

It may be noted that the linear term in (22) is the energy density of a self equilibrated initial stress which produces a localized strain and a surface energy of deformation at a free surface.

The next higher order approximation is constructed in the same way as in the preceding section. First, the fourth order derivatives are retained in the series expansions of the difference operators in the six equations of the type (20). The resulting differential equations are then identified as the Euler equations of Hamilton's variational principle with the potential energy density taken as a quadratic function of the relative displacement and rotation of the two continua and of their strains and first strain-gradients.

Exact, Continuum Form of Lattice Equations

For exponential functions

$$u_j = A_j \exp[i(\theta_1 x_1 + \theta_2 x_2 + \theta_3 x_3)/h], \tag{25}$$

the infinite series expansions of difference operators in terms of derivatives can be summed. Thus:

$$\Delta_j^\pm u_k^{l,m,n} = \left(1 \pm \frac{h\,\partial_j}{2!} + \frac{h^2\,\partial_j^2}{3!} \pm \frac{h^3\,\partial_j^3}{4!} + \cdots\right) \partial_j u_k = \varkappa_j^\pm \partial_j u_k, \tag{26}$$

where

$$\varkappa_j^\pm \equiv \pm i\, \theta_j^{-1}(1 - e^{\pm i\theta_j}) \tag{27}$$

and $\varkappa_j^\pm \partial_j$ is not summed over the repeated index j. We also adopt the definitions

$$\varkappa_i \equiv \tfrac{1}{2}(\varkappa_i^+ + \varkappa_i^-), \quad \varkappa_i^2 \equiv \varkappa_i^+ \varkappa_i^-, \quad \text{(not summed)}. \tag{28}$$

Then, for exponential functions

$$\Delta_i^2 = \varkappa_i^2\, \partial_i^2; \quad \Delta_i \Delta_j = \varkappa_i \varkappa_j\, \partial_i \partial_j, \ (i \neq j) \quad \text{(not summed)}. \tag{29}$$

Inserting (29) in the difference Eq. (6), we have three equations of the type

$$\varrho \ddot{u}_1 = [c_{11} \varkappa_1^2 \partial_1^2 + c_{44} (\varkappa_2^2 \partial_2^2 + \varkappa_3^2 \partial_3^2) + \\ + \tfrac{1}{2} c_{12} h^2 (\varkappa_1^2 \varkappa_2^2 \partial_1^2 \partial_2^2 + \varkappa_1^2 \varkappa_3^2 \partial_1^2 \partial_3^2)] u_1 + \\ + (c_{12} + c_{44}) (\varkappa_1 \varkappa_2 \partial_1 \partial_2 u_2 + \varkappa_1 \varkappa_3 \partial_1 \partial_3 u_3). \quad (30)$$

These are differential equations which yield the complete dispersion relation (3). Similarly, substitution of (29) in the difference equations of the type (20), for the NaCl-type lattice, converts them to differential equations which produce the complete dispersion relation (18).

References

[1] Cauchy, A. L.: C. R. Acad. Sci. Paris **32**, 323 (1851).
[2] Aero, E. L., and E. V. Kuvshinskii: Fiz. Tverdogo Tela **2**, 1399 (1960).
[3] Grioli, G.: Ann. Mat. pura appl., Ser. IV, **50**, 389 (1960).
[4] Rajagopal, E. S.: Ann. der Phys. **6**, 192 (1960).
[5] Truesdell, C. A., and R. A. Toupin: Encyclopedia of Physics, Vol. III/1. Berlin/Göttingen/Heidelberg: Springer 1960.
[6] Toupin, R. A.: Arch. Rat. Mech. Anal. **17**, 85 (1964).
[7] Toupin, R. A.: Arch. Rat. Mech. Anal. **11**, 385 (1962).
[8] Mindlin, R. D.: Int. J. Solids Struct. **1**, 417 (1965).
[9] Green, A. E., and R. S. Rivlin: Arch. Rat. Mech. Anal. **16**, 325 (1964).
[10] Krumhansl, J. A.: Lattice dynamics; Proc. Int. Conf., Copenhagen 1963. Pergamon Press 1964.
[11] Gazis, D. C., R. Herman and R. F. Wallis: Phys. Rev. **119**, 533 (1960).
[12] Mindlin, R. D.: Arch. Rat. Mech. Anal. **16**, 51 (1964).
[13] Cosserat, E., and F. Cosserat: Théorie des Corps Déformables. Paris: Hermann 1909.
[14] Gazis, D. C., and R. F. Wallis: Private communication.
[15] Nye, J. F.: Physical properties of crystals. Oxford 1960.

The Theory of Elastic Media with Microstructure and the Theory of Dislocations

By

I. A. Kunin

Institute of Thermophysics
USSR Academy of Sciences, Novosibirsk

Different phenomenological theories of generalized Cosserat continua have been developed in the well-known works of AERO, ERINGEN, GREEN, GRIOLI, GÜNTHER, KOITER, KUVSHINSKI, MINDLIN, NAGHDI, NOLL, PALMOV, RIVLIN, TOUPIN, TRUESDELL and others. These theories approximately take into account the inner degrees of freedom and the existence of a scale parameter in a medium. From this point of view the classical theory of elasticity may be considered as an asymptotic theory and the above mentioned ones as next order approximations[1].

A different approach based on crystal lattice theory has been developed in the works of the author and his collaborators [1—8]. These works are also concerned with applications to the theory of point defects and dislocations. This paper contains a brief survey of these works. In connection with this approach the reference to the works of KRUMHANSL [9], ROGULA [10], KRÖNER and DATTA [11] should be made.

1. Classification of Elastic Media with Microstructure

A general linear theory of elastic media with microstructure will be considered here with no assumption that the scale parameter is small (strong space dispersion). The existence of an elementary length unit and long range forces induces a non-locality of the theory.

One should distinguish between microscopically homogeneous and non-homogeneous media. For the case of homogeneous medium the elastic operator kernels are of difference type and the equations of motion in the Fourier-representation are algebraic. For the general case, the equations of motion for both physical and Fourier-representations are integral ones.

---

[1] Though the correct transition to the classical theory was not usually discussed.

Media of simple and complex structure will also be distinguished here. In the first case the only kinematic variable is displacement. An additional set of microstrains is introduced for the medium of complex structure. A theory of weak dispersion is obtained for longwave approximations. The integral operators are then replaced by differential operators, and a scale parameter should be regarded as small. And lastly, in the zero approximation all models must be equivalent to the classical theory of elasticity.

2. Quasi-Continuum [1]

This notion is of great significance for the theory. Let $u(n)$ be a scalar or tensor function defined on a three-dimensional lattice with unit cell $\mathscr{A}$ and such that $|u(n)| \leqq C|n|^p$ when $|n| \to \infty$. Let $N(\mathscr{A})$ be the corresponding functional space.

Functions $u(n)$ may be interpolated by generalized entire analytic functions $u(x)$, the Fourier-images $u(k)$ of which are defined in the unit cell $\mathscr{B}$ of the inverse lattice[1].

Formulas (v—volume of the cell $\mathscr{A}$)

$$u(k) = v \sum_n u(n)\, e^{i k \cdot n}, \quad k \in \mathscr{B} \tag{2.1}$$

$$u(x) = v \sum_n u(n)\, \delta_{\mathscr{B}}(x-n), \qquad \delta_{\mathscr{B}}(x) = \frac{1}{(2\pi)^3} \int_{\mathscr{B}} e^{-ix\cdot k}\, dk \tag{2.2}$$

establish a concrete isomorphism between the spaces $X(\mathscr{B}) \ni u(x)$, $K(\mathscr{B}) \ni u(k)$ and $N(\mathscr{A}) \ni u(n)$. Here $\delta_{\mathscr{B}}(x)$ in $X(\mathscr{B})$ is analogous to the ordinary δ-function, and $\delta_{\mathscr{B}}(0) = v^{-1}$ and $\delta_{\mathscr{B}}(n) = 0$ for $n \neq 0$. In the sequel the index $\mathscr{B}$ will be omitted.

Figuratively speaking, an analytic structure is stretched over the lattice and thus it is changed into a quasi-continuum. This permits the application of a well-developed analytic technique for the description of a discrete structure. Each operation in $N(\mathscr{A})$ (not necessarily linear) can be brought into one-to-one correspondence with operations in $X(\mathscr{B})$ and $K(\mathscr{B})$. In particular, the formulae

$$v \sum_n u(n) = \int u(x)\, dx = u(k)\,|_{k=0}, \tag{2.3}$$

$$\langle u \mid q \rangle \overset{\text{def}}{=} v \sum_n \overline{u(n)}\, q(n) = \int \overline{u(x)}\, q(x)\, dx = \frac{1}{(2\pi)^3} \int \overline{u(k)}\, q(k)\, dk, \tag{2.4}$$

$$\begin{aligned} \langle u \mid \Phi \mid w \rangle &\overset{\text{def}}{=} v^2 \sum_{nn'} u(n)\, \Phi(n, n')\, w(n') = \iint \overline{u(x)}\, \Phi(x, x')\, w(x')\, dx\, dx' \\ &= \frac{1}{(2\pi)^3} \iint \overline{u(k)}\, \Phi(k, k')\, w(k')\, dk\, dk' = \overline{\langle w \mid \Phi^+ \mid u \rangle} \end{aligned} \tag{2.5}$$

---

[1] More exactly, $\mathscr{B}$ is a cell with identified opposite faces, i.e. a three-dimensional torus.

are valid. Here $\Phi^+(n, n') = \overline{\Phi(n', n)}$, and the overbar here stands for complex conjugate.

To the discrete convolution $q(n) = v \sum_{n'} \Phi(n - n')\, u(n')$ there correspond

$$q(x) = \int \Phi(x - x')\, u(x')\, dx' \quad \text{and} \quad q(k) = \Phi(k)\, u(k). \tag{2.6}$$

Thus, to any lattice theory we may assign its quasi-continuum representation without any loss of information.

It is natural to generalize the concept of a quasi-continuum admitting a wider class of manifolds for $\mathcal{B}$. So if $\mathcal{B}$ is a sphere, then we have the well-known Debye's quasi-continuum. We can also construct quasi-continua corresponding to laminated and fibrous media. Thus the quasi-continuum can be defined as an aggregate of two objects: the Euclidean space and $X(\mathcal{B})$ over it ($\mathcal{B}$ being a fixed manifold). This definition admits a generalization for the case of non-Euclidean spaces.

3. A Homogeneous Medium with Simple Structure [1, 5]

A simple crystal lattice in the harmonic approximation [12] is considered as a starting micromodel. In the notions (2.4, 5) the corresponding Lagrangian being invariant to the choice of n-, x- or k-representations, is of the form:

$$2L = \varrho \langle \dot{u}_\alpha \mid \dot{u}^\alpha \rangle - \langle u_\alpha \mid \Phi^{\alpha\beta} \mid u_\beta \rangle + 2 \langle u_\alpha \mid q^\alpha \rangle. \tag{3.1}$$

Here ϱ — mass density, u_α — displacement, q^α — external forces. The elastic energy operator $\Phi^{\alpha\beta}$ is defined by the force constant matrix $\Phi^{\alpha\beta}(n - n')$. In k-representation $\Phi^{\alpha\beta}(k, k') = \Phi^{\alpha\beta}(k)\, \delta(k - k')$. The tensor $\Phi^{\alpha\beta}(k)$ can be represented as:

$$\Phi^{\alpha\beta}(k) = c^{\nu\alpha\mu\beta}(k)\, k_\nu\, k_\mu, \tag{3.2}$$

which follows from the energy invariance related to translation and rotation. Here $c^{\nu\alpha\mu\beta}(k)$ is an analytic function expressed through the microscopic force constants $\Phi^{\alpha\beta}(n)$. It is shown that without the loss of generality, one can consider the tensor $c^{\nu\alpha\mu\beta}(k)$ as being symmetrical with respect to the index pairs $\nu\,\alpha$ and $\mu\,\beta$ and the Hermitian related to the pairwise permutation.

The equation of motion in the (k, ω)- and (x, t)-representations is

$$\varrho\, \omega^2\, u^\alpha(k) - c^{\nu\alpha\mu\beta}(k)\, k_\nu\, k_\mu\, u_\beta(k) = -q^\alpha(k), \tag{3.3}$$

$$\varrho\, \ddot{u}^\alpha(x) - \partial_\nu \int c^{\nu\alpha\mu\beta}(x - x')\, \partial_\mu u_\beta(x')\, dx' = q^\alpha(x). \tag{3.4}$$

If the strain tensor $\varepsilon_{\mu\beta}(x) = \partial_{(\mu} u_{\beta)}(x)$ and the symmetrical tensor

$$\sigma^{\nu\alpha}(k) = c^{\nu\alpha\mu\beta}(k)\, \varepsilon_{\mu\beta}(k) \quad \text{or} \quad \sigma^{\nu\alpha}(x) = \int c^{\nu\alpha\mu\beta}(x - x')\, \varepsilon_{\mu\beta}(x')\, dx' \tag{3.5}$$

are introduced, the equation of motion and energy density $\varphi(x)$ can be written in the form:

$$\varrho\,\ddot{u} - \operatorname{div}\sigma = q, \qquad \varphi(x) = \tfrac{1}{2}\sigma^{\nu\alpha}(x)\,\varepsilon_{\nu\alpha}(x). \tag{3.6}$$

Hence $\sigma^{\nu\alpha}$ can be interpreted as a stress tensor whereas (3.5) is the operator Hook's law.

The given equations are also valid for any phenomenological model defined by the Lagrangian (3.1).

Theorem. In a non-local theory of a homogeneous linearly-elastic medium with a simple structure the symmetrical stress tensor and the elastic energy density may always be introduced, the latter being expressed in the usual way through stress and strain.

The Green's tensor $G_{\alpha\beta}(x) \in X(\mathscr{B})$ can be introduced for Eq. (3.4). $G_{\alpha\beta}(x)$ is an entire function for a finite manifold $\mathscr{B}$ and thus has no singularity at $x = 0$. Hence the divergencies existing in the elastic continuum are eliminated.

For the isotropic and weakly anisotropic cases we succeeded in constructing explicit expressions for $G_{\alpha\beta}(x)$.

The transition to long-wave approximations (the case of small k) in a strict sense means that in the series expansion of $c^{\nu\alpha\mu\beta}(k)$ in k a finite number of terms is kept and the quasi-continuum is replaced by a continuum. For the zero approximation we have the ordinary theory of elasticity and for the following approximations a weak dispersion theory. Thus for the centro-symmetrical case, Hook's law in the second approximation is of the form:

$$\sigma^{\nu\alpha}(x) = (c_0^{\nu\alpha\mu\beta} + c_2^{\nu\alpha\mu\beta\lambda\tau}\,\partial_\lambda\,\partial_\tau)\,\varepsilon_{\mu\beta}(x). \tag{3.7}$$

The equation of motion (3.4) coincides in this case with that of the couple-stress theory with constrained rotations (but $\sigma^{\nu\alpha}$ is symmetrical and there is no use in the introduction of the couple stresses). The scale parameter is in this case due to the non-linear dispersion relations $\omega(k)$. In particular, for the Debye model all approximations coincide with the zero approximation. Unlike the strong dispersion case, the theory of weak dispersion when applied correctly, usually gives only a small correction to the ordinary theory of elasticity.

4. A Homogeneous Medium with Complex Structure [2, 3]

As the starting micromodel we have now the complex crystal lattice [12]. This model is generalized for the case of an arbitrary mass distribution in the cell. Instead of individual cell variables (displacements of each particle in the cell) it is convenient to introduce collective cell variables, i.e. centre of mass displacement u_α and microstrains of different orders $\eta_{q\beta}$ (here $q = \lambda_1 \ldots \lambda_q$ — multiindex). Accordingly, the force density q^α and micromoments $\mu^{p\alpha}$ are introduced. The transition from the individual to collective variables is performed with the help of a special biorthogonal basis with the mass density weight. This permits the establishment of one-to-one correspondence between the two sets of variables, which is impossible in the case of the usual power moments. In particular, if a cell contains no more than four particles, then the kinematic variables will be displacement u_α, microstrain $\varepsilon'_{\alpha\beta}$ and microrotation $\Omega'_{\alpha\beta}$, only.

The equations of motion in (x, t)-representation are

$$\begin{aligned} \varrho\, \ddot{u}^\alpha - \partial_\nu\, \hat{\gamma}^{\nu\alpha\mu\beta}\, \partial_\mu u_\beta - \partial_\nu \hat{\chi}^{+\nu\alpha q\beta}\, \eta_{q\beta} &= q^\alpha, \\ I^{p\alpha q\beta}\, \ddot{\eta}_{q\beta} + \hat{\chi}^{p\alpha\mu\beta}\, \partial_\mu u_\beta + \hat{\Gamma}^{p\alpha q\beta} \eta_{q\beta} &= \mu^{p\alpha}. \end{aligned} \tag{4.1}$$

Here I is an inertia matrix; $\hat{\gamma}, \hat{\chi}, \hat{\Gamma}$ — integral operators with difference kernels expressed in terms of microparameters.

Eqs. (4.1) give an exact representation of the starting micromode (without any loss of information). From the phenomenological stand point they describe the most general homogeneous linear-elastic medium with complex structure.

For an isotropic medium with the microrotation Ω' (Cosserat continuum with strong dispersion) we have:

$$\varrho\, \ddot{u} - \hat{\gamma}\, \Delta u - \hat{\beta}\, \mathrm{grad\,div}\, u + \hat{\chi}\, \mathrm{rot}\, \Omega' = q, \quad I\, \ddot{\Omega}' + \hat{\chi}\, \Omega + \hat{\Gamma}\, \Omega' = \mu. \tag{4.2}$$

Here $\hat{\gamma}, \ldots, \hat{\Gamma}$ are scalar operators, $\Omega = \frac{1}{2}\, \mathrm{rot}\, u$.

In the zero long-wave approximation the operators in (4.1) become constant tensors while the next order approximations give differential operators. The weak dispersion theory contains the Cosserat continuum, couple stress and multipolar theories as particular cases[1].

For the medium with complex structure, the dispersion relation exhibits not only acoustic branches but also high-frequency optic branches. Of great significance for the mechanics of continuous media is the domain of acoustic frequencies. Thus in crystals the optic frequencies are of the order of $10^{13}\ \mathrm{sec}^{-1}$, i.e. they significantly exceed those of the sources of mechanical oscillations. Physically it is evident that at low frequencies the main role belongs to the displacement u_α of the mass centres of cells. The analysis, however, shows that there is no small parameter in (4.1) which would permit one to neglect inner degrees of freedom $\eta_{q\beta}$. This paradox is cleared up in the following:

Theorem. In the acoustic frequency domain the equations of the elastic medium with complex structure can always be transformed

[1] The comparison shows, though, that in these theories the orders of approximations in different equations do not coincide.

into equivalent equations of a medium with simple structure with space and time dispersions.

The transformed equations in (k, ω)-representation are

$$\varrho\, \omega^2 u^\alpha - c^{\nu\alpha\mu\beta}(k, \omega)\, k_\nu\, k_\mu\, u_\beta = -\, Q^\alpha, \tag{4.3}$$

$$\eta_{p\alpha} = a_{p\,\alpha}^{\cdot\,\cdot\,\mu\beta}(k, \omega)\, k_\mu\, u_\beta + A_{p\alpha q\beta}(k, \omega)\, \mu^{q\beta}. \tag{4.4}$$

Tensor $c^{\nu\alpha\mu\beta}(k, \omega)$ has $c^{\nu\alpha\mu\beta}(k)$ symmetry and in the zero approximation coincides with the usual tensor of elastic constants measured in the macroexperiments. Eq. (4.3) permits the introduction of a symmetrical stress tensor $\sigma^{\nu\alpha}$ satisfying (3.6). But the Hook's law will also contain integration over time. In the zero approximation $\sigma^{\nu\alpha}$ becomes the usual stress tensor [unlike that of the stress tensor which could be introduced into (4.1)].

These considerations prove that in the acoustic domain Eqs. (4.3) are more adequate for macrodescription of a medium than those of the form of (4.1).

5. Non-Homogeneous Medium with Microstructure [4]

Let us restrict ourselves to the case of the medium with simple structure homogeneous at infinity. The symmetrical stress tensor and an elastic energy density can always be introduced, the latter being expressed in the usual way through stress and strain.

Thus in this case Eqs. (3.6) are valid, but the Hook's law has the form:

$$\sigma^{\nu\alpha}(x) = \int c^{\nu\alpha\mu\beta}(x, x')\, \varepsilon_{\mu\beta}(x')\, dx'. \tag{5.1}$$

Of special interest is the case of local inhomogeneities. They can be described by δ-function perturbations of the elastic operator (this is impossible in the ordinary theory of elasticity). This permits one to obtain an explicit expression for the Green's tensor for the medium with inhomogeneities through the Green's tensor for homogeneous medium. With the help of the Green's tensor the perturbation of the external field induced by inhomogeneities as well as concentration of stresses and forces acting upon inhomogeneities are calculated.

It is interesting to note that this method also permits one to obtain effective approximate solution for a system of inclusions in the usual elastic medium.

6. Inner Stresses in the Medium with a Microstructure [6]

Taking into account a microstructure in the inner stress theory allows an essential extension of the latter's applications making it closer to the theory of lattice defects. The main idea is the introduction of a quasi-continuum and the space dispersion keeping as much as possible to the general scheme of the continuum theory of dislocations.

The field variable now is a distortion $\zeta_{\nu\mu}$. Tensor $\alpha^{\lambda}_{.\mu} = \mathrm{rot}^{\lambda\nu}\zeta_{\nu\mu}$ is identified with the density of dislocations (or quasi-dislocations in the sense of KRÖNER). For the unique definition of ζ we also postulate that $\mathrm{div}\,\sigma = 0$ for the inner stresses. But in this case we have:

Theorem. The elastic modulus operator C for the inner stresses differs, in general, from that for the external stresses by an operator localized in a domain containing inner stress sources. This is exactly the domain where the inner stress tensor may be non-symmetrical. In the zero approximation both operators coincide.

For simplicity, it is supposed below that both operators identically coincide.

It is also convenient to introduce two other characteristics of the inner stress source distribution, namely: incompatibility η and the density of dislocation moments m defined as follows

$$\eta^{\nu\varrho}(x) = \mathrm{rot}^{(\lambda|\mu|}\alpha^{\varrho)}_{.\mu}(x), \qquad \alpha^{\lambda}_{.\mu}(x) = \mathrm{rot}^{\lambda\nu} m_{\nu\mu}(x). \tag{6.1}$$

Then the equations for the inner stresses can be written in the usual form:

$$\mathrm{Rot}\,B\,\sigma = \eta, \qquad \mathrm{div}\,\sigma = 0. \tag{6.2}$$

Here $B = C^{-1}$, $\mathrm{Rot}^{\alpha\beta\lambda\mu} = (\mathrm{rot}^{\alpha\lambda}\,\mathrm{rot}^{\beta\mu})_{(\lambda\mu)}$.

Let us introduce the Green's tensor $F(x, x')$ for inner stresses defined by equations

$$\mathrm{Rot}\,B\,F = \mathrm{Rot}, \qquad \mathrm{div}\,F = 0, \qquad F = F^{+}. \tag{6.3}$$

Thus the representation $F = \mathrm{Rot}\,H\,\mathrm{Rot}$ is valid where the operator $H = H^{+}$ is, however, defined non-uniquely. It permits to obtain the simplest expression for H.

The inner stress tensor σ and elastic energy Φ are expressed through F and H in the following forms.

$$\sigma(x) = \int F(x, x')\, m(x')\, dx' = \int \mathrm{Rot}\,H(x, x')\,\eta(x')\, dx', \tag{6.4}$$

$$2\Phi = \langle m\,|F|\,m\rangle = -\langle \alpha\,|\mathrm{rot}\,H\,\mathrm{rot}|\,\alpha\rangle = \langle \eta\,|H|\,\eta\rangle. \tag{6.5}$$

The choice of the most suitable of these forms is determined by the particular type of the inner stress sources.

For the isotropic and weakly anisotropic quasicontinua with the space dispersion, it is possible to obtain the explicit expressions for the Green's tensors F and H. This permits solution of a number of fundamental problems for the point defects and dislocations in the medium with the microstructure.

7. Point Defects [7]

The point defects of the type of vacancies and impurity atoms are simultaneously the inner stress sources and local inhomogeneities.

For the system of point defects

$$m_{\lambda\mu}(x) = v \sum_i M^i_{\lambda\mu}\, \delta(x - x_i) \tag{7.1}$$

where $M^i_{\lambda\mu}$ is a non-dimensional moment characterising the relative volume or shear distortion in the neighbourhood of the defect at the point x_i.

The force constant change may be described as follows.

$$b(x, x') = b_0(x - x') - v \sum_i b_i\, \delta(x - x_i)\, \delta(x' - x_i) \tag{7.2}$$

where $b_0(x)$ is the kernel of the operator B_0 for the homogeneous medium, and tensors b_i are calculated from the perturbed force constants.

Let $F_0(x - x')$ be the Green's tensor for the homogeneous medium. Then the Green's tensor $F(x, x')$ corresponding to (7.2) has the form:

$$F(x, x') = F_0(x - x') - \sum_{ij} F_0(x - x_i)\, P^{ij}\, F_0(x_j - x') \tag{7.3}$$

where the matrix P^{ij} with the tensor components is inverse to

$$R_{ij} = F_0(x_i - x_j) - v^{-1}\, b_i^{-1}\, \delta_{ij}. \tag{7.4}$$

For $\sigma^{\nu\alpha}$ and Φ according to (6.4, 5) we have now

$$\sigma^{\alpha\beta}(x) = v \sum_i M^i_{\lambda\mu}\, F^{\alpha\beta\lambda\mu}(x, x_i), \tag{7.5}$$

$$\Phi = \tfrac{1}{2} v^2 \sum_{ij} M^i_{\alpha\beta}\, M^j_{\lambda\mu}\, F^{\alpha\beta\lambda\mu}(x_i, x_j). \tag{7.6}$$

These formulae permit us to obtain the elastic fields and interaction forces for the system of defects. In particular, it is shown that the asymptotic value of (7.6) for two dilatation centers in the Debye model coincides with Lifshitz's for two spherical inclusions [13]. Taking into account of the space dispersion introduces an essential correction into the solution. Specifically, under some conditions the interaction force can change the sign at short distances, i.e. the existence of stable pairs of defects is possible.

8. Dislocations [8]

Knowledge of the Green's tensor permits one to take into account the microstructure in the basic problems of the dislocation theory. Thus for stresses and energy of a dislocation loop with the contour L

and the Burgers vector b_α we have:

$$\sigma^{\alpha\beta}(x) = b_\nu \int_L \mathrm{rot}^{\alpha\varrho}\,\mathrm{rot}^{\beta\tau}\,\mathrm{rot}^{\lambda\nu} H_{\varrho\tau\lambda\mu}(x - x_L)\, dL^\mu, \tag{8.1}$$

$$2\Phi = -b_\alpha b_\beta \int_L \int_{L'} \mathrm{Rot}^{\alpha\beta\lambda\mu} H_{\nu\lambda\varrho\mu}(x_L - x_{L'})\, dL^\nu\, dL'^\varrho. \tag{8.2}$$

It is also possible to take into account the influence of the force constant change in the cores of dislocations. For example, the generalized Peach-Koehler force has in this case two components. The first is linear with respect to the Burgers vector and with respect to external field. The second is quadratic relative to the field and does not depend on the Burgers vector.

The plain problem for parallel dislocations can be completely studied. In general, one must solve a system of two equations related to two scalar potentials. But for some important cases (centro-symmetric, isotropic and others) the system decomposes into two independent equations for screw and edge dislocations.

In these cases one can easily obtain explicit expressions for the fields and interactions forces. Under some conditions the interaction forces can change the sign at short distances and the existence of the stable screw dislocation dipoles is possible. Similar to the case of point defects this effect appears for the medium with space dispersion only.

In conclusion we note that the dislocations were also considered for the Debye model by Brailsford [14] and for the medium with space dispersion by Rogula [10].

References

PMM — Prikladnaja Matematika i Mechanika.
PMTF — Prikladnaja Mechanika i Techničeskaja Fizika.
FTT — Fizika Tverdogo Tela.
FMM — Fizika Metallov i Metallovedenie.

[1] Kunin, I. A.: PMM **30**, N 3 (1966).
[2] Kunin, I. A.: PMM **30**, N 5 (1966).
[3] Vdovin, V. E., and I. A. Kunin: PMM **30**, N 6 (1966).
[4] Kunin, I. A.: PMTF, N 3 (1967).
[5] Kunin, I. A.: PMTF, N 4 (1967).
[6] Kunin, I. A.: PMM **31**, N 5 (1967).
[7] Kosilova, V. G., I. A. Kunin and E. G. Sosnina: FTT, No 2 (1968).
[8] Vdovin, V. E., and I. A. Kunin: FTT, No 2 (1968).
[9] Krumhansl, J. A.: Proc. Int. Conf. of Lattice Dynamics, Copenhagen 1963. Oxford 1965.
[10] Rogula, D.: Bull. Acad. Polon. Sci., Ser. Sci. Techn. **13**, N 7 (1965); **14**, N 3 (1966).
[11] Kröner, E., and B. K. Datta: Z. Phys. **96**, H. 3 (1966).
[12] Born, M., and K. Huang: Dynamical Theory of Crystal Lattices. Oxford 1954.
[13] Lifshitz, I. M., and L. V. Tanatarov: FMM **12**, N 3 (1961).
[14] Brailsford, A. D.: J. Appl. Phys. **37**, N 7 (1966).

Interrelations between Various Branches of Continuum Mechanics

By

E. Kröner

Institut für Theoretische Physik
Technische Universität Clausthal

Abstract. The following insights are gained and/or discussed:

1. Mass point and continuum theories have equal rights in the classical mechanics of matter.

2. The main difference in the formulation of dislocation theory by NOLL and that by the former workers lies in the definition of material uniformity; thus it is physically irrelevant.

3. COSSERAT and dislocation theory show many similarities, but also fundamental differences.

4. The strain gradient theories reflect non-local (and dispersive) response characteristics and are successive approximations to the integral non-local theory. Physically observable effects can be predicted.

5. The new developments in continuum mechanics can be classified in a three-dimensional scheme according to the geometry of deformation (Fig. 3).

1. Lattice or Continuum Mechanics?

The fact that all solid and fluid bodies are made up of more or less point-like particles is accepted by the physicists, as is the fact that quantum mechanics provides a correct basis for the prediction of the properties and the behaviour of such a body. Of course, this statement does not imply that quantum mechanics must be a *convenient* basis for calculations about solids and fluids. In fact, if this were so, there would be no need for this symposium at all. We are here, indeed, interested in problems of solids and fluids for which quantum mechanics is *not* a convenient basis. Nevertheless, if it is maintained that the theories to be discussed at this symposium describe nature, they have to be consistent with quantum mechanics and its results.

Acknowledgment. The present exposition has profited from numerous discussions with my coworkers, Drs. B. K. DATTA and E. GRAFAREND, whose work was made possible by generous support of the Deutsche Forschungsgemeinschaft.

For our purposes one result of quantum mechanical calculations is of particular importance: The mass density of any physical body fluctuates with a wave length of the order 10^{-8} cm. No classical continuum theory, no matter how refined, can ever predict a physical body's mass distribution over a distance of 10^{-8} cm: Quantum theory alone is qualified.

This result implies: though the mass distribution over a distance of order 10^{-8} cm is important for the quantum mechanical calculation of the interaction between domains of this size (we call these domains *atoms* though they are not atoms in a strict sense), it has no meaning in any classical continuum theory. Hence, giving the position of all centers of gravity (*nuclei*) of the atoms provides a complete specification of the spatial configuration of the body in the classical approach. In other words: in contrast to an often heard argument it is not a greater limitation to pass from the correct, i.e. the quantum mechanical, formulation of the mechanics of a body to a classical *mass point theory* than it is to use *continuum mechanics*. Of course, in formulating this insight it is implied that no specialisation is made about the interaction of the points defined above. In particular, many particle forces cannot be excluded at this stage.

It follows that continuum mechanics, if it is supposed to describe nature, must be consistent with all the consequences which follow from the rigorous mass point theory.

As a second important result it follows from quantum mechanics that the mass point distributions of many materials have the form of a crystal lattice which may be more or less disturbed. The mass point theory of such materials is known as the atomic lattice theory. Again, its use does not mean any limitation as compared to the continuum mechanics of these materials as long as no additional approximations are introduced.

The preceeding reflections suggest the following rather trivial answer to the question raised in this section: Both continuum mechanics and mass point theory are good tools for handling the mechanics of solid and fluid bodies. Hence it is advisable to use the theory which is more easily adapted to the problem in question.

2. Dislocations and Inhomogeneities

The most important lattice defects are the dislocations. They act as sources of internal stress and are moreover responsible for the most spectacular characteristics of crystalline solids, namely the plasticity.

One of the relevant problems of this symposium concerns the position of the continuum theory of dislocations within general continuum mechanics. W. NOLL has given a classification scheme of continuum mechanics which has attracted great attention. The lines of thought are laid down in the Handbuch article on the *non-linear field theories of mechanics* [1], and, as far as dislocation theory is concerned, more detailed in a yet unpublished exposition[1] [2]. I quote from Ref. [1], p. 88: "NOLL's analysis generalizes and gives a rigorous foundation for the continuum theory of dislocations proposed and developed in various forms by . . . These authors either lay down *a priori* some geometric structure on the basis of considerations of lattice defects in crystals or . . . NOLL's theory, on the other hand, is free from *ad hoc* assumptions . . . he shows that once a constitutive equation such as to define a materially uniform simple body is laid down, the geometric structure in the body is determined. Thus the geometric structure is the outcome, not the first assumption, of the physical theory."

The forestanding statements require an answer. I shall first briefly compare the two standpoints in question for the kinematical part of the theory. As argued in Sect. 1, both the continuum and the mass point view are legitimate in describing the mechanics of matter. I shall here apply the mass point picture, however only for illustration.

NOLL's *configuration* of a body can be characterised as being a distribution of particles (atoms in the sense of Sect. 1) over some region of the physical (Euclidean) space. This distribution can vary from very irregular to very regular. Let us limit ourselves, as NOLL did, to materially uniform bodies, the mainly investigated case.

Standpoint I (old): We define material uniformity by the assertion: the particles of the body are all of the same kind. *This statement has a clear physical meaning and needs no further mathematical formulation.* For a materially uniform body the following conclusion is trivial: If the response of the body is local then the response to equivalent local configurations at any two points $\boldsymbol{x}$ and $\boldsymbol{y}$ in the body is the same.

Standpoint II (new): Here the argument goes in the opposite direction: If the response of the body is local (i.e., if the body is *simple*) and if the response to equivalent local configurations at any two points $\boldsymbol{x}$ and $\boldsymbol{y}$ is the same then the body is (simple and) materially uniform[2]. Incidentally, it is not trivial to extend this definition to the non-local case.

Once material uniformity is defined, one arrives at dislocation theory in both approaches by formulating restrictions on the set of admissible configurations or deformations. Clearly, two sets of restrictions which lead to the same theory are isomorphic. In standpoint I

[1] I wish to thank Dr. NOLL for distributing copies of this at the symposium.

[2] We can renounce here the more elaborate mathematical language of NOLL and also a possible history dependence.

are used the two postulates (A, B): The geometry of the deformation from one configuration to another is affine (A) and metric (B)[1]. In standpoint II are introduced so-called smoothness assumptions: see the lecture of C. WANG at this symposium[2].

The comparison shows that the main non-trivial difference between the two—equally rigorous—standpoints concerns the problem of how to define a body which consists of physical particles of the same kind. Standpoint II is less general because here the body is restricted to be simple, a physical property which *a priori* has nothing to do with the question "dislocations or not?".

I discuss a further difference. NOLL's theory, as far as it is concerned with dislocations, appears to be a theory of a body with *constrained dislocations*. Here the word *constrained* has a similar meaning as in the term *Cosserat theory with constrained rotations*: The latter constraint implies that the kinematical degrees of freedom of the particle's eigenrotation are suppressed, whereas the former constraint refers correspondingly to the kinematical degrees of freedom of the dislocation's eigenmotion to which I also refer as the *dislocation's drift*[3]. In other words: the drift of dislocations is excluded in NOLL's analysis.

In fact, NOLL's exposition in Ref. [2] concerns a *continuous body of class C^p* $(p \geqq 1)$ which is defined among others by the axiom that the deformation from one configuration to another of the body be a

---

[1] One configuration of the so restricted set can represent a perfect crystal (case of the crystalline body). In this case the *material geometry*, i.e. the geometry of the *configurations* (not only that of the *deformations*) is affine and metric. If there is not a perfect crystal configuration in the set, one can always imagine a virtual point lattice to be impressed on the body, as pointed out by myself in 1958 [3]. The set of admitted configurations of these points would be isomorphic to the set of admitted configurations in the case of the crystalline body. This observation provides an answer to a further claim of generality raised by the workers of standpoint II: Dislocation theory in standpoint I is not *a priori* restricted to crystals, though it is convenient to consult crystals for illustration and though one can expect that in all physical applications of dislocation theory the (solid or fluid) body is crystalline.

[2] The way smoothness enters NOLL's theory has confused me for some time. The definition of the materially uniform simple body can be given without any reference to smoothness. It appears clearer to me when the *geometric* postulates of smoothness come in only *after* the *physical* postulates which define the materially uniform simple body.

[3] The eigenmotion (drift) of a dislocation is a motion which changes the position of the dislocation *with respect to the lattice*. On the other hand, if a crystal is deformed in a manner that the position of a dislocation with respect to the lattice is unchanged though its position in *space* has altered, then the motion of the dislocation is not an eigenmotion. Only eigenmotions result in plastic deformation.

(continuous) deformation of class C^p. This axiom excludes from the very beginning the drift of dislocations since this motion implies a non-continuous, namely plastic, deformation.

For evidence, consider the Nye) limiting process: lattice parameter a (= length of the Burgers vector of the single dislocations in a crystal) $\to 0$, total Burgers vector (of all dislocations piercing through a fixed area element) = constant. It is easily seen that in the continuum limit ($a \to 0$) the dislocations remain discrete objects with Burgers vectors which become infinitely small compared to the distance between the dislocations. It follows that even in this continuum limit plastic deformation is fundamentally discrete, the scale of discreteness being related to the distance between the dislocations. This result reflects a physical truth which is confirmed by thousands of observations on slip lines[1].

The following facts give further evidence of the constraint in Noll's theory:

a) This theory does not involve differential equations for the change of dislocation density with time.

b) Noll does not speak of plasticity but of (simple and materially uniform) inhomogeneous *elastic* bodies (so does C. Wang in his lecture at this symposium).

c) In Noll's terminology, dislocations are *inhomogeneities* and the corresponding (Nye) curvature (Fig. 1) is called a *contorted aelotropy*. These terms are suitable when the dislocations are constrained not to

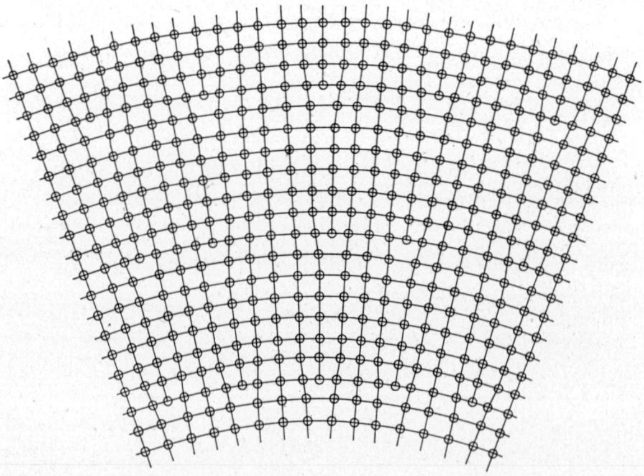

Fig. 1. Nye curvature (contortion) of a crystal lattice. No macroscopic elastic strain is involved.

[1] In constructing a continuum theory of plasticity on the basis of dislocations, it is indispensible that a second limiting process letting the distance between dislocations shrink follows the first limiting process ($a \to 0$).

drift. They appear less adequate when the dislocation distribution changes with time. In fact, in conventional use, the terms *inhomogeneity* and *aelotropy* describe a material, but not its motion.

Except for statical problems one will find little application of the idea of constrained dislocations. Indeed, it is known that even at the low temperature of liquid helium the dislocations in a crystal drift under very small loads. The best known examples of this are the important problem of fatigue and the vibration problem where sound waves of small amplitude excite drift-type vibrations of dislocations.

Of course, the constraint in NOLL's theory can be removed and no strong point of it is made here. Nevertheless, it seems to me that the term *inhomogeneity* has no justification in the unconstrained theory. I fully support the word *contortion* for NYE curvature.

3. Dislocations and the Cosserat Theory

All measurements in an affine and metric geometry can be performed with the aid of an affine connection Γ_{mlk} which is defined by a law of parallelism in such a way that two vectors C_k at two neighbouring points separated by dx^m are parallel if their difference dC_k at the two points obeys the equation

$$dC_k = -\Gamma_{mlk}\, C^l\, dx^m. \tag{1}$$

The affine and metric connection has the form

$$\Gamma_{mlk} = [m\,l,k] - K_{mlk} \tag{2}$$

where $[m\,l,k]$ is the Christoffel symbol taken with the metric g_{lk}:

$$[m\,l,k] = (g_{lk,m} + g_{mk,l} - g_{ml,k})/2 \tag{3}$$

and $K_{mlk} = K_{m[lk]}$ is a similar combination

$$K_{mlk} = -(S_{mlk} + S_{kml} - S_{lkm}) \tag{4}$$

where $S_{mlk} = \Gamma_{[ml]k}$ is the antisymmetric part (with respect to m, l) of Γ_{mlk}. Hence, the metric connection is completely determined by the two tensors g_{lk} and S_{mlk} which in our application, except for a rigid motion, determine the deformation from some reference configuration to the present configuration. The tensor S_{mlk} is CARTAN's torsion. Here, it describes the dislocation density, whereas g_{lk} provides the strain measure (both considered with respect to the reference configuration, usually the perfect crystal).

The dislocation density in a crystal at any time is an observable quantity which can be measured without knowing anything about the past of the body. Hence it is a state quantity in the usual sense of thermodynamics. It is common to define response quantities as the variational derivatives of some energy U (internal, free, Lagrange etc.) with respect to the state quantities[1]. Thus we expect response quantities of the form

$$\sigma^{lk} = \delta U / \delta g_{lk}, \qquad \tau^{mlk} = \delta U / \delta K_{mlk} \tag{5}$$

where the first quantities can be identified with ordinary (symmetric) stresses (at least in the local case). The last quantities have the dimension of moment stresses. They respond to the tensor K_{mlk} defined by Eq. (4). This tensor describes a curvature of the lattice of the form of Fig. 1 [observe that K_{mlk} to be used in Eq. (1) is antisymmetric in l, k]. This so-called Nye curvature (contortion) which occurs without a simultaneous macro-strain is one of the fundamental features of dislocation theory.

The fact that the quantities τ^{mlk} respond to a curvature and have the dimension of a moment stress has lead Günther [4] to his conception of a dislocated continuum as a Cosserat continuum. The idea is, indeed, very tempting. Nevertheless, a complete correspondence has never been established and I am more and more convinced that the difference between Cosserat and dislocation theory is fundamental.

The Cosserat continuum, as I would define it, is built up from particles which possess an inherent orientation. The particles with spin considered in the former work of Amari [5] and also in J. B. Alblas' lecture at this symposium are of this kind. The possibility of spin waves in these materials has been explored by Bloch and others in the late twenties.

In contrast to this picture, dislocations occur in crystals in which the atoms need not possess an inherent orientation. Orientation enters only when the arrangement of the neighbouring atoms is regarded. Hence, although one can speak of an orientation at a point both in the normal crystal and in the Cosserat continuum, the physical situation is basically different. In fact, moving dislocations are not spin waves. Instead, dislocations possess the fundamental ability to produce slip.

As a consequence of these obvious differences the geometry (and kinematics) used in the two cases should be different. S. Amari in his work on the deformation of ferromagnetic crystals has applied Finsler's geometry, which was characterised by Cartan as the geometry of a manifold the elements of which are points equipped with a proper direction (see Amari in Ref. [5]). In the Cosserat continuum the points are equipped with an orientation described by three directions rather than by a single direction. This means that the corresponding geometry

---

[1] Such a proceeding also fits well into the general scheme developed in the lecture of L. I. Sedov and V. L. Berditchevski at this symposium.

should be a modified Finsler geometry based upon three directional vectors. Vectors of this kind are known as directors. These directors have many similarities with the primitive lattice vectors in crystals, a fact which has caused some confusion in the past[1].

Since dislocation and Cosserat theory belong to two different geometries (and kinematics), the new response quantities must also be different. In both cases they have the dimension of moment stresses.

In Fig. 2a is shown an edge dislocation, in Fig. 2b the five central atoms and, by arrows, the type of forces which are necessary to maintain the dislocated atomic arrangement. The forces form two couples with a lever arm of the atomic distance. These couples appear everywhere along the dislocations and nowhere else. They can be considered as singular couple stresses which in the macroscopic description (see footnote 1, p. 334) are smeared out over the volume elements. Hence, dislocation theory is a couple stress theory.

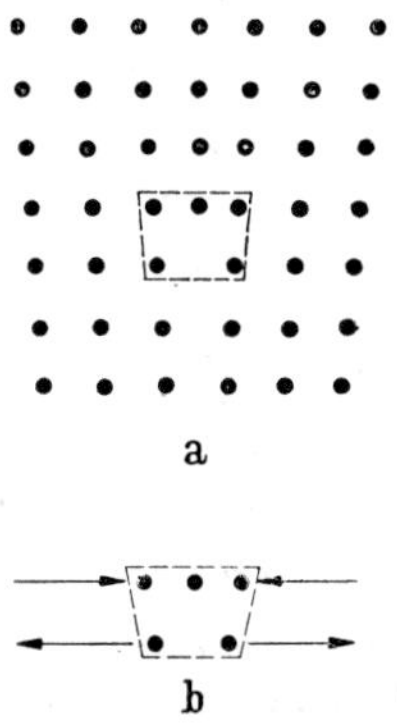

Fig. 2. Couples as response to dislocations.

In contrast to this, the Cosserat theory is not a couple stress theory. In fact, here the rotation refers to one particle and not to two neighbouring particles. This implies that the lever arm is missing. Hence, the Cosserat theory is a spin moment stress theory (or torque stress theory).

Of course, couple and spin moment have a lot in common; thus one more reason for the similarity of dislocation and Cosserat theory. Nevertheless, the difference between couple and spin moment has a fundamental physical implication: It is the balance of the two couples shown in Fig. 2 which keeps the dislocation from slipping along the glide plane. Only when this balance is disturbed will the dislocation start moving. Spin moments, on the other hand, act on single particles and never tend to produce slip. These observations make it clear that couples play a role in plastic flow, as do dislocations.

4. Multipolar and Related Theories

There is a third kind of response quantities which have been called couple stresses. They arise in the so-called strain-gradient or multipolar (of the first kind) theories of which the Cosserat theory with constrained

[1] One should carefully distinguish between directors assigned to single particles and directors which connect neighbouring particles as do the lattice vectors. The modified Finsler geometry cited here is meant to be based on the first type of directors.

rotations is a special case. These theories imply a certain non-local behaviour of the response. The "couples" arise from forces between more distant particles[1].

The strain gradient theories can be considered as successive approximations to the non-local integral theory which was derived from the theory of Bravais lattices by I. A. Kunin (see his lecture at this symposium) and by myself [6]. In the integral theory no couples occur. The response of the body is completely specified by a symmetric tensor σ^{ij} which is related to a symmetric strain tensor ε_{kl} by an integral (or integro-differential) equation (the constitutive law) which, for instance, may be of the (linear) form

$$\sigma^{ij}(\boldsymbol{x}) = C^{ij\,kl}\,\varepsilon_{kl}(\boldsymbol{x}) + \int c^{ij\,kl}(\boldsymbol{x}, \boldsymbol{x}')\,\varepsilon_{kl}(\boldsymbol{x}')\,d^3\boldsymbol{x}'. \tag{6}$$

The strain gradients appear after performance of the integrations when $\varepsilon_{kl}(\boldsymbol{x}')$ is written as a Taylor series around point $\boldsymbol{x}$.

Of course, waves show dispersion in the non-local theory. Beside this, it has been shown in [6] that even in the relatively short range Van der Waals crystals range effects can be found: Two isotropic point defects which have no interaction in the (isotropic) elasticity approach attract each other with Van der Waals forces according to the non-local theory[2]. In materials in which only cohesive forces of sufficient range are present the displacement does not become infinite under the action of a point force. This result may provide some insight into the divergence problems discussed in the lecture of E. Sternberg at this symposium.

It is almost unnecessary to state that non-local characteristics can also arise with respect to moments.

I have not yet mentioned a last big group of theories[3], viz. those in which the particle is no longer considered a point (with or without orientation) but a rather complex structure itself which may consist of sub-particles built up from sub-subparticles, etc. As an example, one may think of a polycrystal, the particles being the grains, the sub-particles being the non-primitive lattice cells and the sub-sub-particles being the atoms. If one likes, additional complications such as

---

[1] The strain gradient theories can easily be derived from the atomic theory of Bravais lattices (see the lectures of J. A. Krumhansl and of R. D. Mindlin at this symposium). E. Grafarend (unpublished work) finds that the higher order geometry of H. V. Craig [7] and of T. Ohkubo [8] can be used adequately in the geometric description of these materials.

[2] This effect is to be distinguished from the "lattice effect" discussed in the lecture of R. Bullough and R. C. Perrin at this symposium.

[3] Most of these theories can be classified under *multipolar not of the first kind* and also under *micromorphic* (see the lectures of R. S. Rivlin and of A. C. Eringen at this symposium).

dislocations and non-local behaviour can be built in. Bodies of this kind show optical branches beside dispersion in their wave spectrum. It is likely that in the extensive literature fully developed geometries can be found which may be adapted to many special situations. In simpler cases, the director geometries may suffice.

5. Geometric Classification of Continuum Mechanics

In the attempt to classify continuum mechanics two principles of order are especially important. They are provided by

a) the geometry of deformation (in the sense explained in Sect. 2),

b) the constitutive laws.

Fig. 3 shows a classification according to the first principle. The scheme is three-dimensional, following the three main developments in the recent years. The additional micro-degrees of freedom occur when the particles obtain some structure (as in the Cosserat continuum).

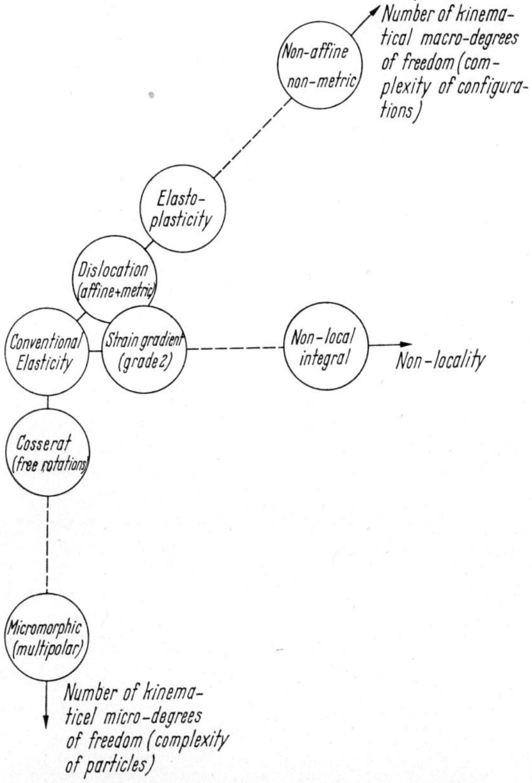

Fig. 3. Geometric classification of continuum mechanics.

The additional macro-degrees of freedom result when the configurations become more complicated (as in the dislocated or in the elasto-plastic continuum.)

One may be in doubt as to whether the third (non-locality) axis belongs to a geometric or a constitutive classification. The non-local theory requires geometric measurements over finite distances. These measurements are different from the measurements on neighbouring particles in the local theory. Hence, non-local behaviour *can* be classified in a geometric scheme (see also the remark in footnote 1, p. 338, referring to the work of Grafarend).

Combinations of the three developments rapidly result in very complicated theories and it appears doubtful if these theories ever can beat the lattice theory. There is no question, however, that some of the new developments offer advantages over the lattice theory in certain problems. For instance, this applies to materials with cohesive forces of some range, in which case often the summations of lattice theory become intractable. It is still an open question whether it also will apply to the final solution of the phenomenological problem of real elasto-plastic materials. Here the great complication comes in that dislocations usually do not form smooth line densities but more complex arrangements. A phenomenological characterisation of these arrangements which determine the work-hardening of the material requires, in addition, to the tensor of dislocation density, further quantities, for instance correlation functions or moments of the dislocation distributions, taken in the sense of statistical mechanics.

Notwithstanding its supreme importance in solid state physics and materials sciences, the dislocation will not be relevant to engineering calculations until the indicated phenomenological theory is sufficiently developed. The way to this aim is still far.

References

[1] Truesdell, C., and W. Noll: in: Encyclopedia of Physics, Vol. III/3, Ed. S. Flügge. Berlin/Heidelberg/New York: Springer 1965.

[2] Noll, W.: Materially Uniform Simple Bodies with Inhomogeneities, to appear in Arch. Rat. Mech. Anal.

[3] Kröner, E.: Kontinuumstheorie der Versetzungen und Eigenspannungen (Ergebnisse der angewandten Mathematik, Vol. 5). Berlin/Göttingen/Heidelberg 1958.

[4] Günther, W.: Abh. d. Braunschweigischen Wiss. Ges. **10**, 195 (1958).

[5] Amari, S.: RAAG Memoirs of the Unifying Study of Basic Problems in Engineering and Physical Sciences by Means of Geometry, Vol. III, Ed. K. Kondo. Tokyo: Gakujutsu Bunken Fukyu-Kai 1962, p. 193.

[6] Kröner, E.: Int. J. Solids Struct. **3**, 731 (1967).

[7] Craig, H. V.: Amer. J. Math. **61**, 791 (1939).

[8] Ohkubo, T.: J. Fac. Sci., Hokkaido Univ., Ser. 1, **11**, 1 (1946).

Properties of Dislocations in Iron — The Importance of the Discrete Nature of the Crystal Lattice

By

R. Bullough and **R. C. Perrin**

Theoretical Physics Division
A.E.R.E., Harwell, Berks.

The continuum model cannot be used to calculate many important physical properties of dislocations. For example, it cannot be used to obtain the energy, the Peierl's stress or the short range interaction between dislocations and other defects. Furthermore, Bullough and Hardy [1, 2] have recently used a harmonic face centered cubic lattice to demonstrate that point defects defined in discrete lattices interact in a qualitatively different way from the analogous continuum defects. This last result is due mainly to the precisely defined spatial distribution of body force (non-local), on the atomic neighbours of the point defect, that can be imposed to create the point defect in the lattice; the absence of real atomic structure with its associated translational symmetry makes such a "discrete" simulation in the continuum extremely artificial. In addition, the discrete lattice approach enables the entire available phonon dispersion data to be used.

Thus, for example, a vacancy in a face-centre cubic lattice is simulated by radial forces on its twelve first and six second neighbours; the magnitude of these forces is calculated from the form of the pair potential prevailing between the matrix atoms with due allowance for displacements of the first and second neighbours. The lattice statics formalism permits the straightforward calculation of analytic expressions for the asymptotic displacements associated with the vacancy and for the interaction energy associated with two widely separated vacancies. A particularly interesting situation then arises when the lattice is made elastically isotropic (by a simple adjustment of the force constants). In this case, whilst the asymptotic displacements agree with the isotropic continuum results, the interaction energy between two widely separated vacancies does *not* agree with the isotropic continuum result (which is identically zero). In fact the interaction energy is proportional to $|R|^{-5}$ and thus, in an *elastically isotropic medium* this interaction is more important than the often quoted $|R|^{-6}$ induced or second order inhomogeneity interaction. The non-vanishing of the first order interaction arises in our treatment because, even though we make the lattice macroscopically isotropic, the microscopic anisotropic nature of the face centred lattice is retained and defined by our particular distribution of body force that actually simulates the presence of the vacancy. It should be emphasised that this result is probably only of formal interest since very few real crystalline materials are exactly elastically isotropic (tungsten is a notable exception) and thus the coefficient of $|R|^{-3}$ in the anisotropic first order interaction will rarely vanish; clearly when it is not zero it will dominate over the $|R|^{-5}$ term above.

The results [1,2] obtained for point defects in a discrete lattice led us to consider the influence of the discrete nature of the lattice on the properties of dislocations. However, the topological nature of a dislocation make it rather difficult to use the Fourier transform methods used for the point defect calculation; thus we have used a large atomic assembly with the relaxation performed in real space. A large parallelepiped of perfect body centered cubic iron lattice is first achieved by allowing a non-equilibrium central force potential to act between first and second neighbour atoms with appropriate restraints on the atoms in the surfaces of the parallelepiped to maintain equilibrium. The potential used consists of a set of splines interpolated to give a continuous and differentiable function. Its precise form is deduced by fitting its slope and curvature at the first and second neighbour equilibrium positions to the linear elastic constants of iron and at short range it is matched to the known radiation damage potential. The potential is arbitrarily made to vanish, with zero slope, at a point halfway between the second and third neighbours. This potential predicts phonon energies in quite good agreement with the dispersion data obtained by inelastic neutron scattering experiments on iron [3].

Various dislocations have been studied using this model [4]. The procedure is to first introduce the topological features of the dislocation by moving the atoms to the appropriate anisotropic elastic positions. When the dislocation is long and straight in one particular crystallographic direction the atoms in the surfaces of the parallelepiped that contain this direction are held in their elastic positions and periodic boundary conditions are imposed across the two surfaces orthogonal to the dislocation. The atoms within the paralleelpiped are then allowed to relax under the above potential and the resulting large set of simultaneous differential equations of motion are solved by an iteration technique. When the relaxation is complete the linear boundary constraints are replaced by responsive forces that simulate the embedding of the parallelepiped crystallite in an infinite elastic medium. The relaxation is then continued until the forces on all the atoms in the crystallite are negligible.

The dislocations studied include the screw dislocation lying parallel to a $\langle 111 \rangle$ direction, the cube edge dislocation and various dislocation loops [5]. In addition the interaction energy between such dislocations and point defects has been obtained and differs considerably from the corresponding continuum values [4].

References

[1] Hardy, J. R., and R. Bullough: Phil. Mag. **15**, 237 (1967).
[2] Bullough, R., and J. R. Hardy: Phil. Mag. (to appear).
[3] Low, G. C. E.: Proc. Phys. Soc. **79**, 479 (1962).
[4] Bullough, R., and R. C. Perrin: Battelle Colloquium on Dislocation Dynamics. Seattle: Harrison, May 1967.
[5] Bullough, R., and R. C. Perrin: Proc. Roy. Soc. (to appear).

A Modernization of MacCullagh's Ether Theory

By

J. A. Brinkman

Science Center, North American Rockwell Corporation
Thousand Oaks, Calif.

MacCullagh's ether [1] may in a certain sense be regarded as a predecessor of the Cosserat Continuum. It was necessary, in particular, for MacCullagh to identify either the electric or magnetic field components with rotations of the ether. It is of some interest to investigate a natural modern extension of MacCullagh's ether theory. We have carried out the generalization of this theory in its linear approximation by (a) extending it from three to four dimensional space and (b) considering a plastic as well as elastic mode of deformation. We consider space-time to be representable as a four-dimensional Euclidean space haviug as its fourth coordinate ict. We consider equilibrium to exist between elastic stresses, dislocations and body forces. To simplify the mathematical treatment we consider two limiting cases, the ideal elastic case characterized by $\mu \ll k$, Y, and the ideal plastic case characterized by $Y \ll \mu$, k. Here μ and k represent the shear and bulk moduli respectively, and Y represents the plastic yield strength of the medium.

As in MacCullagh's model, we identify the electromagnetic field tensor in the linear approximation with rotations:

$$f^{\alpha}{}_{\beta} \propto \omega^{\alpha}{}_{\beta}. \tag{1}$$

In the elastic case the rotations can be further identified with the derivatives of the displacement vector as follows,

$$\omega_{\alpha\beta} \equiv u_{[\alpha,\beta]}, \tag{2}$$

implying an identification of the vector potential with the elastic displacement vector,

$$A^{\alpha} \propto u^{\alpha}. \tag{3}$$

The Nye curvature tensor, in the linear approximation, is related to the rotations by

$$K^{\alpha}{}_{\beta\gamma} = \omega^{\alpha}{}_{\beta,\gamma}. \tag{4}$$

In the elastic case it follows from the integrability conditions on the displacement that

$$K_{[\alpha\beta\gamma]} = 0, \tag{5}$$

from which follow by substitution from Eqs. (1) and (4)

$$f_{[\mu\nu,\sigma]} = 0. \tag{6}$$

In the plastic case Eqs. (6) do not follow automatically since the displacement is not a single valued function of position. Instead, the integrability conditions apply only on $\omega_{\alpha\beta}$:

$$K_{\alpha\beta[\gamma,\delta]} = 0. \tag{7}$$

The Nye relations are

$$S_{\alpha\beta\gamma} = -K_{\gamma\alpha\beta} - K_{\beta\gamma\alpha}. \tag{8}$$

The dislocation node theorem follows from Eqs. (7) and (8):

$$S_{[\alpha\beta}{}^{\delta}{}_{,\gamma]} = 0. \tag{9}$$

The boundary conditions representing the compatibility conditions in an infinite medium in which the deformation is localized are

$$S_{\alpha\beta[\gamma,\delta]} = 0. \tag{10}$$

Combination of Eqs. (9) and (10) gives

$$S_{[\alpha\beta\gamma],\delta} = 0. \tag{11}$$

From Eqs. (8) and (11) one obtains

$$K_{[\alpha\beta\gamma],\delta} = 0, \quad \text{or} \quad K_{[\alpha\beta\gamma]} = C_{\alpha\beta\gamma}. \tag{12}$$

Application of the boundary conditions requires these constants to be 0, yielding again Eq. (5). Thus in the plastic case as well as the elastic case Eq. (6), the half of Maxwell's equations which do not contain the four-vector current, are obtained. The other half of Maxwell's equations,

$$j^{\mu} = f^{\mu\nu}{}_{,\nu}, \tag{13}$$

defines the four-vector current. In the plastic case this turns out to be

$$j_{\mu} \propto S_{\mu\nu}{}^{\nu}. \tag{14}$$

For the elastic case it is

$$j^{\alpha} \propto F^{\alpha}/\mu, \tag{15}$$

where the body force, F^{α}, is given in terms of second derivatives of the displacement by the Navier equation for the four-dimensional medium. The following choice of gauge for the displacement field is necessary:

$$u^{\beta}{}_{,\beta} = \text{const}. \tag{16}$$

In the plastic case the four-vector current is just that distribution of edge dislocations required to stabilize the rotations which correspond to the fields; in the elastic case the current is simply the displacement field produced by the body forces required to stabilize the rotations.

Classical electromagnetic theory does not allow a choice to be made between the elastic and plastic distortion models. It is of some interest, however, to note that a quantum mechanical effect exists which does exhibit a preference for the elastic over the plastic model. The Aharonov-Bohm (A-B) effect [2] demonstrates that the wave function of an electron is able to evaluate $\int A_\mu \, dx^\mu$ along a closed path; in particular, it predicts an observable phase shift for the wave function of an electron which travels along a closed path which encircles magnetic lines of flux in field-free space. In a space-time deformation model which identifies A^μ with the appropriate displacement vector (elastic or Cartan), the A-B effect is correctly predicted by the elastic model, but not by the plastic. Thus, the A-B effect provides strong evidence against the acceptability of all unified field theories which achieve electromagnetism by the introduction of torsion into the space-time connection.

It is of further interest to note that the four-dimensional elastic model allows the existence of electric charge but not magnetic, in contrast to MacCullagh's three-dimensional model which did not forbid the existence of magnetic charge.

In view of the apparent success of the four-dimensional elastic model, it is important to ask what is wrong with it. One of the principal inadequacies seems to be its inability to account for the energy content of the electromagnetic field. The elastic stresses in four-dimensional space-time are in equilibrium with themselves and body forces, but the strain energy associated with them does not appear as the electromagnetic field energy. It is our belief that the resolution of this and other objections to the model cannot be achieved unless one postulates a space-time which is more complex than Minkowskian.

References

[1] MacCullagh, J.: Trans. Roy. Irish Acad., XXI, 17 (1848); see Whittaker,, Sir Edmund: A History of the Theories of Ether and Electricity. London: Thomas Nelson, Vol. 1 (Rev. 1951) p. 142.

[2] Aharonov, Y., and D. Bohm: Phys. Rev. **115**, 485 (1959); also Phys. Rev. **123**, 1411 (1961).

Space-Time as Generalized Cosserat Continuum

By

F. Hehl

Institut für Theoretische Physik
Technische Universität Clausthal

Space-time is characterized through metric and torsion. The metric (torsion) tells us how the mutual distances (orientations) of the structured points of space-time have changed. Correspondingly, energy-momentum of matter fields is coupled to the metric and spin-angular momentum to the torsion of space-time.

1. General Relativity. In the general theory of relativity the space-time continuum is described by the metric tensor g_{ij}[1]. Suppose in space-time there is embedded a matter field with a Lagrangian density $\mathfrak{L}$. Let us alter the metric virtually; then the Lagrangian responds according to the well-known formula[2]

$$\varepsilon\, \sigma^{ij} = 2\,\delta \mathfrak{L}/\delta g_{ij}, \tag{1}$$

σ^{ij} being the (metric) energy-momentum tensor of the matter field. This formula means that energy-momentum, which expresses the dynamical properties of matter with respect to translation, alters the mutual distances of the material points of the space-time continuum via the gravitational interaction.

2. Spin and Torsion. In this picture especially one point seems to be unsatisfactory to us. We know from experience that in general matter, e.g. the neutron, also has independent intrinsic rotational degrees of freedom manifesting themselves on the dynamical level in spin-angular momentum $\tau_{ij}^{\cdot\cdot k}$. Therefore, in the spirit of general relativity, we feel

Acknowledgment. I thank Prof. E. Kröner for many discussions and Dr. B. K. Datta for reading the manuscript. I highly acknowledge financial support by the Deutsche Forschungsgemeinschaft.

[1] $i, j = 0, 1, 2, 3;\ \varepsilon \equiv \sqrt{-\det g_{ij}}$.

[2] One easily recognizes the deep-lying analogy with the following formula of elasticity theory: force-stress = δ (deformation energy)/δ (strain).

justified to assume that space-time should mirror this fundamental property of matter. It is then only too natural to suppose that space-time has a structure like a Cosserat continuum, namely, to assume a certain intrinsic orientation for each point. Through the action of spin-angular momentum these orientations would be rotated relative to each other.

Geometrically we describe this structure curvature through a tensor $K_{ij}^{\cdot\cdot k}$. If the underlying affine connection of space-time is Γ_{ij}^{k}, purely geometrical considerations yield

$$\Gamma_{ij}^{k} = \left\{ {k \atop i\,j} \right\} - K_{ij}^{\cdot\cdot k}. \tag{2}$$

$\left\{ {k \atop i\,j} \right\}$ is the Christoffel symbol of the second kind belonging to the metric g_{ij}. As is easily seen from (2), $K_{ij}^{\cdot\cdot k}$ can be expressed as a linear combination of the antisymmetric part of Γ_{ij}^{k}, i.e. of the torsion tensor.

As soon as space-time is able to respond to spin, we except a relation similar to (1) for the spin. Indeed, we postulate[1]

$$\varepsilon\, \tau_{k}^{\cdot ji} = \delta \mathfrak{L} / \delta K_{ij}^{\cdot\cdot k}. \tag{3}$$

Now spin is as intimately related to the structure curvature, or rather the torsion, as is energy-momentum to the metric.

3. Formalism and Consequences. Taking as essential ingredients (1)—(3), one is able to develop a consistent Lagrangian formalism in a manner like in general relativity[2]. From the invariance of the action function of matter under general coordinate transformations, there follows, e.g., the identity of the usual canonical spin-angular momentum tensor with the dynamically defined tensor (3), showing us the reasonableness of our postulate (3); furthermore, we get the conservation theorems for energy-momentum and angular momentum. With the help of a variational principle we then deduce the field equations, which are straightforward generalizations of Einstein's equations.

Of course this formalism makes sense only, if one is able to deduce certain effects which are observable in experiments. With ordinary matter densities there seem to exist only minor deviations from general relativity. Nevertheless, there is a result of principal importance: on each spinning matter field there acts, besides the usual inertial forces, a force proportional to the Riemann-Christoffel curvature tensor of

---

[1] The analogous formula in dislocation theory reads couple stress $= \delta$(deformation energy)/δ(Cosserat-Nye structure curvature). Cf. e.g. Kröner, E.: Plastizität und Versetzungen. In: Sommerfeld, A.: Vorlesungen über theoretische Physik, Vol. 2, Chap. 9, 5th Ed. Leipzig: Akad. Verlagsges. 1964.

[2] For a recent article on this subject containing also references to former works of Sciama, Kibble, Kröner, and others; cf. Hehl, F.: Abh. Braunschw. Wiss. Ges. **18**, 98 (1966).

the connection (2)[1]. In highly condensed matter, for instance in white dwarfs, there might exist observable effects[2].

Let us add the remark that in the case of the photon field, in the manner one represents it today, spin is not a gauge-invariant magnitude. Therefore there arise certain difficulties for the reported theory which have not yet been removed.

4. Outlook. From general relativity one knows that a space-time dependent metric looks to the light-rays like a space-time continuum with a changing refractive index. If we now have torsion as an additive structure, we take for further development of the theory the following plausible working hypothesis: space-time also possesses optical activity caused by torsion. This would allow us to measure the torsion of space-time. Finally we may remark that there are indications that torsion might possibly have something to do with weak interaction.

---

[1] At least a in first approximation, neutrons, which, as is well known, can be described by a spinning field, fall like balls. But the accuracy of measurement is not too high. Cf. the article of KOESTER, L.: Z. Phys. **198**, 187 (1967).

[2] SCIAMA, D. W.: private communication.

The Cosserat Continuum with Electronic Spin

By

J. B. Alblas

Technological University, Eindhoven

1. Introduction

The mathematical model of the Cosserat Continuum presents the opportunity to consider physical phenomena characterized by internal degrees of freedom. It is the purpose of this paper to apply this model to a series of problems related to the deformation of magnetic materials. The idea to investigate the deformation of a dielectric or magnetic material is not new. Toupin [1] has studied the coupling of an electric field with a deformation field, Tiersten [2] has extended Toupin's calculations on magnetically saturated insulators, while Brown [3] has given a comprehensive exposition of magneto-elastic interactions[1]. The present paper may be considered a recapitulation of some known facts, together with extensions of the basic theory and applications to a number of important problems. It is based upon the modern general theory of the Cosserat Continuum, as given by Toupin [4] but adapted to this category of specific considerations.

In this paper we define at each point a magnetization vector representing the resultant electronic spin and orbital magnetic moment, and consider the dynamical equations for this model. If we assume some kind of saturation, the model may be applied to the deformation of a ferro-magnetic material with localized electrons, as well as to the interaction in a paramagnetic material consisting of ions, surrounded by water ions and placed in an external magnetic field.

The interest in the model lies in the fact that the electronic spin (or the local magnetic moment) is not only the origin of a body force, but also a source of internal angular momentum. The well-known Einstein-de Haas-effect of physics is alien to continuum mechanics. In the paper the ambiguity of the 'spin' concept in continuum mechanics and in atomic physics is partly eliminated.

---

[1] At the Congress the author took notice of an important unpublished paper by Rieder [8] in which related problems are dealt with.

It has to be stressed that this paper only gives a simplified model. More realistic problems concerning magnetic behaviour cannot even be mentioned (cf. [5]).

2. The Spin Continuum

We assume the following form for the Lagrangian L

$$L = T - \varrho F, \tag{1}$$

in which the kinetic energy density T may be expressed by

$$T = \frac{1}{2} \varrho \dot{x}_i \dot{x}_i + \frac{1}{\Gamma} \frac{M_3}{M^2 - M_3^2} (M_1 \dot{M}_2 - M_2 \dot{M}_1), \tag{2}$$

while the directors M_k satisfy

$$M_k M_k = M^2. \tag{3}$$

In (2) M is the (constant) absolute value of the magnetization per unit volume and Γ is the gyromagnetic ratio. The second term on the right hand side of (2) is gyroscopic. A similar form has been presented by Brown [6]. The kinetic energy T satisfies the invariance condition

$$\dot{x}_{[i} \frac{\partial T}{\partial \dot{x}_{j]}} + M_{[i} \frac{\partial T}{\partial M_{j]}} + \dot{M}_{[i} \frac{\partial T}{\partial \dot{M}_{j]}} = 0. \tag{4}$$

The specific free energy ϱF may be split up into a local part and one resulting from the long range magnetic forces. It may be proved [3] that

$$\int_V \varrho F \, dV = \int_V \varrho F_{\text{loc}} \, dV - \tfrac{1}{2} \int_V M_k \varphi_{,k} \, dV, \tag{5}$$

where the local free energy F_{loc} is a function of M_k, $x_{k,\alpha}$, $M_{k,\alpha}$. In (5) the potential φ of the internal magnetic field has been introduced. It satisfies (if there are no currents)

$$\Delta \varphi = -4\pi M_{k,k} \tag{6}$$

while the normal derivative $\partial \varphi / \partial n$ at the boundary of the body has a discontinuity determined by

$$\left(\frac{\partial \varphi}{\partial n}\right)_+ - \left(\frac{\partial \varphi}{\partial n}\right)_- = 4\pi M_k n_k, \tag{7}$$

if the positive side of the surface is outside with respect to the body. A (known) integral representation for φ is

$$\varphi = \int_V \frac{M_{k,k}(\xi)}{r} \, dV_\xi - \int_S \frac{M_k(\xi) n_k}{r} \, dS_\xi. \tag{8}$$

The basic action principle is

$$\delta \int_{t_1}^{t_2} dt \int_V L\, dV + \delta \int_{t_1}^{t_2} dt \int_V H_k M_k\, dV + \int_{t_1}^{t_2} dt \int_V f_k\, \delta x_k\, dV +$$

$$+ \int_{t_1}^{t_2} dt \int_S t_k\, \delta x_k\, dS = 0, \tag{9}$$

where the assumption is made that all variations vanish at $t = t_1$ and $t = t_2$. In (9) H_k is the external (known) magnetic field, the mechanical body and surface forces being denoted by f_k and t_k respectively. The formula (9) only holds for cases where no surface moments are involved. The execution of the variations in (9) leads to

$$t_{ij,j} + M_j H_{i,j} + M_j \varphi_{,ij} + f_i = \varrho\, \ddot{x}_i, \quad \text{in } V, \tag{10}$$

$$\frac{\dot{M}_{ij}}{2\Gamma} = -t_{[ij]} + M_{[i} H_{j]} + M_{[i} \varphi_{,j]} + \left\{\varrho\, M_{[i} \frac{\partial F_{\text{loc}}}{\partial M_{j],\alpha}} x_{k,\alpha}\right\}_{,k}, \quad \text{in } V, \tag{11}$$

and the boundary conditions

$$t_{ij}\, n_j - 2\pi (M_k n_k)^2 n_i - t_i = 0, \quad \text{on } S, \tag{12}$$

$$\varrho\, M_{[i} \frac{\partial F_{\text{loc}}}{\partial M_{j,\alpha]}} x_{k,\alpha}\, n_k = 0, \quad \text{on } S. \tag{13}$$

The stress t_{ij} is defined by

$$t_{ij} = \varrho \frac{\partial F_{\text{loc}}}{\partial x_{i,\alpha}} x_{j,\alpha}. \tag{14}$$

For the derivation of (11) an invariance condition for F_{loc} has been used. The formulas (8), (10)—(13) constitute a system of differential-integral equations of the form proposed by KUNIN at this symposium. The considerations may easily be extended to materials of the second degree.

3. Applications

a) Einstein-de Haas-effect. From (10) and (11) we derive the following form for the torque equation of the whole body

$$\dot{\boldsymbol{D}} + \frac{1}{\Gamma} \frac{d}{dt} \int_V \boldsymbol{M}\, dV = \boldsymbol{L}_{\text{mech}} + \boldsymbol{L}_{\text{mag}}, \tag{15}$$

$$\boldsymbol{D} = \int_V \boldsymbol{r} \times \dot{\boldsymbol{r}}\, \varrho\, dV \tag{16}$$

being the mechanical angular momentum while $\boldsymbol{L}$ represents the moment of the external forces. If $\boldsymbol{L}_{\text{mech}} = 0$, we have

$$\boldsymbol{D} = -\frac{1}{\Gamma} \int_V \boldsymbol{M}\, dV - \boldsymbol{H} \times \int_{t_0}^{t} dt \int_V \boldsymbol{M}\, dV. \tag{17}$$

The conventional Einstein-de Haas formulation is found if the second integral in (17) is either zero or neglectable.

b) Spin waves and resonance phenomena. We assume that a solution of (10)—(13) is known for a strong static field $\boldsymbol{H}^{(0)}$ and consider the case of a small external time dependent field $\boldsymbol{H}^{(1)}(t)$ superposed on $\boldsymbol{H}^{(0)}$

$$\boldsymbol{H} = \boldsymbol{H}^{(0)} + \boldsymbol{H}^{(1)}(t), \qquad |\boldsymbol{H}^{(1)}| \ll |\boldsymbol{H}^{(0)}|. \tag{18}$$

We assume the possibility of linearization and find in first order approximation (and obvious notation)

$$t^{(1)}_{ij,j} + M^{(0)}_j H^{(1)}_{i,j} + M^{(1)}_j H^{(0)}_{i,j} + M^{(0)}_j \varphi^{(1)}_{,ij} + M^{(1)}_j \varphi^{(0)}_{,ij} = \varrho\, \ddot{x}^{(1)}_i. \tag{19}$$

$$\begin{aligned}\frac{1}{2\Gamma}\dot{M}^{(1)}_{ij} = &-t^{(1)}_{[ij]} + M^{(0)}_{[i} H^{(1)}_{j]} + M^{(1)}_{[i} H^{(0)}_{j]} + M^{(0)}_{[i} \varphi^{(1)}_{,j]} + M^{(1)}_{[i} \varphi^{(0)}_{,j]} + \\ &+ \left\{\varrho\, M^{(0)}_{[i} \left(\frac{\partial F_{\text{loc}}}{\partial M_{[j],\alpha}}\right)_0 x^{(1)}_{k,\alpha}\right\}_{,k} + \left\{\varrho\, M^{(1)}_{[i} \left(\frac{\partial F_{\text{loc}}}{\partial M_{j],\alpha}}\right)_0 x^{(0)}_{k,\alpha}\right\}_{,k} + \\ &+ \left\{\varrho\, M^{(0)}_{[i} \left(\frac{\partial^2 F_{\text{loc}}}{\partial M_{j],\alpha}\,\partial M_{s,\beta}}\right)_0 M^{(1)}_{s,\beta}\, x^{(0)}_{k,\alpha}\right\}_{,k}.\end{aligned} \tag{20}$$

This set of linear equations have time-periodic solutions; it is basic for the discussion of resonance phenomena. If $\boldsymbol{H}^{(1)} = \boldsymbol{0}$, solutions may exist which may be considered spin waves. We note that the elementary theory of resonance phenomena is based upon a simplified version of (20) (cf. [7])

$$\frac{1}{2\Gamma}\dot{M}^{(1)}_{,j} = M^{(0)}_{[i} H^{(1)}_{j]} + M^{(1)}_{[i} H^{(0)}_{j]} + M^{(1)}_{[i} \varphi^{(0)}_{,j]}, \tag{21}$$

while, in addition, $\varphi^{(0)}_j$ is approximated by a constant field.

c) Ferromagnetic media. The success of this theory depends on the possibility to construct a form for the free energy. This has been done with the help of the theory of the solid state by Brown [3] for a special model of a ferromagnetic. We refer to his monograph.

d) The paramagnetic solid. Here, we indicate a way to obtain a form for F for a paramagnetic solid consisting of ions, surrounded by water ions and placed in an external magnetic field $\boldsymbol{H}$. The electric field of the water ions, near the origin may be represented at point P, given by $\boldsymbol{r}_1$, by

$$V_0(x_1, y_1, z_1) = \sum_{k=0}^{\infty} r_1^k\, Y_k(\theta_1, \varphi_1), \tag{22}$$

which may be written for the case that the charge distribution on a surface S is given by $f(x, y, z)$ as

$$V_0(x_1, y_1, z_1) = \sum_{k=0}^{\infty} r_1^k \int_S \frac{f(S)\,dS}{r^{k+1}} P_k(\cos\psi), \tag{23}$$

where ψ denotes the angle between $\boldsymbol{r}_1$ and $\boldsymbol{r}$. S is a surface through the water ions. If the body is stressed, this surface will be deformed: $\boldsymbol{r} \to \overline{\boldsymbol{r}}$. Under the assumption that the charge distribution is not affected by the deformation, the potential after deformation is in first approximation

$$V = V_0 + V_1, \tag{24}$$

with

$$V_1 = e_{ij} \sum_{k=0}^{\infty} r_1^k \int_S l_i\, l_j\, f(S)\, dS \left\{- (k+1) \frac{P_k(\cos\psi)}{r^{k+1}} + \right.$$

$$\left. + \frac{P'_k(\cos\psi)}{r^{k+1}} \left(\frac{l_j^{(1)}}{l_j} - l_k\, l_k^{(1)}\right)\right\}, \tag{25}$$

where e_{ij} is the deformation tensor and l_i, $l_i^{(1)}$ are defined by

$$l_i = \frac{x_i}{r}, \quad l_i^{(1)} = \frac{(x_1)_i}{r_1}. \tag{26}$$

The next step is to consider the influence of the disturbed electrical potential (24) on the localized magnetic moment. This step may be performed by standard quantum mechanical methods. Subsequently, an approximate expression can be constructed for the free energy of the ion in the disturbed lattice. This completes the system of Cosserat Eqs. (10)—(14) which may be used to describe mechanical interactions with magnetization for the solid.

References

[1] TOUPIN, R. A.: J. Rat. Mech. Anal. **5**, 849 (1956).
[2] TIERSTEN, H. F.: J. Math. Phys. **5**, 1298 (1964); **6**, 779 (1965).
[3] BROWN, W. F.: Magnetoelastic Interactions. Berlin/Heidelberg/New York: Springer 1966.
[4] TOUPIN, R. A.: Arch. Rat. Mech. Anal. **17**, 85 (1964).
[5] MORRISH, A. H.: The physical principles of magnetism. New York: Wiley 1965.
[6] BROWN, W. F.: Micromagnetics. New York: Wiley 1963.
[7] JONES, D. S.: The theory of electromagnetism. Oxford: Pergamon Press 1964.
[8] RIEDER, G.: Über das mikromagnetische Kontinuum als Sonderfall eines Cosseratschen Kontinuums lecture Saarbrücken 1965, unpublished.

Disclinations and the Cosserat-Continuum with Incompatible Rotations

By

K. Anthony, U. Essmann, A. Seeger, and **H. Träuble**

Institut für Physik am Max-Planck-Institut für Metallforschung, Stuttgart, and Institut für theoretische und angewandte Physik der Universität Stuttgart

A disclination is a lattice defect analogous to a dislocation. It is characterized by a closure failure of rotation rather than of displacement for a closed circuit round the disclination centre. The rotation failure $\Delta \boldsymbol{\Phi}$ equals a symmetry angle of the lattice. The displacement failure $\Delta \boldsymbol{u}$ depends on the starting point of the circuit and increases proportional to the distance from the centre.

Fig. 1 shows two examples of wedge disclinations (rotation axis parallel to the singular line; Volterra's dislocation of the sixth kind), denoted by (+) or (—) signs. They have opposite signs of $\Delta \boldsymbol{\Phi}$ and may

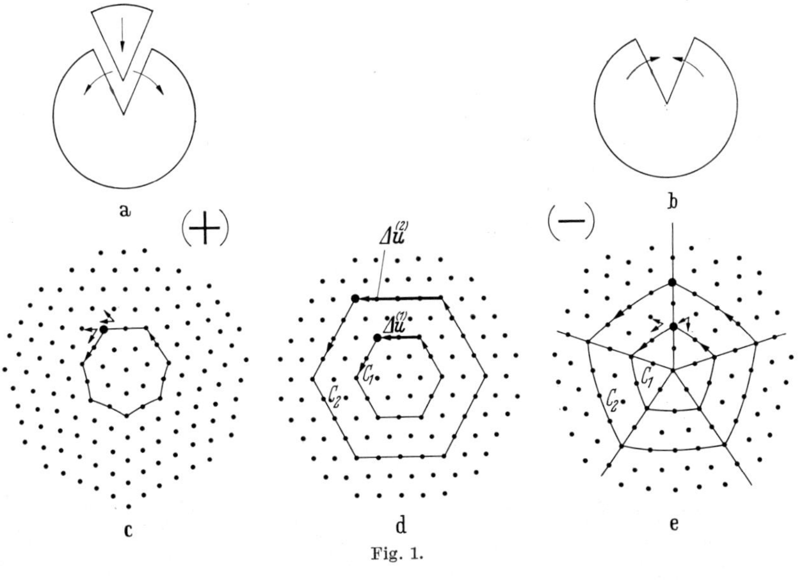

Fig. 1.

be generated by adding or removing a wedgeshaped piece of material and rewelding the material [1, 2]. The wedge angle equals the abovementioned symmetry angle; its minimum value in a hexagonal lattice is 60°. The associated lattice structures with characteristic seven- and fivefold symmetry are shown in Figs. 1c and 1e. $\Delta \overset{(1)}{\boldsymbol{u}}$ and $\Delta \overset{(2)}{\boldsymbol{u}}$ (Fig. 1d) are displacement failures corresponding to the Burgers circuits C_1 and C_2 in Fig. 1e.

Wedge disclinations have recently been found experimentally. Fig. 2 shows a negative disclination in the two-dimensional lattice formed by vortex-lines in the mixed state of type II superconductors [3, 4].

The existence of a closure failure of rotation means that the deformation contains an incompatible rotation field. In order to deal with

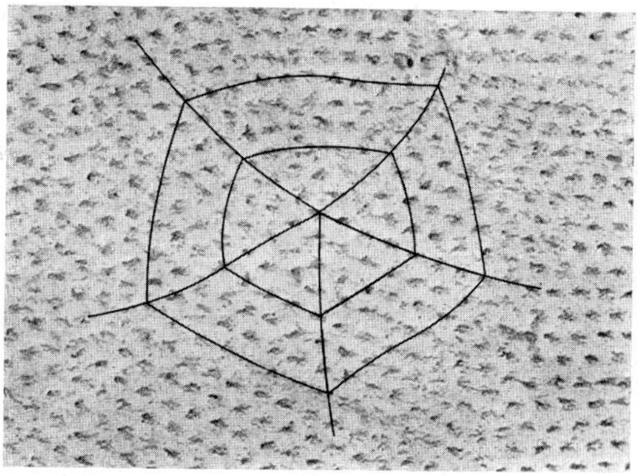

Fig. 2.

such a field we generalize the Cosserat continuum to incompatible rotations. The formulae to be given here are for simplicity confined to the linearized theory. We use rectangular coordinates x_k.

We associate with each volume element of the continuum two independent triads [5, 6]. During deformation the rigid Cosserat triad $\boldsymbol{C}$ undergoes a rotation $\boldsymbol{\Phi}$ independent of the displacement vector $\boldsymbol{u}$,

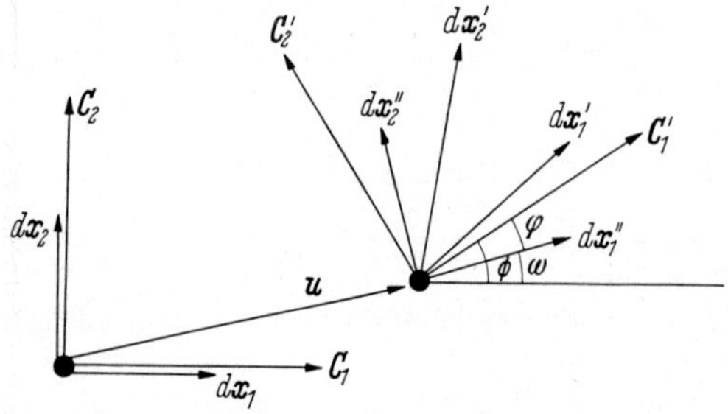

Fig. 3.

whereas the material triad $\boldsymbol{dx}$, defined by the material points in the volume element, is transformed into the triad $\boldsymbol{dx}'$. It undergoes the rotation $\boldsymbol{\omega}$ associated with the displacement field $\boldsymbol{u}$. Furthermore $\boldsymbol{dx}''$ is the material triad in the natural state [7].

Kinematically the Cosserat continuum is defined completely by its curvature tensor

$$\varkappa_{ij} = \partial_i \Phi_j \tag{1}$$

and by its deformation tensor

$$\gamma_{ij} = \varepsilon_{ij} - e_{ijk}\,\varphi_k \tag{2^1}$$

$$= \beta_{ij} - e_{ijk}\,\Phi_k. \tag{3}$$

$\beta_{ij} = \partial_i u_j$ and $\varepsilon_{ij} = \beta_{(ij)}$ are the ordinary distorsion and deformation tensors associated with the displacement $\boldsymbol{u}$. The angle $\boldsymbol{\varphi}$ is defined by

$$\boldsymbol{\varphi} = \boldsymbol{\Phi} - \boldsymbol{\omega}. \tag{4}$$

To describe the compatibility properties of the continuum we introduce the tensor of disclination density

$$\vartheta_{ij} = e_{imn}\partial_m \varkappa_{nj} \tag{5}$$

and the tensor of dislocation density

$$\alpha_{ij} = e_{imn}\,\partial_m \gamma_{nj} + \delta_{ij}\varkappa_{mm} - \varkappa_{ji}. \tag{6}$$

They satisfy the divergence relations

$$\partial_i \vartheta_{ij} = 0 \tag{7}$$

$$\partial_i \alpha_{ij} = -e_{jmn}\,\vartheta_{mn}. \tag{8}$$

Examples:

1) The de St. Venant compatibility conditions

$$\vartheta_{ij} = 0 \tag{9}$$

$$\alpha_{ij} = 0 \tag{10}$$

are fulfilled. We have single-valued vector fields $\boldsymbol{\Phi}$ and $\boldsymbol{u}$.

2)

$$\vartheta_{ij} = 0 \tag{11}$$

$$\alpha_{ij} \neq 0: \tag{12}$$

The rotation field $\boldsymbol{\Phi}$ is single valued, whereas the displacement field $\boldsymbol{u}$ is incompatible. Because of (11) and (3) we can define from $\boldsymbol{\gamma}$ the distorsion tensor $\boldsymbol{\beta}$. Relations (6) and (8) reduce to the well-known relations [7]

$$\boldsymbol{\alpha} = \operatorname{Rot}\boldsymbol{\beta} \quad \text{and} \quad \operatorname{Div}\boldsymbol{\alpha} = 0.$$

3)

$$\vartheta_{ij} \neq 0 \tag{13}$$

$$\alpha_{ij} = 0: \tag{14}$$

---

[1] e_{ijk} = permutation symbol.

Both the rotation $\boldsymbol{\Phi}$ and the displacement field $\boldsymbol{u}$ turn out to be incompatible. As a consequence $\boldsymbol{\beta}$ cannot be defined from (3). For a closed circuit starting and ending at P_0 (Fig. 4) we obtain the closure failures

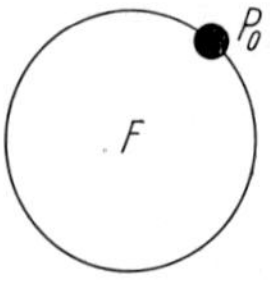

Fig. 4.

$$\Delta \Phi_i = \int_F \vartheta_{k\,i}\, dF_k \tag{15}$$

$$\Delta u_i = e_{i\,mn}\, \Delta \Phi_m\, x_n(P_0) - \int_F e_{i\,mn}\, \vartheta_{k\,m}\, \xi_n\, dF_k . \tag{16}$$

Because of the first term in (16), $\Delta \boldsymbol{u}$ depends on the starting point of the circuit. The isolated disclination is contained in 3) as a special case.

Adding the equations of statics and the constitutive equations to the kinematical concepts stated here, we get a continuum theory for ordinary crystals if we fix the Cosserat triad to the material triad (i.e., $\boldsymbol{\varphi} = 0$). This theory includes displacement and rotation incompatibilities, stress and couple stress fields.

References

[1] Eshelby, J. D.: Solid State Physics **3** (eds. F. Seitz and D. Turnbull) Academic Press Inc., New York, 1956, p. 79.
[2] Nabarro, F. R. N.: Theory of Crystal Dislocations, At the Clarendon Press, Oxford, 1967.
[3] Träuble, H., and U. Essmann: Phys. Stat. Sol. **25**, **373** (1968).
[4] Essmann, U., and H. Träuble: Phys. Letters **24 A**, 526 (1967).
[5] Günther, W.: Abh. d. Braunschweigischen Wiss. Ges. **10**, 195 (1958).
[6] Kessel, S.: Abh. d. Braunschweigischen Wiss. Ges. **16**, 1 (1964).
[7] Kröner, E.: Kontinuumstheorie der Versetzungen und Eigenspannungen. Ergebn. d. Angew. Mathem. Bd. 5 Berlin/Göttingen/Heidelberg: Springer 1958.

721/34/68 — III/18/203